Construction Technology

Analysis and Choice

Second Edition

Tony Bryan

University of the West of England

WILEY-BLACKWELL
A John Wiley & Sons, Ltd., Publication

This edition first published 2010

First published 2005 by Blackwell Publishing Ltd
Second edition 2010

Blackwell Publishing was acquired by John Wiley & Sons in February 2007. Blackwell's publishing programme has been merged with Wiley's global Scientific, Technical, and Medical business to form Wiley-Blackwell.

Registered office
John Wiley & Sons Ltd, The Atrium, Southern Gate, Chichester, West Sussex, PO19 8SQ, United Kingdom

Editorial offices
9600 Garsington Road, Oxford, OX4 2DQ, United Kingdom
350 Main Street, Malden, MA 02148-5020, USA

For details of our global editorial offices, for customer services and for information about how to apply for permission to reuse the copyright material in this book please see our website at www.wiley.com/wiley-blackwell.

Library of Congress Cataloging-in-Publication Data

Bryan, Tony, 1947–
Construction technology : analysis and choice / Tony Bryan. – 2nd ed.
p. cm.
Includes index.
ISBN 978-1-4051-5874-9 (pbk. : alk. paper) 1. Building. I. Title.
TH146.B82 2010
690′.837–dc22

2009038731

A catalogue record for this book is available from the British Library.

Set in 10/12 pt Palatino by Toppan Best-set Premedia Limited
Printed and bound in Malaysia by KHL Printing Co Sdn Bhd

1 2010

Contents

Preface

This book has been prepared for those at the beginning of their career studying with a view to practise in one of the professions concerned with the construction of buildings. It is written for the novice who, with study and experience, will become an expert in his or her chosen field. The more fundamental ideas developed in the text may need to be revisited as experience grows and it is, therefore, hopefully a text that can be referred to in the early years of practice and maybe beyond.

The book takes as its theme the process of choice: what the expert has to know and how the expert could think through the myriad decisions that have to be made about the design, production, maintenance and disposal of buildings. While this involves a process of analysis, the basis of which is universally applicable, the final choice is dictated by context.

The context is set by time and place, the nature of the activity to be undertaken in the building and the aspirations of its benefactor or client. The process of choice is seeking appropriate solutions in a physical and cultural context.

For a text of this nature it may seem to be dominated by words and not illustrations. The contention is that for the novice or the practising professional faced with new ideas and situations the explanation of possible solutions is required. Presenting a final working drawing may be the outcome of choice, but it is the process of analysis with its need for explanations that gives the confidence that it will perform and can be built in the context of the particular building. A picture only paints a thousand words if you already have the words to explain it.

It is hoped that the reader will find that the text tells a good story, that the development of the ideas is logical and easy to follow, but perhaps above all that it is seen as valuable in the world of practice. It is hoped that the text will be illuminating and the illustrations supportive in providing the explanations of both the analysis and the solutions suggested for choice.

The first part of the book focuses on the key areas of analysis that have to be considered to check that the proposed construction will perform (will not fail) and can be built. These two tests are seen as fundamental to the final choice. In order to carry out these analyses this book suggests that it is necessary to visualise not only the physical form of final construction, however outline or tentative, but also, equally importantly, its response to the dynamic conditions under which it has to perform. The key question is 'What if I do it like this?' The answer comes from an understanding of how it works and how it is built, this understanding coming from explanations of why things happen, drawing on knowledge from a number of other disciplines such as science and economics.

Each of the areas of analysis presented in this book is independent, and therefore each analysis has to be separate, yet all have to be satisfactory before the final choice is made. Choice is a juggling act: checking a proposal in one area of analysis may suggest changes that will affect others. Structurally sound solutions may be difficult to build or solutions that will last without maintenance too expensive as a capital investment. The final choice must satisfy all the areas of analysis.

The second and third parts of the book put analysis into practice, focusing on making the final choice. Given the approach identified in the first part of the book, with context being so important, the chapters in Parts 2 and 3 are written from the perspective of making choices for construction in the UK at the beginning of the twenty-first century. In Part 2 the focus is on housing, while Part 3 is concerned with the commercial use and scale of building. While these both take place in the same physical and cultural environments, the way buildings are commissioned and designed varies. For mass housing the choice of technical solutions is made by the developer/builder and this leads to a convergence of detailing and specification for the construction. This current common form of house construction is presented in Part 2. In commercial building there is considerably more diversity in the use and scale of the building, leading to more variety in both the mix of technologies and in detailing and specification. This greater need to consider broad options early in the design and to be clear about the potential for the technical solution is reflected in the approach developed in Part 3.

The final chapter provides a guide to the wide range of published material that is available to develop knowledge beyond this book and particularly to support the process of choice. The book itself gives no direct references. The notion of the framework for understanding developed in the book encourages the idea of seeking relevant and current information in the context of each project. It is hoped that this final chapter introduces the sources of information that will complement the approach developed in this book in order to put these ideas into practice.

This book is written with the conviction that by focusing on the process of choice the range of theory and knowledge that is useful to practice becomes explicit, making the link between knowledge and practice and between understanding and experience clear.

As study, and then practice, unfolds for each individual it is the power to reflect that develops understanding. It is hoped that the framework suggested in this book will help not just the process of choice but also the process of reflection. Readers should take from this text what understanding they can to use it to develop their own good judgement in whatever part they play in the great enterprise of construction and the development of the built environment.

Tony Bryan

Book website at www.wiley.com/go/bryanconstructiontech2e

Contains nearly 200 fully referenced, clear line drawings to download for free, as well as suggested learning activities for lecturers to incorporate into their teaching programmes.

Part 1
Analysis

1 The Framework for Understanding

This opening chapter outlines a framework to help develop an understanding of what has to be known in order to take decisions on how we should build. The framework suggests a way of going about selecting construction and identifies the knowledge that is required to make the choice. It is the framework that will be developed and used throughout the book.

Process and knowledge

This book provides examples of construction, showing how we currently build, and provides an introduction to the understanding necessary to explain how the construction works. These two types of knowledge are both vital to making decisions as to how we should create and maintain buildings in the future. This book sets this knowledge in the context of the process of making choices for the construction. In practice, a great deal of knowledge exists of the types of construction we might use and the materials and details that might be specified. There is, however, the need for any proposed construction works to answer the questions 'Will it fail?' and 'Can it be built?' Clearly it requires the answer to these two questions to be 'No, it will not fail' and 'Yes, it can be built'. These categorical no and yes answers may be difficult to give in practice. There often needs to be some analysis to develop a level of confidence with which to make the choice. The amount of analysis required determines the level of understanding and experience needed to provide the evaluation of the suggested solution before the decision is taken to put the final proposal into practice.

It is this basic ability to make a suggestion of how a building might be constructed and then to carry out an evaluation asking the questions whether it will fail and whether it can be built that is developed in this book. The evaluation points to changes in the suggestion, which after re-evaluation will lead, through a series of refinements, to the specification and details to be adopted.

This requires knowledge of what potential solutions might look like, with current practice and precedent as the major sources for an initial suggestion. It requires an understanding of what is necessary to be specified in order to describe the proposed construction in sufficient detail to carry out the evaluation. The ability to carry out the evaluation, on the other hand, is dependent on an ability to 'see' the proposal working in the dynamic systems of the physical and social conditions in which the building is to be built.

This process of choice and the way it leads to the identification of the knowledge required is shown in Figure 1.1.

It will be shown later that the process of choice is not just an analysis of the behaviour of physical performance. This evaluation of a proposal's response to the dynamic physical, chemical and biological systems of nature is vital, but building also takes place in a social and cultural context. There will be an imperative to ensure that the resources and know-how are available to manufacture and assemble the construction. This will require knowledge of the available industrial systems. Further, it will be necessary to ensure that the cost of the solution

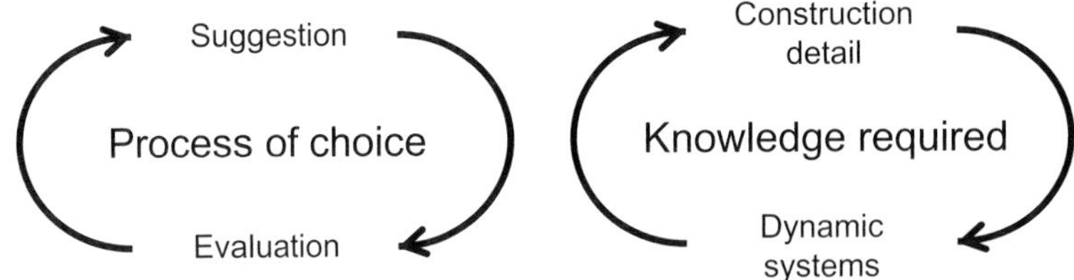

Figure 1.1 Relationship of process of choice to knowledge required.

is monitored and that its social and environmental impact is audited. These will demand an understanding of both the economic and social systems in which the building is to be created.

The initial suggestion

If the process is suggestion and evaluation, and the starting point is suggestion, it is necessary to know how we make the initial suggestion. How do we make that first best guess?

In most cases the suggestion is informed by precedents. It is necessary to have knowledge of current solutions and how they perform in practice. Something, somewhere, has been done before that gives clues as to how the new solutions might be formulated. In times where change is limited the particular circumstances will have been faced a number of times in the past and successful solutions will have evolved. The well-tried and tested solution needs only to be suggested and, with some evaluation to ensure the circumstances have not materially changed, can be immediately adopted. If the circumstances are changing, current solutions may still be the best starting point. They represent not only a sound basis in performance but also an existing base of resources and know-how to manufacture and produce the materials and details. Evaluation may modify but not fundamentally change the solution. This, over time, gives rise to a number of general forms from which specific solutions can be derived.

In some cases, perhaps where there are increasing user demands, or the structure of industrial practice is changing, then similar or related practice may have to be investigated. On the rare occasions where suggestions have to be derived from little or no existing previous work, a more fundamental understanding of the behaviour of the construction will have to be applied.

While experts approach making suggestions based on their knowledge and experience, it is, in many ways, not necessary to know much at all to make suggestions. It is the evaluation that requires the expertise. Novices or casual observers may make suggestions that may be hailed as brilliant observation, but in truth they have no way of knowing whether that particular suggestion is workable or not. It will take the expert to spot its potential and prove its worth through evaluation.

The power of the expert to spot potential is probably associated with the ability to carry out a rapid, approximate evaluation before subjecting the suggestion to more rigorous and explicit analysis.

Carrying out the evaluation

The heart of the success of the process of technological choice lies in the ability to be able to carry out an evaluation. The need to carry out a series of analytical exercises determines the knowledge required and manner of its application. It requires a level of understanding to answer the question 'What if we built the building as proposed?'

While the suggestion and the ultimate solution describe the construction in what appears to be a static detail, the evaluation of the suggested construction has to describe a dynamic system of behaviour ('Will it fail?') and a production process ('Can it be built?').

The process starts with the client's brief, the design then being devised from the requirements of those who commission the works, and as a response to the social and physical context in which the building is to be built. It is against these criteria that failure will be judged. There will be many possible ways to construct a building to achieve performance expressed in the client's brief. The criteria for choice of the technical solution come from the identification of the function of the parts and how they contribute to the function of the building as a whole.

The dynamic behaviour of the construction responding to changing conditions has to be understood, anticipating the modes of failure. Some specification of performance has to be established and then the suggestion has to be tested in the mind to assess the risk of failure within agreed design conditions.

It is easy to see technology as only the final construction, an assembly of components and materials, the building as a static object. However, making the choice of what construction should be adopted has to be rooted in an understanding of the construction as a dynamic system responding to changes in conditions and open to a failure to perform.

It is necessary to be able to visualise not only the physical construction but also its behaviour under the conditions it will have to endure in its working life. Both of these are of equal importance. If the suggestion is not visualised correctly, its behaviour under analysis may be misinterpreted. If the dynamic system of behaviour is not visualised correctly, a flawed proposal may be adopted. Visualising the building as a dynamic system will involve identifying the flows and transfers that take place both within the building and through the construction itself. These ideas are explained in Chapter 7.

Technological choice demands skills in these areas of visualising the object and the systems, and then the conceptual manipulation of the systems acting on the construction to predict behaviour and assess the risk of failure.

If the two basic questions to ask of a proposed solution are whether it will fail and whether it can be built, the criteria for choice come from an understanding not only of the potential dynamic flows and transfers when the building is in operation but also of the resources available. The performance of construction can only be realised if the resources are available to construct the building. This relies on the manufacturing and assembly possibilities but will also include the existence of design expertise and the options for maintenance and disposal. Knowledge of available techniques and know-how for production is crucial to the choice of the final construction if it is to be successful in reality as well as on paper.

As a design concept emerges, it is necessary to question what construction solutions may be used to fulfil the design requirements. It is then necessary to question whether this solution is available with current technology and resources within any environmental, cost or time requirements. The resources available make the design a reality. Technology stands between the design of the building and the management of the resources.

Choices can be made that may extend current production and design knowledge, but this must be recognised, and any costs involved in prototypes or training and the risks involved must be accepted before the final choice is made.

The design and resources available condition the choice. These two areas are shown in Figure 1.2. These are the two areas that have to be understood before any analysis can be started leading to the final choice of construction.

Physical and social context

Figure 1.2 also indicates that before any choice for a specific building can be made it is necessary to understand something of the context in which it will be constructed. There needs to be some knowledge of the conditions that exist when and where the building is to be built, possibly with some assessment of how conditions may change in the future.

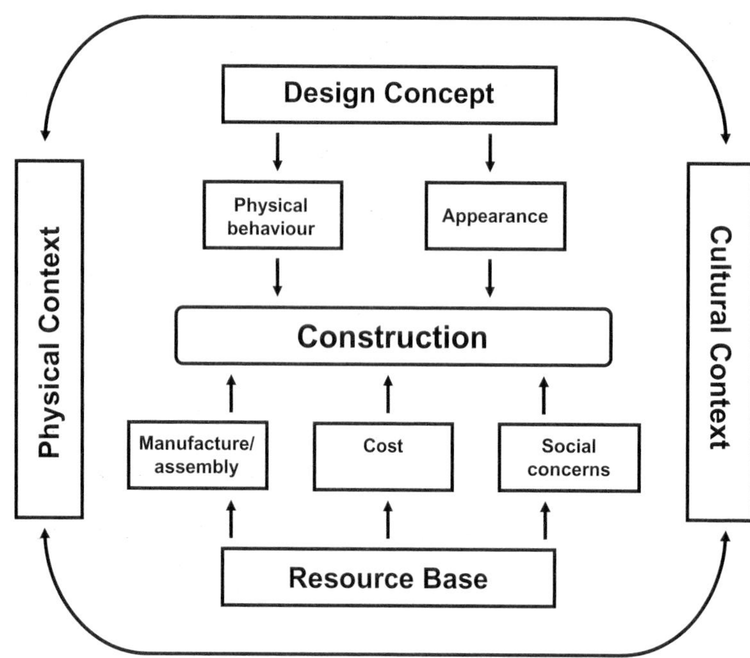

Figure 1.2 Framework of analysis.

Construction takes place within, and has an impact on, the world in which it is undertaken. Some description of this world is necessary for both technical and socio-economic analyses. The world can be represented as a series of contexts, both physical and cultural. The physical context includes nature and climate. These exert forces that act on the building; they provide the raw resources and may be adversely affected by the construction and operation of buildings. The physical context includes the surrounding development and the need for the design to respond to existing buildings and spaces.

People and their social, economic and political systems create the cultural context. It needs a response to local, national and even global needs based on beliefs and fundamental world views of the relationship between individuals and between society and the other components of the natural world.

The interaction between these two sets of contexts has been brought into focus at the beginning of the twenty-first century by the movement for sustainable development. The realisation that development cannot continue without consideration of its impact on the natural environment as well as established economic and social considerations calls for new knowledge associated with materials choice, energy use and waste disposal aspects of our chosen construction.

There are, therefore, dangers in seeing these two sets of contexts as separate. However, it is still useful to consider them as having different dynamics, as, generally, in order to resolve technical questions, it is necessary to have an understanding of the physical context, while to evaluate the chance of a solution being successfully applied requires an understanding of society, its economic and political systems.

The basis of analysis

Having identified the design concept and the resource base as providing the criteria for choice and having recognised the need to understand the physical and social context in which we build, it is possible to identify five areas of analysis as shown in Figure 1.2.

The design concept is the translation of the physical and social needs for the operation of the whole building into a scheme that identifies the function and performance of each of the parts. In order to achieve this, the design concept has to articulate both the arrangement and appearance of the spaces and the technical contribution each part of the construction will make to the creation and maintenance of the internal conditions.

Two tests indicated in Figure 1.2 have to be applied to a suggestion to see if it complies with the design concept:

- Does the *physical behaviour* of the construction provide a building fulfilling its functions to the required performance level?
- Does it provide the right attributes of *appearance*?

The test for physical behaviour involves three separate areas of analysis:

- Creating environments
- Under load
- Over time

Three tests indicated in Figure 1.2 have to be applied to a suggestion to see if it is achievable with available resources:

- Can it be produced, including *its manufacture and assembly* and the subsequent processes of maintenance and disposal, with existing skills and know-how in a reasonable time and at the required quality?
- Are the resources available at reasonable *cost*?
- Will it be compatible with the *social concerns* that currently exist?

These resources are both natural (from the environment) and social, so both aspects have to be part of the evaluation. It is only possible to use resources that nature has provided. However, the level of economic development in a society will provide the capacity to process materials, develop the skills and machinery to work them, the intellectual capacity to undertake design and the provision of capital to invest in the enterprise of the construction itself. Both the technical properties of materials and the existence of the knowledge to exploit them plus any environmental impact of their use must be analysed in evaluating a suggested form of construction.

These seven areas of analysis (physical behaviour being three) that have to be undertaken are, by and large, unrelated to each other and each needs its own tools of analysis and knowledge base to be successfully applied. Carrying out one analysis may indicate the need for a change in the suggestion, but the change that is chosen may make the analysis of a previous aspect invalid. It is not until all the relevant areas of analysis are shown to be satisfactory that the suggestions can be adopted as the solution.

These seven areas of analysis have potentially to be undertaken for all aspects of the construction, from the overall structural system to the finest detail such as the screw that fixes the final fitting to a wall. The work in this is clearly overwhelming if it has to be carried out for every building that is constructed. Current practice can inform much of the process of choice. One of the most important decisions that experts take is, of all the thousands of choices that have to be made to fully specify the construction, which of the areas of analysis for which parts of the building are significant. Where, if analysis is not thought through, is the greatest risk of failure?

It is not clear how experts make this judgement. One possible explanation is that while looking at a suggestion they make many rapid checks against the seven criteria. From their knowledge and experience they take a view on what is well within failure limits and where the risks of failure are higher.

Knowledge needed for choice

It is now possible to start to identify the knowledge that is required and the areas of understanding that have to be developed in order to carry out a full evaluation. Although used in various combinations in the different areas of analysis, it is possible to put forward a tentative list of types of knowledge that will be required:

- Setting the performance levels as the criteria against which the evaluation will be made.
- Defining the conditions under which performance has to be achieved.
- Determining the fundamental behaviour of the construction that would lead to failure to meet performance requirements.
- Identifying the materials' properties that will govern the behaviour that could lead to failure.
- Thinking through the behaviour of the particular combination of materials and details under the conditions envisaged.
- Identifying the process of manufacture and assembly together with the resources required to produce the building within quality, time, cost and safety limits.
- The cost in economic, social and environmental terms of using a particular form of construction.

The process of evaluation requires the building to be conceptualised in a number of ways. While the final building will be seen as a physical construction, 'bricks and mortar', it is necessary to perceive the building in a number of other ways when carrying out the evaluation. Initially, the construction has to be seen as fulfilling a set of functions with associated performance levels. For evaluation, the building has to be viewed as a series of physical systems responding dynamically to changing conditions. The building then has to be seen as a series of production operations with the resources required for the realisation of the design. All these will have economic, environmental and social implications that have to be understood. When all these can be mastered, choices can be made associated with an assessment of risk and with some confidence of a successful building.

Summary

1. The process of choice is one of suggestions and evaluation that requires knowledge of the physical construction and the dynamic systems in which it will be built and will operate.
2. How a suggestion is generated, and the extent to which a formal evaluation has to be undertaken, depends on the scale and nature of any changes from current practice required by either the design or the resources available.
3. When there is a changing demand for buildings and rapidly developing technical knowledge, the role of experience changes from predominantly reproducing known solutions to the integration of experience into a process of analysis to explore the possibilities of failure in the modified or new solutions before they are built.
4. The design and the resource availability define the criteria for the evaluation. This has to be set within a physical and cultural context.
5. The suggestion will probably be based on precedent and this leads to general forms from which specific solutions can be derived.
6. The evaluation needs to be carried out through seven areas of analysis: appearance, behaviour creating environments, behaviour under load, behaviour over time, manufacture and assembly, cost and social concerns.
7. This will involve constant reference to both production and design. The concern is for what will not fail and can be built.

2 Building Purpose and Performance

Technology is used to mediate between the world as it is and the world as we want it. This chapter expands this idea, focusing on the users and their activities, introducing the functions expected of the building and how these have to be translated into a process of choice for the construction of each part.

Activities, space and construction

Buildings are often characterised by their overall purpose or function – house, factory, hospital, prison, etc. – and are often immediately recognisable although their architecture may vary. To understand the form of the building, it is necessary to understand the day-to-day activities that make up the function for which the building is commissioned. The construction will have to play its part in ensuring the well-being of these activities. The notion of well-being is normally associated with the people who use the spaces, but the building will have to protect the equipment being used and the storage of items associated with the activities. Well-being of people also needs to include the requirements of the owner and the wider public, as buildings are social institutions and create the surrounding spaces.

The majority of spaces in buildings have to accommodate people and consideration of their well-being becomes a dominant aspect of the performance of the construction. In some cases, such as the rooms for the early computers and specialist storage areas such as vaults and cold stores, the equipment housed in the building or the storage conditions dominate the requirements. Designing for the well-being of people is essentially concerned with health, safety and comfort. Health should be understood as meaning both physical and mental health, both being a function of physical and social conditions created by the building.

The owner may see a purpose beyond the simple support of activities, to include image and status. The surroundings in which the activities are undertaken have social significance and must form a feature of the design.

Each society over time evolves common building forms for each purpose type that become recognisable within that society because of their size, arrangements of internal spaces and social statement. With this emergence of building form there is a need to develop appropriate construction solutions. This process is dealt with in more detail in Chapter 3, but first this chapter explores the complex range of performance issues. These will be established predominantly by the users' activities, although later by the owner and then by government in the form of legislation on behalf of society as a whole.

Introduction to performance requirements

In order to explore the range of performance issues it is useful to engage in a mind experiment. The situation to be imagined is one where civilisation as we know it does not exist and the task is to provide shelter. The scenario is one of few resources of materials, labour or knowledge. There is also little development of an economic system, so the whole design and production process has to be undertaken by the users. The mind experiment starts by asking which performance the user of the building would seek to satisfy first. The experiment can then be extended through time, assuming increasing resources and socialisation, introducing additional requirements that would add value and utility and improve well-being.

By undertaking this mind experiment it will be possible to introduce the needs of the user and hence the functions that he or she will require the building to fulfil. As the user is also, at least initially, the designer and builder of the shelter, he/she must decide on the construction. The experiment will, therefore, also start to introduce the ideas and knowledge that will be needed to make decisions on the construction of the building itself. All the issues and factors introduced here are still relevant today.

In the beginning

Given no immediate risk from attack, we can speculate that people would organise their technological endeavour in the following way. Thought would have to be given to the range of activities of everyday life that should be housed in the shelter. Initially this is likely to be very few, perhaps only sleeping, with most activity being undertaken outside depending on the prevailing climate. This will provide a definition of size and any division of the internal space that needs to be made. It would be immediately recognised that these requirements for size and shape would be limited by the possible construction available. Some decision would have to be made as to how much effort should go into developing construction to match aspirations or use existing resources to satisfy immediate needs.

The priority in terms of establishing functions for the building fabric would be to remain dry, a condition fundamental to the health and comfort of people and the storage of many natural products. This would be accompanied by the requirement to be warm, but it is possible that this will initially be achieved in combination with other technologies. These would include clothing and the use of fire, which at this stage may or may not be associated with the construction of the shelter.

As soon as any construction is contemplated the builder is immediately faced with achieving a stable structure and making some judgement on how long it may last. Although buildings may be erected for environmental control, their very existence demands considerations of stability and reliability if they are to be deemed satisfactory by the user. This will require the identification of possible modes of failure. At this early stage this would include how the construction might leak or fall down. This understanding of the ways in which performance may be lost becomes crucial to directing effort to significant aspects of the solution. This would also lead to some questioning of what maintenance may be necessary and what parts may have to be renewed to continue the performance until the construction had to be abandoned or demolished to make way for a new building.

It would also become necessary to have some knowledge of the environmental conditions in which the construction has to maintain performance. This knowledge would be required to consider not only the environmental control but also the stability and reliability. These three become the key physical performance issues. External climatic conditions such as frequency and volume of rainfall will need to be assessed for the weatherproofing function. Damaging wind speeds and direction would need to be assessed for stability. It would, however, soon become clear that wind is a major factor in weatherproofing. Some of the most arduous weatherproofing conditions come with wind-driven rain. It is not possible to say what aspect of the environment needs to be known to assess durability until some choice of the basic materials is made. Specific decay agents in the environment degrade different materials. Linking materials to their generators of decay is just one more question that has to be answered in the quest for knowledge to make sensible choices for the construction.

This assessment of external climate would soon reveal that often the more extreme conditions of open land are ameliorated by local natural features, that exposure can be modified by the position and orientation of the building which will limit the demands on the construction and/or give better internal conditions.

The next steps

As the number of activities carried out in the building increases, the provision of good air may be the next priority for improvement of the construction. This is probably the next requirement for health and comfort after dryness and warmth.

Seeking a solution that would improve the quality of the air would probably involve ventilation since there is an abundance of good air outside. It would have to be recognised that the air outside may also be cold and that this will make the requirement for warmth more difficult to achieve. This may mean that a solution for good air may only be possible when air temperatures outside are sufficiently high or air quality inside is particularly poor. Assuming that the active technologies of building services are not available, modification would be necessary in the existing construction. It would be necessary to use natural ventilation driven by wind pressures around the building. Openings would have to be introduced to ensure the exchange of air. Their position and size would have to be considered. It would be possible to introduce a level of control by providing shutters. The openings could eventually be as sophisticated as chimneys, but history tells us that these often come much later.

It is important to note that the containment of the bad air is a direct result of employing construction to remain dry and warm. The air outside the building is almost certainly good. This realisation that the construction itself can create undesirable conditions should not go unnoticed. Changes introduced to fulfil new performance demands may change conditions that have always been satisfactory. This happened towards the end of the twentieth century when a requirement for energy saving was satisfied by the introduction of insulation. Under some environmental conditions this led to condensation that has a deleterious effect on the basic requirement of keeping dry. This led to health problems for the users of the buildings.

The introduction of new performance requirements may also introduce failures in the construction that have previously not occurred. This happened in the condensation example, where dampness caused not only health problems but also a deterioration of the construction materials themselves. In this mind experiment it is possible to speculate on a similar problem with the improvement in internal air quality. Given the lack of social organisation to produce more processed materials, the construction would almost certainly be made of relatively unmodified organic materials. In this case it might be speculated that if the poor air had a large component of smoke it would have permeated the construction. This would have effectively fumigated the organic materials and kept infestations or decay-promoting fungi and insects at bay. Improving internal air quality may change the conditions within the construction itself. This may allow significant numbers of some of the natural agents of decay to thrive, thereby reducing the life of the construction.

More physical requirements

This mind experiment continues on the assumption that no major threat to security arises. This would inevitably lead to a diversion of resources and may change the priorities of individuals or society in seeking improvements in their buildings, although it may also lead to new types of buildings, such as defensive works.

Given peaceful social development, the next priority may be difficult to identify. It may be that the infestation of insects and small animals that will be attracted to the improved internal environment becomes a problem. It may be that in the harsh realities of life, and in the absence of any realisation of the health risks, these infestations would be tolerated. Measures to eliminate the larger animals, if not already considered as a danger, may now become important. At some time, however, infestation control would have to be included in the list of functions required of the construction to improve well-being. This is part of the more general requirement to keep the enclosed space clean. Internal surfaces that are dust-free, smooth and wipeable become desirable attributes where cleanliness becomes important to the health and comfort of the user.

Having gained good clear air, more activities could be carried on indoors. It is, therefore, possible to imagine an increasing need for light to improve comfort levels and further extend the range of activities in the building. The choice is between natural and artificial means of lighting. Natural would involve openings, while artificial would only demand a safe place to support what would almost certainly be an independent light source. Although independent of the building, it would have an influence on it. Perhaps the major influence it would have is in the increased risk from fire. While it may not be possible to improve the fire resistance of the whole building at this time, it would be sensible to employ non-combustible materials at least where the light source is to be mounted.

Here we see the increased use of technologies independent of the building, seen by the user as an asset to the well-being of the activities, influencing the required functions of the construction itself.

Another requirement is for some measure of noise control. This would, of course, only become a concern if the external level of noise were somehow interfering with the overall sense of well-being. It may become a problem when the building has separate rooms for specific activities. Noise from one activity may well be incompatible with the conditions necessary for another. Noise is a particular problem of civilisation, and especially for those societies that have advanced technologies. It also seems that it has a cultural basis in attitudes towards individual rights, privacy and intrusion.

The technical considerations become more complex

It is useful to view the need to consider noise performance from a different perspective. It may be that there is a need to modify the sound between outside and inside, yet the construction used for the other requirements already has sufficient, if not excess, ability to perform as a sound barrier. With low performance expectations the final form of construction is dictated by only a few of the functional requirements. As more functions are identified and the level of the required performance increases it becomes difficult to provide simple construction solutions. Composite solutions are necessary with different materials to satisfy different functions at the levels of performance that are required.

The analysis of exactly how each part of the construction is interacting in providing the overall function becomes more difficult. The danger comes when changes are suggested. It is necessary to ensure that the change does not make a previously satisfactory performance critical so that the construction fails to perform, in an unexpected way. The trend towards lighter partitioning in housing has, by reducing the weight and changing natural vibration frequencies, reduced the sound transmission resistance. In the past this was normally more than satisfactory when partitions were thicker or made of denser material. The lighter, cheaper solution may become critical in some situations where the soundproofing performance of the heavier solution had always been taken for granted.

The need for composite construction leads to the search for materials with properties matched to the specific functions. Processing or manufacture of materials allows these properties to be more accurately achieved. This not only changes the ability to gain performance but also influences the production resources required.

Materials and manufacturing methods may also be changed to provide more economic buildings. The innovation may not be driven by performance but in order to seek production and cost gains. This may lead to performance failure in previously satisfactory solutions. Those responsible for the choice of construction must re-evaluate the possible performance failures if construction is changed for production or cost advantages.

Taking storage and equipment into account

The environmental requirements that have been introduced are those necessary to maintain the well-being of the people who use the building. These can be summarised as:

- Dryness
- Warmth
- Cleanliness
- Light
- Quiet

Observations of this list will identify that, with only slight modifications (e.g. warmth needs to be more generally interpreted as temperature control), it could equally well be applied to maintaining the well-being of the equipment and stored items. The required internal conditions for storage and equipment may be different, even hostile to life as in the case of cold storage. They will still demand the same type of functions of the construction. There would, however, be changes in performance level due to the change in specification for the internal environment, even though the external condition may be the same. In the example of cold storage, levels of insulation would be different.

More significantly, the equipment and stored items, as well as the living occupants, will contribute to the internal environment with noise, smell, heat, fumes, etc. It may be that the construction is called on to moderate these effects by isolating storage from living areas by partitions. Independent technologies in the same room such as cooling of food may be applied using ideas such as evaporating water from a porous pot or a refrigerator. However, independent technologies may influence internal conditions. In the example of cooling food the porous pot releases vapour and the refrigerator releases heat.

The dynamics of these changes in the internal environment due to occupants and independent technologies in the building must become part of the analysis. Sealing up a building to reduce heat loss through air changes can lead to a build-up of vapour (if only from people breathing), which makes condensation more likely.

Although the list of environmental requirements generated for people will cover most circumstances associated with the equipment and storage in the building, it may in some situations be incomplete. Some modern equipment has environmental operating requirements, such as the need for anti-static precautions, that may not currently be considered necessary for the well-being of people. What has to be achieved is a balanced solution to ensure the well-being of both the participants and the processes associated with the activities for which the building is required.

What should be considered tolerable?

Setting performance levels involves considering what is tolerable. What should be considered as normal limits to be maintained in the internal conditions? What external conditions should be allowed for in the design? These questions have to be answered in terms of risk and consequences. Extremes that happen rarely but have serious consequences to the well-being of the activity or even threaten life and property may well be taken into account. The more frequent events, even with lesser consequences, are also more likely to be taken into account. This is not independent of the economic resources available and social attitudes. In many parts of the world, earthquakes will occur, but the construction to minimise the risk to life is seen as increasing costs beyond that to be extended to all sections of society. Similar circumstances have existed in the past with fire and building. Increasing wealth, effectively distributed in society, will affect what is considered tolerable, reflecting more demanding performance expectations.

Generally, wide limits would be accepted for solutions at the beginning of our mind experiment, but as societies develop they seek to extend the range and level of demands. This often includes a requirement for a greater level of control as it defines a narrower band within which performance is deemed satisfactory. These trends place more demands on the solutions and failure is more likely.

Some extremes that happen rarely have to be considered if they have serious consequences. One risk that comes from the internal activity that generates extreme conditions has already been raised. It is the issue of fire within the building. At the beginning of the mind experiment, while escape for people would be required, it is unlikely that there would be any expectation on the part of the user that any of the building would remain after a fire. Although fire is now rare in modern buildings, its consequences are such, particularly in risk to life, that it is currently deemed necessary for the construction to continue to maintain a selected range of performances both during and after the fire. This would almost certainly include stability, although insurance requirements may extend the performance range.

External climatic extremes such as wind, flood, lightning and even earthquake may be considered. Design for flood may be sensible if the building is sited by water, for lightning if a tall building and for earthquake if the building is in an unstable region of the world.

New threats created by increasing sophistication of solutions are now emerging. Draught-free homes created to conserve heat may suffer from a build-up of radon gas if built in an area of the country where it occurs naturally in the rocks below the building. The chemical contaminants in the soil on sites used for redundant industries threaten both the occupants and the fabric of the building itself.

Introducing social requirements

Having established the requirements for health and physical comfort, it is now necessary to consider the social aspects. In reality the consideration of performance for more social aspects does not come only when physical requirements have been fully satisfied. Human endeavour very quickly becomes a social activity and there grows a need for social organisation, establishing cultural norms, provision for leadership and political activities and an economic system for the exchange of goods and services. Buildings play their part in this process not only to house the activities but also in that they reflect, affirm and even become instruments for change as the cultural and economic order develops.

As well as the issues of status and image, to be discussed later in this chapter, two other functions associated with well-being arise when society starts to develop. These are the requirements for security and privacy. They are mainly associated with mental health, although the security risk, if it becomes a reality, may involve physical danger. The point at which people would take these functions into account would be dependent on the nature of the society that develops. They are not as predictable as the physical needs and may well have to be addressed before the desired level of physical comfort and safety is achieved.

Security requirements will be dependent on the existence of a perceived threat. If the risks are seen as sufficient, technological ingenuity, time and resources will be devoted to combating the threat if social devices such as the rule of law or treaties cannot be achieved. Threats come from both inside and outside a society, each of which pose different challenges to the technological skills of the builder. In the extreme, in buildings such as bunkers, the security threat dominates the design as the technology of survival becomes pitted against the technology of destruction.

The requirement of privacy is much more rooted in the cultural attitudes to community. It is cultural interpretations of what activities should be shared and those that should be carried out in private that dictate the need to introduce this performance into buildings. Performance levels of privacy can be identified associated with the senses, visual and aural being the two main requirements of the building influencing both the external envelope and normally demanding some division of internal space.

The functions of the building that have a social origin often translate into aspects of physical performance already introduced. Security makes structural demands on the fabric. Privacy determines many of the sound transmission requirements of construction.

Influence of production resources

As more functions are included and more materials are introduced, production resources and know-how have to develop. The relationship between the need for a building and the availability of the materials, skill and tools to produce it is always a determining factor in the choice of the construction. Access to production knowledge and expertise limits the pace of development of new solutions to achieve new performance. However, developments in materials and production can also suggest different ways of achieving similar performance.

This mind experiment has taken the perspective of the purpose of the building and requirements of the user. It is easy to forget that the means of realising that performance may need as much thought and ingenuity as the design and detailing of the fabric and structure of the building. In many old buildings it is easy to see how they work but often difficult to see how they might have been built without appreciating the tools and skills available at the time. In many innovative buildings, detailing is often determined as much by manufacturing or erection procedures as it is by satisfying performance. The availability of manufacturing and assembly resources also determines the cost. The need to develop production procedure affects the time to construct and risk associated with a particular solution. All these contribute to the final quality of the building and its success judged by performance.

Considering cost and value

Throughout this mind experiment the question of cost and value will have to be considered. Buildings are usually costed in terms of money, but there are also considerations of cost to the environment and on the impact they may have on the society itself. The resources available to any society are limited and decisions have to be made as to how these should be allocated. As part of the culture these are the political decisions involving both the vision of the type of society that should emerge and the preferred methods of achieving it. It is from this cultural process that demand is initiated, the cost of resources determined and value assigned.

Buildings will only be commissioned when the cost of a building that performs to an adequate standard is considered good value. If the costs are too great, decisions have to be taken on eliminating functions, reducing performance or changing production methods. These changes have to ensure that the building is still considered as value for money. Assigning this cost requires that at least a broad outline of both the construction and the production methods is known. At some point in the process the decision to proceed to full design and production based on estimated cost has to be made. For buildings that can be produced using well-developed current forms of construction this decision can be made very early with some certainty. For a project where there are few precedents, possible solutions may have to be investigated before cost can be established with sufficient certainty to take the decision to proceed. It is, however, always a requirement to set cost limits as early as possible and then develop the final solution with these costs in mind.

Building for status and image

Status and image will at some time in the development of society start to influence the requirements of the building. This may be reflected in the size of the space provided over and above that strictly required for the physical satisfaction of the activity itself. It may be associated with the quality of the materials used being acknowledged as expensive. These expensive materials must have the ability to perform their physical tasks without failure, but they could have been replaced by a less expensive alternative. These requirements are often then overlaid with issues of art and fashion and symbolism. The influence of these should not be underestimated, but a full discussion of their nature and form is inappropriate in this text.

The pressures of status and image are not just related to statements of wealth and position. We all need to make statements and feel comfortable about our place in society related to the prevailing cultural norms. It has to be recognised that these cultural norms can be highly regionalised, being associated with tribes or nations, although improved communication seems to reduce this variability. We all care about what our buildings look like and the environment they create. Using and being associated with buildings that match the psychological demands of the activities carried out in the building is very important to our mental well-being.

These issues become associated with aesthetics, the visual and deeper, even spiritual, quality of beauty of both natural and manufactured articles. These are the aspects of the world that satisfy our feelings rather than our physical needs. Buildings intrude into the rural landscape and create the cityscape. Their shape, form, colour and ornamentation have a significant impact on our lives.

Control from society and legislation

All social groups have cultural norms of behaviour that limit the actions of individuals and impose sanctions for transgressions. This eventually has to be formalised into legislation, some of which will start to impinge on the choices that may be made for construction. Direct legislation is normally associated with public health, limiting the use of scarce resources or the position and appearance of whole buildings. Once manufacture and assembly start, there are laws associated with the health and safety of operatives and people in the surrounding area.

All these will have to be acknowledged and satisfied in the final choice. If the choice of construction is made in the sprit of responsibility within the cultural norms, it should not dominate the solutions chosen. Legislation is for the protection of those who use the buildings and should, therefore, be easy to accommodate into a process whose objectives include satisfying social and individual needs.

In reality the law on these matters can be slow to change and may be drafted such that its interpretation becomes a matter of opinion. These factors can spur or inhibit technological change, but it is clear that in an advanced society those responsible for the choice of construction have to be aware of the law and its interpretation.

There is much legislation that will indirectly influence the choice of construction, although predicting and understanding the links are more difficult. Contracts become necessary between the organisations that undertake different parts of the building process. This legislation affects decisions on the construction and what risks they carry. The knowledge and experience that the organisation has then influence the choices made for the construction of the building.

Clients, users and change of purpose

Another complication that arises as society develops is that individuals who become the owner/occupiers commission fewer buildings. It is now possible for those who commission the work to be an investment company who pass on the task of operating the building to another company for use by yet another group of people. This does not nullify the idea that satisfying the needs of the user should govern the performance of the construction solutions. It is the provision of space that ensures that the desired activities can be carried out efficiently and effectively by the user that gives value and utility to a building. Those who commission the building may have objectives such as investment, but these will generally only succeed if the purpose of the building and its performance are satisfactory.

Even when relatively simple relationships between client and initial occupation exist, the purpose of the building is likely to change with time. These can be quite dramatic changes such as churches to flats, but most are much more subtle. Changing practices in medicine have made many old hospital designs inefficient and the general electronic information revolution has changed the basic activities carried out in many offices. Increasingly the operators of buildings (as opposed to the users) see the building as part of the business's efficiency. This gives different value to maintenance, alterations and refurbishment and makes the building a more integrated part of the facilities as a whole. This leads to the questions about the whole life costs. More expensive initial options may be better value over the life of the building, reducing operating costs and increasing efficiency and well-being.

Environmental impact and sustainability

This long-term view of the building leads to wider questions about environmental impact. Construction has always had a local impact on the environment; it takes resources, changes ecosystems, consumes energy and creates waste. The society in both our mind experiment and now in reality has grown more prosperous; infrastructures have improved, allowing the growth of manufacturing and materials processing to serve an expanding population with ever higher performance expectations. The environmental impact of technologies chosen for individual buildings can now contribute to global consequences.

This demands yet more analysis, with new knowledge required in order to make choices about the technology we employ for our buildings. It will lead to different solutions being favoured, innovation and risk. The story continues.

Summary

1. The purpose of the building is to support the activity to be carried out in the building using both physical and social criteria.
2. Both physical and social functions of the building lead to physical performance requirements covering environmental, structural and reliability aspects of the construction. These requirements lead to performance expectations and hence definitions of failure.
3. Choice of construction can only be made in the knowledge of the environments in which the construction operates and an understanding of the mechanisms or actions that lead to failure.
4. The other factors to be considered in making the choice of construction include production resources and know-how, cost, legislation and the need to consider the whole life of the building and its environmental impact.

3 Common Forms – Specific Solutions

This section develops the idea that in any society there is a recognisable set of common building forms. These common forms will be drawn from a range of generic types that are then worked into various specific solutions. The generic type and constructed form are two distinct ways of identifying the construction that is adopted for a building. The conceptual generic type is characterised by its arrangements of components and action in achieving performance. The physical constructed solution has defined materials, size and shape of components, joints and fixings. This construction solution becomes specific to an individual building, although the building may be repeated many times and the detail may work in many buildings.

General and constructed forms

In any society at any time there is a pattern of needs that supports a recognisable series of common building forms. For each of these building forms there will be an interrelated set of construction solutions to cope with the size of the building and the resources affordable for each. Construction solutions are derived from a limited set of generic types. Technological choice often starts with the selection of an appropriate general form that is then developed into a detailed construction solution for a specific building.

The ability to describe a building in these two forms, the general form and constructed form, at the early concept stage in the design of a building is most important. It should be the big idea from which the individual details can be developed so they work together to ensure the performance of the building as a whole.

The emergence of general building forms

It appears that the prime determinant of the construction of a building is related to space. The users' demands will translate fundamentally into spaces and their size and interrelationships will suggest the scale and shape of the building. At the domestic scale small cellular spaces lock together, while many commercial or public buildings require more voluminous spaces, often with smaller associated spaces such as toilets, to form the building as a working unit.

Construction solutions are directly related to this demand for space. Most significantly it influences the structure, but these choices have inevitable consequences on the solutions available for the construction of the enclosure and other elements of the building.

Currently there is a wide range of spatial requirements both on plan and in the height required that leads to the use of a range of structural forms. Three general groups of structural form can be identified:

- Loadbearing walls
- Skeletal frames
- Long-span roofs

Each of these will create a range of building forms, some examples of which are shown in Figure 3.1. Each building form will be made up of a combination of a range of structural

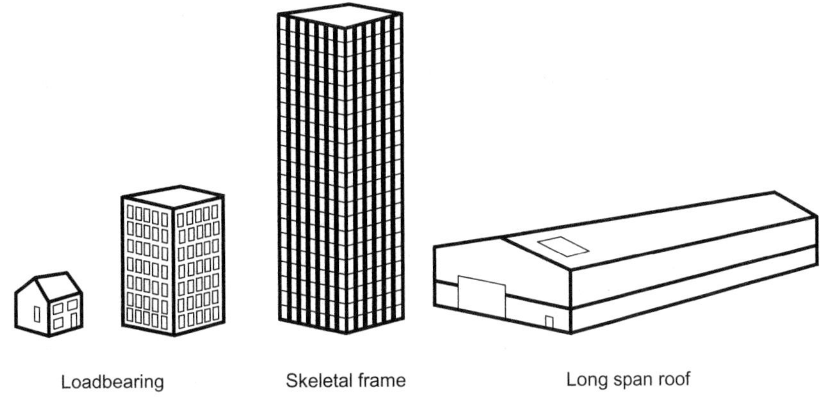

Figure 3.1 Typical form associated with types of structure.

elements. An example of a generic structural element is the beam. With its structural action in bending it appears in cellular loadbearing construction as joists for the floor and a lintel over an opening, in skeletal frames as a main structural member and in the long-span roof in a portal frame. This generic type has been adapted for specific solutions using different materials to achieve the required characteristics: great strength, small deflections, small section size, low weight, good looks, low cost, fast erection. Each of these characteristics will have a greater or lesser significance in a specific building. In a house the joist will most likely be timber; in the skeletal frame the beam will generally be steel or reinforced concrete. The portal frame is most likely to be made from steel, but for a building of the right use, scale and size it could be made from reinforced concrete or even from timber in a laminated form.

While the ability to enclose the space is the primary determinant of the generic form, design and resource conditions will lead to a dominant economic detailed solution. This leads to two ways of thinking about a building, each having it own vocabulary to describe the solution. Early in the selection of the construction it is helpful to think about the elements of the building in generic terms. This can then be developed into a full description of the constructed form, including specification of materials, joints and fixings.

Generic elemental – domestic, loadbearing walls

One widespread use for the loadbearing wall form is the house. Described in their elemental form, houses in the UK at the beginning of the twenty-first century employ a semi-permeable (cavity) construction for all external walls to provide (with internal cell walls) a stable dry box. This box will be roofed with a tiled covering supported on a trussed rafter structure taking the loads directly to the external walls. This means that any internal loadbearing walls need not extend to roof level. Suspended floors will be joisted, often taking advantage of internal walls to limit spans, making some ground-floor partitions loadbearing. Ground floors may be ground supported or suspended floors depending on the site and ground conditions. Foundations are most likely to be strip footings unless the ground conditions are very poor. This form is shown in Figure 3.2.

Many houses built in the first half of the twentieth century and still in use were not constructed like this. Many have solid permeable walls and have roofs made from individual structural members (rafters, purlins, struts and hangers) which, when acting as a whole, require internal wall support. Later in the twentieth century the width of terraced houses reduced and cross wall construction, where only walls

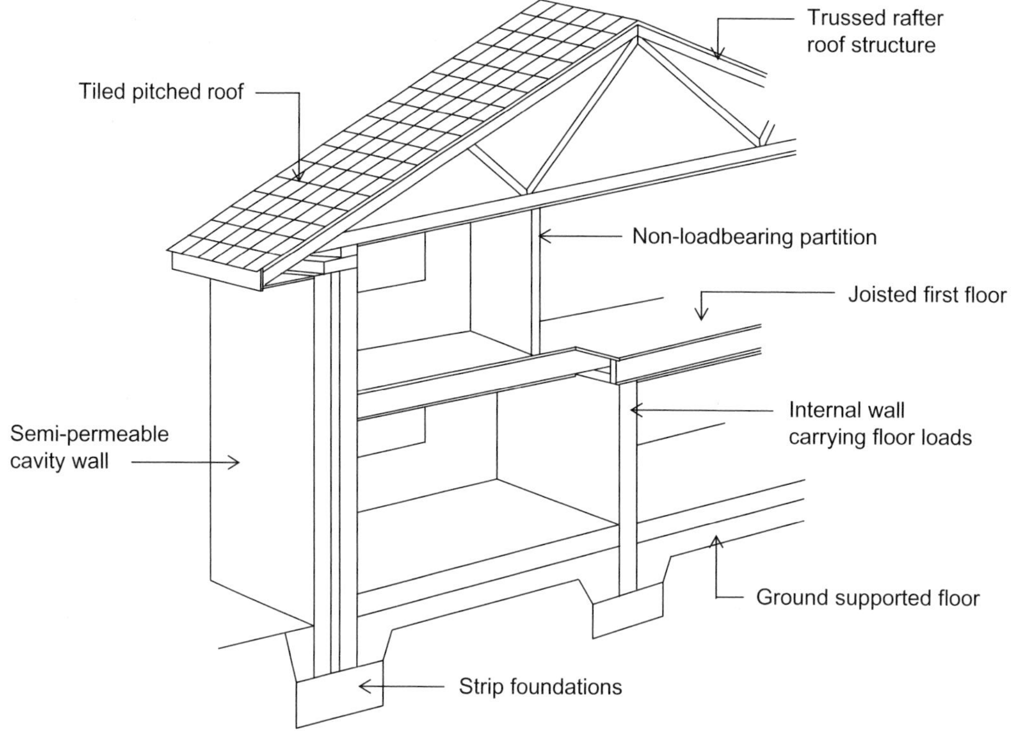

Figure 3.2 Generic description of house construction.

between the houses are loadbearing, became an option. This lost the stabilising effect of the box but allowed a range of less expensive external back and front walls to be chosen.

As the loadbearing wall solution is developed into multi-storey flats other elemental forms have to be employed. Floors with greater sound and fire performance requirements become more economical as slabs, not joisted, even though the spans might not have changed significantly. The loadbearing walls may become prefabricated, taking advantage of the repetitions floor by floor to gain a production advantage.

As this arrangement is developed into multi-storey blocks, the alternative structural form of the skeletal frames will become an economic alternative, even though the plan layout still shows walls dividing the building into small spaces with limited floor spans that made the loadbearing wall option viable.

Constructed form – domestic, loadbearing walls

None of the generic descriptions talks about the material, the shape or size of the components or the joints and fixings employed. They give the elements general names and describe them in terms of their general form or how they perform. For the constructed form, details have to be worked out to ensure economic sizes and manufacturing techniques. From this a recognisable physical constructed form will appear. The constructed form of the generic elements shown in Figure 3.2 is given in Figure 3.3. This shows concrete single-lap interlocking roof tiles supported on timber battens over felt fixed to timber-trussed rafters jointed with galvanised steel gang-nail plates. The external walls are masonry, although timber frame is a viable alternative. The wall may be constructed of brick outer skin and block or timber framed

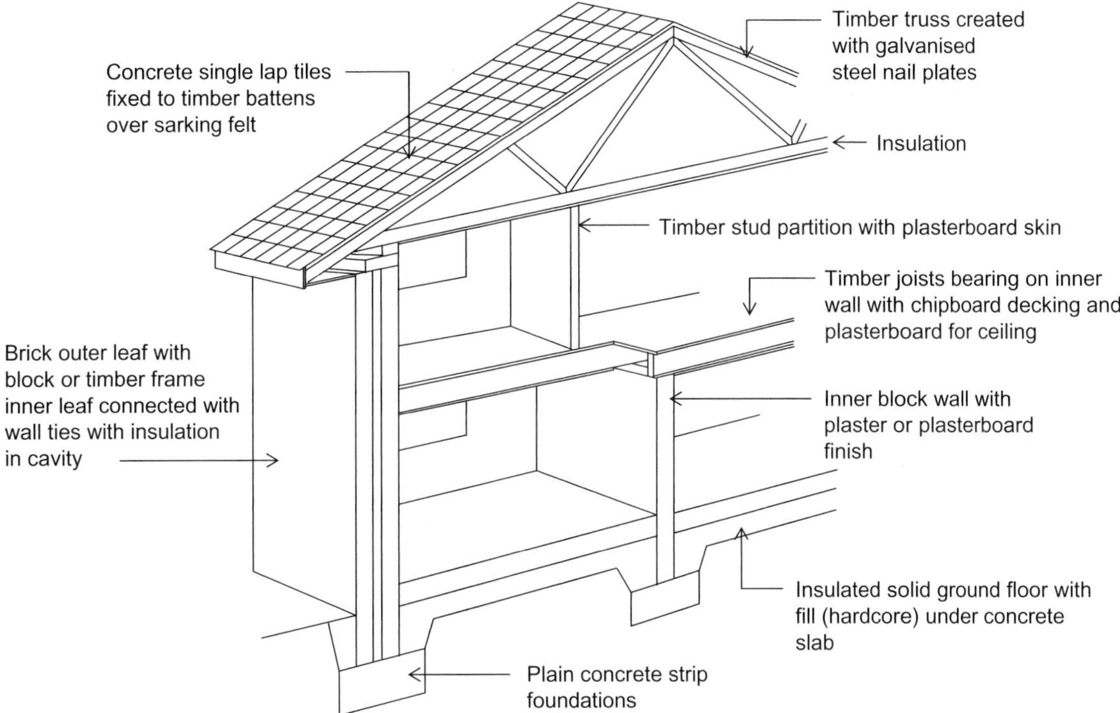

Figure 3.3 Constructed form description of house construction.

panels for the inner leaf. This is unlikely to meet the requirements to limit the passage of heat. The most cost effective way to limit heat loss in the winter condition (loss of heat from warmed inside air) is by insulation incorporated into the wall. In summer the design condition is heat gain from outside. As this gain is predominantly from direct radiant heat from the sun, shading or thermal capacity (not shown in Figure 3.3) may be better actions than insulation.

The floors are timber joist supported by resting either directly on the inner leaf or on galvanised steel joist hangers and finished with chipboard sheet flooring. As the ground-floor partition walls carry the joists they are constructed with concrete blocks. However, as the trussed rafters only transfer load to the external walls the first-floor partitions only need to be timber stud with plasterboard finish, as long as they can provide appropriate levels of sound insulation. Plasterboard has been used for the ceilings and as a finish for the downstairs blockwork, although gypsum plaster may be used on any block walls in the house. The external wall would have to be modified below ground and the foundation would be a strip of plain concrete.

The size of each of these components has to be chosen. Common components, such as the brick or the standard range of windows, have been developed in appropriate sizes to work with the domestic scale and performance levels. The size of some components, such as the trussed rafters and floor joists, has to be designed to the size and use of the building.

In the same way as houses in the past have been different to the construction shown in Figures 3.2 and 3.3 there is no reason to suspect that either generic forms or the physical solutions will endure into the future. They will evolve and develop to meet new design needs and resource conditions.

Other general forms and current practice

Buildings using skeletal frames and long-span roofs show a greater variety of both generic elemental types and physical construction solutions. To try to catalogue them here is not appropriate. However, a few examples serve to introduce these forms as both generic and physical entities.

The skeletal frame structure has beams, columns, slabs and some means of ensuring overall stability. This basic format can be repeated to generate large buildings both on plan and in height. These structures will, however, need internal structural columns or walls, thus imposing some limits on clear internal space. Where this structural form has no current rivals is in its ability to achieve great heights with relatively open internal spaces. The physical constructed form is generated in either steel or reinforced concrete. The reinforced concrete can be either precast or in situ. These materials can then be mixed between the elements. Most steel solutions have steel beams and columns but concrete floors and either concrete or steel wind-stability provision. Reinforced concrete structures normally have reinforced concrete members throughout, but the development of flat slabs as spanning structures has eliminated the need for beams.

Long-span roof structures are used for industrial, retail, storage, exhibition, sports venues and other activities that need large floor space under the roof. They are normally only single-storey, although they may require considerable internal clear height. Some buildings used for these purposes are not large, and so although termed long span because of their potential the generic forms may be used for buildings of modest scale.

Long-span roofs have perhaps the greatest variety of generic elemental forms to be developed into construction forms. Both plane and three-dimensional structures in flat and curved forms have been developed in steel and reinforced concrete and even timber and aluminium.

Both skeletal and long-span structures have to be enclosed. The enclosing fabric can normally be supported and restrained by the structure and there is a wide range of options for the external walling. Again, generic forms such as cladding or facings can be identified, with constructed solution in materials such as metal, concrete, glass or brickwork being chosen for the specific building. The different structural forms tend to use different enclosure solutions. In long-span structures the building often requires less openings, and reducing the weight of the fabric allows economic gains to be made in the structure. This leads to sheet cladding in metals with heavier forms such as brickwork used only around the building at ground level, where it can be supported directly on the ground. In the multi-storey skeletal form the interaction of the enclosure and the structure plus the need for performance often lead to more complex construction.

Building services

To illustrate the idea of the general form and the specific solution, descriptions have been given for the construction of the structure and enclosure elements. Buildings also include a group of technologies collectively known as building services. It is not the intention of this chapter to introduce their general forms and possible solutions. This is not because they are any less significant in supporting the everyday use of the building and the well-being of the occupants. Each building will have to be resolved with a mix of structure, enclosure and service systems to meet the performance expectations of the building. Hopefully in this chapter the point about general and specific forms has been made with the examples of structure and enclosure. The general forms and possible solutions for services will be introduced in later chapters in the book.

Uncertainty and risk

Where change is slow, there is the opportunity for current detailed solutions to be adopted with little modification from one similar building to another. There is always danger in assuming two building to be the same, but if the general form is robust and not being applied close to a failure cusp the solution can be suggested in detail without much analysis, with little risk.

As the pace of change increases, initial choice may deviate from currently detailed and even generic forms. The initial suggestion becomes based on an understanding of generic forms that then need to be thought through in detail. Initially this may only need some alterations in materials or detailing, but as demands become more divergent from common practice then alternative elemental generic forms in new combinations may be more appropriate.

In making any technological choice it is important to be aware of the uncertainty and risk in adopting a particular solution. The technical risk is one of failure in performance. As with any risk it must be weighed against the consequences. Consequences that lead to loss of life always require careful attention. Where the consequence is interruptions in the operation of the building, the severity of the loss will have to be evaluated against the risk of failure.

Whatever the pace of change, the choice of construction is based on an understanding of the action of the generic form and the performance of the detailed design and availability of resources to construct the specific solution.

Summary

1. Elements and components have a limited range of conceptual generic forms based on their action in achieving performance.
2. Detailed construction solutions have to be chosen for specific design and resource conditions and are therefore ultimately specific to an individual building, although the building may be repeated and one detail may work in more than one building.
3. Within each society a range of common building forms and economic construction solutions arise which normally form the basis for choice for subsequent construction solutions.
4. Particularly in times of change, the process of choice has to include analysis that recognises the uncertainties and risks for each individual building in both failing to achieve design performance and acknowledging the reality of available resources.

4 Construction Variables

This chapter explores the variables that define construction: materials, shape, size and spatial relationships. It introduces the reality of deviations in assessing whether the components will fit together and perform. The chapter concludes with a section on the specific aspects of jointing and fixing as the process by which the parts of the building can contribute to the behaviour of the whole.

The variables of construction – the outcome of choice

It is important to recognise at the start what aspects of the physical construction can be varied in order to achieve the desired performance of the building. Identifying what can be varied also identifies what has to be chosen and specified if an unambiguous solution is to be defined.

Choices have to be made for the materials to be formed into components that have a defined shape and size. Each component has then to be given a spatial relationship with other components together with the joints and fixings necessary for them to act as a whole building. This simple identification of the variables is represented in Figure 4.1.

Defining the final physical arrangements of the construction involves a specification for each of the component parts of the building, including any joints and fixings, detailed in assemblies of components to create the final building. This specification will have to include:

- The material of which components are to be made and the quality of that material
- The shape into which the material has to be formed
- The size of each component, giving dimensions sufficient to define the shape
- The relationship of each part either to an absolute position defined by a reference framework or relative to another part of the construction or the surrounding environment

This specification determines the space that the components and their joints will occupy. It may be in competition for this space with other parts of the construction and/or the usable space of the building itself. The choice of construction cannot be confirmed until this has been checked and it has been established that there is a unique space for each part of the building.

Joints and fixings also need to be identified by their component parts, their material, shape and size and spatial relationships, as they themselves are part of the assembly. These assemblies are then connected (where they are often identified as junctions) into the whole building. Any components associated with junctions also have to be specified.

The process of choice is about defining the components, their materials, shape and size and their spatial relationships, including the joints and fixings necessary to complete the building. The focus of technological knowledge is taking decisions about these variables to create a building that is safe and has utility. This is a technical choice that has to be made following the creation of a design concept and coming before the management of the production process. Yet it is a choice that has to be made in the full knowledge of the design requirements and the available production resources.

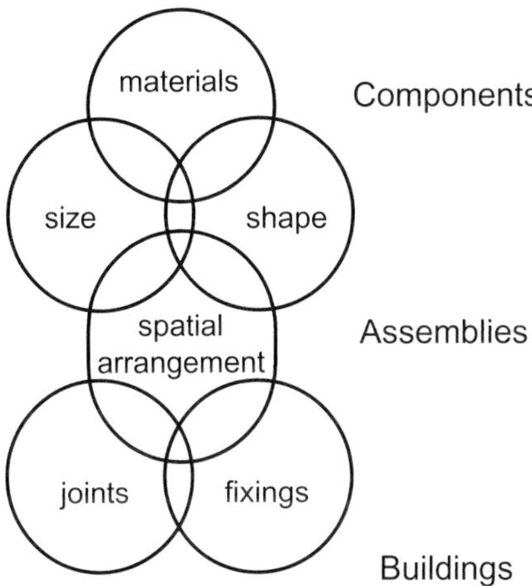

Figure 4.1 Variables or outcome of choice.

Choosing materials

Specifying materials is a process of matching the properties of the materials to the performance requirements. This is not, however, independent of size and shape, for all materials have all properties to some degree or another. We only choose to exploit certain properties when their value is such that, in economic quantities and at sizes and shapes that can be produced, transported and fitted within the space available, they will perform at the required level.

Buildings are perhaps the largest manufactured items widely used throughout all societies in the world. They demand relatively high quantities of materials at reasonable cost and it is, therefore, not surprising that, historically, materials used for everyday buildings are those with high bulk, locally available and with limited process costs. Increasing economic wealth and its attendant higher levels of infrastructure allow the use of materials requiring greater processing off site and less need to rely on local materials. With economic development comes the possibility for transfer of production activity to factories remote from the site, which in turn gives opportunities to use a wider range of materials than a wholly site-based process.

The materials that have been dominant in building in the past are timber and stone, with lime, limited amounts of iron for fixings, lead and glass. Fired clay for bricks and tiles has been used for some time in the UK, but up to just before the turn of the nineteenth century these would have been only used locally, particularly in the conurbations where there was a sufficiently large market close to the kiln site. In rural areas such as Devon, soil was still being used for agricultural cottages up to the end of the nineteenth century.

Perhaps the two materials that have made the greatest impact on the size and scale of everyday building have been steel and concrete. The superior strength properties have allowed greater loads to be carried on smaller sections but in particular have introduced the possibility of smaller, efficient spanning members such as beams. This opened the way to large, comparatively voluminous building to a wider group of building clients. In the last half of the twentieth century there was an increasing number of materials introduced into buildings. More metals such as copper and aluminium now are used, particularly in services. Plastic has made a considerable impact but only where its relatively high cost can be justified by thin-sectioned components. Processing and manufacturing developments have offered some old materials new opportunities, with techniques such as laminating timbers.

As well as increasing the range of materials, developments have been made that offer more grades with greater quality control of the more traditional materials, allowing even finer choices to match performance with cost. All this demands increasing knowledge. Evaluating the more recently formulated materials is more difficult for many have not yet stood the test of time. More materials introduce more possibilities for incompatibility so even well-tried and tested materials may fail in new combinations. Adverse reactions between materials and between materials and the environment may threaten durability in ways that have to be imagined before an analysis can be undertaken.

In imagining failures it is possible to argue that all failures are ultimately materials failures.

It is also possible to argue that the failure of the material is always the consequence of a previous sequence of events and therefore never the direct fault of the material. The usefulness of this argument possibly lies in accepting that both are true. In considering the behaviour of the material, it is necessary to identify the significant events in the life of a component when a coincidence of conditions may lead to failure.

The events that lead to failure will determine which properties of the material will become critical. Significant events may well be during manufacture, assembly or during the life of the building. Understanding the production process as well as the behaviour of the finished building is vital, for the most demanding event in the life of a component may well be in storage, handling or as part of a partially erected building. An example of this is a precast, reinforced concrete pile where transporting and in particular driving induce entirely different stresses in the materials than when it is under load as a foundation.

Materials – health, safety and the environment

While it is the properties of a material that may identify it for use in a building, there are reasons why it may not be preferred or even excluded from the ultimate choice. Cost and the availability of local manufacturing techniques have already been introduced, but aspects of health, safety and the environment also have to be considered.

There are many reasons why a material may be hazardous to health, particularly during the manufacturing and assembly processes, and these will be discussed in more detail in Chapter 13 in the section on health and safety. However, materials have also to be evaluated against their impact on the environment and their contribution to sustainable development.

Society is reassessing its relationship with the natural environment and the demands and stresses we place on it. Much of the concern is associated with our development as a technological society. This new thinking is bound to play a major part in the future in the way we build buildings. Such concerns as energy use, biodiversity, waste and pollution can be directly affected by our choice of building materials.

In assessing a material for its environmental impact it is necessary to trace its entire history, often known as cradle-to-grave analysis. We can only use the materials provided by nature. Any material that finishes up in a building will have been part of an ecosystem that will be altered by the removal of the resource we require to make and shape our components. In many cases the effect will be local and nature will have the capacity to restore the balance in a short amount of time. However, as the volume of development increases and infrastructure becomes available to exploit resources on a large scale, the restorative powers of nature may be severely tested. The impact becomes more than local to the extent that the exploitation of the earth's resources begins to have global implications.

This is the fear at the beginning of the twenty-first century. Concerns associated with resource depletion, loss of biodiversity and global warming are all affected by decisions to use certain materials in buildings. This is not just in the extraction of materials but also in their transport, for the business of building materials is now part of global trade. There may be a preference for a renewable resource if this can be identified.

Processing and manufacture use energy and produce waste and pollution. In an attempt to compare the impact of producing materials we use the idea of embodied energy or embodied carbon. It is the sum of the energy or carbon used to extract and process materials normally calculated up to the delivery to the site gate. Where components are delivered to site, the energy or carbon used in manufacture may vary even if they are made of the same material. This is discussed in more detail in Chapter 15. Assessment of embodied energy or carbon allows not only comparisons between materials but also the evaluation of using recycled material where process energy may be less than that of the original conversion. Waste and pollution also pose a major problem in that returning the by-products of processing to nature may place a burden on the environment it is not capable of dealing with. Disposal strategies are a major concern in evaluating the environmental impact of the choice of materials.

Disposal is of concern not only at the stage of manufacture and assembly but also at the end of the life of the building itself. Specifying materials where the recycling opportunities are limited and the disposal and pollution costs may be high will have to be another part of the analysis of the environmental impact of the choice of construction.

Choosing shape

The shaping of components can bring great benefits in terms of performance. It can give strength by corrugating comparatively thin materials and can provide ornament and finish to satisfy appearance requirements. It has a particular role to play at joints and at points of fixing, where shaping such as grooves and rebates can be the major contribution to achieving function. Shaping is, however, dependent on manufacturing processes that require skills, tools and machinery. This adds cost to the component that has to be justified by its utility or a saving in material. It is also dependent on the material chosen, not only for its ability to be worked with current process skills and technology but also for its ability to be worked to fine enough deviations (precision) to achieve the allowable tolerances required for performance.

Early construction methods exhibit the characteristic of minimal working or shaping of the raw materials. Exploiting shape comes with increased resources where skills and manufacturing technique become available. Shaping is achieved either by forming the shape directly from a formless material or by working from simple shapes available from the natural form of the raw material. In some cases these are combined where a crude shape is formed and then refined by removing material to work towards the final form.

Shaping has to be related to an appropriate manufacturing process. Techniques such as casting, rolling or extruding can shape a formless material such as concrete, clay or hot metal. Cutting or machining can then reduce materials to refine or modify the shape. Parts can be added by fixing additional pieces to make complex shapes from simpler forms. Simple forms can also be bent, pressed, beaten or forged into more complex shapes. These processes introduce a range of materials properties associated with workability. They may make a material with all the properties for performance unusable or at least too expensive to use.

Many of these processes are essentially factory based; the range of methods that can be undertaken on site is still limited in most cases by individual skill and tools, leaving pouring, cutting and attaching with some bending as the predominant shaping processes.

Choosing shape and appreciating its performance value and production cost is a major consideration in technological choice. It has particular significance in the issues of deviations and fit, to be discussed later in this chapter. Where parts with complex shapes have to fit together, there will be particular demands on accuracy that limit the ability of the assembly to accommodate distortions. Hence, while shaping of individual components leads to many opportunities in jointing and fixing, this must be considered carefully in terms of the components' relationship to the assembly of which they form a part.

Choosing size

So far it has been established that the level of performance required from the construction leads to the initial choice of material based on a match between the function and a property that can be exploited economically. It has also been established that, for each material, certain processes such as cutting or shaping may be available to make best use of the material and facilitate assemblies that function as whole elements of the building. It is now necessary to establish dimensions sufficient to define the shape, allow manufacture and ensure assembly. In the process of choice it may be that there is a limit to the size of the component that will determine the choice of material and/or the shape. This limit may be associated with the need to fit together with other components or there may be manufacturing/handling limitations. It may be that visually the size of the component confounds the design vision.

It is helpful to see the establishing of dimensions in three phases working towards a manufacturing or working specification:

- Gross size – based on performance
- Coordinated size – based on incorporation with other components
- Working size – based on jointing, fixing, deviations and movement

The gross size will be determined by performance. It represents a minimum size for waterproof materials to keep out the rain or for structural elements to support the loads. It will, therefore, identify the scale and proportion to the solution and may even show at an early stage that a solution is not possible with that material with the chosen shaping in the space available for the construction itself.

If the gross size appears to be sensible to achieve performance economically and within the space available, it is then necessary to look at the dimensions in terms of the detail and how the parts of the construction will go together. This is common, for instance, where the major component is incorporated into brick or blockwork. This is illustrated in Figure 4.2 for the sizing of the depth of a concrete lintel over an opening in brickwork. Structural calculations may show that the lintel only has to be 125 mm deep. However, the co-ordinating depth of a brick course is 75 mm (65 mm for the brick and 10 mm for the joint), so the lintel would have to have a coordinated size of 150 mm.

Establishing coordinating sizes becomes particularly significant in more industrialised and prefabricated buildings that are often

(a) Gross depth based on strength performance

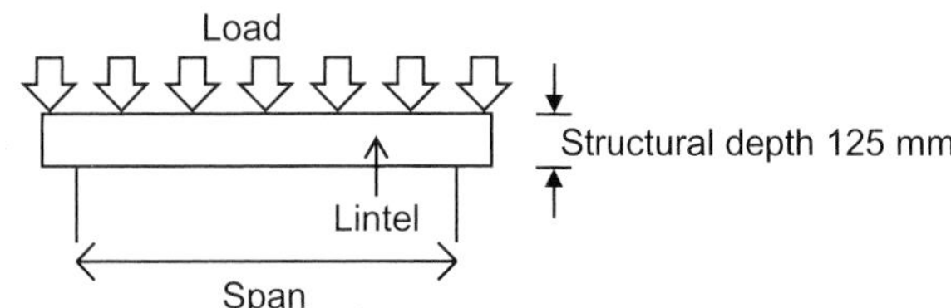

(b) Co-ordinated depth based on brick courses

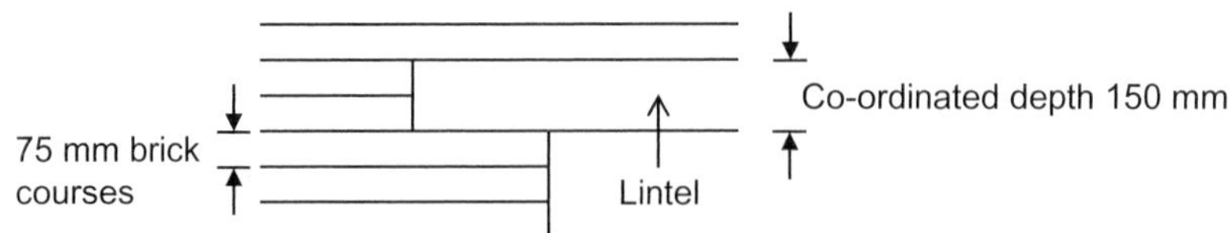

(c) Working depth based on joints and deviations

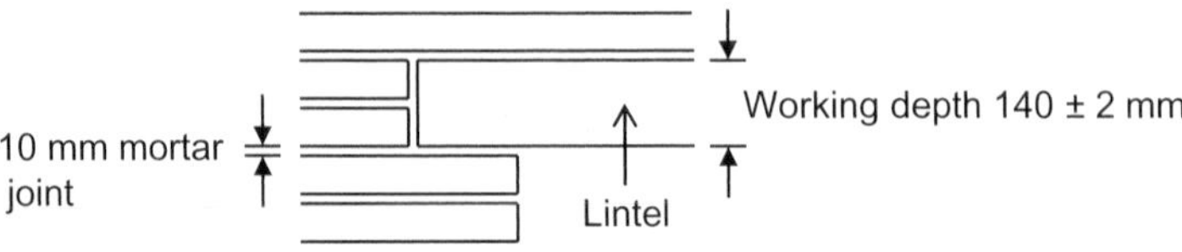

Figure 4.2 Sizing depth of lintel.

designed within a dimensional discipline around a module, leading to a basic or modular size being established. The most common module is likely to be 300 mm. When selecting factory-made components available from manufacturers' catalogues, the economics of production means that they are often only available in a limited number of sizes. Although these will be based on a module most likely to maximise their potential for integration with other components, they do limit the choice without incurring additional cost.

The coordinating size does not take into account:

- The need for space for joints and fixings
- Allowance for deviations and movement

These have to be considered together, as in many cases the joints will have to allow for the deviations and movement, and the fixings will have to accommodate both deviations and movements if the assembly is to fit together on site and maintain its integrity in operation. Consideration of these aspects will lead to a working or manufacturing size.

Joints are a necessary part of the assembly process to ensure fit in that they can take up deviations in components. Fixings are required to hold assemblies together and these also take up space. Sometimes the joint is also the fixing. The material of the joint and the fixings dictates their own dimensions (with its own deviations). If the fixing is in addition to the joint and this cannot be accommodated completely within the component or in the joint, space has to be provided for the fixing itself. This will have to include any access required to complete the fixing on site. In some cases thought has to be given to the space necessary to remove the fixing for repair or maintenance.

In the example of the lintel the joints in the brickwork are 10 mm. This will be a mortar joint that will allow for deviations in the brick. It is now possible to establish the working size for the lintel at 140 mm. For a full specification of the depth it is necessary to be clear that in the manufacture of the lintel any deviation from the 140 mm will have to be limited as these also have to be accommodated in the 10 mm mortar joint. Figure 4.2 suggests an allowable deviation or tolerance of ±2 mm. It is now necessary to ensure that the material can be manufactured or worked within these deviations or there is a risk that the lintel will not fit with the brickwork.

Working sizes also have to take into account any allowance for dimensional changes that may take place during the life of the component. These are known as movements or inherent deviations, as they are inherent in the materials chosen, and these are discussed in more detail in Chapter 12.

Dimensional deviation and fit

When establishing a working size and thinking about how assemblies might fit together, it is necessary to think about three things:

- The size of the space into which the component has to fit
- The size of the component and its position in the space
- The gap that will be formed between components for the joint/fixings

Generally a component can be said to fit if it can be positioned within the space so that the gaps allow for effective fixing and jointing. During construction, all of these – the space, the component and its position and the gap – will vary: they will all need to have tolerances established.

In what is often thought of as traditional construction, trade processes control fit. Procedures such as building in and cut to fit allow for dimensions to be set by the construction itself as the work proceeds. It requires judgement as to how much reduction should be made for fixings, jointing materials, gaps or movements, but these are part of the craft tradition. It is not a question of analysis but one of common practice. One person can take all the deviations into account at one point in the assembly procedure in one judgement.

With the use of components and the development of prefabricated construction, building in and cutting to fit is no longer an option. There has to be some understanding of these deviations as a basis for analysis to decide on final working sizes before manufacture. With elements of the construction being formed from many components this may have to be based on a statistical analysis of the deviations that can be achieved in manufacture and assembly and will be closely associated with the detailing of the joint.

Consideration of the production process will soon reveal an inability to reproduce the specified dimension for the space, component or the joint in practice without deviation. This is not a reference to people making mistakes, the detection of which is a management problem, but an acknowledgement that materials shaped with the most skilled hands, even assisted by machines, will produce components and assemblies that will deviate from the specified figures, and this is a technical problem. These deviations are induced by the chosen production processes. They will be within limits dependent upon the type of material, the level of skill and assistance provided by tools and machines plus the conditions under which the individual has to work. It is possible to identify these induced deviations or characteristic accuracies from a particular production process.

The particular dimensions that are actually achieved for the space, the components and the joint will vary within a range of deviations, most of which will be close to the target value with a decreasing probability of larger deviations being produced. This probability follows a pattern known as a normal distribution, as shown in Figure 4.3. This tells us that you cannot predict the size of the space, component or gap in advance, only the probability that it will be within certain limits. This deviation from the specified value is also true for the specification of the quality of the material.

Because these production deviations are a reality, it is necessary when evaluating the dimensions of components and assemblies to consider how such deviations will affect the performance. Some limits will have to be placed on these induced deviations, for, if too great, the performance will be impaired. These limits, or allowable deviations, are known as tolerances, or what that particular assembly can tolerate before performance may be affected.

(a) Process producing limited deviations

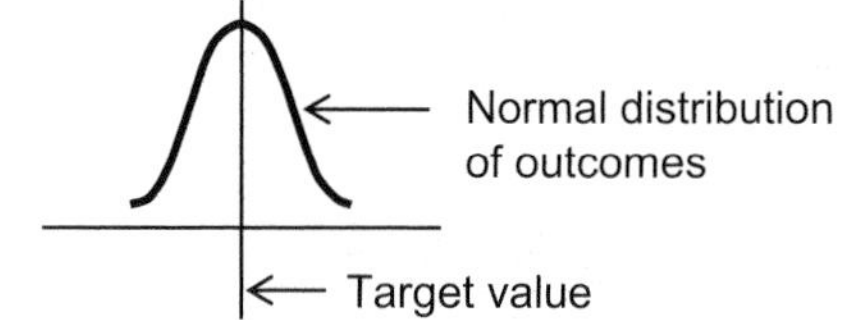

(b) Process producing wider deviations

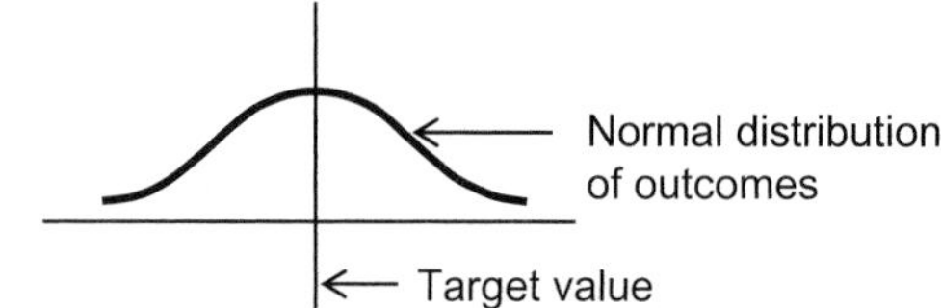

Figure 4.3 Identification of achievable deviations.

(a) Specification as average

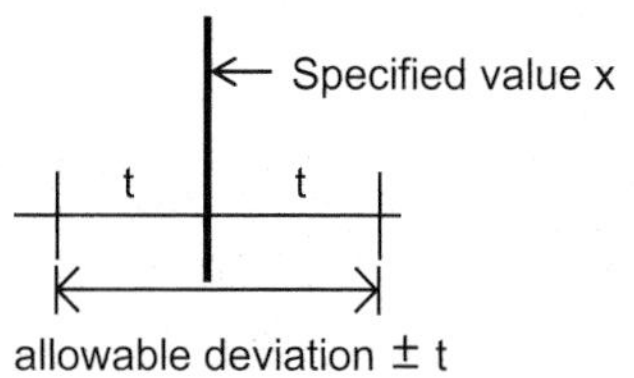

(b) Specification as maximum

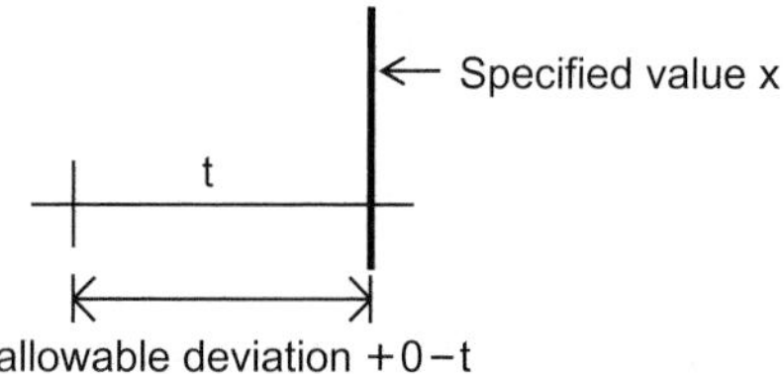

Figure 4.4 Specification of allowable deviations.

Tolerances are quoted as a range within which an assembly will successfully perform. They are limits where it is deemed that components installed outside these limits will fail, or at least risk failure in the assembly. They are represented graphically in Figure 4.4. They take the form of a ± value from the specified size. While most specified sizes are set as an average where equal deviations on either side can be tolerated, some specifications are maximums (or minimums) allowing deviations in only one direction.

Knowing the probability distribution for the chosen production process and having an allowable tolerance set by the detailing allows an estimation of the probability of fit. If the allowable tolerance is greater than the deviation that will be achieved for the great majority of the components from the chosen production method, the probability of assemblies failing will be small.

It would be usual to ensure that the detailing chosen can tolerate the deviations inherent in current production processes. However, in some designs it may be necessary to consider revised or even new production processes that can achieve the required tolerances, but the cost of this development and the risks involved must be appreciated before the proposal is adopted.

It is true to say that before a solution can be specified with complete confidence an analysis of both allowable tolerances and inherent deviations (characteristic accuracies) must be made. This will involve an understanding of both the behaviour of the construction under working conditions and the outcomes of the production process before a solution can be said to be a viable proposition against performance criteria.

Types of dimensional deviations

The previous section looked at the relationship between inherent deviations and the tolerances that can be allowed to ensure fit and operational performance. In defining the component, its space and the gaps that will arise for the joints and fixings, it is important to realise the ways in which deviations can distort as well as just affect size and position.

Figure 4.5 shows a simple rectangular solid component, but this could also be the space available for the component or the space representing the gap between components for the joints and fixings.

Variations in size can be seen as linear deviations affecting the three principal dimensions.

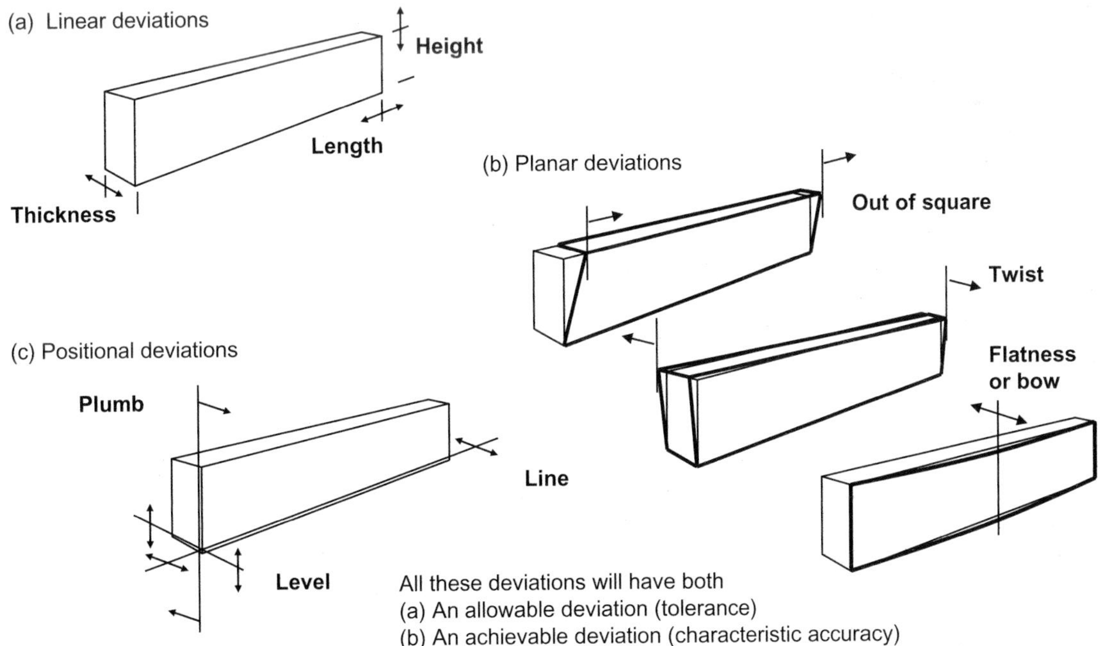

Figure 4.5 Possible dimensional deviations.

However, these may not be the same at all points across the component and this will cause distortions or planar deviations of square, twist or bow. The third set of deviations is positional associated with the line, level and plumb aspects of site control that ensure the spatial relationships are maintained.

Spatial relationships

There are two ways of defining relationships that can be employed to specify position:

- Referenced – specifying a position in space for each individual part within a common reference framework
- Relative – maintaining a positional relationship with a number of other parts of the construction

The use of each will depend on the scale of the building and the stage of construction. Since the process of construction is predominantly building one part after another, dimensional deviations can become cumulative if relative positions alone are used. It may be that fit in parts of the building may become difficult to achieve within their own allowable deviations because previous specifications did not take into account the possibility of knock-on effects. Components can move further and further away from the intended spatial arrangement that the relative dimensions were intended to maintain. This is despite the fact that all previous dimensions have remained within specified deviations.

For construction where cut to fit and building in can be employed, spatial relationships can be controlled to maintain the detailing of the parts and the joints between them. By working within tolerances, parts can be adjusted for the specific circumstances. In this way the all-important joints and fixings can be produced in the required position and within the acceptable limits of their allowable deviations. In modern construction the process is more the assembly of manufactured components of predetermined size. The preformed parts tend to be built one on top of the other. The possibility for dimensional creep or moving outside spatial constraints may be high if too much of the specification is relative.

It is, therefore, common to specify most of the early stages of the construction to a reference framework. Sufficient of the construction has to be specified and controlled in this way until enough real construction exists so only a few parts have to be added to complete that section of the work.

The frameworks developed to control position in this way are usually grid lines on plan and finished floor levels on elevation. These are illustrated in Figure 4.6. In both these cases the 'lines' as they appear on plans and sections are actually planes that 'box up' space and from which positions can be specified.

The control 'box' should not be too small, because the positioning of finishing, fixtures and fittings can be relative to the major construction so long as they have been well controlled by referenced dimensions. For this reason the grid system chosen is usually associated with the structural layout for the building. Every effort would be made to keep it simple, with repeating dimensions to avoid mistakes in both design and production phases.

It is these grids and floor levels that would be the basis of site dimensional control. For all but simple building this would usually involve the use of surveying equipment, tapes, levels and theodolites. Advances in this type of equipment including electronic distance measurements particularly associated with theodolites in total station equipment have led to the use of coordinates for site control.

Joints and fixings

Where two or more components or elements of a building meet, there is a need to consider how they should be joined. Where one or more of these parts requires support or restraint from the others, there is a need to consider how they

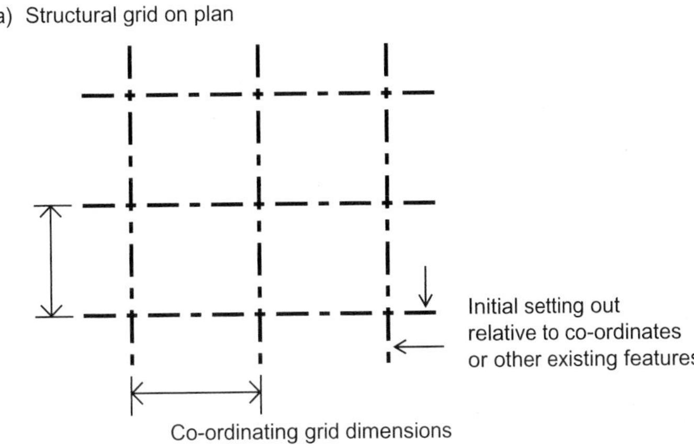

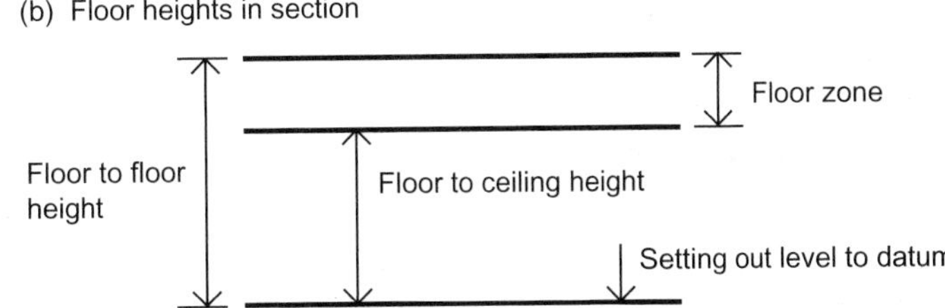

Figure 4.6 Dimensional control framework.

should be fixed. Joints and fixings are the all-important details that:

- Hold the construction together
- Allow it to be assembled
- Ensure the performance of the fully erected building

Like components, they are made of materials, and have shape and size. They have to work within gaps or their own space between components. They only work within certain limits of that size (allowable tolerances) and can only be produced within limited accuracies (induced deviations). They then have to be positioned within specific limits. They are, in fact, subject to the same considerations of performance and production criteria as the components they join and fix together.

Joints and fixings arise to facilitate the creation of the whole from a series of parts. They are, therefore, crucial in the overall behaviour of the building. They influence its stability, its ability to accommodate movement and provide the continuity of performance of the elements of the building of which they form a part.

They also play a major part in facilitating the assembly process, particularly where many of the components are manufactured off site, when the joints and fixings become the focus of the site activity. Buildability, safe working and skill requirements become a matter of handling the components and the completion of the joints and fixings.

In common detailing it is often difficult to separate out what is a joint and what is a fixing. Fixing implies a support or restraint function with load transmission capabilities. Joints tend to describe the whole detail where materials, components or elements come together. The joint may have many functions such as weathering, fire, sound or movement and may not include a structural requirement

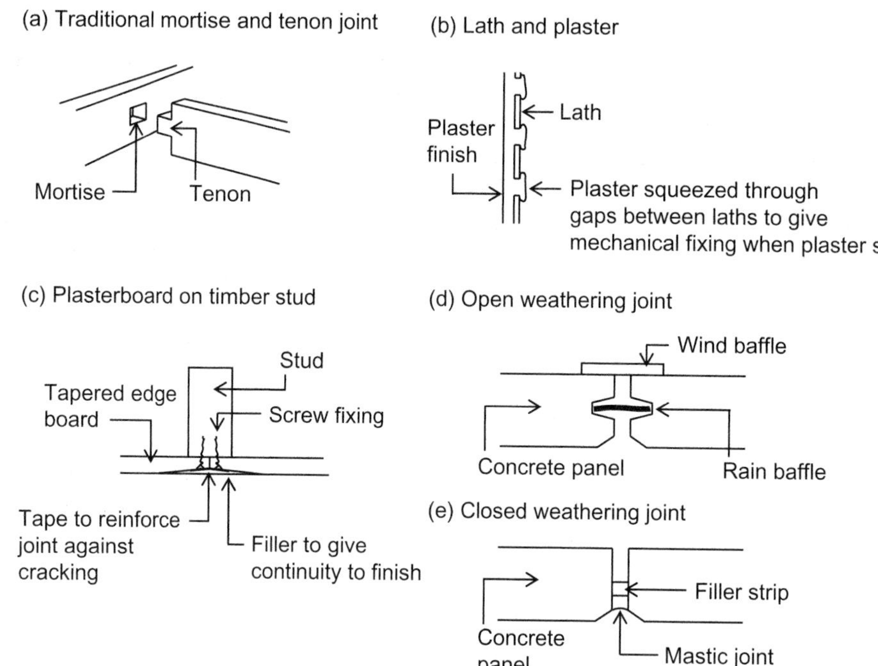

Figure 4.7 Joints and fixings.

that implies the action of a fixing. In some cases components may be fixed independently and are stable but need to be jointed to achieve continuity of function across similar components. The particular requirement of a joint to isolate movement in a building needs special attention. Fixings may also have to be chosen to allow movement in at least one direction or the joints may have to be created totally independently of the fixings.

A number of joints and fixings are shown in Figure 4.7. In traditional timber jointing, such as the mortise and tenon, the detail is entirely based on shaping the components to be joined, which then becomes the fixing of one part to the next. Although glue might have been used, often no other components were necessary. The joining of plaster to a wall also needs to provide fixity. With old plaster systems, laths allowed the wet plaster to form a shape, giving a mechanical key, while modern plasters can achieve fixity with adhesion alone. The jointing of plasterboard to studs is achieved with a mechanical fixing, a nail or screw.

Joints are often subject to performance requirements other than fixing, such as appearance, weatherproofing or fire. The joint to be made between the sheets of plasterboard in the stud system shown in Figure 4.7 needs to take into account movement that may cause cracking of the jointing material, spoiling the appearance. Joints have to be reinforced with tape to resist these forces. Cladding panels on the external face of a building are normally fixed to the structural frame independently, being supported and restrained with a range of fixings such as bolts and angles. The joints between panels then function to allow for deviation, both inherent and induced, and provide continuity of appearance, weathering and other functions of the cladding system itself.

Joints between structural members have a particular significance since the behaviour of the joint determines the distribution of stress in the structural system as a whole. The detailing determines whether the joint can rotate or is stiff (fixed). This determines the way the forces are transmitted from one element to another.

More detailed analysis of this aspect of jointing and continuity between members is provided in Chapter 12.

The function of joints and fixings to facilitate assembly needs to consider the issue of deviations. They have to be detailed to work within acceptable deviations, but they can also be used to absorb or even out deviations to maintain the position of components. Joints and fixings that can maintain local allowable deviations ensure the achievement of fit throughout the building.

There is now a wide range of fixings, the exact form being determined by the nature of the forces being transmitted, the part to be fixed and the structure to which it is being attached. Its form will also be determined by the need for accuracy and adjustment in the positioning of the part to be fixed. Options include welding, gluing, nailing, screwing and bolting and may involve brackets or plates. Specifications must include shape, size and materials as well as the pattern or position of fixings. The choice is also influenced by economics, which can be significantly affected by the tools required and the influence of deskilling associated with the overall construction specified.

Performance and fit depend on the careful consideration of the joints and fixings. Problems are created and resolved around such details. Careful choice and specification of the materials and components has to be matched by a similar attention to detail on joints and fixings if the building is to work as a whole.

Summary

1. In making a choice for the construction it is necessary to specify the materials, shape and size, and spatial relationship, of the parts.
2. Specifying materials starts with identifying appropriate properties for performance. This then has to be linked to the availability of the material in an appropriate form with reliable supply within acceptable deviations. Health, safety and environmental considerations are also significant in making the choice.
3. The ability to shape materials brings great benefits in using the properties of the materials efficiently in stable configurations, particularly at joints and fixings. Shape is particularly important to appearance and in the fine detail that ensures performance. Shaping, however, requires tools, machinery and skills that must be available at reasonable cost.
4. Size has to be specified to define not only overall dimensions but also the required shapes. The gross size is identified by the basic performance requirements, while the coordinating size is determined by the need for the parts to work together. Given the need to join and fit parts together, the coordinating size will have to be modified to a working size that will then need to have deviations identified to ensure the detail can be assembled and will perform.
5. Each detail of construction will have to be evaluated for its ability to be assembled within these deviations and still perform. It is not possible to produce perfect construction, so it is necessary to establish what deviation could be allowed or tolerated without loss of performance.
6. Specifying and controlling spatial relationships can be to a common reference system or by positions relative to other parts of the construction. Given the existence of deviations, if a reference system is used there will be less chance of cumulative deviations, reducing the probability of failure due to dimensional creep.

7. Joints and fixings allow the parts to act as a whole and the assembly process to be achieved. They influence stability, the accommodation of movement and provide continuity of performance in the elements of the building as well as the building as a whole.
8. Fixings imply a load transfer function, while jointing involves many functions which may, or may not, include a structural role.
9. In facilitating the process of assembly, jointing and fixings have to be chosen that can accommodate deviations in order to maintain spatial relationships and the performance of the building as a whole.

5 Defining Conditions

This chapter identifies the fundamental importance of the creation of both physical and social conditions to the choice of construction. It establishes the need to specify internal requirements, define external conditions and then to identify the contribution of the activities of the occupants and the construction itself to the internal conditions. These conditions will be ever changing and therefore the whole system must be seen as dynamic, a series of flows seeking equilibrium. The chapter concludes by identifying how this process can be applied to the choice of internal elements, such as floors and partitions, and introduces the idea of microclimates within each element of the construction itself.

Physical and social conditions

It could be argued that buildings are fundamentally concerned with providing physical construction to create the conditions for the well-being of those who use them. This statement would raise in most people's minds the modification of the natural environment: from the weather conditions outside the building to a less variable and more comfortable and safe environment inside. This is clearly a primary technical role for the construction, and therefore a clear understanding of the external prevailing conditions and a clear specification of the required internal conditions are of primary importance before any process of choice can begin.

Increasingly the external environment is being significantly modified by human activity by noise and air pollution that any understanding of the external environment must be more than just the natural weather conditions. It becomes necessary to understand the impact that producing and operating a building will have on the local, regional and even global environment.

The case for the need to understand the interaction of physical conditions is easily understood, but solutions chosen only against criteria of modification of the physical environment are not sufficient. Buildings are created in a social context in which they serve social functions that influence performance requirements. This simple idea is shown in Figure 5.1, which illustrates that the user's interests in modification concern both the physical and social worlds. If an activity can be carried out effectively within the limits of prevailing external conditions, and it is socially acceptable to do so, there will be no requirement for a building in any form.

It is important to remember that buildings have an impact on social conditions with both the internal and external spaces they create. While an understanding of social behaviour is beyond the scope of this text, it is well established that building may be part of social problems but can also be part of solutions. This is recognised in many aspects of urban design in areas such as crime and community safety.

The social context also determines the resources that become available for building. There is a need to understand the influence of the economic, political and cultural environments on the choice of construction set against its cost and its value in its wider social interactions. This has an influence in the specification of the required internal environments. The expectations of the user and the variations in

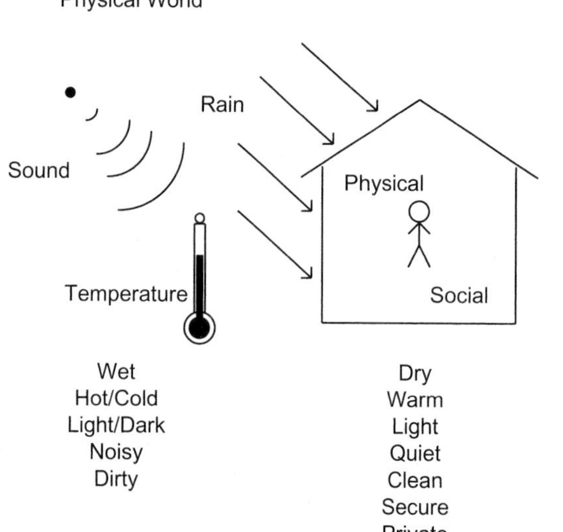

Figure 5.1 Physical and social environments.

internal environment deemed acceptable are set against these wider social standards and the economic possibility determined by the amount of income clients are prepared to spend on their buildings.

Activity and construction-modifying environments

It is clear that construction modifies the external environment to create the desired internal conditions. However, the activities within the building will themselves contribute to the internal environmental conditions. Cooking generates moisture, heat and smells; machinery generates heat and noise; and cleaning activities generate waste. They will in turn, if detrimental to well-being, have to be controlled, which may increase the technical demands of the construction.

It also has to be appreciated that the construction chosen may itself modify the environment. This is particularly true of the services in building that often create heat, fumes and noise, but it can also be true of the passive construction. Poorly detailed insulation to resist the passage of heat to provide the thermal environment with lower energy inputs can induce condensation and create damp conditions. Finishing materials can give off toxic fumes in fire. Both are examples of the choice of construction affecting the achievement of the desired internal conditions.

The dynamics of the system

Although a stable internal environment may be the objective, it is clear that the external conditions and the levels of activity within the building will change in a variety of ways. This will produce constantly changing conditions. The whole system can be seen as a series of flows and transfers seeking equilibrium, as shown in Figure 5.2. This dynamic makes the process of control of the internal conditions a significant aspect in the process of choice.

Predicting the possibility of exceeding design limits in dynamic systems needs an understanding of both the variability of the conditions and the response of the construction providing the modifying effect. Indeed, a major consideration in choosing a solution may well be its ability to respond in sympathy with the expected magnitude and speed of changes in the environ-

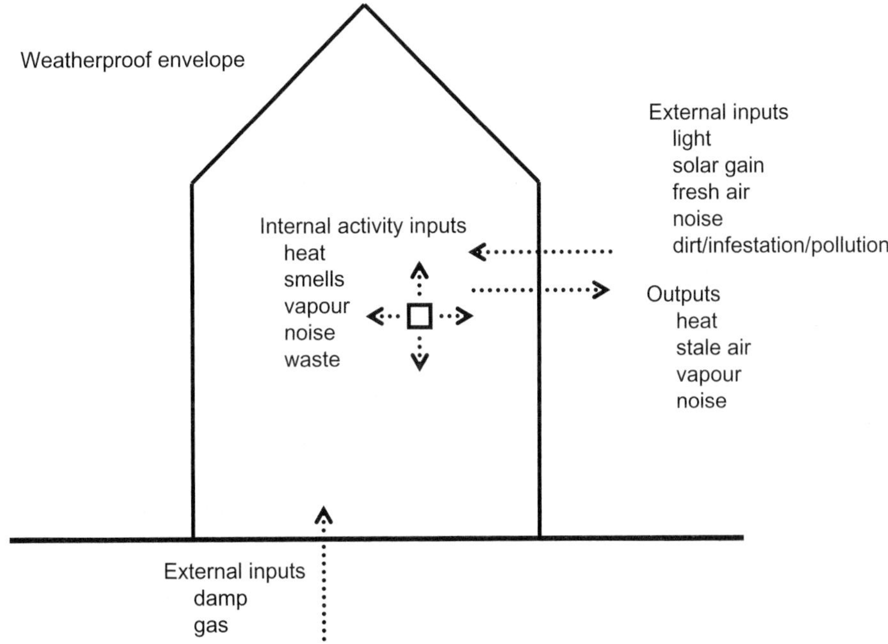

Figure 5.2 Environmental flows – a dynamic system.

ment. This has been recognised, for example, in the advantages of high thermal mass for building in limiting the use of energy for cooling on hot days to limit the variations in the internal temperature between day and night.

To enter into an analysis of the behaviour of the construction it is necessary to define the following:

- Target internal conditions and acceptable limits within which the activity can operate
- Design limits in the external environment and their pattern and variations
- Inputs to the internal environment from the activity and construction including the response to changes in conditions

Target internal conditions

Defining the target or ideal internal conditions is a question of physiology and psychology in the case of people, engineering in the case of machines and spoilage processes in the case of stored material. Aspects of the environment important to the well-being of life, machines and materials have much in common. Temperature and dryness are often the most significant to the well-being of all three, although ideals and limits may be different. In cold stores the temperature is clearly dictated by the stored food. Any people or machines that also have to operate in the store have to be protected by other technologies (e.g. clothes in the case of people). In addition to temperature and dryness, consideration has to be given to aspects such as humidity, light, noise and air quality. The recent concerns over air quality, associated with hitherto unrecognised contaminants which occur in very low concentration, such as radon, illustrate how new aspects of environments can come into play as we strive for more demanding and less variable conditions.

People, machines and materials do not, however, need precise environmental conditions and can operate within limits. Indeed, in the case of people, constant, unvarying conditions can cause fatigue and some variations are desirable. Specification of internal environments is greatly influenced by social expectations and the resources clients are prepared to devote to the building. With low levels of resources available for the construction, wider limits of environmental conditions may have to be accepted. In the early factories of the industrial revolution, concentration on machine efficiency often made noise and dust levels far from ideal for the workforce.

Social expectation is often now embodied in law to ensure clients make resources available to maintain healthy and safe environments within their buildings. Many regulations and advisory standards now set design limits for comfortable, safe and healthy environments.

Establishing external design conditions

Establishing external conditions, both climatic and social, is clearly a different task from setting internal conditions. It is necessary to understand the broad patterns of events, with the probability of extreme conditions being identified. The broadest climatic patterns can be identified by global classifications. The fact that the UK lies in a temperate zone gives an overall picture of conditions. However, regional and even local conditions may well determine the pattern of variations and extremes. The South-West has the fewest frosts, and the highest winds occur along the west coast of Scotland. Variations also exist between urban and rural areas where the pollution and the modifying influence of existing developments can have a significant effect, an example of which would be traffic noise. All these may influence the orientation of the building on the site and the choice of materials and detailing to ensure the required performance is achieved. This would also be true of the social conditions, indicated by broad patterns such as inner city, suburban or rural classifications. These may give an indication of the social conditions, but often much more local knowledge will be required if the performance of the construction is to be achieved in practice.

The external conditions need to be understood in order to evaluate most aspects of performance. The need to define climate for environmental control is straightforward: temperature variations and sun paths for heating and cooling requirements, rain and wind conditions for weathering to ensure the building remains dry. However, external conditions also need to be known for structural performance and for durability. Both snow and wind place loads on the building structure but in different ways: sunlight fades colours and embrittles plastics; saturated bricks subjected to frost may disrupt under the internal pressures created during the growth of the ice crystals.

It will be noticed from the above examples that combinations of climatic factors are often the worst design conditions. Wind and rain can be the most testing conditions for water penetration; snow and wind may give the worst loading conditions; wet weather followed by frost causes damage to brickwork. Defining single factors may not reveal the most testing conditions. The pattern and variations in the climate have to be understood well enough to visualise the significant combinations as well as individual extremes.

Establishing variations and extremes requires knowledge of conditions beyond that of the daily cycle or the annual seasons. Extremes do not occur every year but at longer intervals. Some of these seem to be identified with long-term cycles or trends and some seem more random. Since most buildings have design lives measured in decades, cycles, trends and random extremes that happen over these timescales need to be investigated.

For each stage of the analysis (environmental, structural and durability) it is necessary to visualise the most testing set of conditions to maintain performance. For durability this will be dependent on the materials chosen, for the agents of decay are specific to each material. Each set of conditions will then need to be defined in the most appropriate way. Knowledge of annual rainfall gives a broad indication of the magnitude of the problem faced in choosing the construction for a building in that area but it will not give the whole picture. Occurrence during the year and some measures of intensity are required. Rain that occurs in a single season of the year at high intensity poses different problems from rain that occurs in light showers throughout the year. Intensity over a period measured in minutes will determine the size of the drainage system, while lower intensity over some hours, particularly if associated with high winds, may determine the detailing of weathering to the walls and roof. In another set of circumstances, prolonged light rain may penetrate porous materials sufficiently to cause dampness on the inside of construction well able to resist a sudden downpour.

While this rainfall example is directly related to the physical climate, the principle of visualising the most testing conditions is true for other performance requirements. Changes in social environments, say with crime and terrorism, also need to be established in some way that allows a design solution to maintain performance of the internal environment even with the perceived levels of threat. Crime figures may allow some quantitative analysis of probability, timing (day or night) and mode of entry, which allows choices to be made as to how best to maintain performance.

Since the external conditions are variable, it becomes a design decision as to which limits are to be taken into account. Generally, design involves establishing the probability of the occurrence and a level of 'value' on the disruption, danger or cost of the consequence of the building failing to perform. Rare but devastating or life-threatening events may be given similar importance to common occurrences that cause inconvenience or a lowering of operational efficiency. It requires the client to assign value to any loss of performance associated with prevailing social and economic conditions. It may also be subject to the law, for in many cases loss of performance poses a danger that the client may not value as highly as society in general. Over the years this has been deemed to be the case with loss of life in a fire and this has led to extensive regulations.

Yet another aspect of the analysis of proposed construction that will require knowledge of the external environment is its manufacture and assembly. Conditions influence the accuracy that can be achieved, leading to variations in the achievable deviations. Protection to both operatives and the partly completed work will be influenced by the condition. Efficiency and hence cost will change with conditions. The way the conditions are defined will again vary, depending on the purpose of the analysis. Number of hours of daylight or frost days per month may now be the most appropriate way to seek the information about the external environment in order to identify the effects on the decision to specify works to be completed on site in the winter.

Social conditions also influence decisions on the production process. The labour market, local, regional, national and even international, has a major influence on the availability of skill and the cost of labour. Site security will be influenced not only by local crime rates but also by legal and social obligations to keep a safe site that protects both operatives and the general public.

Inputs from the activity

For each aspect of the activity, the types of outputs and the quantity of each will have to be established. People within a building will produce varying amounts of heat, moisture vapour, other products of exhalation, odour, etc., depending on the activity being carried out. While some of these may be helpful in, for example, heat gains in winter to maintaining temperature, others will militate against clean air. In another example, the contribution of moisture vapour can have a significant influence on the formation of condensation if ventilation rates are not sufficient. Machines can produce noise and heat and there are concerns over the electromagnetic fields emitted by electronic equipment. While most stored material is inert and contributes nothing to the internal environment, this may not be true if the material stored is organic. Bulk storage of straw can, under some conditions, generate sufficient internal heat to catch fire.

As with the external environment, broad patterns of these internal contributions may be established, but questions of variations in duration and rates of activity will have to be determined. Designers often look for the single figure to represent inputs from the activity, but if variations are great or change rapidly it may be necessary to analyse the dynamics of the change to match the response of the building more carefully. If data are not available for particular activities, studies will have to be carried out to determine the rate of net inputs into the environment for appropriate construction solutions to be developed.

Inputs from the technical solution

Clearly the effects of the construction and the services cannot be defined before the construction and services are chosen and must, therefore, become part of the process of choice itself. Problems associated with the building in this respect have to be foreseen and minimised at the beginning of the process by the broad strategies developed rather than by resolving problems at the end in the specification and detailing.

Deciding on broad strategies involves decisions on all the performance requirements and is therefore a careful balancing act. This includes the balance between passive construction and the active services. With increased demands for quality in the internal environment with smaller variations, active services are difficult to avoid. Unfortunately it is the services that are most likely to generate outputs that affect the internal conditions and use energy. Some input can be used to advantage. Heat recovery from exhaust air can reduce the size of the heating system and reduce overall energy use. However, the noise from the fans necessary to circulate that air may be transmitted through the ducts. This in turn will require decisions on the use of silencers or sound absorbers in the ducts. The materials for these now have to be chosen with care as fibres from the sound absorption material may be picked up in the air stream and carried to the rooms the ducting system serves. Designs of building that can take advantage of natural ventilation generated by wind and other pressure differences have considerable advantages, but this has to be part of the overall design concept and not a question of detailing.

Most construction materials are inert and therefore add nothing to environmental conditions. Moisture from wet construction at the early stages of occupation of the building has been largely overcome with the use of dry forms of construction.

Choice of interior elements of the building

So far the analysis of internal and external conditions has focused on the inside and outside of the building. However, buildings are rarely used for a single activity. Although there will be an overall purpose for the building, fulfilling that purpose for the user will almost certainly involve a series of interrelated but separate activities. In specifying internal elements, floors and partitions, in the building the notion of an external environment may refer to another internal space used for a separate activity. In this case the internal/external labels are interchangeable.

Problems posed for the construction to solve will depend upon which way the undesirable elements of one environment will affect the second. The task of the construction depends upon which way the flow disturbing the desired conditions will take place. It may become the task of an internal partition to modify sound transmission in one direction and heat in the other if it is between an unheated factory area and an office.

Inside the construction itself

Although apparently not of immediate interest to the user, the process of modifications of environments across construction creates intermediate or micro climates within the construction itself. They can become highly significant to the running cost or life expectancy of the construction and that is clearly the concern of the building user or operator. Perhaps the best example of this is the formation of interstitial condensation causing damp when it is apparently dry on both sides of the wall or roof. The wetting of insulation will immediately reduce its effectiveness and, if persistent, may create the condition for rot or corrosion, leading to premature failure.

The analysis of these intermediate conditions that rarely reach steady state is another task for the technologist seeking to provide satisfactory, continuing performance throughout the range of environmental conditions that may prevail.

Summary

1. Building is concerned with providing physical construction to create environmental conditions within the building that are modifications of those prevailing externally. These modifications are required to fulfil both physical and social functions.
2. In providing physical construction, the building will have an impact on both the physical and social environments and these must be understood and be part of the analysis.
3. Both a target value and an acceptable range have to be identified for a number of parameters to define the internal conditions to ensure a comfortable, healthy and safe environment. These will depend on the activities being carried out in the building.
4. The activities within the building generate inputs that will modify the internal conditions. If these inputs take these conditions outside the desired range, the building will have to provide the necessary control to maintain the required internal environment.
5. Under operational conditions the interplay between changing external conditions and inputs from the activities and the construction itself must be seen as a dynamic system: a series of flows seeking equilibrium.
6. In addition to understanding how internal conditions are to be maintained it is important to understand how the choice of solutions will impact on the environment in general.
7. Defining the external conditions, both climatic and man-made, is necessary not only for the environmental modification role of the construction but also because it loads the structure and deteriorates the materials of which the construction is made.
8. Because external conditions influence so many aspects of the analysis of a proposed solution, the same aspect of climate, say rainfall, may have to be defined in a number of ways depending on which part of the analysis of a solution is being carried out. Aspects of climate are also important when deciding on appropriate methods of manufacture and assembly.
9. Internal elements of construction often separate areas of different activity requiring different internal conditions. In this case the definition of the external environment is another but different set of internal conditions.
10. When modifying environments, the construction has different conditions on either side, which means that the conditions within the construction are neither internal nor external. Identifying what the micro environment is within the construction and its rate of response to change will have to be analysed.

6 The Resource Base

This chapter introduces the four key resources required for the successful creation of buildings. Materials are identified as the primary resource, being associated directly with the forming of components and the construction of the final building. The second is the knowledge and skill held by people undertaking design and production in order to exploit the properties of the materials. Production equipment allows the application of materials in forms, sizes and shapes not possible with human effort alone. Finally, money as a resource has to be deployed by clients and the players in the construction industry.

Introduction

Resources are normally associated with the activities of management, organisation and control. They have often been quoted as materials, people, machines and money that have to be employed in the construction process. In management terms these four resources represent the inputs for the construction of the building which need to be obtained, deployed and controlled. For the technical choice of the construction the understanding has to be different from that of management, but the idea of four basic resources is a good starting point to think about the evaluation of a suggested construction solution.

A knowledge and understanding of resources becomes central to the analysis of buildability, cost and social interactions in the evaluation of the suggested solution. The whole concept of buildability is centred on the capacity to successfully obtain resources to achieve the required specification that will realise performance. Costs arise only from the deployment and control of resources, the price of which are determined by the prevailing economic and political conditions. Society has an interest in the impacts on the physical and social environments of the use of resource for the process of construction as well as the newly created resource of the building itself when in occupation.

Materials – the primary resource

Since materials are one of the basic variables of the final construction, they can be seen as the primary resource. They are identified as a fundamental aspect of choice for the construction because of their properties and behaviour to achieve performance, as discussed in Chapter 4. However, there is also a need to understand our ability to bring the raw material to a finished state in a number of component forms suitable for incorporation into a final assembly. It is bringing the raw material to make components, followed by fixing in the final position, that determines all other resources required.

The use of a material is, therefore, dependent not only on its potential properties but also on the availability of resources to process, manufacture and assemble the building. It is the existence of (or at least the economic and political will to release) the resources to bring the material to its finished state that determines the success of a chosen construction form or detail.

In societies where resources are not readily available, construction is often limited to relatively unprocessed materials of local origin. The properties of the material may be difficult to predict, with variations in the quality of the material brought to the site. This means either a process of selection giving a high level of waste (often economically unacceptable) or the design has to assume only limited performance from the material, leading to larger sections. Many of the developments in processing materials seek to limit the variation as well as enhancing the basic natural properties.

The consequences of an increased manufacturing base and infrastructure are not only a guaranteed quality but also a greater range of materials. The existence of these new materials does not, however, lead directly to an ability to incorporate them into a design with the expectation of successful construction. There is always a need to be able to detail with the confidence that the required skill to work the materials for assembly into the building is available.

In addition to understanding the properties of a material for performance, its workability properties have to be known. Very few materials are usable in the form in which they are found in nature. There are whole industries concerned with the processing of materials upon which buildings rely for the availability of components. In some cases, such as timber, this is mainly careful conversion so the potential materials properties are maintained. In other cases, such as plastics, formulation and processing can determine a wide range of properties, although only within the basic behaviour established ultimately by the long chain molecular structure of this group of material. In both the cases of timber and plastics, not all types have similar workability. Straightness and closeness of grain for timber and the basic distinction between thermoplastic and thermosetting plastics, for instance, influence how they can be used in practice.

The ease with which it is possible to perform operations on materials is key to choosing size and shape for components as well as possible fixing methods to provide the assembly. The ease with which each material can be cut, bent, poured, extruded, etc. without loss of its useful properties determines the manufacturing and assembly methods. This in turn determines the economics of the construction.

This may lead to a range of materials being used for similar components. Bricks, although usually associated with clay that is moulded or extruded when wet and then fired, can also be made of concrete, which relies on chemical reactions that take place at ambient temperatures. While this gives a product with many similar properties, some are different. Clay bricks expand in early life, while concrete bricks shrink; clay bricks can be dimensionally and visually irregular, while concrete ones have fewer deviations. These are endowed by the properties of the material; they are not a basic attribute of the component.

The materials we use are not limitless and their conversion and processing have an impact on both the built and the natural environments. As discussed in Chapter 4, this has perhaps been the major realisation of the late twentieth century and its influence on the choice of materials for building is likely to be profound. Concern for the impact of extraction, depletion rates, energy for conversion, pollution and waste and ultimate disposal leads to cradle-to-grave analysis, showing the true implications for the use of each material on the environment. Materials with all the right properties and that are workable with the skills available may not be chosen because of their environmental impact.

Knowledge and skill – the human resource

Many animals can be identified as builders. A few higher primates seem to have the rudiments of using tools. Humans alone seem to have the abilities to take this to the level of complexity we observe in the world today. This intellectual power when applied to the development of the skills to design and build is the second great resource we have to create buildings.

It is the development of knowledge and skill that determines not only the genesis of ideas about how we might build but also our confidence to try. Human activity is predominantly social, and increased social complexity demands a greater range and purpose for its buildings. Given a sympathetic cultural environment, increased economic activity releases more resources to technological activity, which will be reflected in the methods used for building. Perhaps more significantly, the cultural environment determines the speed at which progress in technological development takes place. If developments move too far ahead of knowledge and skill, there is an increased risk of failure.

Every time a new idea is put into practice it is a trial that may fail due to an error in our understanding of the material or the behaviour of the construction. Technology derives a great deal of its knowledge and know-how from practice as well as from the sciences. It is an assessment of this total knowledge and skill that has to be available before the risk inherent in the choice of construction can be evaluated. In a complex social environment this knowledge and skill is unlikely to be available in any one person or even a small group of people. Communication between those with different aspects of technical understanding becomes increasingly important.

This resource, characterised as knowledge and skill, is identified in social organisations involving people. The operatives, designers and managers are most directly involved, gaining their knowledge from their education and training and then from their daily experience. Material and component manufacturers also generate and accumulate information and knowledge that is vital to the process of incorporation into the building. Behind all this lies the research and development activity often not directly related to the construction industry. This feeds in knowledge that can be called upon by practice to improve the chance of sound decisions and lower the risk of failure.

This knowledge is fundamentally, but not exclusively, technical. Design is a key ingredient in the choice of construction to provide fine buildings but is of limited value without an understanding of the technical issues involved. Organisational and motivational skills are vital to the manager but can be misplaced and ill-directed if the technical issues are not well understood. The skill of the operative to use tools cannot be isolated from an understanding of technical matters, particularly where new construction is being introduced. The judgements being made by the operative on, say, deviations (tolerances) are not just about the knowledge of the workability of their material but may have consequences for subsequent operations of assembly or the long-term performance of the building.

It is the capacity of each individual to make technical judgements combined with their individual skill that is the key to identifying the availability of appropriate resources. It is the technical understanding that has to be shared and communicated so that each individual may make sound decisions in order to limit the risk in the overall construction process.

Production equipment – the technological resource

With the right people with the knowledge and skill to select, process and work the materials into effective components and assemblies, it is possible to produce buildings. These buildings would, however, be made of a small number of materials and would be limited in size. While hand tools will be associated with the operative, it is the use of production equipment, both temporary works and plant and machinery, that expands the technological possibilities. This is discussed in Chapter 13.

Although often seen today as the way to greater productivity and economy, production equipment was originally a resource that allowed solutions to be chosen that could only be built if the temporary works or machine itself could be conceived and constructed. It formed a chain of technologies that had to be available before a particular construction could be realised.

In the early days of large construction, such as cathedrals and early civil engineering works, these production technologies would have been designed and built by those responsible for the construction of the works themselves. They were often associated with lifting or simple materials processing. Manual skill would still be employed to shape and finish most of the components and fix them into their final position.

Although this production equipment is still fundamentally about breaking away from the limitations of the strength of an individual with only hand tools, the availability of these technologies may now make operations safe and economic and therefore become the preferred choice as they lower cost or increase quality. Knowledge of production equipment at the design and detailing stage becomes significant if safe and economic solutions are to be adopted.

This process of mechanisation is not just associated with the site assembly process. Behind the expanding choice of materials lie developments in processing plant. Behind the expanding choice of components lie mechanised factory processes. The technological chain that allows choice of construction for the building itself is now extensive.

Plant and machinery has long been available for processing and lifting materials but often by mechanical advantage still using human power. Developments have been greatly influenced by alternative sources of a power supply. Individual power units can now be devised for very large and very small machines, making the availability and versatility of plant and equipment very great. However, most machines still require human knowledge and skill to guide and operate them. Perhaps the most radical change in the relationship between the choice of construction and machinery will come with the introduction of machines that take over the operator functions. With the application of microprocessor technology to automation of construction equipment, the conventions of the material–labour relationship may be supplanted by material/machine protocols. This can be seen in the use of computer numeric control (CNC), now available for structural steel fabrication, which can take its information directly from computer-held design information.

The increasing use of power often involves the burning of fossil fuels, and this will have an environmental impact that needs to be taken into account in the choice of construction.

Money – the enabling resource

Money is fundamentally different from the other three resources in that it is not used directly in the construction of the building. It is not technical in nature and is therefore the most easily transferred to other areas of economic activity. Its use for building has to be seen as good value by those who are willing to deploy it. In the past, patronage for great building has been directed towards status and the association with fine things. More typically today it is about return on capital and this determines much of the ethos in which construction is undertaken.

There are two major groups of individuals, or more typically businesses or corporations, for whom the deployment of money must be seen as good value.

The first are the clients who commission the building. They have requirements that, starting with the function of the whole building, can be translated through the design to the function and then the performance of the parts so that technical choices can be made. If the building can be built for the money the client is prepared to allocate to the construction, the client is likely to judge the building to be good value.

The second group who have to have money as a resource to place at the disposal of construction are the professionals and the contractors who employ their money in the technical expertise and know-how of design and construction. Their attitude towards money, particularly the value of long- and short-term investment, is crucial to the technical potential of the whole industry. If the human resource is just seen as employing people from the job market then this can be viewed as short-term investment. However, if, as is suggested here, the resource is not people but knowledge and skill then the investment is long term and its costs must be acknowledged. The willingness to deploy money in education and training fundamentally influences the technical capacity of the industry to deliver value for money to the clients as well as the long-term potential for their own money investment.

There is a third institution, government, that interacts with the investment and value-for-money decisions of the first two. Grants and tax regimes change the decisions of clients as well as of the industrial players. There are also implications of the government's role in education, training and research and development to the investment attitudes of the industry. Not least is the role of the government as client, spending taxpayers' money on buildings. Not only can this be a sizeable part of the construction industry's workload but also, as the client, government can influence the way construction is organised.

Money is not a fundamental resource in the technical potential of a particular construction solution. This potential lies in the materials, the individuals with expertise and the production equipment available. However, money plays a part in determining the levels at which these are available for construction and ultimately whether construction, however technically superior, will be applied.

Summary

1. There are four resources that have to be available for the successful creation of a building: materials, knowledge and skill, production equipment and money.
2. Materials are the primary resource, being directly related to the final construction of the building. Chosen for their properties to fulfil performance, they also have to be workable to make components suitable for incorporation into the building.
3. The level of development in a society determines the ability to process and shape materials as it determines the growth of knowledge and skill, the application of machines and the availability of money.
4. Knowledge and skill gained from both training and experience become invested in people. Analysis involves the quality of the understanding held by all the parties involved and the quality of the communication between them.
5. Production equipment in the form of temporary works and plant and machinery allow for the development of solutions that break out of the limits of individual human strength. The production equipment itself has to be conceived and constructed and hence forms a technological chain to provide the ability to use processed materials, manufactured into components, which can be erected on site.
6. Money, although not used directly in the technical solution, enables the available resources to be brought together so long as society, the client and those who seek to earn their living in the construction of buildings see the endeavour as good value.

7 Design Concept

This chapter establishes the relationship of the design concept to the choice of construction, suggesting an approach that considers the flows and transfers through the building, the fabric and the services in order to identify the function of the parts of construction from the function of the building as a whole.

Linking design to construction choice – flows and transfers

It is not the intention of this chapter to presume to discuss either the theory or the practice of designing. This chapter seeks to make the link between the design and the choice of construction. The design will be an interpretation of the client's brief in defining the size and arrangement of spaces and their interrelationships, making them work in terms of their feel and environmental conditions for the operation of the building. The design will also have to respond to the physical and wider social context incorporating the requirements of regulation and the need to achieve sustainable development. This intention has to be worked right through into the specification and details to ensure the whole design concept is achieved.

When seeking a link to technological choice, it is necessary to be concerned not only with spatial arrangements but also with flows and transfers in, out and around the spaces as well as through the construction itself. Design is often seen as three integrated areas: architectural, structural and environmental, each with not only their own concern for space and aspects of the construction but also considerations for flow and transfers.

Designs therefore have to establish not only an arrangement of spaces, both internal and external, but also a way of defining this series of flows and transfers between them. It is by establishing the flows and transfers that the dynamic behaviour of the construction expected during the operation of the building can be visualised. This allows an initial suggestion of the generic forms of construction that may be appropriate. However, perhaps most important, it allows for the identification of the function of the parts of the building and their contribution to the function of the building as a whole.

Flows and transfers take place in one of three ways. The first is in and around the building itself. Perhaps the most obvious from our experience of buildings is the flow of people and the transfer of objects around the building. This is of direct concern to aspects of the design that considers the arrangement of usable space. This has to be extended to considering emergencies such as fire, where the extraordinary direction and volume of people have to be defined. However, movements in and around the building will also involve air carrying moisture, heat and odours. As a process of natural ventilation this flow has to be planned and may require openings other than those for the movement of people and objects.

The second way flows and transfers take place is through the construction fabric itself. This is of direct concern to the choice of structure and enclosure elements. Although the construction is often characterised as passive, the behaviour is dynamic in response to these flows and transfers. There are energy flows associated with creating environments, normally heat, light and sound, as well as transfers of forces through structural elements. These flows also include the physical transfer of gases and liquids through cracks and permeable materials.

The third way is through building services. These are purpose-designed systems that are characterised as active because it is their function to transfer commodities around the building. They may be carrying energy either in a medium such as heat in water or directly with electricity in wires. Water and drainage systems have the function of transferring clean water to points such as sinks and basins and then disposing of the soiled water safely. Signals and data have to be transferred in communication systems. As active systems they need control. They need to be turned on and off, up and down, and this distinguishes them from the passive fabric, where flows and transfers are determined by the properties of the materials and size of the components.

The size and interrelationship of spaces, the fabric of the building and the building services and the interchange and transfers between them have to be incorporated into the design concept. The design intention has to be made clear so it can be taken right through the process of choice of the technical solutions.

Concept design – broad options

Architectural, structural and environmental considerations all need a conceptual design that can identify some general construction forms or systems early in the design process which can be worked up into specific solutions later. Designs are made reality by technical solutions that can be conceived, along with the resources to put them into practice. Inevitably most designs are worked out with practical technical solutions in mind. This does not mean complete detailed solutions. It is more likely to be associated with the identification of broad technical options, indicating the materials and sizes and even some details involved and how they would work with the other elements of the building. In some cases the demands of the brief from the client may lead to designs that cannot be achieved with immediately identifiable technology. This can be seen in the early sustainable designs using natural ventilation and with night purging schemes to reduce the energy consumption of buildings. This requires a deeper exploration of the general forms and the way they might work as the design concept is turned into practice.

All buildings pose challenges somewhere between copying the technology from the last building to developing new and novel solutions for general forms. Perhaps the key understanding in this process is just how difficult it is to achieve a technical solution to an evolving design. There is no direct relationship between novel designs and technological difficulty. The radical design may be easily achieved with existing technical knowledge and experience.

Physical performance and appearance – commodity, firmness and delight

The design concepts must allow for the function of each part of the building to be identified with the level of performance required if the building is to function as a whole. The move from client's brief with its language and images of the operational nature of the building to a definition of the technical requirements of the parts should emerge from the design concept expressing the intention for both physical performance and appearance.

Concepts of appearance are a matter of taste, style and image. They involve aesthetic judgements and rely on responses conditioned by cultural and social influences. Appearance should not be interpreted too narrowly for this aspect of design. It involves not only the finish of the surfaces that can be seen but also the sense of space and the appearance of the space itself to the beholder. It is as much (if not more) to do with the feel of the space defined by the building as by the strict look of the building itself.

The business of providing a working building is a matter of physical performance. Concepts have to be based predominantly on the realities of the physical world and the laws of nature, although the arrangement of physical space influences our psychological well-being. They involve physical, environmental and structural judgements and have in the recent past called on knowledge from ergonomics, science and engineering. They require an understanding of the physical behaviour of the construction when subjected to the forces of nature.

Most definitions of 'good design' would argue that both the working performance and how the building appears should be satisfied in order to attribute any judgement of quality to a building. The Roman architect Vitruvius expressed this in the first century BC as 'commodity, firmness and delight'. Commodity and firmness refer to the building's ability to accommodate the user's activities and to its structural integrity. These have been referred to earlier in this text in terms of utility and safety. Both of these require attention to the building's physical performance. Delight, on the other hand, is about the way individuals feel about the building. Without all of these, the building is left wanting in some respects. All are legitimate objectives to be set for the technical solutions to achieve. However, since the nature of these objectives and subsequent evaluation have to be made in different ways, the distinction between the physical performance and what is being referred to in this text as appearance (delight) seems appropriate here.

Appearance and physical performance differ in the way judgements of success are made. The response to aspects of the appearance will be based on a human, cultural, even spiritual, response expressed as emotions, if they can be expressed at all. The response to the physical performance will be physiological and psychological, expressed in feelings of comfort, and a mental state. There will almost certainly be an overlap between the two. Both the appearance and the physical performance in aspects such as safety and privacy influence a sense of psychological well-being.

Function of the whole, function of the parts

The design is the translation of the client's physical and social needs for the operation of the whole building into a scheme that must be able to identify the function and performance of each of the parts. The conceptual themes that unify the appearance and working of the building have to be able to identify the role of each part of the building in achieving the performance of the whole.

The design will have to identify not only the function but also the performance level. For example, the environmental design will determine whether a wall will have a function to resist the passage of heat. The design will have to determine the required internal condition and then will have to suggest the relationship of heat inputs to retention rates and the role of services and passive construction in maintaining the design conditions. Given the general concerns for energy saving, most designs look for retention of heat to balance low energy inputs, which means a high-performance requirement from the wall in resisting the passage of heat.

Having established the basis of the function and performance levels of the elements, assemblies and components of the building, it is now possible to undertake the analysis necessary to evaluate proposed construction specifications and details.

Summary

1. There has to be a way to link the design concept formed from the client's brief to the process of technical choice of the elements, assemblies and components of the building to ensure that the design concept and intention are carried right through to the specification and detail.
2. The design has to determine not only the arrangements and interrelationships of the spaces themselves but also the flows and transfers required between the spaces. This will allow the function of the parts to be established from the function of the building as a whole.
3. The design is most likely initially to be developed with some ideas about the broad options, from which the specification and detail could be developed to ensure the full design intentions are achieved.
4. Good designs require both physical performance and qualities of appearance (delight), which have to be addressed in the technical analysis but have to be considered separately.

8 Appearance

This chapter suggests the ways in which appearance may be interpreted from the design concepts and then introduces how these influence the physical construction and the space that may be available for the technical solution.

Visualising attributes and response

It was suggested in Chapter 7 that 'appearance' is perhaps an inadequate word to convey the full meaning of this requirement of the design concept that has to be achieved through the specification and detailing of the construction. The words 'feel' and 'delight' were also used, but even they would need to be explored further if they were to be successfully translated into the choice of construction.

It is necessary to think about the attributes that will influence the appearance of both the space and actual physical presence of construction. There is a need to judge the response of people to a suggested solution. The response will be conditioned by cultural and socialising influences. The response required may be an aesthetic judgement concerning taste and beauty. It may also be required to convey some message about the purpose of the building or the status or beliefs of the owner. These would have been part of the brief and should be present in the design concept. It is then important to evaluate the extent to which the suggested solution evokes these appropriate responses.

Space and context

The requirements of appearance will be primarily judged by our response to the experience of the building and the spaces it creates. Buildings divide space: internal and external, public and private, illuminated and shaded. These not only must inform and enhance the function of the building itself but also should respond to the context of site and surroundings. These are deep responses and involve judgements with cultural and social overtones. This generates a paradox in that, while they may be the most difficult to articulate objectively, at the time of choosing the solution it is the attribute that is most easily 'measured' by the users of the building.

Since this book is predominantly technical, it is not intended to discuss this analysis in any detail. It may, however, be of value to examine one or two of the issues that contribute to the formulation of this design intent from the brief and the context in which the building is to be placed. The client's brief has to be translated into appropriate spaces arranged to mimic the activities in both scale and interdependence. In achieving this the design intent has to encompass such ideas as style, image and aesthetics as well as context.

The need to satisfy the requirements of appearance also influences the space available to the construction since it can only occupy the zones between internal space and external form. The design concept can be confounded not because a construction form cannot be conceived and created to perform but because there is not sufficient space for it to exist without it adversely influencing the aesthetics. All these are the ideas that are briefly explored in the remainder of this chapter.

Style

Historically many cultures have developed different styles of buildings. To the Greeks the analysis was predominantly geometrical. Early in the twentieth century modernists saw buildings as representing order and clean living. Later in the century many buildings can be seen as expressing the overt use of new technologies. Everyday building, where the form has evolved driven by local conditions, is often referred to as vernacular.

What is perhaps interesting in this historical perspective is that, however different the cultural origins, there is considerable agreement that, in their own way, all of these buildings can be of great quality when the style is carried through to the specification and detailing. It is perhaps the unity of concept, with careful attention to ensuring it is maintained right through to the details in the construction, that is important.

Copying style is often not successful. There does seem to be a need to thoroughly understand the origins of a style, to be driven by the same unwritten forces, and to appreciate the objectives it satisfied for the owner of the original buildings. It also seems important to have the appropriate resource base. The very quality of the materials and the understanding and skill of the craftsmen which developed in the exploration and development of the original style will be missing in any attempt to reproduce or copy a style in another place at another time.

Image

Another aspect of any design concept is the image it intends to portray. This is associated not directly with the activity but more with the social and cultural purpose of the building and the dreams and aspirations of the owner or occupier. It involves issues of status and statements about social and commercial position. It puts forward the social (or individual) beliefs to which the owner/occupier pays particular regard.

This holds true if the owner is not an individual but corporate or governmental, although often in this respect the image is the creation of one individual, most often the head of the corporation or government. Commercial, business and political beliefs dictate the appearance of buildings. Compare the civic pride of the Victorian town halls with the bureaucratisation of local government manifest in the public building of the late twentieth century, or the sense of security in banking halls when the business was holding clients' money with encouragement to deposit, compared to the open layout when the banking business became predominantly credit and lending. Both these examples demonstrate a considerable shift in the importance of space and the choice of materials and detailing.

Although the purpose of two buildings may be the same and hence the functions of the building remain constant, issues of image may determine the spatial arrangement and materials quality that significantly influence the technical solutions. It may call for larger or smaller spaces, which will influence the possible choices of structural solutions. Image determines aspects that will influence the choice of quality materials and have considerations for the lifespans of components. Materials may be chosen not only for properties of appearance but also for their ability to be worked to fine limits or their qualities when weathered.

Image can become stylised into building as the major vehicle for corporate image. Once this state has been reached, many design decisions are removed from the process. While spatial questions and best use of site will still have to be considered, the building may become a series of standard details, materials and colours with only limited scope to influence the design. If this is carried through to solutions based on prefabrication and off-site processes decisions on design are even more limited. This can be seen in franchised retail outlets, particularly those associated with fast food at the end of the twentieth century.

Often the ideas of image are not directly translated into the spatial arrangement and the appearance of the building. The image of the 'green' client, 'caring for the planet', may introduce, through the brief, consideration for choice of materials and details in providing the physical construction of the building. This fact may or may not then be used overtly by the client at a later date depending on whether the image is significant to the client's business or whether it is just part of the client's self-image as part of his/her own personal beliefs and aspirations.

Aesthetics

Beyond the issues of style and image, yet often incorporated within them, lie the pure ideals of aesthetics. Concerned with the recognition of beauty, it is a quality sought in much art, of which architecture is a recognised form. It is concerned with feelings generated within the beholder when confronted with beauty. It may not be too strong to suggest that the experience can be spiritual. The human response in this case may defy analysis, but the emotions it creates, and the influence it has on well-being, would be shared by all of similar taste that experience the building.

Response to context

Buildings are experienced not only from inside, associated with the activity of the building itself, but also by a wider public from the outside. In addition, the view from inside a building can be an extension of the internal space itself and change the experience of the internal space. Buildings contribute to landscape and cityscapes, in which they may enhance or detract from the sense and feel of the space.

Some aspects such as orientation to provide logical access and movement between inside and outside or to provide the best use of the natural environmental conditions such as sunlight are relatively easy to identify, analyse and justify. Other aspects such as scale, massing and the sympathetic use of materials are often not so easy to identify, particularly in the context of a statutory planning framework.

Buildings can become the subject of public debate locally and in some cases even nationally. Such debate often centres on controversial responses of liking and disliking, sometimes the building itself, but often its relationship to its surroundings. These debates are normally about the shock of the new or the tedium of the widespread. It is often not clear whether the building itself lacks design content or if it is its position and surroundings that are in contention. What does seem to be clear is that both need to be agreed before the building is to attract universal acclaim.

Physical space for the construction

The bulk or slenderness of the construction has a profound effect on its appearance. The size and scale of the actual fabric of the building and the detailing will have to be in sympathy with the intended style, image and aesthetics as well as being significant to context. This is true for both the internal spaces and the external appearance of the building and its contribution to its surroundings.

Much of the expression of appearance is about aspects such as line and surface shape as well as colour, texture and light of the actual surface. These are spatial limits to construction as well as definitions of the surfaces themselves. The construction has to be contained within a zone that remains when internal space and the external form have been defined.

This limit to the size and arrangement of the construction has to be part of the analysis. It may be that in order to achieve the required level of physical performance in an economic manner the size and shape of components will exceed these limits and intrude into the visual field of the onlooker, interfering with the required appearance.

Moreover, these spaces or zones for the physical construction often have to contain a number of elements of the fabric and services required for the building. Planning for the space for the elements of construction can become as significant as planning for the space for the occupants.

The approach to analysis

Image, aesthetics and response to context are not subject to the same type of analysis as the physical performance expected of the construction solutions. They do, however, influence the choice and must be part of any evaluation of a suggested solution. The analysis is not a dialogue with the laws of nature but a dialogue with the perceptions of taste and cultural norms. When these are not well defined, design is often seen to lose its way and levels of controversy over buildings grow.

Normally much of the construction is hidden from view and therefore space availability is the only limit associated with appearance. Clearly some aspects of construction, referred to as the finishes, are specifically required to contribute to appearance. In some designs the natural aesthetics of form of the components is used as part of the design concept. In these cases, shape and size and in particular joints and fixings become part of the generation of a desired appearance.

Some designs employ decoration and appendage to create the desired qualities of appearance. These parts of the construction may play no other part in fulfilling function other than appearance. Their justification lies entirely in the contribution they make to the response they invoke, giving value to a particular form of style and aesthetic expression. No less than any other part of construction, choices have to be made about the materials, shape and size plus some means of fixing. This is still a technical challenge, for the parts still have to have sufficient strength and need to maintain their effect over the years. In some forms of decoration, such as sculpture, the artist determines material size and shape; only fixing may become part of the choice for the construction.

The ability to generate a building that fulfils these criteria would be associated with a creative act in the best traditions of art. Historically this may have been more associated with craft, where the emergence of design came from experience with the materials, rather than an analysis of the creative arts themselves. This represents a technological tradition where design, detailing and construction were more closely related in the everyday choice of technical solutions for buildings.

Summary

1. Appearance is a major aspect of the design, associated with evoking response to the building conditioned by social and cultural norms.
2. It concerns the generation of space as well as the physical surfaces and ornamentation, and is associated with the external appearance in response to context as well as internal design.
3. Appearance can be associated with the creation of style and image as well as aesthetic consideration.
4. This makes demands not only on the choice of materials and details but also on the space that is available for the construction between the internal design and the external appearance.

9 Analysis of Physical Behaviour

This chapter is the introduction for the three chapters that follow. It establishes the rationale for the titles of these three chapters and draws out the common approach to analysis that is associated with understanding physical behaviour. It considers function and performance as the outcome of the physical behaviour, which, linked to the need to prevent failure, leads to the need for an explanation of how the construction responds when working under design conditions.

Visualising conditions and response

Given that a proposal for the construction has been suggested, however tentative or outlined the description, it is possible to start to think about its physical behaviour. There are three primary areas of behaviour that are of significance to the success of the building, namely:

- Behaviour creating environments
- Behaviour under load
- Behaviour over time

The analysis of physical behaviour is predominantly a technical activity, since it is undertaken after the function of the parts has been defined and after the conditions under which it has to operate have been established. It varies from an analysis of the appearance of the building in that it is not based on a social or cultural response but on a response to the condition of the physical environment. The analysis of physical behaviour must be based on an understanding of the laws of nature, the forces involved both in states of equilibrium and in the dynamic behaviour during changes in conditions.

Function and performance – a technical aim

This analysis of physical behaviour takes as its criterion for success the idea of performance. Every part of the construction fulfils a function. It has a reason for its existence and a part to play in contributing to the function of the building as a whole. While the client or user determines the function of the whole building, the functions of the parts are determined by the chosen design concept.

Having established a design and assigned functions to a part of the construction, it is necessary to identify a level of performance. It is against these levels of performance that analysis can be undertaken and the failure modes identified. The language of performance levels will vary, but the common idea is that the performance statement must be testable. Examples would be that beams must carry load without exceeding a determined deflection, walls may have to resist the passage of heat with a specified level of insulation and no dampness should appear on any internal surfaces of the external elements of the building.

Preventing failure – a technical objective

Failure is a loss of performance. Failure follows when the response of the proposed construction does not meet performance levels. If the beam were too small or made from a material that were too elastic and had not been shaped or fixed to overcome these shortcomings, it might deflect too much. The beam has failed to perform. If the loads were larger than predicted and the beam deflects too much, the origins of this failure are in the prediction of the design condition.

The analysis of physical behaviour requires three areas of understanding:

1. Identification of all the performance criteria
2. Careful consideration of the source and nature of the operating conditions
3. Understanding of the mode of failure of the construction responding to working conditions

This requires an ability to imagine the ways in which the construction might physically fail to achieve performance under the design conditions. The clues to this should be in the performance specification itself, for example excessive deflection in beams under loading conditions or excessive heat transfer through the walls under winter environmental conditions. It will also require a method of predicting how much deflection or heat transfer will occur to check that behaviour will remain within performance limits. The clues to this will come from visualising the internal action of the construction under these conditions in order to determine the response.

Future predictions play a part in this analysis, and this is often difficult, given the expectation of longevity for most buildings. Just establishing higher performance levels for some expected future requirements is problematic, for higher performance levels will have cost implications. Variations in conditions with time, including accidental events, need to be predicted with some probability, since designing for all imaginable circumstances also has considerable cost implications.

These considerations for the future add to the difficulties of prediction. They may be included in the brief but, if not, can be part of a technical analysis in so far as the sensitivity to error or change in conditions can be carried out. A solution that is sensitive to change may be less attractive if the future is not certain. This adds an additional dimension to the analysis. How close to failure is the solution when under working conditions? Will small changes greatly increase the probability of failure?

This prediction of sensitivity can be part of the analysis undertaken by visualising the response of the construction under the expected conditions. Understanding of the response of the construction to the external conditions allows the estimate of behaviour under working conditions and any capacity it may have to withstand unexpected or deviant conditions.

The role of observation and science in the analysis

It is suggested that an understanding of the ways of nature is necessary to provide the descriptions and determine the behaviour, and hence the performance, of the construction.

Experience can offer direct observation of construction both during production and under working conditions. In unfortunate cases experience can offer the opportunity to observe failure. If no failure is observed in a form of construction, it can be repeated in similar conditions and be expected to achieve similar performance levels, although this observation tells us nothing about the sensitivity of the construction. If failure is observed, some explanations are necessary to determine what changes are required to achieve performance the next time. Requirements for higher performance, operating in different conditions, or a change in the construction for production for economic gain also require explanations. These explanations have to be in the form of a description of the response of the construction in resisting failure.

For those engaged in technological activity the naturally enquiring mind is likely to seek such explanations. It is possible that over the past centuries, individuals have provided their own explanations in order to develop construction in a systematic way, refining solutions and occasionally making larger leaps of understanding to new forms of construction. Whatever the scale of change, these explanations have given confidence in perusing new ideas that would achieve performance with a limited (or at least considered) chance of failure. It is interesting to speculate whether the great cathedral builders had their own explanations to visualise behaviour.

Science provides a framework for exploring the laws and forces of nature. It provides a method of investigation from which explanations of the natural world can be verified. These explanations are useful in explaining the behaviour of construction. Science does, however, have its own language, which can be daunting to many. The knowledge derived from science has two major powers, one of explanation and one of prediction. Only the second, that of prediction, really needs to use the language of science. Explanations can be reduced to everyday language for they are only descriptions of the way the laws of nature operate. While it is the explanations that help understanding of the way in which construction may be functioning (action), prediction becomes necessary to establish a level of performance.

The approach to analysis

The analysis will be based on visualising the dynamic actions and processes involved. This will be achieved predominantly with verbal explanation or word pictures. Description of the mechanisms involved seems to offer the most widely applicable approach to understanding technology. All this can be greatly assisted by diagrams and mathematical modelling on the one hand and by experience on the other, but both these seem to need a verbal articulation, not only for a personal understanding of behaviour but also to provide an explanation to others of the basis on which the solutions have been adopted.

The aim should be to build a picture in the mind's eye first of what the solution will physically look like (the suggestion) and second of the actions that determine the response that describes the potential failure under working conditions (the analysis). Providing a description of both physical construction and then the response to the construction against certain criteria, which if not satisfied would result in the construction being deemed to have failed, is the underlying approach used in this text.

Four basic areas have been established and need to be understood to complete an analysis:

- Conditions expected through the life of building
- Failure or loss performance
- Proposal or suggestion of the construction (components and assembly)
- Response of the proposal to the condition

The chapters that follow provide the basis for this visualisation. It involves a dialogue between the proposed construction and the response of the construction to the conditions in which it has to be built and operated in order to limit the risk of failure.

Summary

1. Performance identifies the level to which the construction has to function under the design conditions and becomes the criterion for any evaluation of success.
2. Failure is a loss of performance at some time in the service life of the building under a predicted set of conditions.
3. Failure follows the action of construction. Action is the internal response of the chosen construction under the conditions in which it has to perform.
4. Visualising the conditions under which the construction has to function and the expected response of the construction chosen provides the basis for an analysis to evaluate satisfactory performance.

10 Physical Behaviour Creating Environments

This chapter takes in turn each of the aspects of environment that are important to the well-being of individuals and their possessions in the building. These have been identified as dryness, warmth, light, acoustics, cleanliness, security and privacy. In achieving these environments the building must be seen as a system employing both active services and the passive building fabric. The overall performance requirements will depend on the severity of the external conditions and the specification for the internal working conditions. The design will have to assign the role and contribution (function and performance) of each part of the construction to the creation and maintenance of the required environment. Once the contribution to performance for any part of the construction is known, it is possible to undertake an analysis of the response of construction under the working conditions to evaluate the chance of failure. For each of the aspects of environment, the dynamic response of the proposed construction can be visualised, based on an understanding of the law of nature and hence the action or mechanisms that may lead to failure.

Aspects of environment

This chapter is concerned with the aspects of physical behaviour of the construction that ensure the internal environment required to support the activity carried out in the building. These activities require aspects of both social and physical environments to be considered to ensure the well-being of the occupants and their possessions. All aspects of the physical environment will rely on the physical behaviour of the construction to ensure a satisfactory internal environment. However, not all aspects of the social environment will require a contribution from the physical behaviour of the construction. Space and appearance is not considered in this text as a physical behaviour and has been discussed in Chapter 8. Some aspects of the social environment such as privacy and safety do rely on some physical response from the construction. In the case of safety, the physical condition may be associated with loading and will, therefore, require the analysis outlined in Chapter 11. The range of physical and social environments is given in Figure 10.1.

The building as a system

Before evaluating the behaviour of the construction it is necessary to assign functions to each part and define the required level of performance that each part has to achieve to ensure the balance of flow and transfers to create the conditions required.

A definition of the required environment does not say anything about the contribution to be made by the various parts of the building. Even when the social and physical worlds outside the building have been defined, all that has been established is the total task of modification.

Before functions can be assigned to the various parts of the building it is necessary to develop an overall view of how the required environment is to be created and maintained. There are three basic groups of technologies available to fulfil these functions:

- Passive fabric
- Active fabric
- Active services

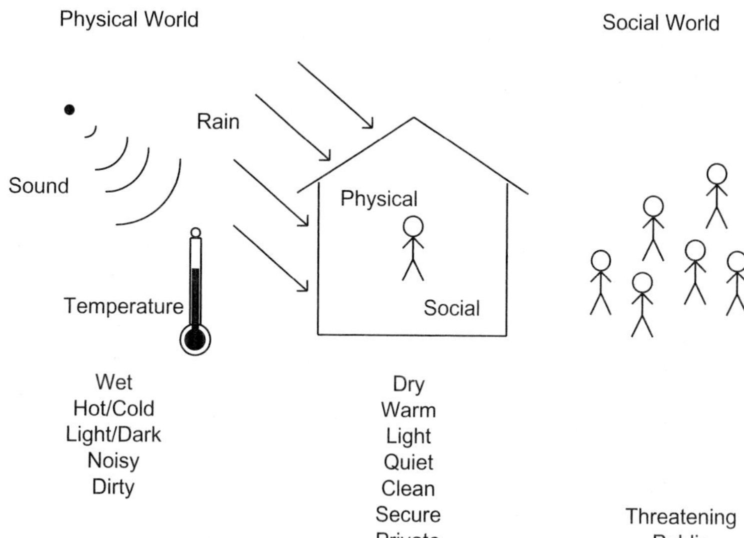

Figure 10.1 Physical and social environments.

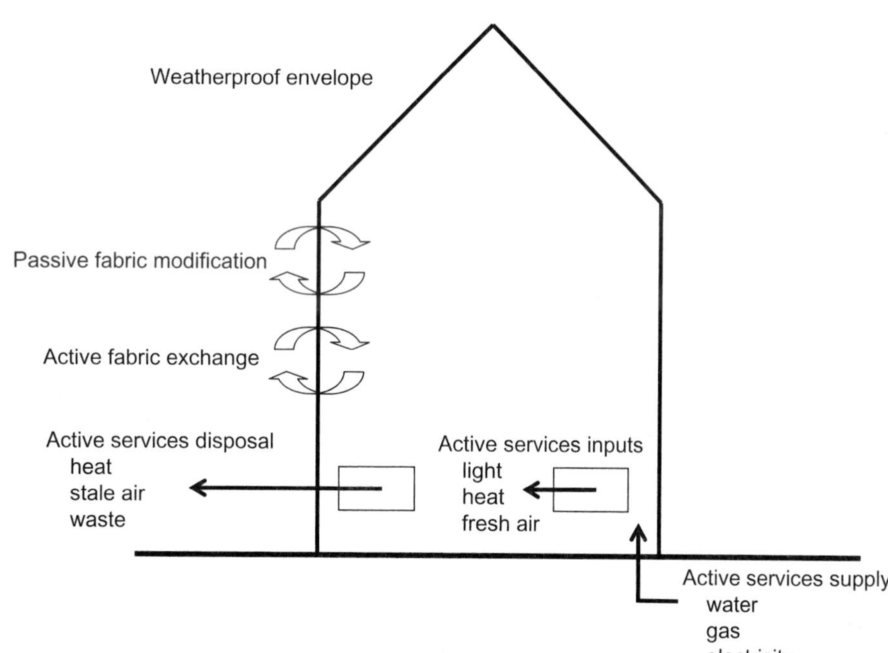

Figure 10.2 Creating internal conditions – possible technologies.

These are shown in Figure 10.2. It could be argued that the supply and disposal services are not all strictly environmental services, but all contribute to the provision of the active creation of the environment even though they contribute to other operational services as well. Water, gas and electricity, for instance, are often categorised as utility services.

Many of the aspects of environment can be achieved with different combinations of these three types of technology, for example the use of artificial lighting to supplement natural lighting provided by windows or roof lights together with the colour of walls. Decisions on this balance are influenced by initial costs, running costs, energy use and the control arrangement to be incorporated. With an increase in emphasis on the environmental impact of buildings, the combination of building design and active fabric solutions can reduce energy use from the active services.

Even when this balance between services and fabric has been established it is still possible to assign the functions of the fabric in a number of different ways. Some functions are fairly obvious, such as the roof covering's function of weatherproofing to create and maintain a dry internal environment. For others, alternatives exist. In providing a warm environment thermal insulation may be chosen to maintain temperatures in order to limit the need for heat inputs. This can become a function of either one or all of the main elements of the building, roof, walls and ground floor, each with various levels of performance contributing to that part of the strategy for maintaining internal temperature levels.

When the balance between services and fabric has been established, it is possible to start to develop details and specifications. The approach to each of the three types of technologies is different, generating a different set of decisions for each.

Passive and active fabric

Where the creation of the internal environment depends on the contribution of the fabric, the process of choice is focused directly on the variables identified in Chapter 4.

If passive fabric is to be employed then the primary concern will be for:

- The properties of the materials
- The dimensions, usually thickness
- The joints and fixings

For example, if sound transmission is being considered and mass is to be used to achieve performance then, as will be seen later in this chapter, the density of the material will determine the thickness required, so long as the joints do not support flanking paths and the fixings provide sufficient rigidity.

Where active fabric such as windows are being used, the focus becomes the joints and fixings to ensure the operation of the moving parts. The material, shape and dimensions of the active component will be determined more by their strength and reliability in use, although they may also have a passive contribution to make that will require analysis, as suggested above. They may have to be considered for appearance, as these components are almost certainly visible.

For active fabric there will also have to be some form of control so that their effect can be brought into use when required.

Active services

The final choice of components and materials in services systems is not as straightforward as the direct choice that associates the specification of the fabric with the environmental conditions. Their choice has become a design discipline in its own right. They are systems within the system of the building as a whole. Figure 10.3 suggests a way of considering the active environmental systems and their integration into the overall building design.

The outlet for the services are required to link the user with either a supply or a disposal system, as shown at the bottom and top of the diagram. Taking a heating system as an example, a radiator may be provided as an outlet but has to be linked to a heat source. In this case the active distribution would be via heated water contained in pipes, most probably using a pump to provide a mechanical force. The radiators would be located in each room, probably under the windows. The boiler to provide the heat would need to be located to make fuel and flue connections plus with consideration of appearance and as a possible noise source. The pipes may be routed through the floors and either within or

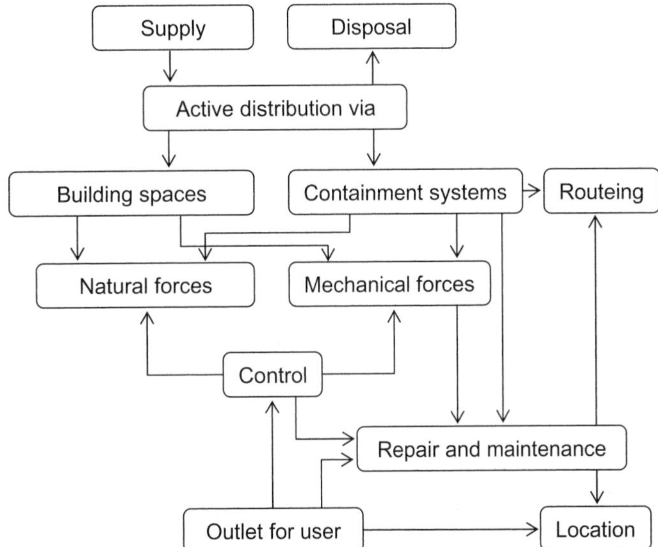

Figure 10.3 Aspects of active services systems.

Analysis

on the walls of the building or through specifically designated ducts or voids. This would have to include an access point to allow for repair and maintenance. In another example, a natural ventilation system could use the building space as the distribution system, using natural forces of wind and thermal gradients to move the air from the user, to dispose stale air at high level and bring in fresh air (perhaps tempered with heating) at low level.

Figure 10.3 also shows that, like the active fabric, these active services systems require control. When the internal conditions change, some control mechanism will have to be invoked. This control can be manual, relying on the user to identify unsatisfactory conditions and then being offered some way of re-establishing the required environment. For example, opening a window as long as fresh and relatively cool air existed outside the building could control a stuffy or over-warm room. It would include leaving the occupants to switch on lights when they considered it too dark. The alternative is to devise ways of providing automatic control, such as the use of a sensor to respond when design limits are exceeded and send a signal to a part of the system, which can start a chain of events to bring the environment back into design conditions. In the example of the heating system in the previous paragraph, this may be a room thermostat that can switch on the boiler and the pump in a central heating system when the temperature in the room (more correctly at the thermostat) falls below a predetermined limit.

Assigning environmental modification functions

Table 10.1 shows the aspects of the internal environment that may have to be created and maintained. These are each then identified with the major aspects of the external environment that will influence the achievement of modified conditions. The final two columns indicate typical functions that could be assigned to parts of the construction to achieve the modification required. For most of the aspects of modification, both passive construction and active services could be employed with different functions. For example, to create dry internal conditions it is necessary to understand patterns of precipitation, the conditions of ground water and the presence of vapour in the air. From this, waterproofing can be assigned to the fabric and water disposal to the building services.

Table 10.1 The role of fabric and services.

Internal conditions to be established	Significant aspects of the external environment	Possible functions of fabric	Possible function of services
Dry	Precipitation Ground water Vapour in air	Waterproofing Vapour control	Water disposal
Warm	Air temperature Solar gain Wind Humidity	Resisting the passage of heat Thermal mass Ventilation	Heating Cooling
Light	Sun path Overcast sky	Transmission Reflection	Room Task Mood Architectural
Acoustic	Natural sounds Pollution	Transmission Absorption	
Clean	Dirt Pathogens Fungal and plant life Animal life Air pollution	Exclusion Filtering Cleanable	Removal Filtering Disposal
Secure	Social threats Act of God Fire	Resistance Separation (barriers)	Detection Alarm
Private	Cultural and social conventions	Separation (visual and acoustic)	

Rarely does one part of the construction act alone in fulfilling one of the environmental functions and rarely does it contribute to one of the functions only. Elements of the building such as walls and roofs may contribute to many of the functions. Elements are assemblies of sub-elements, such as the external skin of a cavity wall or tiling on the roof, where a primary function is often identifiable. Choice often centres on these sub-elements, for they are normally identified with a major component or material where the focus of the functions is also often fairly clear. For example, an external wall probably plays its part in all the environmental functions. Even major sub-elements such as windows have a range of functions. Primarily for light and view, these may be used successfully for ventilation but cause problems with solar gain and security. As finer division of the sub-elements is considered, down to single components, then function is often clear. Within a wall, the damp-proof course (DPC) has a very well-defined function of damp control. Analysis at this level has to keep sight of the function of the building as a whole, while defining the function of the components and materials of which the sub-element will be made. This division and refinement of the environmental functions is illustrated for a roof in Figure 10.4.

Assigning performance levels for environmental modification

Assigning of a function only identifies the part to be played by an element, sub-element, component or material in creating and maintaining the required environment. Assigning a performance involves defining a level of attainment. While it is unusual for a building's design to be so radical as to cause the functions to be unclear, the new demands of clients and regulations often lead to a change in performance. All the aspects required to set a performance level have already been introduced:

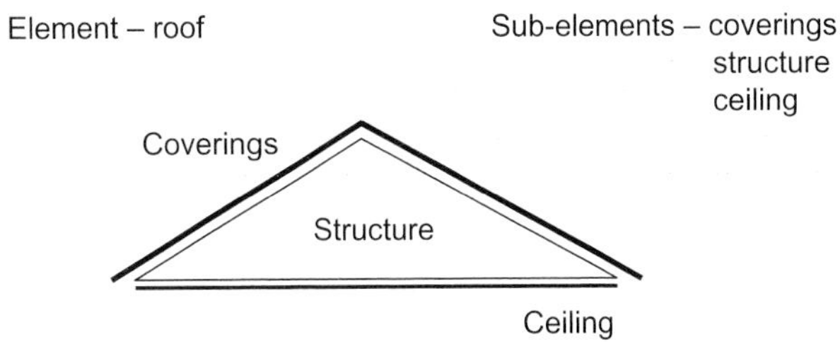

Contribution of sub-elements to environmental performance

Coverings	Dry	Waterproofing
	Clean	Infestation exclusion
	Secure	Fire spread
Structure	None	
Ceiling	Warm	Resistance to passage of heat
	Light	Reflection
	Clean	Dirt exclusion

Figure 10.4 The role of sub-elements of the roof.

- The range of what will be considered normal external conditions and what extreme or accidental conditions it may be worth designing for
- The ideal internal conditions, the limits that would still be considered satisfactory and any control mechanisms that are intended to be included
- The proportion of the task of creation and maintenance of the environment to be assigned to that part of the construction

Thinking through the conditions

The analysis of each aspect of the internal environment will be considered in turn in the rest of this chapter. The focus will be on the analysis of the fabric, although the choice of services systems will be introduced.

Following the general format for thinking through physical behaviour established in Chapter 9, each section will consider:

- The origins and nature of the environmental conditions
- The nature of the failure linked to performance
- Possible types of solutions
- Analysis of the response involved to achieve performance

A dynamic description of the environmental conditions to be modified and the mechanisms that determine the response of the construction to the expected environments in which it will have to operate should be carried out. This should be based on a visualisation of the flows and transfer taking place through the construction.

The dry environment

Substances exist in a solid, liquid or gaseous state. Unlike nearly all other substances, water can exist in all these states within the normal climatic range of temperatures and pressures. This creates particular problems for buildings trying to maintain dry environments. An excess of liquid water is undesirable, but a limited quantity of gaseous water (vapour) in the air is important for comfort, providing levels of humidity. In visualising the behaviour of construction, the possibility of a change of state has to be considered. Perhaps the most important example of this for dry environments is the change from vapour to liquid in the depositing of condensation.

As well as having an undesirable effect on health and comfort, water is a major source of deterioration of construction materials (see Chapter 12). Much detailing has the dual function of protecting the environment and the fabric of the building. In this respect, visualising the dynamic interaction between water and the construction is common to the analysis of both environmental control and durability. However, the analyses of the two need to be considered separately since different actions of the water become significant, as, for instance, in the change of state from liquid to solid with frost action, significant for durability but of little concern in considering environmental conditions.

If the requirement is for a dry interior then the clear failure can be characterised as leaks, where water enters the building through the external fabric and openings or from failure of the building services. However, much of the detailing of buildings is concerned not only with the free entry of water as leaks but also with excessive moisture in the fabric, known collectively as damp. The dividing line between these, both as observable actions and in the appropriate detailing, is often fine, so they are normally considered together.

Maintaining a dry interior (absence of dampness rather than lack of humidity) requires the control of moisture as both liquid and vapour. As a liquid, the source of water is normally external and hence the external fabric has a major functional role in damp control. As a vapour, sources are predominantly internal so both the design of the external fabric and the overall process of ventilation are used to prevent the incidence of condensation.

When considering the analysis and choice of the fabric to ensure a dry interior, it is usual to see three major threats:

- Penetrating damp
- Rising damp
- Condensation

Among others are blown light snow entering ventilated openings and subsequently thawing and leaking from water-carrying services.

To visualise the behaviour and analyse the ability of the construction to fulfil the function of damp control requires knowledge of the origins of the moisture, the forces acting on it when it comes into contact with the construction and the pathways then available for it to reach internal surfaces.

Predictive methods are available to model vapour and temperature patterns in order to assess the risk of condensation. Penetrating and rising damp assessment is still largely a matter of traditional detailing; maintaining barriers to the progress of dampness into a building by applying tried and tested methods. Testing of prototypes to wind and rain penetration is possible, particularly associated with components systems such as windows or claddings. This prototype testing is especially significant when new principles such as rain screening are applied because predicting the behaviour of water can still be a matter of trial and observation.

Each of the major threats will be considered in turn, with descriptions of behaviour, identification of pathways and of the major detailing techniques that can be employed.

Penetrating damp

Rain is the major source of water that interacts directly with the external fabric of the building, creating the possibility of penetrating damp. Three major actions can then be identified, as follows:

1. The water that comes into direct contact with the outside of the building will soak into permeable construction where pore networks form continuous capillary pathways through the material.
2. Any water running over the surface of the building is primarily influenced by gravity

moving it down the construction, but other forces also act on this surface water, particularly when it meets joints and projections. The water may be drawn into small cracks by capillary action. Wind that accompanies the rain will create pressures around the building, affecting rates of movement of the water across the surface and penetration of the fabric, particularly at open joints. Even the 'skin' formed on the surface of water exerts a force, known as surface tension, that not only holds water on the surface but also can cause it to migrate, upside-down, under horizontal surfaces beneath overhangs. Visualising this highly complex set of forces acting on water on the wetted surfaces of buildings is particularly important at the detailing of joints between components and at junctions between elements.

3. The majority of the water that falls on the surface of the construction will arrive at an edge. If it is a free edge, the water will fall off mainly under the influence of gravity. However, its behaviour at these edges will also be influenced by wind and surface tension drawing some of the water back under horizontal overhangs. It is important to make sure this water is removed from edges effectively so it does not have the chance to accumulate and soak the construction more than that expected from the original rainfall.

In reality the distinction between the first action (soaking in) and the second (surface water interacting at joints) is not so clearly defined. Most construction, such as roof tiling and brickwork, is not monolithic. It is not just one material where the properties of the material will be the only influence on the rate of soaking in. Although they may be considered as one skin, they are made up with joints, which in the case of brickwork are made with a second material: mortar. The behaviour of soaking in is now dependent not only on the permeability of the brick and the mortar but also on the reaction of the water at the joint. Some of the forces identified in the second action (interaction at joints) come into play. In fact, in brickwork with cement mortar, where drying shrinkage breaks the bond between brick and mortar to create fine capillary paths, water penetration can be observed at the joints long before it penetrates the brick itself.

The depth of penetration of the water soaking into the material depends primarily on two things: the cycle of wetting and drying and the permeability and thickness of the construction. Prolonged periods of wetting with only short periods of drying in between pose the greatest threat. If the construction is not specified thick enough, or barriers to halt penetration are not provided, moisture will appear on internal surfaces. The three basic construction forms to overcome this direct transfer through the walling material itself are shown in Figure 10.5.

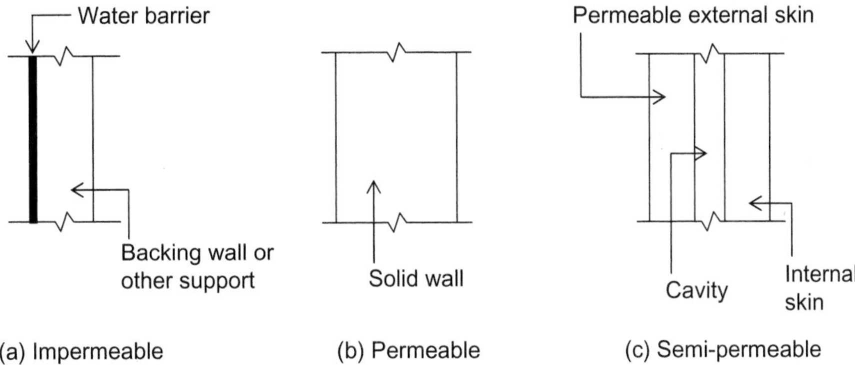

Figure 10.5 Generic forms of walls to resist penetrating damp.

All of these forms can be exploited. At the beginning of the twentieth century most house construction would have had solid walls, but by the end of the century the changes in materials, particularly the use of cement mortars (see explanation in previous paragraph), had led to the almost universal adoption of cavity construction.

The water that does not soak in or is not drawn in will move down the face of the building. If the surface is impermeable then a considerable quantity of water can accumulate. There is a danger of damp penetrating at any time when the water meets joints or projections. The expected pathway and potential quantity of water has to be visualised for each situation of joint and projection.

The action of capillarity at a filled joint has already been identified between the brick and its mortar if the bond breaks and fine hairline cracks are formed. This is the scale at which the mechanisms have to be visualised. Pathways through the fabric are often very fine and tortuous, with damp appearing on the inside, some distance from its entry on the outside. Other joints are open with no filling material. If these joints are narrow, they may support capillarity if a fine continuous open gap is formed. However, two other forces will assist water penetration at gaps or joints: gravity and wind. These may be significant even where the gap size would not support capillarity. Not only may wind change the surface water's movement directly, but also the pressures that it builds up around the building may blow water through joints or even pump water up behind detailing, such as flashings.

Two other forces that may act on the water when at joints or projections are kinetic and surface tension. Kinetic energy is present in the falling rain. It is the force that causes it to bounce on hitting the surface, so apparently sheltered areas may become wet. Kinetic effects are relatively weak, but surface tension has the potential to cause problems at overhangs and projections. On the underside of projections, which may also be considered to be sheltered from wetting, surface tension will hold water on the underside of the projection where it will migrate towards an apparently protected joint and enter from below the projection. The basic forces acting at joints are shown in Table 10.2.

Penetration at joints can be aggravated when they occur at projections such as sills. Water will collect on the top surfaces of a projection, causing high levels of soaking or additional water to be drawn in through joints. Identification of the damp pathways and the detailing of damp-proofing at the sill of a window are shown in Figure 10.6. This figure illustrates the use of jointing materials (putty), filled joints (water bar), barriers (DPC) and, on the underside of the sill, the simple device of the groove or drip detail. In the drip the water collects and gains sufficient mass to break the surface tension

Table 10.2 Forces potentially assisting water to enter at joints.

Force	Direction of action	Effect
Capillarity	Any	Water travels through fine cracks and pore networks
Gravity	Downwards	Water enters any opening or crack with downward path
Pressure differences	Any	Above ground wind • Driving rain • Holding water in joint • Moving towards lower pressure • Pumping behind flashings Water pressure below water table in ground
Kinetic	Rebound	Localised to area of impact
Surface tension	Any surface	Movement at underside of projections Holds film of water on surface and on ledges

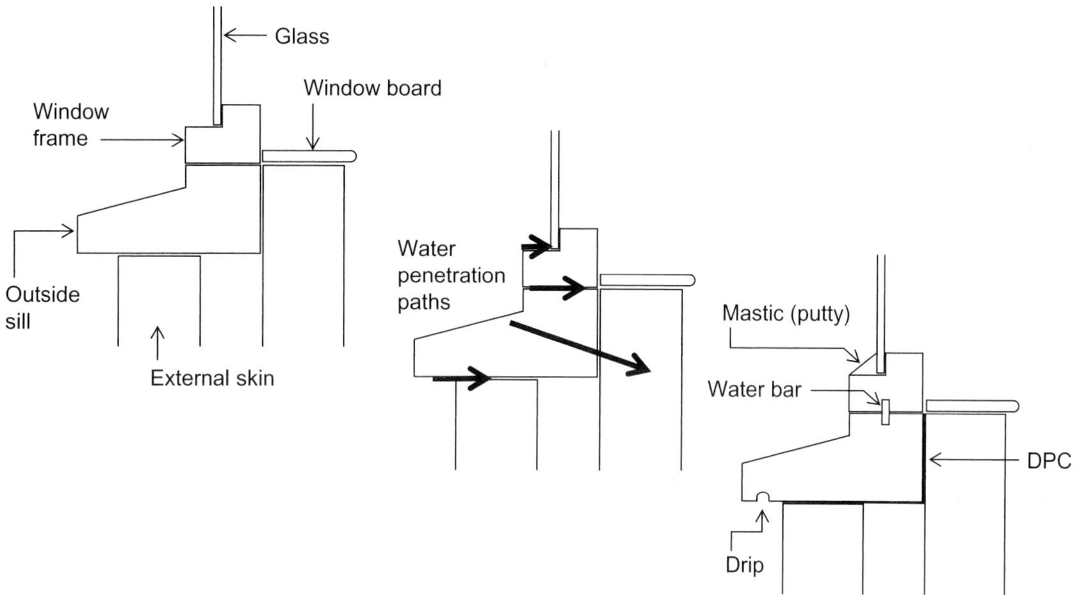

Figure 10.6 Pathways and details to resist damp penetration at sills.

when droplets fall from the underside before reaching the joint under the sill.

The distinction between absorption from prolonged wetting and direct penetration through joints and projections has particular importance when thinking about aspects of external climate and rainfall in a particular locality. Any building subject to heavy, driving rain may need to be detailed differently to one subjected to prolonged drizzle. The total average annual rainfall may be the same.

Finally, it is necessary to consider the water that is shed from the building, particularly where the roof discharges water at the top of walls. If not intercepted, gravity will take it towards the ground. It is not a direct threat if it falls directly to the ground, but it may fall or be blown onto the surface of the construction, soaking materials to an extent not envisaged by direct exposure to the rain itself. If it reaches the ground without contact with the construction, it may still splash (kinetic forces) onto the fabric at ground level and increase wetting and soaking of materials close to, but above, the ground. The rest of the water reaching the ground will either directly soak into the ground or collect and flow away on any flat, hard surfaces. Intercepting this water at the edge of the roof is normally the function of gutters and drains.

Rising damp

The threat from rising damp is ever present in temperate climates. Moisture is found in the ground at virtually all times of the year. While the greatest threat may be during wet periods when the moisture content of the soil will be high, the moisture persists in all but the surface layers, even in dry periods. This is even above the water table. If a building is built below the water table, and comparatively few are, then the building is under an additional threat of water pressure assisting entry through the fabric of the building.

The major action associated with rising damp, as the name implies, is a transfer of moisture up into the building against the pull of gravity. The water travels through the fine network of connected pores in permeable materials by capillary action. This rise within the pore system in the materials can reach

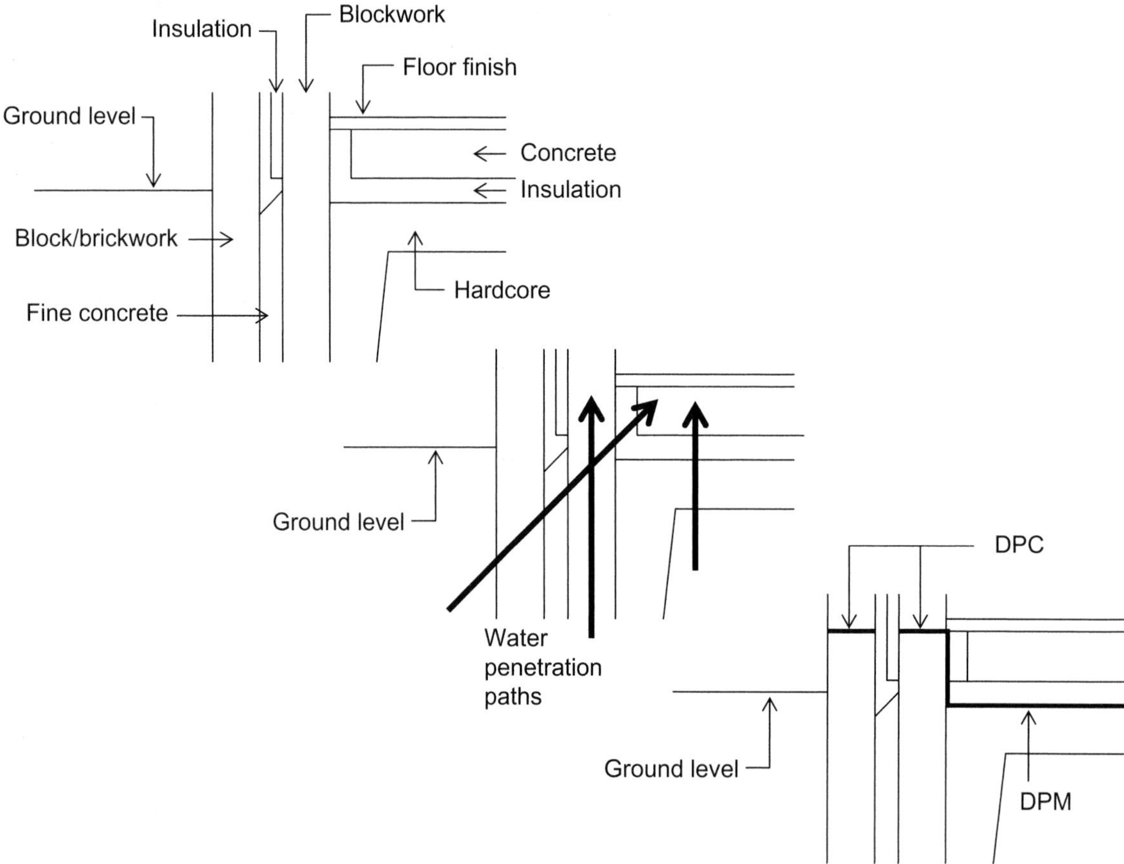

Figure 10.7 Pathways and details to resist rising damp.

considerable height above the source of the moisture in the ground.

It is not normally economic to use anything other than permeable materials for the major components below ground. The usual detailing solution is to provide an impermeable barrier known as a damp-proof course (DPC) in walls and damp-proof membrane (DPM) in floors. Identification of the damp pathways and the detailing of damp-proofing at the junction of the ground floor and the external wall of a house are shown in Figure 10.7.

The detailing and care in construction to ensure continuity within and between these damp-proof barriers become extremely important. Given that the major action is capillarity, which by definition exploits the finest pathways, this indicates that any lack of care in detailing or assembly can have a disproportionate effect. Once damp-proofing is breached, the materials on the dry side absorb the moisture and transfer of the moisture will continue to internal surfaces.

Condensation

The descriptions so far have been of the possible pathways threatening a dry interior, essentially from outside, through the fabric. This is, however, not the only source of internal dampness. Primary among other sources is condensation. The moisture for condensation does not enter the building as water but as vapour, and most is generated within the building through everyday activities. This vapour not only comes from cooking and bathing but also is generated by the act of breathing, leaving almost all activities as generators of vapour which, as a gas, is carried away by any process of ventilation occurring in the building.

Vapour is not a threat to the dryness of the environment (although excessive vapour causing high humidity can be uncomfortable); as long as all the vapour can be held in the air it causes no damp problems. There is, however, a limit as to how much vapour can be held in the air, dependent on a number of factors, primarily temperature. If vapour increases at constant temperature or temperature decreases with constant vapour the result may be condensation or the appearance of liquid water. This is most commonly seen on the internal surfaces of window glass, which will probably have one of the lowest surface temperatures in the room. Another vulnerable area is at the internal corners between wall and ceiling and behind furniture where ventilation has least effect. Here the surface temperature may be higher but so will the vapour concentrations. As these surfaces, unlike the window glass, are permeable the condensation will soak in and the first sign of a problem may be mould growth.

The temperature at which condensation occurs varies depending on the quantity of vapour in the air. The less vapour in the air, the lower the temperature at which condensation first occurs. The temperature at which condensation will occur for a given air–vapour mix is called the dew point temperature, where the air is said to be 100% saturated or to have a relative humidity of 100%. If the temperature continues to fall (or more vapour tries to enter the air), more condensation will be produced, as the air can be no more than 100% saturated with vapour.

As well as occurring on surfaces, condensation can be a problem within the materials themselves. Most of the materials of which construction is made are not only water permeable but also gas (and therefore vapour) permeable. The two are, however, not directly related. Gas permeability of a material is a different property to water permeability. When air inside the building is warm, it is probably holding more vapour than cold air outside. This difference is sufficient to cause vapour from inside to migrate through the construction towards the outside. As it is warm on the inside and cold on the outside, heat will also be flowing out across the construction, the temperature of each layer of material becoming colder towards the outside. This drop in temperature becomes particularly large across any insulation. Insulation material does not normally hold up the vapour being transferred, so this may create the conditions for condensation to occur within the construction. Because it happens inside the materials, it is known as interstitial condensation.

Water is now present within the construction. If vapour continues to flow and temperatures remain low, the quantity of condensate increases. The presence of water within the materials may threaten the durability of the materials and may change other properties on which the function is depending: principally their insulation and air permeability values. Depending on the detailing and where the condensation occurs, it may create dampness on the internal wall. This will depend on there being a pathway for the condensate water back towards the inside face of the construction. Similar forces and mechanisms create these pathways, as discussed in penetrating dampness. The water will travel in capillary paths within porous materials or through cracks. Gravity will work on the water (hardly at all on the vapour), draining it through the construction where it may find a pathway in.

If it does not find its way into the building it will, like absorbed rain, have to dry out from the outside surface, where wind, rising temperatures and the sun will dry the surface, creating the conditions for a migration of absorbed water to the surface or to evaporate it back to vapour. This may be a slow process and condensate can remain within the construction for considerable periods.

Detailing, including the position of the insulating layer along with ventilation and vapour control layers, can limit the risk of condensation and this is discussed in the section on the warm environment that follows.

The warm environment

In temperate or cold climates it is natural to see the function of the building as creating and maintaining a warm environment. In warm or hot climates the function may be characterised as providing a cool environment and this will lead to very different strategies and will therefore need a different analysis to the one described here. In hot climates thermal mass and white exteriors are likely to be employed rather than insulation, which is common in temperate climates. This section will focus on the temperate climate winter heating condition, although the summer condition will also be considered.

Feeling warm is not just a case of the air temperature. It is dependent on radiant effects, cooling air movement (draughts or breeze) and humidity.

Radiant effects are a process by which heat is lost (or gained) from the body by radiant exchange. This is independent of air temperature and only requires a cold (or hot) surface to be in direct line with the individual (or another object) for heat to be exchanged.

Areas of glazing can be associated with these effects. The rate of radiant exchange is proportional to the temperature difference, and this is likely to become large if the glazing itself is directly exposed to sunlight (heat gain) or a clear winter night sky (heat loss). It is likely to be uncomfortable if activities have to be carried out too close to the glazing. Under these circumstances some form of shading, blinds or curtains may have to be used, either externally to keep the surface at air temperature or internally to cut out any direct radiant exposure from the surface to the individual. Placing warm radiators beneath cool windows will also help mediate the radiant exchange experienced by the individual if the temperature of the cool glazing is not too extreme. Working very close to a radiator can also be uncomfortable due to radiant exchange.

Air movement will come from ventilation, where if the rate of airflow is too high a cooling effect will be felt by evaporation from the skin. This will be discussed in the section on ventilation below. High humidity reduces the efficiency of skin-cooling evaporation, making us feel less comfortable in high air temperatures even with high air flow rates.

The function of the passive fabric of the building in maintaining a warm environment within the building is complementary to the function of the active environmental services systems. It is necessary to see any thermal design to be integrated between the fabric, heating and ventilation arrangements of the building. This strategic decision is also concerned with issues of energy use, clean air and running and maintenance costs. Whatever the overall contribution in each case the fabric will fulfil the same function, only the level of performance will vary.

In a temperate climate, the usual initial concern is the loss of heat through the fabric of the building under winter heating conditions. The more heat that is lost through the fabric by transfer and ventilation, the more heat has to be provided by the heating system. This reflects the wider concern associated with energy consumption and running costs. There is also increasing concern with larger commercial building regarding cooling in summer, dealing with high internal heat gains and reducing solar gain so that less cooling is required. The role of the fabric under these conditions is different from the winter heat retention role. This will, for instance, require greater rates of ventilation, the use of shading on the external face and exploiting thermal capacity rather than lower ventilation rates and insulation of the building skin required for winter conditions.

To consider the external skin of the building it is necessary to understand how the construction responds when a temperature difference exists on either side of a wall, roof or floor. The fabric will almost certainly consist of various layers performing different functions,

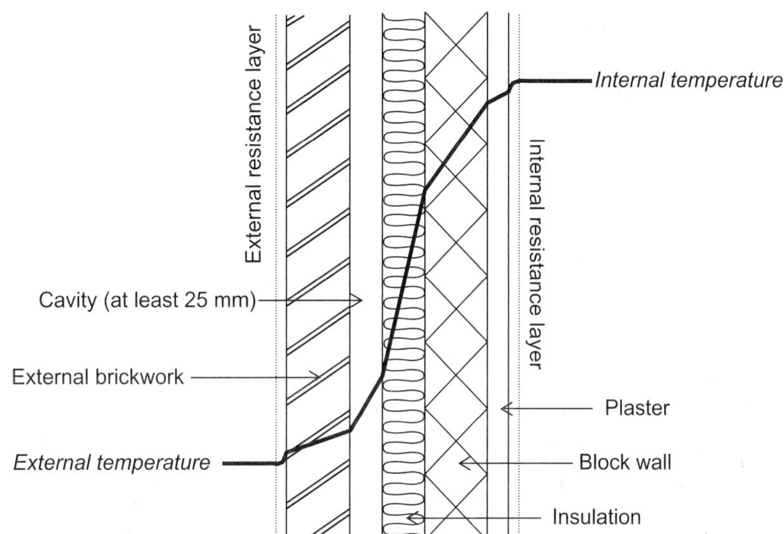

Figure 10.8 Typical temperature drop across insulated cavity wall.

only one of which may specifically be to enhance thermal performance. However, it is the behaviour of the fabric as a whole that needs to be considered.

It is possible to visualise the materials and layers of construction from the point of view of either ease with which they transmit heat or the resistance they offer to the passage of heat. Generally the description that follows will think in terms of layers offering resistance. The function is resisting the passage of heat. Layers with high resistance have materials normally thought of as insulating materials.

The simple view of fabric heat loss

The simplest view of heat loss through the fabric considers the system of inputs and loss to be in equilibrium. It assumes the temperature on each side to be constant and the rate of transfer to have reached steady state. It also assumes that all heat moves directly through the fabric from one surface to the other. The transfer of heat through the fabric is, under these circumstances, dependent on the resistance to heat transfer offered by the layers of the materials chosen, including any cavities, plus the layers of still air held at the surfaces of the construction.

Steady state is the simplest way to visualise the heat flowing through the construction and it illustrates the fundamental influence of the construction. Given a temperature difference on either side of the fabric, heat will flow from the high to the low temperature. The quantity of heat that passes through the fabric, in a given time, the rate of heat flow, will depend on the resistance of the layers, including surface air layers and cavities of air. The greater the resistance of the whole construction, the less heat passes per unit of time. It is not possible under steady state for more heat to be flowing through one layer than another.

The second influence of having layers with different resistances is the way in which the temperature changes across the construction from the warm to the cool side. This is dependent on the relative resistance of each layer. This is illustrated in Figure 10.8. Layers contributing a large proportion of the resistance to heat flow (the insulation) will experience a large proportion of the temperature drop across the construction as a whole.

Not only is steady state the easiest way to visualise heat flow, but once the resistance of each layer is established it is the simplest condition for which to provide calculations. The rate of heat loss is expressed as a function of the

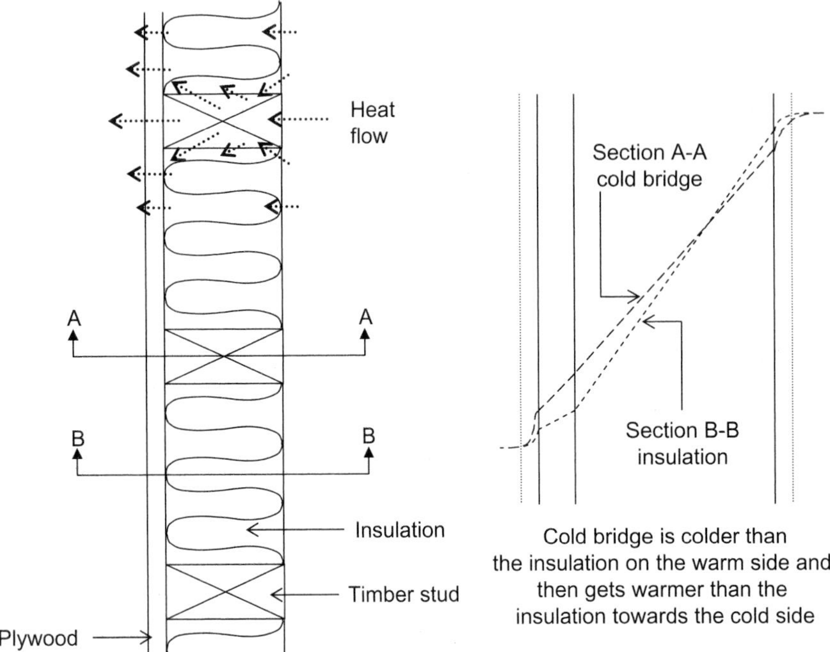

Figure 10.9 Effect of studs as cold bridge in timber frame wall panel.

whole construction, combining the effect of all the layers into one figure representing the transmittance of the complete construction. This is a single transmittance value known as the U value.

Heat loss calculated in such a way is, however, based on a description of behaviour that is rarely observed in practice. While the internal temperature may be relatively constant, the external temperature throughout a heating season will vary a great deal. Deciding on a design condition for external temperature becomes crucial to the value of the estimates provided by such calculations. Chosen too low and heating systems will be over-designed for most of the heating season, while chosen too high and conditions of comfort may not be achieved for a large proportion of the days when heating is required. Current design practice in the UK is to take 0 °C as the external design temperature.

The accuracy of any estimates calculated in this way is sensitive not only to the choice of design conditions but also to detailing. The assumption was made that all heat moves directly through the fabric from one surface to the other. If the resisting layers remain continuous over a large area, this assumption is reasonable. However, whenever the layers are interrupted by structure or openings this assumption must be questioned.

Cold bridging

Figure 10.9 illustrates the effect on heat flow through construction where the studs in a timber framed wall panel do not provide a single layer. The studs are said to be providing a cold bridge by the diversion of the heat so it does not pass directly from one side to the other. This will also occur at junctions of floors and around window openings. Heat will flow naturally towards the lower temperature, distorting the temperature profile. This thermal bridging has the effect of losing more than a simple heat loss calculation would suggest.

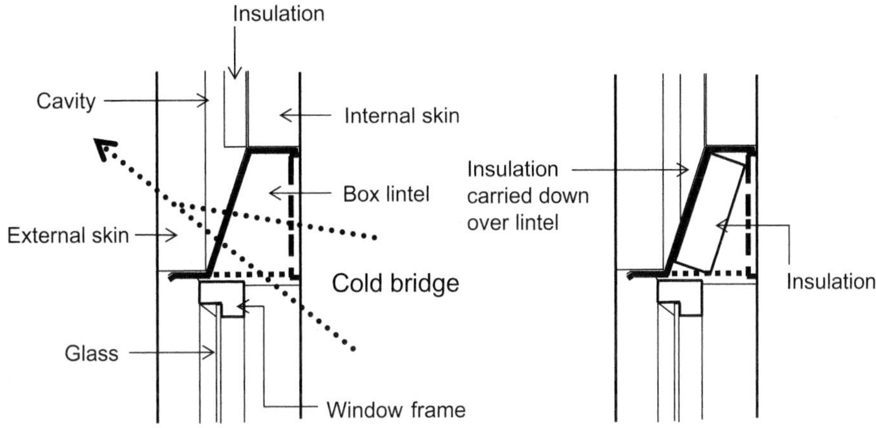

Figure 10.10 Cold bridge and details to reduce heat loss at window head.

Hence where a building has little continuous high resisting fabric with frequent penetrations or openings, the detailing to reduce this effect needs to be considered. The potential bridge at the head of a window and the detailing to minimise the effect are shown in Figure 10.10.

Non-steady-state heat exchange, capacity and diffusivity

While a building is being heated, the internal temperature will be relatively constant. The outside temperature will vary much more, even on a daily cycle. Not only does this involve the air temperature but also surface temperatures will be directly affected by radiant exchange by direct exposure to the sun (gain) and to clear night-time skies (loss). Surfaces, particularly if dark in colour, will absorb heat if directly exposed to the sun and lose heat if directly exposed to a clear winter night sky. Surface temperatures can by this mechanism reach values significantly different from air temperatures.

Buildings may not be continuously heated. Intermittent heating may be necessary to match the occupation pattern. Under these circumstances both internal and external temperatures are changing, often in the opposite direction, with internal temperatures rising just as external temperatures are falling.

In order to visualise this dynamic thermal behaviour it is necessary to think about how rapidly the temperature profile in the fabric changes in response to temperature changes at the surface. The thermal properties of the materials and thickness of each layer will determine the speed with which the fabric will respond. This needs not only a measure of resistance or insulation value of the layers but also one of thermal capacity. Capacity is a measure of the amount of heat necessary to create a given rise in the temperature of the material (or for the heat to be given up during an equivalent fall in temperature). The rate at which the temperature within the layers changes will be quick when there is little resistance to heat flow and little thermal capacity. With increasing resistance or capacity of the layers, changes in temperature within the materials will be slower. This measure of the speed at which the internal temperature profile changes is a function of both resistance and capacity and is known as thermal diffusivity. Generally, materials offering high resistance are low density with large volumes of air entrapped in the materials, while high thermal capacity materials have higher density and hence a lower resistance to heat flow.

Analysis of response and capacity becomes more important than insulation in climates where the large changes occur between night

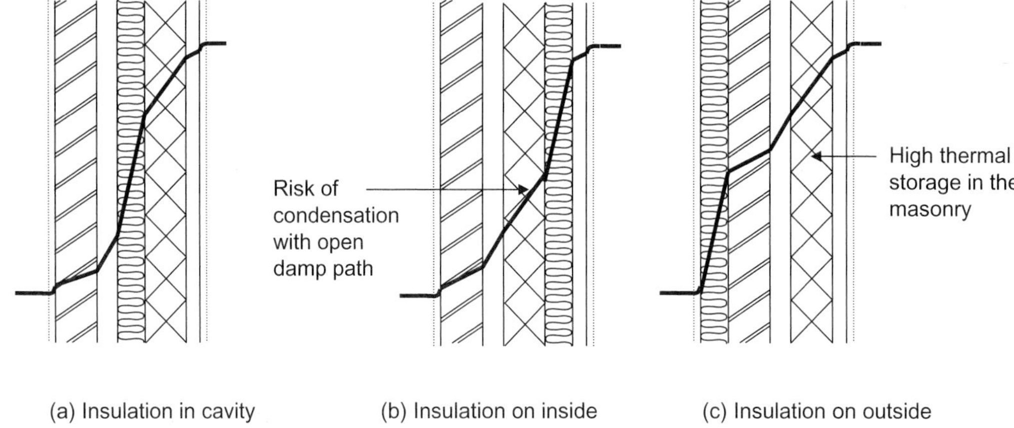

Figure 10.11 Temperature gradient with position of insulation in cavity wall.

and day, such as in desert areas, rather than between summer and winter, as in temperate climates. Hence, heavier, thermally massive materials are used. This is similar to the summer cooling designs in temperate climates where the fabric is used to store the heat generated during the day, which is then purged at night.

Analysis of response is also important where buildings have to be heated for short periods, when thermally light building may be appropriate. Another strategy would be to have a heating system that was chosen with radiant heaters that heat the individual directly without heating the air so that the fabric has little role to play in feeling warm other than reducing draughts.

Position of the insulation

When only considering heat loss and taking the simple view of steady state conditions, the position of the high-resisting, insulation layer is of no consequence. When the behaviour of the fabric is being analysed for any purpose where temperature profile or response time is important, the position of the relative resistance offered by each layer may become significant.

If the insulating layers are near to the inside of the building (warm side) then the outside layers will have a smaller temperature rise in order to achieve steady state. In these circumstances not only is steady state approached relatively quickly but also there is little heat stored in the fabric even if constructed with relatively high thermal capacity materials. Often more significant is when the insulating layer is on the warm side, in which case there is an increased risk of condensation with the lower temperatures that arise on the cold side of the construction.

Figure 10.11 shows the temperature profile for walls with the insulating layer in different positions. The converse is, therefore, that insulation on the cold side reduces condensation risk but may increase warm-up periods where intermittent heating is required.

If the condensation risk is high then additional detailing will be required. This will normally take the form of either a vapour control layer or the ventilation of cavities to take away the vapour migrating across the construction. Ventilation can be particularly effective in roof construction.

Ventilation

Maintaining a warm environment is affected not only by the choice and detailing of the fabric but also by the amount of ventilation. The direct loss of warm air from a room under winter conditions accounts for a significant proportion of the heat loss. Ventilation involves air exchange and therefore airflow. If the speed of this airflow becomes too great, it can be called a draught, increasing heat loss from the body so we no longer accept the environment as being warm even if the air temperature has not changed. Under summer conditions the increased speed of airflow may be called a breeze and is welcomed in order to maintain a comfortable thermal environment. In temperate climates, ventilation is the main means of maintaining a comfortable thermal environment under summer conditions. Air conditioning, being expensive to install and having high running costs, is restricted to a limited range of buildings in these climatic zones.

Ventilation, however, is not just a consideration for maintaining a thermal environment. It also plays its part in two other functions. First, it removes water vapour, which reduces the risk of condensation and hence dampness in the building. Second, it removes stale air, maintaining a clean environment, keeping down the level of airborne pollutants and odours to a safe and comfortable level. This will be covered in the section on the clean environment, for there is no doubt that an exchange of air is necessary within buildings, and some means of achieving a controlled level of ventilation has to be considered.

For ventilation to take place it is necessary to have two conditions. There needs to be a pressure difference to move the air and the presence of openings to let air in and let air out. This assumes a supply of air that is suitable to replace the air in the building. If this is not the case then cleaning by filters or dehumidifying may be necessary, but this will not be considered here.

The need for a pressure difference and openings is true for both natural and mechanical systems. The way they are provided and the means of control are, however, very different. They may be hybrid approaches, the simplest of which is mechanical extraction to increase (or at least control) the pressure difference in order to dictate the exit opening but leaving the fabric to provide the inlet openings. Even with a full mechanical system, the fabric must be detailed and built so it is not too leaky, as this could influence the performance of the mechanical system.

The part played by the fabric in the provision of ventilation is one of providing openings. These are not only those specified for this purpose such as air bricks or chimneys but also the cracks that are present around opening windows and doors. Sealing these cracks with draught excluders is often seen as a cost-effective way of saving heat. This may be true, but it may also exclude a level of ventilation necessary to maintain a dry and healthy environment. Windows with draught seals should be provided with some other ventilation process such as trickle vents if there is a danger of low ventilation rates in the building.

Pressure differences created by the wind occur naturally around buildings. Typical pressure distributions around a building are shown in Figure 10.12. Air will enter any openings on

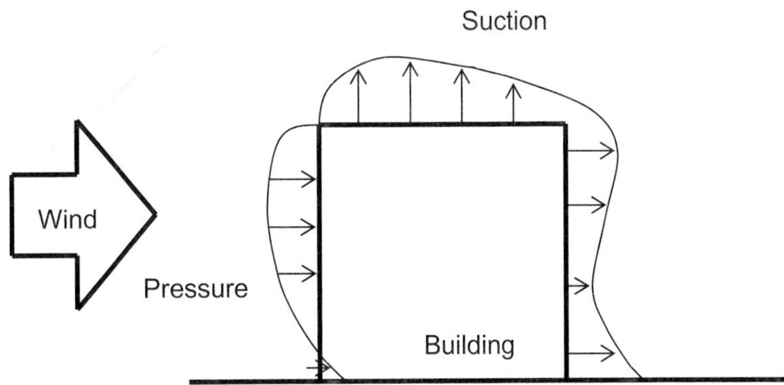

Figure 10.12 Typical wind pressures around buildings.

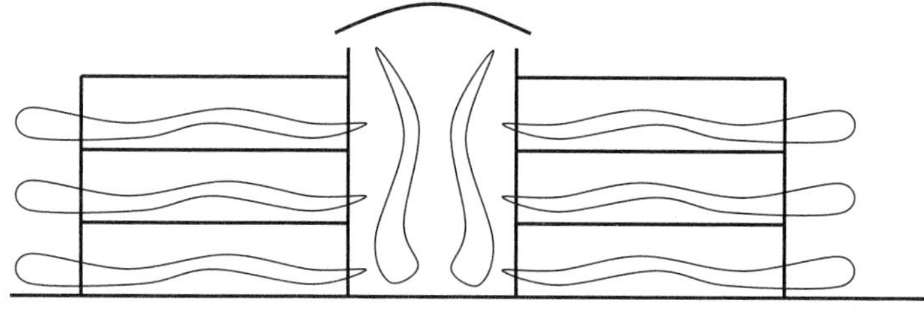

Figure 10.13 Natural ventilation using atrium or internal street.

the side of the building experiencing positive pressure and be drawn out from the side with negative pressure.

A second process that can be harnessed to create natural ventilation is the stack effect. This does not induce a cross flow of air but a vertical air movement. It relies on a temperature difference with warm air at the bottom and cooler air above, but it will be assisted by wind if the upper opening is in a negative pressure and the lower opening in a positive pressure region on the outside of the building.

Since the primary force inducing natural ventilation is the wind, it is clear that control has to use openings in the fabric and that this control will have to account for a variety of conditions experienced locally by the building. Local conditions can be highly modified by adjacent buildings, which may have a profound effect on the levels of ventilation achieved in practice.

Visualising these processes demands an analysis of wind pressure and temperature differentials as well as the pattern of opening provided by the fabric. These can be highly complex and particular to an individual building. Mathematical modelling, wind tunnel and salt bath trials can be used to inform this process of visualisation for building using natural ventilation. This can be a major strategy in this important area of the building's performance as it can greatly reduce the energy use in a building.

The design of the building as a whole has to be considered in order to harness the forces in natural ventilation. One way this can be achieved is with the use of the atrium or street (generally an atrium along the length of the building). The basic design and the natural ventilation forces are shown in Figure 10.13. This design can also take full advantage of natural lighting, again reducing energy use throughout the life of the building.

The light environment

To fulfil the majority of its functions the enclosure of a building will almost certainly use materials that do not transmit light. This means that if the user is to make use of natural light, specific consideration will have to be given to getting light into the building. Since most buildings are occupied during the hours of both daylight and night-time, provision has to be made for artificial lighting as well as natural lighting. These two may be separate in that the building is to be designed to operate during the day without the use of artificial light. Alternatively there may be an element of permanent artificial supplementary lighting for daytime activity. This does have considerable running cost and energy implications as well as considerations for the consistency of level and quality of lighting provided. One further issue in lighting is that, over and above providing general levels of light, there may be a need for the provision of task lighting or dramatic or mood illumination, which may also be provided by either artificial or natural lighting, depending on circumstances.

Natural lighting

The simplest way to allow light into a building is to leave empty openings in the external enclosure, but this is unlikely to be satisfactory given the other functions of the walls and roof. The materials available for enclosure that transmit light are limited, being primarily glass and plastics. Increasing the thickness of these glazing materials reduces the amount of light transmitted, limiting them to relatively thin sheet applications.

These materials will have limited strength and are therefore unlikely to be able to carry structural loads from the building, although they will have to carry wind loads to the structure. They will most often be framed to provide the strength required. As single sheets, the glazing materials will transmit heat easily, giving high heat loss in winter and high gain in summer and will provide limited sound resistance. At the same time they must remain waterproof and are often used to control ventilation by providing sections which can be opened. They may have limited ability to provide security and privacy.

These components, such as windows and roof lights, become sub-elements of walls and roofs of which they form a part. They have to contribute to the performance of the element as a whole, but it has to be accepted that they will have different performance levels. These components may also be made of very different materials from the fabric that surrounds them, and the detailing of the fixings and jointing to the main enclosure materials has to be thought through independently and yet be complementary with the consideration of the elements as a whole.

In addition to all these considerations the size and shape of windows has a significant effect on the appearance of a building. They are an integral part of most styles of architecture, often playing a major part in giving the style its distinctive character.

Against this background it is clear that the solution to providing natural light may often be a compromise with the other functions of the building. Its own importance in function and design terms varies greatly, depending on the use of the building and the designer's interpretation of its importance to the design. The use of multi-layers of glass or plastic components is now common and recent development in materials has given these multi-layer transparent components improved thermal and sound performance and even strength and fire-enhanced properties to help security functions.

What is clear is that the energy use in electrical artificial lighting is high and therefore the provision of high-quality natural light can play a large part in achieving sustainable development.

Source of natural lighting

Being dependent on the sun, the quality, intensity and time during the day when natural light is available are highly variable. These vary with latitude and with orientation. In the northern hemisphere light from the south is generally brighter than from the north. It does, however, vary in intensity, often causing uncomfortable or disabling glare rather than even illumination. Light from the north is more consistent and reliable but colder and rarely dramatic. This made it suitable for factories and warehouses where roof lights on the steep pitch of the characteristic saw-tooth roof-line were orientated to face north. For other purposes, variety and qualities of brightness and warmth may be more appropriate, which, with some control for glare, is best provided from the south.

Natural light also varies with the time of year and the weather conditions. The hours of daylight vary with the seasons, and the light levels and the angles from which the sun can be directly viewed will vary. This will give a quality of light that varies with the seasons but also means that shading may have to be variable to match the angle and direction of the sun. The cloud cover affects the quality of the light. The highest level of light is available with no clouds covering the sun, whereas storm clouds can all but eliminate the light, creating near-night conditions during the day.

With all this variety it is unlikely that performance levels can be assured throughout the year. Bright light may need to be shaded, while storm clouds may dictate that artificial light is needed during the day. This variety is greater in temperate rather than in hotter climates, where the sun is more predictable and remains higher in the sky. In buildings designed for these hot climates windows may be small and/or be permanently shaded by shutters or verandas facing the main path of the sun.

A variety of natural light is not necessarily undesirable. Changing levels of intensity and colour give information about outside conditions. They keep occupants in touch with natural conditions, marking the passing of the natural cycles of the day and the seasons. In addition the light received brings with it a view from outside. Windows, and perhaps to a lesser extent roof lights, allow occupants to look outside to see what is going on. In many cases, landscape (or cityscape) can be designed to provide spaces seen as visual extensions to the internal space. This gives vistas and panoramas that enhance the activity for which the building is designed. The increased cost of offices on upper floors of skyscrapers and penthouse suites is based in some measure on the view they provide.

Despite the technical difficulties of providing the performance of the enclosure where lighting openings are to be included, and the probable additional capital and running costs, windows and roof lights are judged to be good value (if not essential) in the provision of views, light and the influence they have on the appearance of the building. This makes the technical challenges particularly acute.

Penetration of natural light to internal spaces

Given the prevailing external lighting conditions, the quality and quantity of light reaching the internal space are controlled by the size and shape of the openings, the opacity of the glazing material and the shading or other control mechanisms provided. The proportion of light then available penetrating into the space depends upon the area and shape of the room, both its width and depth. Lighting levels reduce with increasing distance from the windows, but penetration can be enhanced by the careful choice of shape and position of the openings, particularly the height of the head of the window.

Lighting levels from natural light are also influenced by reflections from objects or surfaces, particularly towards the back of the room. Surfaces that are shiny or light in colour will reflect a greater amount of light, raising the level of light at the working surface. In the same way as it is possible to provide shade from the direct glare of the sun at the window it is possible to provide reflectors to enhance the levels of light incident at working positions within the room.

Predicting lighting levels from natural lighting is now well established. Data on the external conditions and calculations that take window shape, depth of room and the reflection of surfaces into account are available to predict the level and reliability of natural lighting into a building.

Artificial lighting

For any building that is to be in use after dark the need for artificial lighting is clear. The provision of this lighting will assume that there is no light available from outside. Windows left uncovered will then act to transmit the internally generated light out and the view will then be available from the outside of the inside, affecting privacy and security. Heat loss will also increase as the temperature falls at nighttime. Given these circumstances, curtains or blinds become associated with window design.

After dark, not only is lighting for internal spaces lost but also the exterior of the building loses its illumination. This has implications for its appearance and for security. These two may require different types of lighting. They may need different levels of lighting, colour and modelling requirements. However these are resolved, there is often a need to provide sources of artificial lighting outside as well as inside the building.

If levels of natural lighting are assessed to be inadequate in the daytime then artificial lighting may be seen as a solution, even during the day. This is particularly a problem in deep building where the middle of the space is some distance from the external walls. In single-storey buildings (or the top floor of a multi-storey building) this can be solved with roof lighting. In very deep multi-storey building the use of light wells or atria can bring light to a central space, making light available to areas some distance from the outward-facing walls. In other circumstances artificial lighting will have to be used, although this has energy and running cost implications over the lifetime of the building.

For many buildings the daytime use of artificial lighting is only associated with limited periods when the external conditions are poor. In these cases the lighting designed for night-time use may be switched on. However, even if general lighting levels with natural lighting are satisfactory, specific task lighting may still be required and provided by artificial sources.

Sources and control of artificial lighting

The source of natural light is the sun, often obscured by cloud, so its quantity (level of illumination) and quality (colour, warmth, etc.) are variable dependent upon the time of day and the seasons of the year. With artificial light sources, the parameters of quantity and quality can be chosen. Since the case for artificial lighting also has to consider energy consumption, it is important to be able to identify the energy use for a given output since this also varies greatly between the different sources.

The demands made for lighting both internally and externally vary greatly and hence a wide range of sources has been developed. Originally powered by oil or gas, artificial lighting is now almost exclusively powered by electricity. The initial characteristic of the light is governed by the way the electrical power is converted into light. Whether the light is generated by the incandescence of a solid material such as tungsten halogen lamps or the excitation of a gas such as the fluorescent tubes, it is this conversion process that determines the basic colour and the power consumption of the source.

This light can then be harnessed to best effect by the fitting and positioning of mounted units. Directional spot lights get their power and colour from the light source, but their directional capacity comes from the way they are housed and even focused, with lenses to minimise scatter, as with theatrical lighting. Colour can also be added to white light with filters. For general lighting, the fluorescent tube, mounted in a reflective box, or luminaires with a diffuser on the front to give an even light with limited glare and then mounted in the ceiling at relatively regular spacing provide an overall level of illumination with some control over colour with the choice of tube.

More general control over individual rooms or sections of large areas is achieved by switching the electricity on or off or dimming to reduce the level of light emitted. In many situations this control can be manual by providing switches for the occupants to operate. For energy saving, some level of automation should be considered. Often based on movement sensors to identify occupants, automatic switching will turn off lights when rooms are not being used and turn them on again when people return. This might have to be set to leave low levels of illumination if safety might be compromised. Light level sensors can be used to switch on lights as levels of natural lighting reduce. It is particularly effective to bring in banks of lighting parallel to the windows to supplement the natural lighting at the back of the room without using lights close to the windows where natural lighting is still satisfactory.

As with natural sources, artificial light received at a point in a room is not just in a direct line from the source but also arrives from reflections from surfaces. The strength of these reflections is dependent on how close the surface is to the source and on the texture and colour of the surface itself. This can be used either to increase the level of illumination (quantity) or to create effect (quality) when used in conjunction with the source, its housing and its mounting position.

The predictive methods and the data available for the sources concerning the levels of general illumination are well developed. Both predictive models and data on reflective surfaces and the manufacturers' values for the sources and the luminaires are freely available and reliable. Even predictions of loss of efficiency with time, energy usage and heat generation are possible with the current data available. Architectural lighting and effect or mood lighting are less susceptible to calculation and need a measure of imagination, but some calculation and computer modelling can give clues to comparative effects especially in terms of power and the interpretation of light and shade for soft or dramatic effects.

The acoustic environment

Sound can be considered either as wanted or as unwanted noise. Sound that is wanted can be enhanced or degraded by the fabric of the building, particularly by the finishing and the shape of the room in which the listener is located. Most wanted sound is generated within the space occupied by the listener, most often associated with talking or music. Some wanted sound comes from outside the room. Often sounds from the natural environment need not be totally eliminated since they bring an awareness of the wider world and even pleasure. Like a view or the changing quality of natural light from a window, they limit any sense of isolation that can be psychologically undesirable. Very low background noise conditions the ear so it picks up small noises that become intrusive. The lack of any external sound can also be disturbing as sound is used to identify threats. In modern building the possibility of natural threats is now supplemented by a number of alarm systems provided within the building and it is necessary for these to be heard by all the occupants. Fire and intruders alarms are the most obvious and these must be audible in all parts of the building when activated.

An increasing number of modern sounds are now considered unwanted. They are categorised as pollution and as such are now being given much attention in reduction at source, but this still often leaves excessive noise that has to be reduced through careful choice of materials and detailing of the construction.

Characteristics of sound

Sound has two characteristics as identified by the ear: frequency (pitch) and loudness. Frequency determines how high or low we perceive the sound to be. If too high or low, it becomes inaudible to humans, although some pets can hear higher frequencies, and some low-frequency sound does affect other organs in the body and causes discomfort, but these sounds are very rare. Combinations of frequencies can be harmonious, as in music, or carry information, as with speech. Groups of frequencies that are disharmonies or carry no information are often called noise and, while useful as alarm signals, are generally considered undesirable.

Loudness is associated with volume. Because it has energy or power, it has the capacity to cause pain if it excites the mechanism in the ear to excessive movement. At the other end of the scale, quiet noise can be irritating, since it is frustrating if you think it is carrying information that you cannot obtain, or it is continuous and begins to intrude on your consciousness.

All sound, whether wanted or unwanted, reaches the ear through vibrations set up in the air around us. These vibrations from an airborne sound can also travel through the construction, and hence the provision of a barrier will only modify and not eliminate noise. Further, impact sound created by striking the construction itself, such as footsteps on a floor, will create a sound not only within the room but also transmitted to other areas by the direct connection of the structure and fabric of the construction. The transmission of the sound around the walls and floors directly between source and listener is by means of flanking paths. These noise sources and transmission paths are shown in Figure 10.14. In addition, mechanical plant has the potential to create noise, and this can be passed through the building via the services duct and pipework systems.

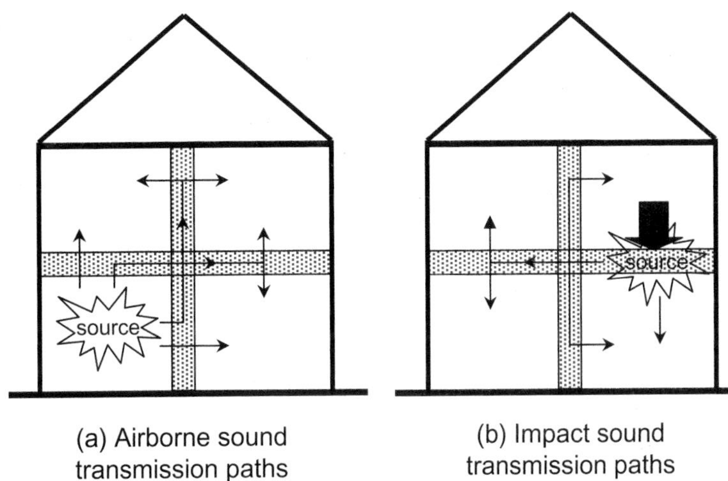

Figure 10.14 Sound source and transmission paths.

Sound created within the space

The sound will travel out from the source and there will be a natural decay of its power as it moves away. If in a direct line with the source, the listener will receive the sound directly. The sound will continue through the air in the room, striking other objects and eventually the floor, walls and ceiling that make up the room. Sound reaching openings in the floor, walls and ceilings will pass straight through. The sound that strikes objects or the surfaces in the room will be partly absorbed and partly reflected, depending on the nature of the surface itself. The reflected component will return into the room, where it may be received with a reduction in power and fractionally later than receiving sound direct from the source. These reflections will continue to be heard by the listener until the power in the reflection is too low to be audible. The quality of the sound is, therefore, highly dependent on the acoustic (absorption and reflection) characteristic of the objects in the room and the surfaces. However, since the sound loses power with distance, and takes time to return as a reflection, the size and shape of the room are also important. The time each individual sound takes to decay to become inaudible is known as the reverberation time. With highly reflective surfaces in a large room reverberation times will be increased.

For both speech and music it is important to have a quiet background; little or no noise from outside the room and little or no generation of sound within the room other than either the speech or music. For both it is necessary to have a good strong direct path from source to listener, but for reflections and reverberation time the requirements for speech and music are very different. For speech a strong early reflection reinforces the sound, but later reflections interfere with the clarity of the separate sounds. Reverberations as low as a twentieth of a second are required. For music it is the later reflections that enhance the sound, enveloping the listener in the sound field. Now reverberations of one to two seconds are required. This enveloping also requires the room to have a certain volume to maintain a smooth and prolonged decay of the sound.

There is one other phenomenon that can occur if a sustained sound with a dominant low frequency is coincident with the dimensions of the room. This is most likely to arise from vibrations from outside the room, which will be considered later, but the characteristic is associated with the room itself. If the room dimension is a function of the frequency (wavelength) of the sound, a standing wave may be generated sustaining that frequency.

It is clear that the major influence the construction can have over the quality of the room acoustics is associated with the reflection/absorption of the finishes, having taken into account the influence, particularly absorption, of the objects (including the listener) in the room. In general terms, hard, flat surfaces are highly reflective. As the surface becomes progressively softer and more textured, absorption is increased. Soft, fissured or even re-entrant surfaces are sound-absorbent as the energy in the vibrating air becomes dissipated within the cavities of the irregular surface.

Airborne sound made outside the space

Increasingly, airborne sound is unwanted, but, as has already been pointed out, total sound isolation may not be a sensible performance requirement. The sound will move out from the source as before. If it were in the open air it would be modified by wind and other atmospheric disturbances, but no such disturbance would be experienced if the sound were created in another room. The sound is again going to be influenced by objects it meets, being either absorbed or deflected, but at some point it will impinge on the dividing construction, such as walls and floors.

An opening in the construction will not change the sound; it will enter (or leave) the room unaffected. It is important to appreciate the significance of the size of openings in transmitting sound. Small openings have a large effect, considerably reducing the sound reduction performance of a barrier, however well detailed.

The walls and floors of the building will act as a barrier to sound. The construction will reduce the loudness but not equally for all frequencies. The action of the construction in reducing the noise is two-fold: it dampens the vibrations due mainly to the mass of the construction and it isolates the vibrations if discontinuities (layers of resilient materials or cavities) can be introduced in the detailing. Hence the overall performance of a wall, floor or roof in attenuating sound will depend on the openings, mass and discontinuities and any sound-absorbent materials in the cavities. Sound will also be carried deeper into the building by other elements attached to the initial barrier along flanking paths, making edge detailing particularly important. Figure 10.15 shows possible soundproofing details for a timber joisted floor, illustrating how mass, discontinuities including abortion in cavities and edge details can be integrated into a basic structural floor.

Care has to be exercised in assessing the performance of cavities. If the cavity is too small it has little effect as the vibration energy is transferred directly across the cavity. This is known as acoustic coupling. It happens in standard double-glazing, where the gap between the glass panes is small and therefore the soundproofing effect is almost entirely due to the additional mass of the second pane of glass. For cavities to be effective they have to be approaching 200 mm or larger.

While the potential sound reduction lies in the basic construction of the floor or wall, the part played by edge conditions and flanking paths means that the actual transmission loss when the building is in use may be less. Transmission loss of the proposed construction can be measured in the laboratory to provide a sound reduction index R_w normally between the frequencies of 30 and 315 Hz. A value of 40 dB reduction will render loud talking inaudible. This value helps to select appropriate specification, but these laboratory-based values may not be achieved when incorporated into the final construction on site, depending on the detailing and any variability arising from the site operations. It is possible to measure on-site performance, where the transmission loss R'_w will be quoted.

Sound generated by striking the fabric or from mechanical services

Sound is created by directly setting up vibrations in the construction itself. These vibrations will be much stronger than those set up by airborne sound reaching the fabric of the building. Flanking paths created by continuity between elements of structure and fabric are important

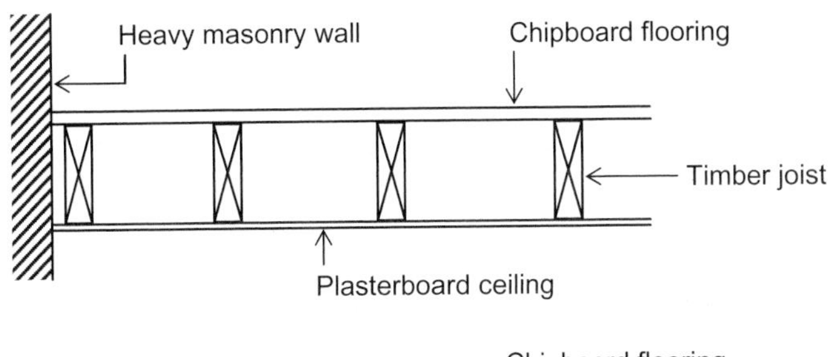

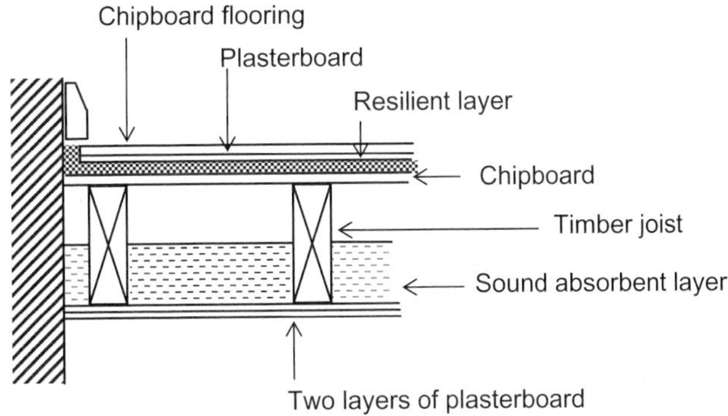

Figure 10.15 Timber joisted floor and details to provide sound proofing.

since if the vibrations can travel across junctions between elements and via joints and fixings the noise will penetrate deeper into the building. Mass will again play its part in damping the sound, but rigidity and natural vibration or resonance of the components sympathetic to the frequency of the noise may become important. Discontinuities and edge detailing are particularly effective in limiting the transmissions of these sounds.

Striking hard surfaces creates the greatest sound, not only in the room but also in setting up vibrations throughout the structure. Soft surfaces dampen the original impact and absorb the airborne sound that results.

There are particular problems associated with the pumps and fans of the services. The continuous rotation generates noise, but the movement of the rotation may also set up continuous vibrations in the construction to which it is attached. This may be the fabric of the building supporting the pump or fan or it may be the pipe- and ductwork to which it is attached. Isolation or discontinuities have to be used to mount the pumps and fans and connect them to the distribution system.

Methods of prediction

Definitions of a comfortable acoustic environment are well established for a wide range of activity including spaces with both general and specific acoustic demands (e.g. offices and concert halls). There is a great deal of empirical measurement of the characteristics of the materials commonly used in construction, and predictive calculations based on theory can utilise these values to predict performance in a wide variety of circumstances. Modelling, originally physical but increasingly on the computer, can be used in the design of specific spaces.

For the predictions to be fulfilled much care is necessary in the design detailing and the production phase. Identifying isolation discontinuities and in particular edge fixings to elements of the construction are of prime importance. During site operations, fixings and edge details need particular care. Gaps at junctions and the rigidity of connections can be crucial in achieving acoustic performance.

The clean environment

As with all aspects of environmental performance it is necessary to be clear about what activities are being carried out in which part of the building. Acceptable standards of cleanliness will have to be defined for the occupants but may also have to be considered for the materials and equipment to be housed and used within the building. For all these aspects of the activity, cleanliness will have to be defined against two complementary sets of criteria: those that would be detrimental to health or operational requirements and conditions that would be socially defined as unacceptable.

The socially defined criteria for cleanliness may not be detrimental to health or operational efficiency. They will, however, contribute to the social and psychological health of the occupants. The distinction between the health of occupants and the efficiency of operations is also important. Definitions of cleanliness will have to be identified against ideals of hygiene, the control of pathogens harmful to life or health. They will also have to be defined for the materials stored and equipment operated in the building. Levels of dust and dirt may not be detrimental to life but may interfere with the operation of equipment.

Another way of considering cleanliness is to identify the possible undesirable components of the environment as physical, biological or chemical. The term 'clean' is most associated with the absence of dust or dirt. These particles will be of various sizes and possible composition. They may be inert minerals but are also likely to include matter of biological origins (dead skin cells). They may also be chemically reactive if associated with, say, industrial waste. Particles are most commonly associated with surfaces but can be suspended in the air. The air, however, also carries other potentially undesirable components. Gases can mix with the air and odours can be carried in the air. The source of both gases and odours can be either natural or manufactured, and while some can be a health hazard, some are only a nuisance. Some gases, such as carbon monoxide produced from faulty heaters or the burning of fossil fuels, have an immediate deadly effect, while the naturally occurring gas of radon from some ground conditions only shows its effect after many years. Other gases such as methane and carbon dioxide from the degradation processes in the ground such as landfill sites that are now being used for development are not just a health hazard but also introduce the safety risk of explosion. In addition to particles, gases and odours, it is now possible to detect appreciable amounts of electromagnetic flux, the long-term health effects of which are still under investigation.

It is clear that surfaces and the air are two major areas of concern in providing a clean environment. Water is also a medium that has to be clean. Once it has entered the building at an appropriate quality, avoiding cross-contamination becomes a priority in the design of the water distribution systems.

Surfaces, air and water are all open to inspection. Occupants and equipment come into contact with them. However, within the construction or in unused or not often visited spaces there is the possibility of infestation. Varying in size from cracks in the construction to whole loft spaces, there exists the potential for infestations of animal and plant life. Infestations of insects, rodents or birds are all possible, as are bacteria in some locations. Plant life is most likely to be mould or other fungal growth, as these often do not need light to maintain growth.

Infestations of this type not only are undesirable against the notion of a clean building but also can be harmful to the fabric of the building itself. Control of these infestations is important for the durability of the construction as well as maintaining a clean environment for the occupants and the operations of the building.

It is, therefore, necessary to undertake an analysis of the air in the building, the surfaces of the construction visible and in contact with the operations in the building and the infestation potential of the construction and unused spaces in the building.

Clean air

It has long been recognised that particles in the air, particularly from smoke, result in a high incidence of respiratory disease as well as making surfaces, including clothes and skin, dirty. Similarly it is recognised that dampness in the air is not desirable and that limits to humidity are sensible. Although not as much of a problem as foreseen by early medical predictions, some disease is carried in the air, although often associated with water droplets suspended in the air, and breathed in with the air itself. Smell is another component of the air that, if allowed to concentrate, can become unpleasant, although it is not really a health hazard in itself. It is now being recognised that very low levels of chemical and ion (radiation) concentrations in the air can also be injurious to long-term health or have a low-level debilitating influence. The full implications of many of these are not yet understood and hence safe levels have not yet been established. It appears, however, that as we 'clean' our environment we improve health but then become susceptible to thresholds of other components in the environment that then redefine our levels of performance.

This general process of demanding new levels of cleanliness has also been true where the activity in the building has been manufacturing. This can be seen at its most obvious in the electronics industry, where some processes are now housed in areas called clean rooms.

Clean air is maintained primarily through ventilation. Most harmful components carried by the air are generated from within the building itself and the process is one of replacing the contaminated air with 'fresh' air from outside. Increasingly the air outside also contains contaminants and, therefore, some measure of processing of either the stale internal air or the incoming outside air has to be undertaken. Typically this is the process of filtration with a variety of processes used for physical, chemical and even biological cleansing.

The process of ventilation, both natural and mechanical, has been described earlier in this chapter. The necessity of a pressure difference and openings to create the air movement is the same, but the rates of ventilation may be significantly different. It is necessary to identify not only the types of contaminants but also the rates at which they build up and the threshold levels at which the air is to be considered no longer clean. Like all aspects of the clean environment, the performance level of clean is not absolute. There are only acceptable levels or thresholds of contaminants, many of which cannot be detected by sight or smell and require monitoring in other ways. Traditional methods of leaving the occupants to open windows when the room becomes stuffy will still be satisfactory in many situations. However, contaminants such as radon are tasteless and odourless and therefore have to be detected by other means. As a gas from the ground, its control will require gas-proof layers similar to the damp-proofing discussed earlier in this chapter.

Cleanable surfaces

The specification of surfaces is not that they shall be clean, for all new materials are clean. The requirement is that they shall be cleanable. This demands that the nature of contaminants must be known, as must the means that will be used to clean them. It is possible to speculate that the widespread use of fitted carpets has been dependent on the invention and affordability of the vacuum cleaner.

Taking the lifetime cost to the user of a surface, it is more than likely that more will be spent on cleaning a surface than on its initial purchase. In the domestic situation, reducing the time needed for cleaning may be seen as quality of life; to a commercial undertaking this will inevitably mean money.

The method of cleaning will be dependent on the contaminant. It is possible that some cleaning agents may be aggressive to some materials and these will have to be excluded from any possible specifications. Much cleaning associated with hygiene will be water based and hence impervious surfaces will have to be specified. The specification of surfaces with respect to cleanliness will only fail if they cannot be cleaned in the manner prescribed. Lack of cleaning in use does not constitute a failure of the specification.

Infestation

Many infestations are unpleasant rather than a threat to health. Infestations occur where either plant or animal life breeds and lives within the fabric or within voids created by the detailing or within the furniture and stored items, particularly food. The scale of infestation and the size of the animals vary considerably, as does any definition of what might be unacceptable. Generally, in the developed world, very low levels of infestation are tolerated. Performance specifications would wish to exclude the larger animals such as birds, mice and rats and most forms of insects. Some life sharing a building is, however, inevitable. What is shelter to us provides shelter to many other life forms unless the choice of material and detailing preclude them.

The first line of defence with animals (including insects) is exclusion of the mature adult. Although this is possible with animals and larger insects, it is often not possible with the smaller species and bacteria. In the case of plant life it is the seeds or spores that have to be excluded, and this can prove very difficult. Openings that lead to internal, dry cavities have to be detailed and constructed with care to ensure minimum dimensions to exclude adults. Larger openings have to be protected with screens where infestation is possible.

If exclusion is not an option the key to stopping infestation is to understand the life cycle of the potential infester and the conditions necessary to sustain each stage of its life. Removing food sources may deter an infestation, but for most infesters it is only shelter and protection that they gain from the use of the building, which allows them to breed and hence establish an infestation. If it is difficult to limit the conditions suitable for habitation, positively injurious conditions to the health of the infester may have to be created. While this may be part of the specification, as with wood preservatives, it is more likely to be arranged by the user during occupation of the building.

So far, infestations and breeding patterns have been considered. In some cases it is necessary to exclude adults from the actual rooms, not because they would cause an infestation but because they would be a nuisance or health hazard in themselves. This would include flying insects, either disease carrying or able to inflict injury. Similar detailing is necessary, although this may be contrary to requirements for ventilation.

Our knowledge of the animals and plants that are likely to seek shelter within our building is well developed and much traditional detailing takes the elimination of infestation into account. However, we have to be ever vigilant in that poor detailing or poor construction may lead to colonisation and infestation.

The safe environment

The need for safety in a building arises from the perception of a threat. Understanding the nature of the threat determines the performance required of the construction, the nature of failure and hence the choice of materials and detailing. Threats can be seen as arising from natural events, fire and from within society, where they are associated with security.

Natural threats

Many of the natural threats have been discussed in other sections of the text. For example, infestation and the presence of landfill gases such as methane are discussed in the section on maintaining a clean environment. They were seen as a direct threat to health rather than to safety, although clearly some safety measures are designed to prevent injury and damage to health. This distinction is arguable, but this text has chosen to discuss these health issues with the general requirements for a clean environment. Irrespective of how the distinction is made, the choice of materials and detailing will be the same, and these were discussed in the previous section and so will not be covered again.

Natural threats not yet discussed come from what is often termed an act of God. It includes the natural extremes of climate, such as high winds and excessive snowfall, and other natural occurrences, such as earthquakes. This threat can be identified with the performance of stability and strength. These natural threats exert forces that the structure and fabric have to withstand without failure. Processes of structural analysis and design are now well established, even though the natural forces involved are often complex and dynamic. The dynamic nature of the forces means the solutions may have to be carefully worked out in terms of the response of the structure and fabric of the building. This is discussed in Chapter 11.

An extreme of climate is flooding. Practical steps that can be taken to hold back water inundation in the choice of construction fabric are limited. Infrastructure projects in coastal and river defences for low-lying areas may limit this threat. Often location and the initial choice of levels for the development are the most effective measures to take against this threat.

The threat of fire

Fire can be a natural threat, but its probability is much higher in the built environment than in the natural environment. The source of most fires in building comes from the activity within the building or a malfunction of the services, particularly electrical. It is also associated with the possible threats from society itself when associated with arson. However it starts, it is seen as a major safety issue in the built environment and its analysis concerns the whole building. It has an influence on most aspects of the construction. There is now much regulation concerning fire safety design and this shows the level of threat that is perceived in modern buildings. Fire is a major threat to life, but there is also a financial implication in the destruction of the property, and therefore the design of a building often takes both into account.

Many of the requirements for construction are now laid down in codes, standards and regulations. When operating with any highly codified rules, it may be difficult to apply them to new design ideas. There becomes a need to return to an engineering approach. This has always been difficult given the complex environment of a fire, but computerised modelling can help in the evaluation.

Whether making choices by using published standards or by fire engineering techniques, there is a need to understand the progress of a fire in a building in order to make informed choices.

Like any environmental design, it is necessary first to understand the conditions in which the construction has to perform. The progress of a fire in an enclosed space follows a fairly consistent pattern. This is shown in the typical fire curve illustrated in Figure 10.16.

For a fire to start at all there has to be combustible material (fuel), oxygen and an ignition source. Fire in a building is really a series of fires starting at various time intervals in each room and each following the basic development on the fire curve. Fire in each space requires ignition and then the amount and type of fuel and the rate of supply of oxygen will determine its progress.

In order to understand how fire grows and spreads, it is important to remember that heat can be transferred by conduction, convection and radiation, and each will be involved to a greater or lesser extent depending on how well the fire is developed. These transfer processes are important in both the growth of the fire in each space and the generation of the ignition source for the spread to other parts of the building.

It is also important to know that the objects and materials that combust give off not only smoke but also flammable volatiles, such as gases, which burn in the presence of oxygen, generating flames. This adds greatly to the heat evolution.

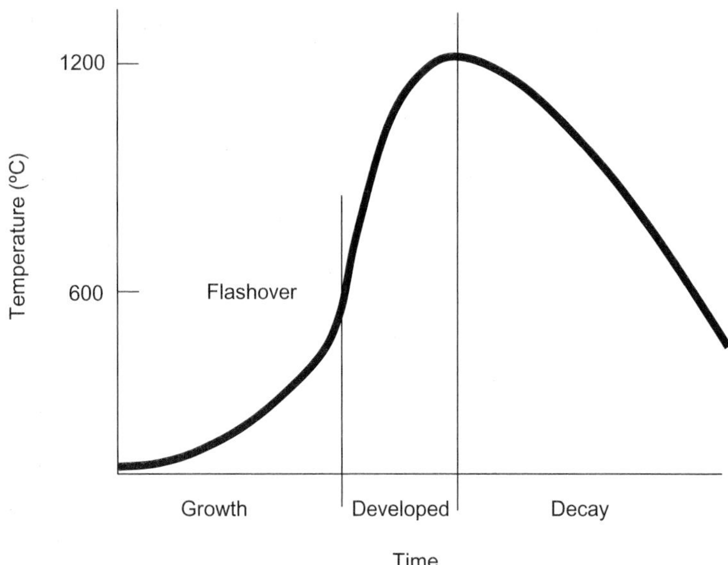

Figure 10.16 Typical fire time–temperature curve.

Fires normally start small. Only objects in contact with the original ignition source start to burn. Gases, heat and smoke will be given off. These will rise in convection currents, heating other possible fuel sources, which start to give off gases, and these all collect under the ceiling, where the surface area of this flammable material becomes much greater. If there is still oxygen to support the fire, when the flames reach the ceiling and the temperature is great enough the gases will ignite. The resulting increase in radiant heat will now cause ignition in most of the remaining flammable objects and materials (particularly surfaces or linings) in a very short space of time. This point occurs at around 600 °C and is known as flashover; the fire is then fully developed. If there is insufficient oxygen the gases will not ignite, but the objects that are on fire will continue to smoulder and give off heat. The introduction of oxygen as the door is opened will cause the sudden ignition of the gases, causing the phenomenon known as backdraft.

The temperature in the space now rises rapidly and the potential for the fire to spread becomes much greater. Containing the fire at this point depends on the fire resistance of the elements such as walls, floors and the building structure.

By following the development of the fire as described in this way it is possible to identify the technical solutions that could be used to make sure people escape and that the integrity of the building is maintained throughout the fire.

Fire prevention

It is clear that the best defence against fire is to prevent ignition. Identifying the possible sources of ignition and isolating them from a significant supply of fuel will accomplish this. If the activities or processes undertaken in the building produce heat, they should be carefully designed so as to be isolated some distance from any combustible material. These can often be easily identified as heat-producing appliances and flues. Minimum distance of timber components from flues is a good example of fire prevention. However, another major threat is the malfunctioning of any energy distribution service such as electricity and gas. Malfunctioning electrical systems are a common cause of fires. If the electrical current arcs it provides a direct source of ignition at high temperature, causing immediate ignition of surrounding fittings or construction. In extreme risk areas, low-voltage electrical supply may be specified to minimise the risk of electrical fires.

Ignition from accidental sources such as discarded cigarettes is difficult to control by strategies involving the choice and detailing of the construction. However, treating fabrics with fire retardants has had major success in limiting the start of fires from such accidental sources.

Fire detection and escape

If a fire does start then there should be a maximum chance of it being dealt with in the initial phase of the time–temperature curve. This involves early detection and fast extinguishing. These two functions are most often associated with the services provided in a building. Both detection and alarm can be either manual or automatic. The simplest means of detection and alarm that has been used is the provision of a bell (or triangle) to be rung by hand by the person who detects the fire. Alarm may still rely on an individual activating a building-wide alarm system automatically by breaking a glass disc in front of the alarm switch. Increasingly an automatic detection system will be specified using heat and smoke detectors to set off the alarms.

Having detected the fire and raised the alarm, two activities have to be considered together: the evacuation of the people and the provision for fire-fighting.

Given that the fire has been detected during the growth phase indicated on the time–temperature curve, heat will still be relatively low, although smoke may build up very quickly. If the fuel is chemically complex, the smoke may be acrid or even toxic. Escape has to be a priority. The majority of deaths in fires are from asphyxiation by hot smoke rather than burns due to the heat.

However, with very early detection it is possible to smother a small fire. Traditionally this would have been done with manually applied material such as sand or a fire-blanket, but nowadays a fire extinguisher would be used. There is a link here with the operational management of the building. The contents of all extinguishers are not the same. Products of water, foam, powders and CO_2 are available for different types of fire; clearly water should not be used on electrical fires or even fat fires. If initial fire-fighting equipment is to be provided then personnel have to be trained in its use and the equipment checked regularly.

Automatic initial extinguishing can be achieved with sprinkler systems (or their gas equivalents as long as evacuation can be achieved first). These include a detection system that automatically releases the extinguishing medium, water in the case of sprinklers. In sprinklers each sprinkler head is its own detector (heat expands a liquid in a glass stopper which breaks and allows water to escape from that head only), so extinguishing can be as local or widespread as the fire. Even if the system does not extinguish the fire, it gains valuable time in that it considerably reduces the production of both heat and smoke, prolonging the initial phase on the time–temperature curve and the opportunities for escape. It does, however, require a large volume of water if the control is expected to be widespread as in large open warehouses. Here the public water supply may not be able to deliver at the required rate. Large storage tanks with pumps may have to be provided to maintain flow should the sprinklers ever have to be used.

As soon as detection and alarm have been established evacuation should commence. Again, the link between construction provision and occupation management is very clear. The design of the building must provide the means of escape, while it is the responsibility of the occupation management to ensure that the routes are known and maintained unobstructed.

The design of the building for means of escape is initially a case of planning routes and determining the size of these to ensure a speedy and panic-free evacuation. Everybody should perceive that they have an equal chance of escape. Initially anybody in the space in which the fire starts has to escape. They have to be able to move to areas that are on the escape routes directly linked to the evacuation point. Here they may meet people from an as yet unaffected part of the building also heading for the evacuation point. While most of these escape routes will utilise the everyday corridors and stairs, additional provision is often necessary due to the fact that everybody would be using them at the same time. It is also necessary to provide alternative directions of escape so individuals do not have to go through the fire to reach evacuation points. The distances and size of these routes have been determined empirically and are now provided in codes and regulations, but computerised modelling can also provide predictions of behaviour in these early stages of evacuation.

These escape routes need protection. They need protection from heat and smoke if escape is to be effected. This protection will be provided primarily in the choice of materials and detailing of the construction that encases the route and ultimately in the stability of the structure of the building itself. In order to think about the properties of the material chosen and the appropriate detailing it is necessary to see how a fire spreads from its initial ignition to other parts of the building.

Building fabric fire performance

Fire development and fire spread are the two areas where the fabric of the building makes most of its contribution to fire design. The floor, wall and ceiling coverings, or linings, influence the development of the fire, and then fire resistance of the elements, both structural and non-structural, influences the spread. The role of the fabric can be identified as significant in the following two broad areas:

1. Lining materials
2. Fire-resistant elements
    - Internal wall and floors
    - Structural members
    - External walls and roofs

While linings are significant in the initial stages of the growth of a fire, within each space the fire-resistant elements are important as fire barriers and maintaining structural stability. Fire barriers also slow the spread of smoke within the building and, in some cases such as compartment walls and floors and in protected shafts, contain the fire within sections of the building. In the design of party walls and external facades the aim is to ensure that the fire does not spread to adjacent property. Structural integrity needs to be maintained throughout the fire, where the question of its potential performance in any rebuilding might be considered.

Fire performance of linings

The choice of linings can affect the time available for escape by slowing down the growth of the fire, but they may also contribute toxic gases and/or flammable volatiles, which may add to flashover gases or create hot droplets that may be a source of ignition. The properties of the materials for fire performance for the linings to the floors, walls and ceilings can now be identified. Combustibility, surface spread of flame and fire propagation all indicate how quickly, if at all, the lining will assist the development of the fire. Smoke and droplet (burning particles) tests along with measurements of contribution to flashover gases give a complete picture to evaluate the choice of lining against fire performance criteria.

Fire resistance of elements

Having chosen materials for the linings, the next concern is the spread of smoke to escape routes and the fire igniting material outside the room. Controlling spread is the role of the fire-resistant elements. Performance is not just dependent on materials' properties but the design of the whole element, be it wall, floor or structural components. This is recognised in that while we talk about lining materials we also talk about the fire resistance of elements. Whole assemblies have to be tested, including all the detailing, joints and fixings (and loading for structural elements), for a performance rating for fire resistance.

Fire-resistant elements may have the role of a barrier to smoke and fire and/or be a part of the building structure, the collapse of which would not only allow the spread of fire but also cause physical danger to occupants escaping and members of the rescue services.

There are three ways a fire-resistant element could fail and allow a fire to spread:

- Resistance to collapse (stability)
- Resistance to fire penetration (integrity)
- Resistance to the transfer of excessive heat (insulation)

The stability of a structural element may be affected if the fire reduces its load-carrying capacity to the point where the element collapses under the load it has to carry at that stage in the fire. The integrity of a barrier may be lost if the element cracks, shatters or just burns away. This includes jointing materials that may burn or melt, or fixings that fail under the influence of the fire. The insulation of a barrier must be sufficient to keep the other side of the element from reaching temperatures that will cause ignition. More correctly, 'insulation' should be seen as heat diffusion. High thermal mass or endothermic reactions (chemical changes that absorb heat from the fire) will have a similar effect of maintaining the other side at temperatures below ignition for the duration of the fire, even though in thermal resistance terms the material of the element may be considered a poor insulator.

Internal walls and floors can contain fire and smoke to control the spread of a fire. Not all of these barriers need the same degree of resistance to the passage of fire, even within the same building. The performance requirements will depend on two factors. First, the nature of the occupancy including the total number of people in each area and their mobility at the time at which they have to use the escape routes. This will determine the actual time it will take to evacuate everybody to safety. Second is an estimate of the speed of growth and ultimate ferocity of the fire as it follows the pattern illustrated in the time–temperature curve. This will determine safe evacuation periods.

Another consideration that will determine the fire resistance of elements is the size and configuration of the building. In order to limit the fire resistance requirements for individual elements it is possible to design the whole building as a number of fire compartments separated by compartment walls and floors designed to contain the fire completely. While the specification of these elements may be particularly onerous, the elements themselves do reduce the demands on the other elements inside the compartment as the risk to both people and property is significantly reduced.

Fire may spread more easily at openings in the floors and walls acting as fire-resistant elements if not correctly detailed. Openings are potential breaches in the integrity or insulation of the element. Two types of opening can be identified: doors (and opening windows) and spaces for services to pass from one area to another.

Doors need particular attention, as they have to contain both heat and smoke from escape routes. Not only must the door itself have properties of integrity and insulation but also the gap around the door is particularly vulnerable in spreading the fire to the outer surface of the door. The dimension of the closing edges and the coverage offered by the rebate become a particular aspect of detailing. Closing edges on double doors may have to be fitted with intumescent strips. These materials expand on heating, sealing gaps and ensuring integrity and offering an insulating layer against heat transfer.

The small spaces created around the services passing through walls and floors need to be filled, or stopped, with a fire-resistant material to maintain the integrity of the element.

The services are in fact often carried in voids either above the ceiling or below the floor. The convenience of these voids is that the services may pass unhindered above or below the rooms independent of the partitions between rooms. In fire design these are known as cavities and have the potential to spread smoke and fire rapidly to other parts of the building. There is a need to provide barriers in these cavities to prevent the spread of both smoke and fire. These cavities should not extend across compartment walls or through compartment floors.

Another danger is spread via vertical shafts. These shafts are provided in buildings for circulation (stairs and lifts) and for services. Once breached by fire, they will act as flues and transmit heat, smoke and volatile gases to higher floors very quickly. The fire resistance of these shafts becomes an important aspect of the fire design. Protected shafts between compartments have to have walls with high resistance ratings and the openings such as doors and service entry points have to be fully protected. Since the stairs are probably escape routes, preventing the intrusion of smoke may be a major performance requirement, possibly by the introduction of positive pressure.

So effective are vertical shafts and cavities that even the cavities in cavity wall construction provide a pathway for the rapid spread of fire to higher floors. This leads to the need to provide barriers in these cavities if the internal skin has limited fire resistance.

As well as establishing barriers to control the fire it is necessary to ensure that the structure of a building does not collapse or damage the barriers for the duration of the fire. There may also be some consideration of the integrity of the structure at the end of the fire for its use in any rebuilding work.

Structural materials in fire

Elevated temperatures change the properties of many materials, including strength, so stability of load-carrying elements may be lost without any apparent physical fire damage to the element. It is necessary to understand that structural sections exposed to fire may not burn, but the rise in temperature may cause expansion and/or a significant loss of strength. The section may deflect and even collapse, as it is no longer strong enough to carry the loads.

The mechanisms are different for the major structural materials. Steel has a high thermal conductivity and low thermal capacity, which leads to a rapid rise in the temperature when exposed to fire. There is, therefore, a rapid loss of strength as steel has lost something like 50% of its yield strength at 600 °C, well within the temperatures reached in a building fire. Steel can be fire engineered but is usually protected from the heat. Casing or spraying the steel with either an insulating material or a material that can absorb the heat can achieve this. Absorbing can be achieved by either high thermal capacity or endothermic reactions, where chemical changes in the materials use the heat of the fire. These casings and spray-applied materials must remain attached and not crack during the fire. Other strategies that have been used include filling the structure with water to absorb the heat or putting the structure on the outside of the building, but these options have not gained widespread application.

Reinforced concrete has a high thermal capacity so, unless the sections are thin, the rise in heat during a fire is slow other than near the surface. If the surface cracks or spalls, or if the steel reinforcement that is near the surface is allowed to gain heat, the section may fail. This focuses attention on the concrete providing cover to the reinforcement. The aggregates in the concrete and the depth of cover are the specification items of significance to the fire resistance of reinforced concrete.

Timber has a completely different mechanism associated with achieving fire performance. Timber burns, but the charred products of combustion remain attached to the section and insulate the unburnt timber from oxygen and heat. Burning slows considerably as the fire progresses. If unprotected structural timber is specified oversize to account for a charred layer, the timber will carry the load for the duration of the fire but will need to be replaced in any rebuilding.

Fire performance of external elements

Having specified internal barriers and the structure so that they perform during a fire, there remains one other major threat. If the fire breaches the external envelope, it not only will gain a source of oxygen but also may spread between adjacent properties. There are two ways that this spread between properties could happen. The first is spread along external surfaces to adjacent buildings, particularly along roof surfaces. The second is directly from areas of the facade breached by the fire. While spread along external surfaces implies that the buildings are connected, breaches in the facade imply a distance to the adjacent building. If this space between the buildings is large enough, the risk of spread is eliminated altogether. Up to this point, at least some of the facade must act as a fire-resistant element. The closer the buildings are together, the greater the area that must be fire-resistant. The codes and regulations give guidance on how much unprotected area can be allowed against notional boundaries between buildings.

Specification of fire resistance

All these performance requirements need to be specified in terms of time. There remains the problem that while evacuation can be established in real time the actual time an element will hold back the fire depends on the ferocity of the fire. For comparable performance between elements, testing must use a standard fire, one with a predetermined time–temperature curve. Hence the times quoted for fire resistance are not real time but standard test times. The required fire resistance has to be chosen by taking into account both the required evacuation real time and the ferocity of the fire defined by its time–temperature curve.

Fire-fighting provision

There remains one other area of consideration for the design of the building, that of the provision for fire-fighting by the professional fire service. Earlier discussion covered the provision of equipment for fighting the fire by the occupants early in its development. Once a fire starts to develop, this equipment has limited value. It becomes necessary to give access and provide services such as hose reels, wet and dry risers and foam inlets for the use of the fire brigade. This will require consideration to be given to the access of fire appliances into the grounds of the building and, possibly, on large buildings, even specialised personnel access in the provision, for example, of fire lifts. Like so much fire provision, this is a matter for initial design, but it also has to be managed 24 hours a day for the life of the building.

Threats from within society

Some threats come from the society within which the building is designed to operate. Like the natural threats (acts of God) and fire, they are threats to both person and property. In most politically stable societies direct threats to the person from strangers are relatively low unless associated with a threat to property. These threats are most usually associated with crime, possibly including the extreme of terrorism. They are characterised by the means individuals are willing to use to act out the threat. These are complex sociological problems and it has been suggested that the quality of the built environment can play its part in reducing the threat as well as providing defence against it.

The nature of the threat is highly dependent on the activity to be housed in the building. For the most part it is concerned with the intrusion of an external threat, but there are situations where the threat is being housed (e.g. prisons) and the risk is to the activities being carried on outside the building.

Whichever of these is the case, the approach to design is likely to follow that identified for fire: deterrent or prevention, followed by detection and alarm, followed by the provision of resistance to the forces of the threat in order to limit injury to persons and damage to property. Unlike fire, however, the perpetrator of the threat has reasoning and therefore defining any set pattern of attack is difficult. For instance, there is for each perpetrator a fear of detection and an assessment of the risk of being caught.

Assessing the threat involves assessing determination and the means the perpetrator is willing to use. Petty crime against households may be deterred by the provision of good locks and simple common-sense occupation management to stop opportunistic crime. The determination to burgle any specific property is low and the means that will be used will not involve large forces. At the other end of the scale, terrorism is characterised by having specific targets, high levels of determination and access to considerable force to inflict the damage necessary to breach the defences offered by the fabric and structure of the building. An interesting third example would be a cat burglar, who may target specific objects, plan and execute the intrusion with great determination but operate with very little force. Devising strategies and the performance of the construction solutions chosen is dependent upon understanding the mind and means of the perpetrator.

Personal accident

The main focus of the analysis of a safe environment has been the possibility of external threats, but there is also the need to reduce the risk of personal accident while using the building under normal operational conditions. The evaluation of the potential for accidents in the building is the same as the approach taken to health and safety on construction sites. It starts with the identification of hazards and an assessment of the potential to do harm. When the chance of an incident causing harm is established, the risk can be assigned. In the operation of most buildings safety equipment is not acceptable, so the hazards should be eliminated in the design. This determines much of the dimensioning and detailing of ramps, staircases, balustrades and handrails. Safety issues are implicit in specifying fixings for fittings where failure during use would cause harm. Another example is specifying floor finishes that are not likely to become slippery when wet. In addition to reducing the risk of personal accident, materials should not be chosen that might be injurious to long-term health.

The private environment

The origins of the requirements for privacy are the conventions of society. Unlike safety, the needs do not arise from the existence of a threat but from a sense of what is right and wrong, what is acceptable or unacceptable social behaviour and organisation. To understand the needs of privacy it is necessary to understand the ways of the society itself and the beliefs of its people.

The concept of the family dictates much about the form of dwellings. Management style and organisational structures will determine much about the form of commercial buildings. This is linked to the concern for image and status, which was considered an aspect of appearance but is associated with other aspects of the performance of the division between spaces to carry out separate activities.

The feeling of privacy comes not from the absence of the sense of impending physical danger but from the close proximity of others invading so-called personal space, even if no physical barriers exist in a room. However, social convention often requires physical separation, involving limiting the attention of others seeing or hearing or in any way being aware of the activity associated with the use of that space.

As such, the major requirements for privacy are acoustic and/or visual separation. The need for privacy adds to the performance requirements of the enclosing elements of the building. Although predominantly walls, it includes the construction of floors, ceilings and service distribution routes, where cross-talk can be a problem.

The analysis and evaluation of the performance of visual separation needs no complex theory or testing other than that of direct observation. Predicting acoustic performance is less direct and has been considered earlier in this chapter.

Summary

1. The aspects of the environment that need to be considered have been presented as dryness, warmth, light, acoustics, cleanliness, safety and privacy. The analysis of these seven areas is the focus of this chapter.
2. The building is a system of active services and passive fabric whose role and contribution (function and performance) is decided in the design.
3. For each of the aspects to be considered it is necessary to have a dynamic description of both the external conditions and the response of the construction to those conditions.
4. Each aspect of the environment that could lead to failure in the proposed construction must be analysed to ensure the performance of the whole.

11 Physical Behaviour Under Load

This chapter recognises the fact that, while all parts of construction are subject to loads, even if it is only their own self-weight, there is a specific need for a building to have a system of structural members jointed together to transmit the loads to the foundations. It is necessary to determine the way the loads are applied and how the support is provided for each structural member together with the way they are jointed to one another. This will establish the pattern of internal forces that will generate the stresses in the materials chosen. It is suggested that visualising the distorted shape of the structural members will indicate the pattern of internal forces. The system must contain sufficient members and be jointed in such a way as to remain stable, particularly against horizontal forces such as wind loads. With knowledge of the stress–strain relationship of the material this pattern of internal forces gives the clues to the efficient shape of the structural members and the economic size.

Forces, external and internal

As soon as any part of a building is created it will be subject to forces even if only from its own weight, and therefore has to have some strength and adequate fixing. There are some elements of the building, however, whose function would be recognised as part of its structure. They maintain the shape of the building and transfer loads from one part to the next to reach an existing stable support, normally the ground, which has to take the accumulated loads and therefore becomes part of the structural system.

These external forces, or loads, are predominantly generated by the weight of the construction and the objects and people inside the building. However, there are also environmental loads, such as wind and snow, which create forces on parts of the building exposed to the elements that have to be transferred successfully to the ground.

These external loads create internal forces within the structural elements. The nature of these forces and the mechanism by which resistance can be provided become the focus for the analysis of the physical behaviour under load. These elements of structure have to ensure the building remains stable, has limited distortion and does not break. Instability, distortion and breaking are the possible failures that will have to be identified, and a level of performance will have to be established for each.

The initial design task is to identify a system of structural members and supports that, when jointed together, provide a stable structure against the loads that will be applied to them. Loads applied to one member in the system have to be transferred to a support through some form of joint. The load being transferred through the joint effectively places a load on the support that becomes the next member in the system. This load path has to be traced through the structure. Each member and joint must be capable of resisting the accumulation of loads being transferred through the structural system.

It is important at this initial stage to ensure that the structural system has all the members necessary to ensure its overall stability. The initial effect of the loads will be to displace or rotate the members to which they are applied. This can be resisted at the joints between structural members, but often no such resistance exists. It is not until the additional stability members have been identified that the overall structural system has been devised.

Once the structure has sufficient joints and members to achieve overall stability, the loads will appear to distort the structure. This distortion arises because the externally applied loads induce forces inside the members and joints that stress the materials of which the structure is made. When materials are stressed, they undergo dimensional changes, known as strains, that cause the members to distort. As stress increases in the material it will ultimately cause the member to break.

Understanding structural behaviour and undertaking the design of structures is concerned with establishing stable structures against the external forces created by the loads and the analysis of the internal forces and the stresses the loads induce, along with the strains that arise causing the distortions.

The outline above follows the general approach to the analysis of physical behaviour established in Chapter 9:

1. The loads define the conditions under which structures have to perform.
2. Failure or loss of performance can occur if the loads
   - disturb the structure, making it unstable
   - distort the structure too far so it cracks finishes, transfers loads to non-loadbearing members or looks and/or feels unsafe
   - destroy part of the structure by breaking members or connections at the joints.
3. The proposal or suggestion is for an arrangement of structural members with choice of materials and approximate sizes to support the loads.
4. An evaluation of the response of the proposed members to the loads is by analysis of the distortions and the distribution and size of the forces within the structural members to see if they will maintain the loads without collapse.

It is now necessary to look at each of these stages in more detail.

Loads and loading patterns

Loads on the building come from the weight of people and the objects in the building, the weight of the building itself and from the environmental conditions to which the building will be subjected. What they have in common is that they all apply forces to the construction, which have to be resisted if the building is to continue to remain stable and then not distort excessively or even break.

The most usual way to classify loads is as follows:

- Dead loads – the weight of the construction itself
- Imposed loads – from the activity within the building and environmental loadings from snow
- Wind loads – an environmental loading from the wind

Before the characteristics of each of these types of loads can be appreciated it is necessary to identify what needs to be known about a load before it can be evaluated as a force on the building.

The following represents a definition of a load:

- Magnitude or size
- Distribution
- Direction

Magnitude is the total weight or force to be taken by the structure, while distribution identifies the pattern on the member. Distributions are often seen as a single point, uniformly distributed or possibly moving and impact loads. Direction completes the specification. Most loads on a building act vertically downwards under the influence of gravity; some will act at other angles (thrust at the bearing of an arch), with some, such as the wind, even acting vertically upwards under certain conditions.

It is now possible to see the basis of the classification. Dead loads are permanent loads where their size, distribution and direction can be established with some certainty. Imposed loads are variable so can be identified with less certainty. While magnitude can be established within certain limits, the patterns of loading need to be established to identify design loading conditions. There may be some objects in the building, such as machinery or storage arrangements, that are so much larger and more permanent than any of the other components of the imposed load that they would be treated separately, with their position defined with the characteristics of a dead load. However, the majority of imposed loads can be brought together as static uniformly distributed load defined by the notional use of the building (e.g. domestic or office).

Wind loading is also variable. Its inherently dynamic nature coupled with its variability in intensity and direction make a definition of size, distribution and even direction problematic. As wind speeds up as it flows over and around building it creates low-pressure areas, resulting in uplift forces, particularly across roofs. If vortices are created at edges, even greater suction forces can occur. In order to provide a definition of the wind load (and hence the forces on the construction) that can be used in a structural analysis certain simplifications have to be made to obtain an equivalent static load. These simplifications must, however, ensure that the effect predicted is at least as demanding as the effect that will be experienced by the real structure under the complexity of the real environment. Procedures for this are now well established and published in codes and standards.

Another condition that has to be allowed for is the possibility of accidental loading causing a local structural failure that leads to a more general failure of a major part of the building. This can lead to a disproportionate collapse in that the final proportion of the building affected is disproportionate to the local nature of the original failure.

Some buildings, however, will have to be considered for other accidental loading, such as impact from aircraft for high-rise buildings and blast for security buildings or those deemed to be at high risk from terrorist attack. The natural world can also create loadings from ground water or earthquake. This serves to highlight the importance in performance specification of defining the conditions that should be designed for, either as the limits considered as normal or in terms of extreme yet potentially disastrous conditions. However, the majority of buildings are only designed for dead, imposed and wind loads plus measures against disproportionate collapse. Designing for extreme loading in most buildings would not be seen to have value in the inevitable extra building cost, if the probability of them occurring would be very low.

Basic structural members

A limited number of types of structural members can be employed in buildings. They act either to span around spaces or to provide support or stability to hold the spanning elements in place to generate the spaces within the building.

While the common names for these structural elements – walls, columns, beams and slabs – give some clues to the position in the building and their basic form, understanding their structural behaviour needs closer analysis:

- Where the loads are applied
- What support is provided (to make stable and stop disturbance)
- How the stable member distorts

By visualising the distorted shape it is possible to identify the distribution of internal forces that would break the member.

Figure 11.1 illustrates perhaps the most straightforward structural members: the strut and the tie. The strut (most commonly identified as a column) is loaded and supported in such a way that it experiences compression, identified from visualising the shortening caused by the loads. By contrast, the tie is loaded and supported so it goes into tension, indicated by visualising lengthening.

Figure 11.2 illustrates the beam, a spanning member where the loading and support pattern causes the member to bend. Two beams are

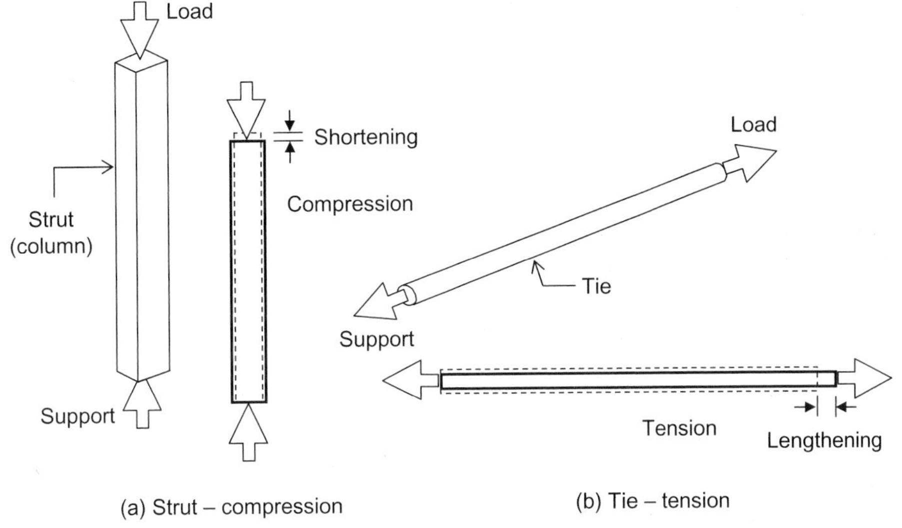

Figure 11.1 Struts and ties.

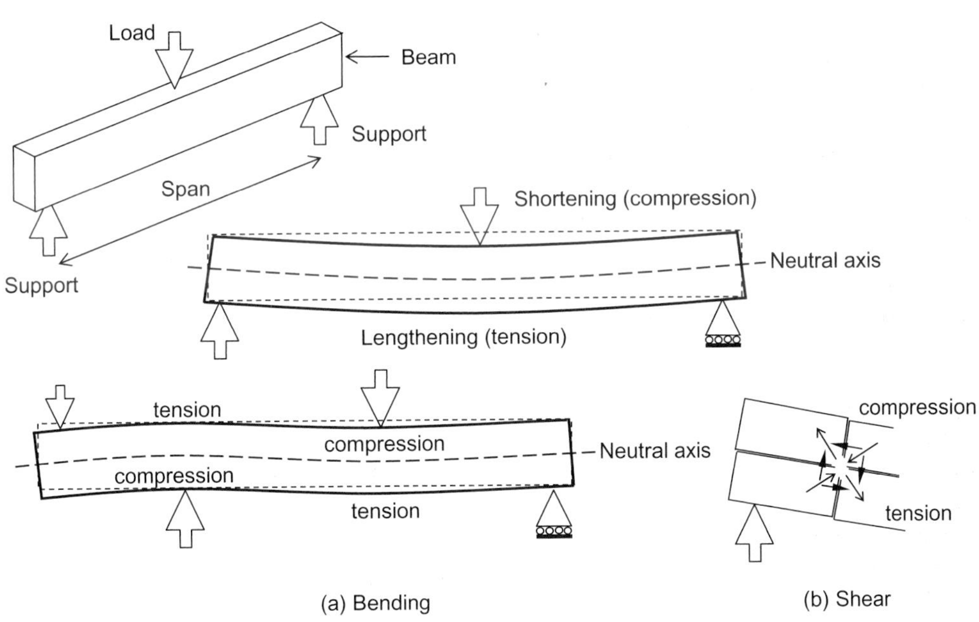

Figure 11.2 Beams.

shown in Figure 11.2, each with their distorted shape, indicating that where the material in the beam shortens it experiences compression and where it is stretched it experiences tension. The maximum lengthening and shortening will occur in the top and bottom of the beam, with no distortion being experienced along the neutral axis. All bending members have this pair of tension and compression forces acting together across the depth of the section. All bending members have a second pair of tension and compression forces, known as shear. This

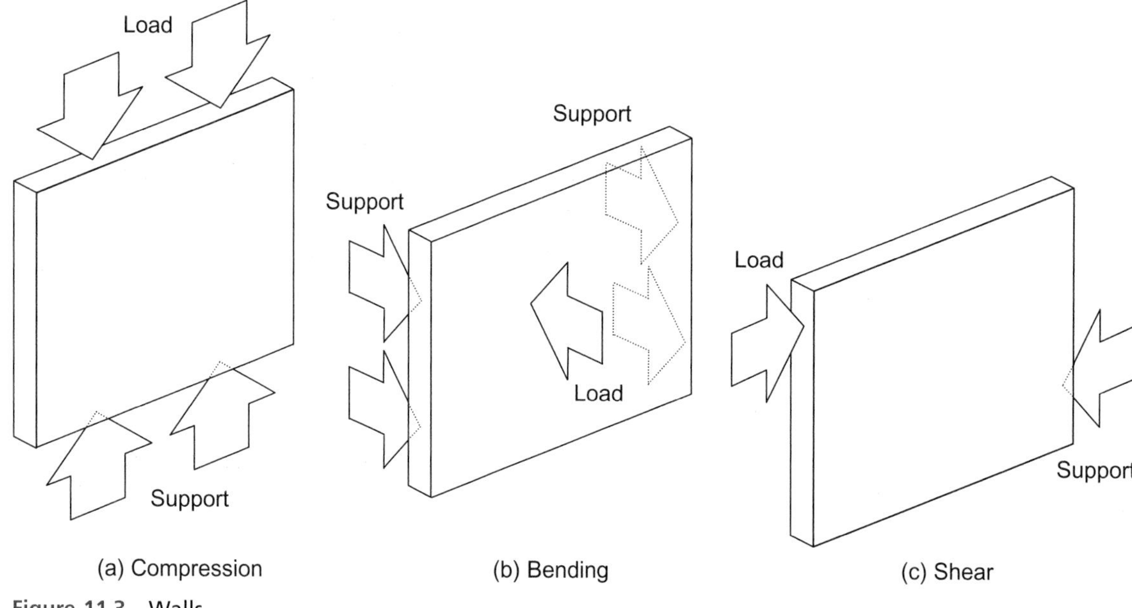

Figure 11.3 Walls.

time the forces are greatest at the neutral axis of the beam where they act diagonally across the material of the member.

Struts, ties and beams have direct compression, tension and bending forces respectively. The other two major structural members, walls and slabs, can have different load and support patterns giving different sets of internal forces. Figure 11.3 shows how a wall can be in compression when taking loads through the building, but may be in bending if resisting horizontal forces such as wind on the facade or ground pressure as part of a basement structure. The shear-loading pattern is created when a wall is used to stabilise the racking effect of wind forces on building structures, as discussed later in this chapter.

Similarly slabs can be subject to different internal forces depending on the loading and support arrangements. Figure 11.4 shows that this is mainly a distinction between suspended and ground-supported slabs. Suspended slabs will be in bending. The illustration shows support on only two edges, creating a one-way span, but it is possible to support the slab on all four edges, creating two-way spanning, to reduce sections or offer potentially greater spans. Ground-supported slabs carrying uniformly distributed loads will experience compression from the applied loads. When slabs are placed under columns and walls as foundation, the forces they experience depend on the ratio of the depth to the width of the foundation. If the foundations are comparatively wide, the foundation must act in bending for the load from the column or wall to be transferred evenly to the ground. When the slab becomes narrow, the predominant force experienced by the foundation is a punching shear.

Lateral distortions

Figures 11.1 and 11.2 show the shortening of the strut, lengthening of the tie and deflection of the beam as distortions that arise as a direct result of the application of the load and in the direction of the load. These are not the only distortions that occur. Lateral distortions may be experienced in slender members where compressive forces are experienced. These are known as lateral distortions because the member bends in a direction at right angles

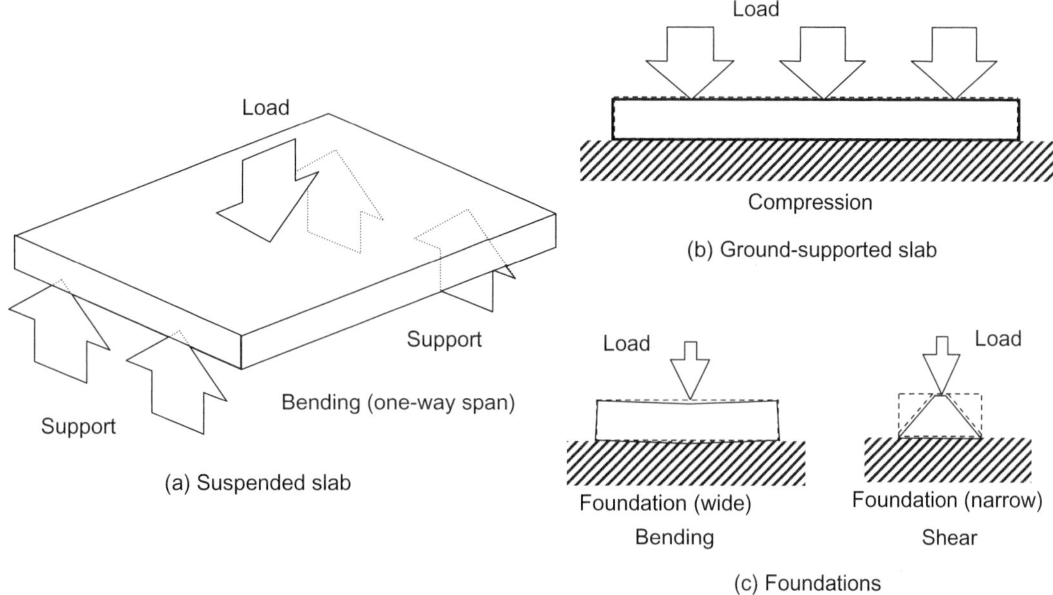

Figure 11.4 Slabs.

to the direction of load. Slenderness, as the name suggests, is primarily a function of the height (or length) to width ratio, but it is also dependent on the way the member is connected with the building structure. The effect of the connection will be explained later.

Observation of the internal forces that develop in the members discussed in Figures 11.1 to 11.4 shows that compression zones develop in all types of members except ties. However, not all of these members become slender in a direction at right angles to the load. Again, by observation, perhaps the most obvious potentially slender compression member is the strut and the potentially slender compressive zones that arise in beams. The lateral distortions, known as buckling, which can occur in struts (column or wall section) and beams, are shown in Figure 11.5. In columns the buckling induces bending as the whole section is in compression. However, in beams the compression zone buckles while the tensile zone remains straight. This induces a twisting action known as torsion. The whole lateral distortion effect is known as lateral torsional buckling.

Perhaps less obvious is the possible buckling associated with shear. In thin web sections in beams the principal shear compression force shown in Figure 11.2 may develop buckling. In thin shear walls, shown in Figure 11.3, the buckle will occur across the diagonal, giving a potential bending failure similar to that of the buckling strut.

It is important to realise that the buckling distortion does not replace the distortions of the direct loading, but when it occurs the structural member experiences both. The two sets of internal forces develop together. For instance, in a column already experiencing compressive forces the buckling adds bending, increasing compression on one side but reducing the compression where tensile bending forces from the buckling develop.

Eccentric loads

In identifying all the distorted shapes and internal forces so far it has been assumed that the loads have been applied and support has been provided at the centre line (on the axis) of the

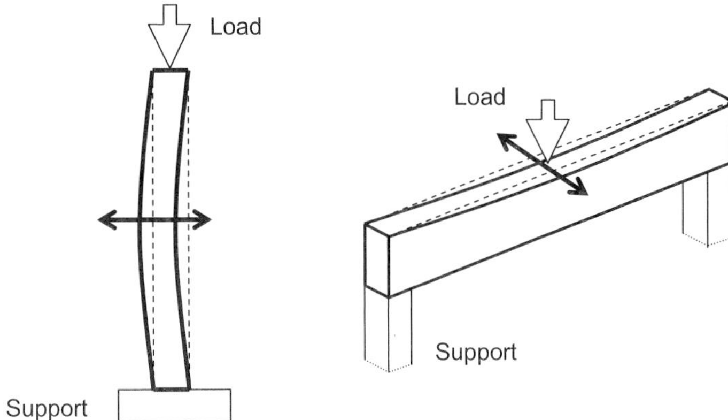

Figure 11.5 Lateral instabilities.

member. These are known as axial loads, acting through the centre of struts and ties and along the centre lines of beams and walls. Although it is desirable for the economy and efficiency of structural members to be axially loaded, it is not always possible. Loads that act off the axis are known as eccentric loads, eccentricity being a measure of how far the load acts away from the axis.

The application of eccentric loads may cause the member to overturn if the dimensions and/or connections are not sufficient for it to remain stable. However, assuming the member is not disturbed in this way, eccentricity creates a further distorted shape and therefore changes the distribution of internal forces. In columns and walls they induce bending. In beams if the ends are restrained from rotation the effect is torsion, similar to lateral torsion buckling.

Like lateral distortions the additional distortions caused by eccentricity do not replace the direct loading effects but add to them. The effect in the column is, therefore, as described for buckling. Already in compression, the bending forces will increase compression on one side and the bending tensile forces will relieve some of the compressive force on the other. Because eccentricity induces a bending effect, this is likely to be the direction in which buckling occurs, given that the member has the same slenderness in all directions.

Curving (and folding) structural members

The members discussed so far have all been straight or flat. Curving the member may increase the manufacturing and assembly cost but does provide structural advantages and may therefore be worth considering. The structural advantage is associated with spanning members. To develop sufficient resistance to span in bending, the material in compression and tension from the direct bending effect (i.e. not the shear) has to have sufficient separation to develop the moment of resistance. This makes beams and slabs deep if the spans become too large. This increases the depth of the structural members and, unless some of the solid section can be removed, increases the self-weight, adding to the dead load to be carried by the beam or slab.

Advantages can be gained from curving both in the direction of the span and at right angles to the span, but for different reasons. Consider the members in Figure 11.6. These members that curve across the span are known as arches and suspension cables. Two significant aspects should be noted. First, they require both a horizontal as well as a vertical support and, second, the internal force becomes pure compression and tension respectively: they are effectively curved struts and ties. This allows materials

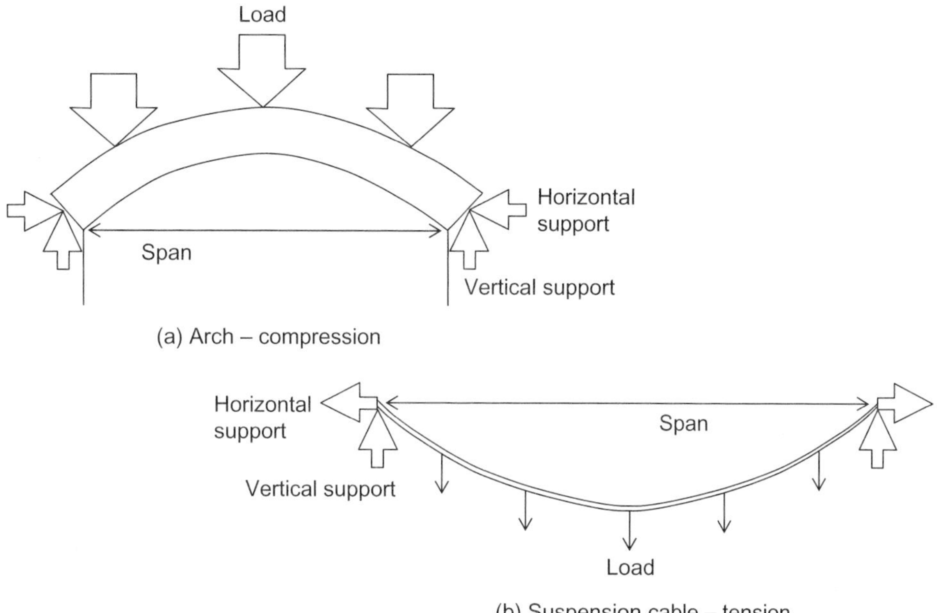

Figure 11.6 Curved structures across the span.

and/or components naturally strong in either compression or tension, but not both, to be used in the construction. Less material is needed and more material options are available. When slabs are curved like arches, they are known as shells and where allowed to hang suspended they are known as nets or fabric structures (tents).

Curving and folding at right angles to the span gains a different advantage. It gives the depth that is needed to act in bending from thin sections. The most common use of this technique is for roof sheets or decking, as shown in Figure 11.7. Thin sheet materials can span many times that of the flat sheet form. Again, less material is used, but there is a manufacturing cost associated with the need to form and control the folding or curving. Curving in this direction is still a bending action and therefore the supports are only required to take vertical forces.

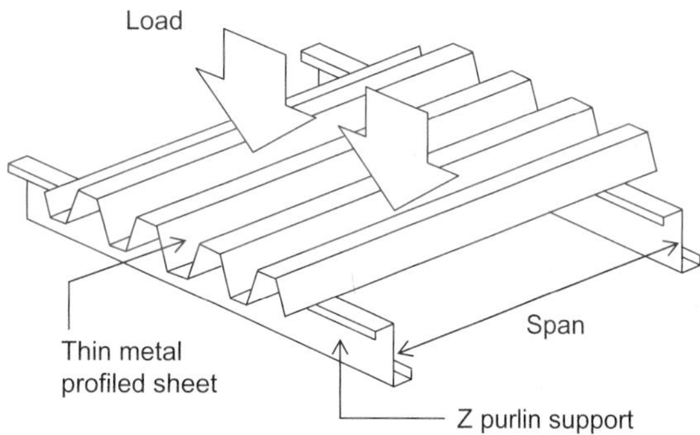

Figure 11.7 Curved (folded) in the direction of the span.

Structural connections

In the analysis of the five basic members shown in Figures 11.1 to 11.4 there is an implicit assumption about the way the member is connected to the support. All the distorted shapes assume that the member can rotate if it bends and can shorten or lengthen freely in response to the distorting effect of the loads. Whether this can happen depends on how the connections are detailed.

There are advantages to restricting the movements at the connections, particularly that of rotation, to induce different distorted shapes to redistribute the internal forces to reduce the size

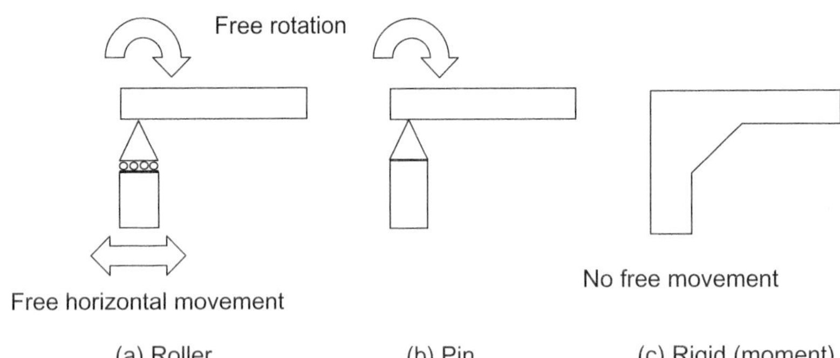

Figure 11.8 Structural action of connections.

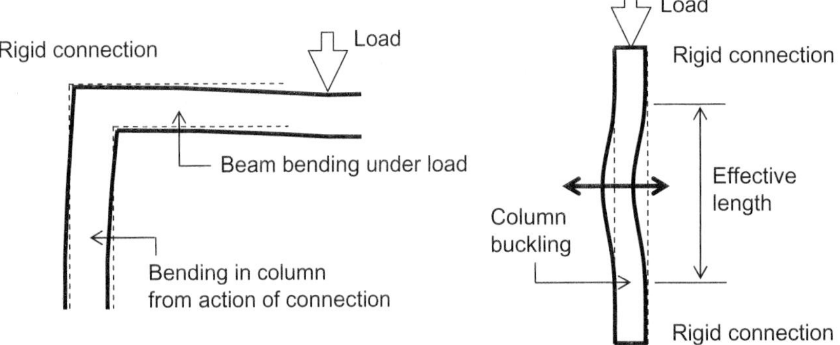

Figure 11.9 Effect of rigid connections.

of the member. This does sometimes increase the manufacturing and assembly effort and therefore has a cost to be set against the saving in section size.

In order to think about the effect of restricting movement at the connection on the potential stability and distorted shape of the member, three 'pure' actions can be identified and are shown in Figure 11.8. They are shown diagrammatically as they are idealised structural joints. The detailing to gain pure action is often difficult and expensive. The roller joint allows both rotation and freedom to move against the shortening (or lengthening) effect. Pin joints allow rotation and rigid (fixed or moment) joints allow neither.

Figure 11.9 shows the effect of making the joints rigid on two of the conditions already considered. The first is making the joints rigid at the end of the beam. The support now has a rotation (or moment) action to resist. If the support were a column with no balancing force from the other side, for instance from a similar beam with similar loading and similarly fixed, the column would go into bending. However, the internal forces in the beam are redistributed along the beam and the maximum force in the beam is reduced, but bending forces appear at the supports where there were none when the joint was assumed to be a pin. Tensile forces appear at the top of the beam at the support, the balancing compressive force being at the bottom. This redistribution of the bending along the length of the beam reducing its size is the basis of the portal frame, which can achieve long spans economically where a simply supported beam could not.

The second example in Figure 11.9 is the effect on buckling in a column where the connections are fixed. Because of the moment

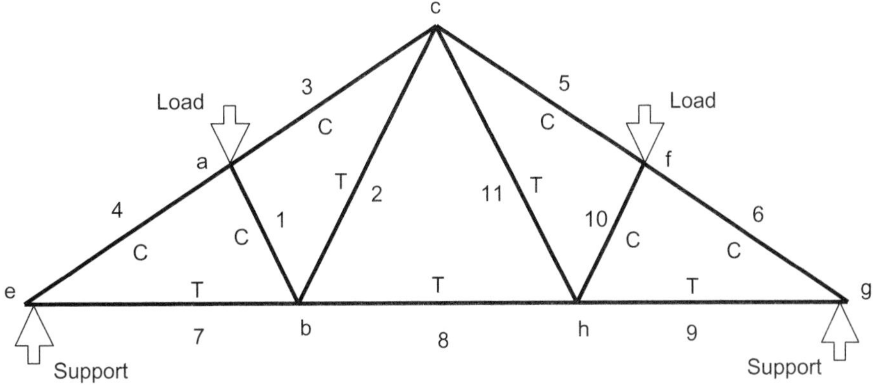

Figure 11.10 Pin-jointed frames – the truss.

resistance experienced at the end of the column, the effective length of the column is reduced and hence the slenderness is reduced. This column will take greater loads before it buckles.

In reality, most connections are detailed for ease of manufacture and assembly rather than to ensure a pure structural behaviour. They give these freedoms of movement to a greater or lesser extent and therefore the real instabilities and distortions may not be exactly those assumed, and different internal forces will develop. However, most simple connections behave near enough to a pure form (normally pinned) for the assumption to be safe to use with the practical detailing. If there is doubt about the degree of fixity (often known as semi-rigid joints), an estimate of the effect has to be established so that a more realistic distribution of internal forces in the members can be calculated. If the structural advantage of making the pure behaviour outweighs the additional cost of manufacture and assembly, details have to be devised to ensure the pure behaviour is achieved.

Grid members – pin-jointed frames

All the structural members considered so far, flat or curved, have been solid. Curving across the span was worth considering for bending members to generate pure forces (compression or tension) and reduce the dead loading that occurs as beams and slabs are designed to span greater distances. Pin-jointed frames are another way to achieve the same ends. This is a very versatile idea and many forms of these frames are used, both as plane frames (having little width, like beams) and acting in three dimensions like slabs, both flat (double-layer grids) and curved (single-layer grids as domes) structures.

The truss is one of the simplest forms of the pin-jointed frame and will serve to demonstrate the principles. It is shown in Figure 11.10. This is a spanning member with loads and supports shown and all pinned joints are assumed. Taking each member of the truss one by one, it is possible to trace the load and support of each to determine its distorted shape. The load applied to node a is directly transferred into member 1, which gains its support from node b. This will shorten member 1, putting it into compression. It has to be designed as a strut (note the load will be axial but it may buckle). At node b, forces are transferred into member 2. Consider the effect on member 2. A downward force is experienced, which is supported at node c. Member 2 is stretched: it is in tension and therefore acts as a tie (note not subject to buckling). Forces at node c now pull down on members 3 and 5, pulling them together, rather like the action of an arch. This puts compressive forces into these members that are continued through members 4 and 6 to nodes e and g respectively. This generates a need to support a

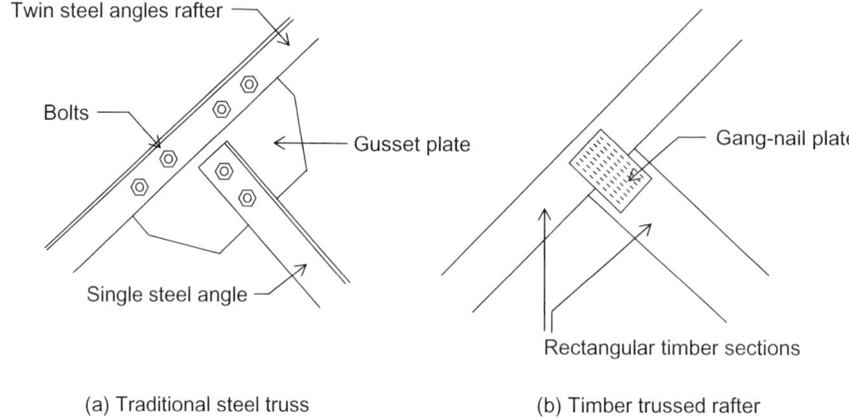

Figure 11.11 Detail of joints in pin-jointed truss.

horizontal force at e and g, and members 7, 8 and 9 give mutual support to e and g. These all become ties. By the same logic, member 10 is a strut and 11 a tie. All the forces are resolved within the frame as either struts or ties. The supports now only experience vertical loads, assuming the connection of the truss to the support to be a pin.

This analysis is based on the assumption of the truss acting as a pin-jointed frame. There is, however, a need for the connections to be practical while being able to assume that the structural action is that of a pin joint. Figure 11.11 shows a typical detail in steel and in timber of the connections at node a. The illustration shows that in reality members 3 and 4 are all one piece, but in this case the relatively low stiffness of the members themselves means that the assumption of pin jointing is safe.

Building structures – wind stability

When considering structural members, so far the loads have all been vertical, essentially from the dead and imposed loads on the building. Actions of forces other than vertical have been identified, but these have come when members transfer their loads to other members as the loads find their way to the ground. When visualising a whole building structure capable of carrying these vertical loads, it may not be stable against wind loads. Only when the whole building structure is identified can the effect of wind loads be considered.

Wind loads can act vertically. One of the most destructive aspects of wind to the fabric of the building is the negative, uplift forces on roof structures as the wind speeds up as it is forced to go round and over the building. Where roofs are over large open spaces, the wind can get into the building and create positive pressures under the roof, increasing the effect of the uplift forces. However, the major initial concern to the overall structure safety is the horizontal forces acting on the sides of the building. These can be both positive and negative forces, often being positive on one side and negative on the other, having a cumulative effect on the structure. One further aspect of the wind is that, while it may have a prevailing direction, it will sometimes blow from other directions and therefore the building must be designed to withstand wind from all directions.

Wind or overall stability of a building is a major design criterion. Additional members may be necessary to stop wind disturbing the structure. These members and/or connections allow the forces to be distributed through the structure and safely transmitted to the ground.

The major structural systems used for buildings are loadbearing walls and framed structures, which also include most of the long-span roof solutions. The predominant instabilities

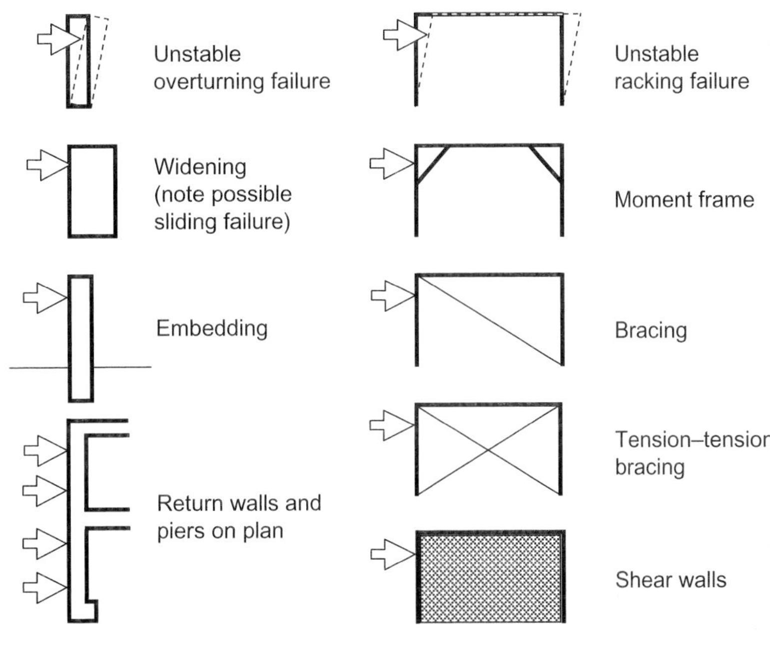

Figure 11.12 Stability of walls and frames.

and the measures to overcome them vary between these two structural systems. However, there are only a limited number of strategies to overcome these overall instabilities, which can be used alone or in combinations:

- Increasing dimensions (against overturning)
- Changing joint type (from pinned to fixed)
- Providing additional support members (buttressing, bracing, etc.)
- Changing shape (curving, folding, etc.)

Some of the alternatives used for walls and frames are shown in Figure 11.12. The major instability failure for walls is overturning. Widening and embedding are options most likely to be used in retaining walls resisting the horizontal forces from ground pressure rather than from wind. They have limited application in the superstructure against wind. Walls above ground are almost certainly likely to be slender and will rely on return walls and piers (buttressing) plus the stiffening effect of floors.

In frames the instability is racking. The frame can rely on moment (fixed) joints between the structural members, but this often is an expensive option. For frames with pin joints, bracing can be used, although it is possible to utilise other parts of the construction, such as internal walls or lift shafts, as shear walls (core stability) or the external hull either braced or with moment resistance from the structure and/or facade. It is not necessary to stabilise every frame. As long as a sufficient vertical section of frames is stabilised the whole building will be stable. The number and position of these stabilised vertical sections will vary, but structures must contain at least one section in order to stabilise each direction and at least one other section to resist a rotational collapse of the whole building. An example of this is shown in Figure 11.15 and will be discussed in more detail in Chapter 25.

Whatever additional support members are used these then become members with their own connections, distortions and internal forces. Consider the single diagonal bracing shown in Figure 11.12. Depending on wind direction, it may be either a strut or a tie. It has to be designed as a strut as it is almost certainly slender, and this makes it susceptible to buckling. The benefit of tension–tension bracing, as its name implies, is that the members only need be designed as ties, with only one tie acting at any one time, depending on the wind direction.

Stress–strain and the choice of material

Visualising the sets of internal forces in structural members can be considered without any reference to the size, shape or materials of the structural members. However, it is an understanding of the type and magnitude of these distorted shapes and internal forces in the members and joints that gives clues to the design of shape and materials that may be possible within economic sizes and manufacturing process.

In order to proceed with the analysis it is necessary to propose the shape and size of the structural member. The internal forces can now be associated with the cross-sectional area of the member over which they will act. From this it is possible to determine the stress that will be generated in the material. The stress is determined by the internal force acting over the proposed area given by the size and shape of the cross section of the member. So far, the illustrations have shown solid cross sections. For beams this has been rectangular, but the distribution of stress across the section suggests an 'I' or box section could be used. This removes material from areas where the bending forces reduce towards the neutral axis, using less material and reducing weight, but it changes the stress distribution across the section. Whatever the cross-sectional shape the stress distribution is independent of the material chosen; it is only a function of size and shape.

The choice of size and shape, however, cannot be established without knowledge of the material's structural properties, mainly (although not exclusively) associated with the stress–strain relationship. It is now that it is possible to ask questions about the two key failures associated with structural performance:

- Will it break? This is known as the collapse criterion and is directly related to the strength characteristics of the material. For a safe solution the allowable stress in the material must be greater than the stress that will be generated by the internal forces over the area of material proposed.
- How much will it distort? This is known as the serviceability criterion and is related to the strain characteristics of the material. The distortion must be limited in order to maintain visual lines and ensure finishes do not crack or become detached. The structure must also not feel live (bouncy) when subject to the imposed loading.

In fact the simple question of whether it will break for the collapse criterion to be satisfied is not enough. In the interests of limiting the risk of collapse while designing an economic structure it would be more accurate to ask, 'How close is it to breaking under working conditions?' The design must incorporate a factor of safety. This will be greater the more sensitive the design is to variability of the loading and/or the materials' properties. How this is allowed for depends on the approach taken to the design and is beyond the scope of this text.

For every material there is a relationship between the stress experienced and how much it strains, known as the e value. This relationship is shown in the stress–strain curves depicted in Figure 11.13 where the curves for concrete and steel are shown. The stress–strain relationship identifies the stiffness of the material, how much it strains with an increase in stress, which makes some materials more difficult to use in building to gain the full structural benefits. It may not be possible to take some materials to their full working stress without the strain induced by that stress leading to excessive distortions. Generally, building structures are made from relatively stiff materials such as masonry, timber, concrete and steel.

There is another important question in considering the collapse criterion. Not 'When will

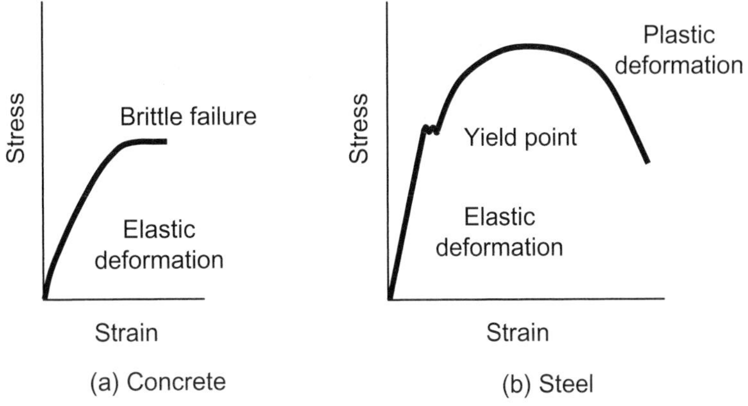

Figure 11.13 Stress–strain curves.

it break?' but 'How will it break?' This is another characteristic of the stress–strain relationship of the material that can be identified from the curves in Figure 11.13. It is the distinction between brittle and plastic failure. The curve for concrete shows that failure is sudden: it fails without any visual warning; it is a brittle material. The curve for steel shows that beyond a certain stress, known as the yield point, the rate of distortion is much greater as the stress increases, so visual warning of failure is given; it is a plastic failure. A second characteristic is that releasing the load after the yield point has been passed will not return the material to its original shape; it is deforming plastically as opposed to elastically.

Once the stress–strain relationship of the material is known, it is possible to evaluate the suggested shape, size and material specification for each member, ensuring that the stresses induced by the forces on the member are not excessive and the strains induced by the stress do not cause unacceptable distortions.

This process of checking suggested member size and shape against induced stresses and materials properties is the basis for structural analysis and design.

Structural analysis and design

Having established the loading pattern and devised a structural system of members and joints that is inherently stable, it is possible to identify the structural behaviour of each joint and member and determine the pattern of forces that will arise in each. Knowing the expected behaviour of the structure and materials under stress, it is possible to suggest appropriate cross-sectional shapes, sizes, materials and details for the structure.

It is necessary to evaluate the performance of these suggestions to see whether they can offer sufficient resistance without breaking and within limits of distortion. Initial proposals can normally be made from observations of past construction combined with an understanding of structural theory and design methods. Theories of mechanics and strength of materials support engineering knowledge. These theories are used to devise mathematical relationships that predict forces, stresses and materials behaviour sufficiently accurately to be relied upon in design. They must, however, be based on an initial ability to visualise the behaviour of the structure and the materials involved. If the engineer does not imagine behaviour and then suspects that behaviour as being a possible failure mode, the calculations necessary to check the design may not be undertaken. Many structural failures can be attributed to this error of judgement.

Once behaviour of the stable arrangement of members is visualised and the most likely failure modes identified, calculating the magnitude and distribution of the internal forces induced in the members is known as the process of structural analysis. Choosing and evaluating the shape, dimensions and materials for the members that will safely resist these forces to limit distortion, local instabilities and ultimate loss of load-carrying capacity is the process of structural design. Design also includes the detailing of the members and joints to ensure the behaviour assumed in the analysis.

Movements and structural behaviour

So far the analysis has assumed that the only movements or change in shape or dimension from those originally formed are the distortion of the members under load. It has also been assumed that these will be able to take place freely, without any restraint from other parts of the building other than their connections one with another. In reality there will be other movements caused by changes in environmental conditions changing the temperature and moisture content of the materials. These will not only change the dimensions of the structural members themselves but also may affect all the other parts of the building depending on the detailing of the joints and fixings.

When these movements take place, they may affect the structural system in one or more of the following ways:

- Restraining this change in dimension will change the magnitude and distribution of the internal forces in the member itself.
- If the member itself has to restrain these dimensional changes from other parts of the building, this will act as additional loading and hence change the magnitude and distribution of the internal forces.
- If movements elsewhere in the building result in a change in the position of some support in the system (most likely ground movements), major redistribution of the internal forces may occur (often determined by joint rather than member behaviour).

Every effort should be made to predict these movements and to detail the joints and connections to allow free movement. If this is not possible then the effects must be evaluated as additional loads and hence levels of stress.

An additional complication is that some materials, such as concrete, continue to distort with time even though the load remains constant. This phenomenon, known as creep, means that the effects of these dimensional changes may develop at some time in the future, leading to failure some time after the building has been occupied. A more comprehensive analysis of movements is covered in Chapter 12.

One important aspect of movement in the design of a building structure is that of the relationship between any differential settlements in the foundations and the ability of a building structure to move with these settlements without distress. Each overall design of a superstructure (members, materials and joints) will have an associated flexibility that will determine how much differential settlement they can accommodate between adjacent foundation supports and this will be a key performance criterion for the foundation design.

Stresses in the ground

All the analysis so far has been to establish a stable arrangement of structural members, where the task has been to suggest the shape, dimensions and materials to be used and then check that the internal resistance is sufficient to carry the loads safely without distortion or breaking of any part of the structural system. The ground on which the building is founded is also part of the structural system but cannot be considered in this way. It too will have to be checked for stability (subsidence), distortion (settlement) and breaking or collapse failure. However, unlike any of the structural elements discussed so far, the ground cannot be specified, it has to be investigated.

The aim of the investigation is to determine the information necessary to check out the ground's ability to carry the loads without excessive settlement or collapse failure. However, it is not possible to determine the

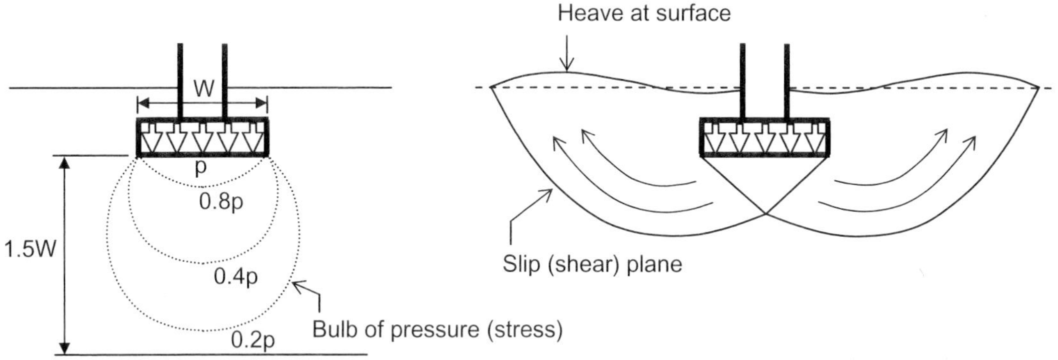

Figure 11.14 Stress in ground due to foundation loading.

scope of the investigation or type of information required without some idea of the pattern of forces that will have to be resisted if the soil is not to fail. In this respect it is the same process as for all structural members, but the visualisation of the behaviour needs to identify the shape of the stressed area. This is very different from the structural members considered so far, since there is now no obvious edge or boundary to the soil as there is for a structural member.

The two diagrams shown in Figure 11.14 help to show the way the soil will be stressed and explain both distortion (settlement) and collapse behaviour under a bearing foundation. To understand settlement, the diagram showing the bulbs of pressure (Figure 11.14a) gives an idea of how the stress will build up in the soil. This shows the volume of soil that will strain, leading to an ability to predict settlement.

There is a simple classification of soils into cohesive (broadly clays) and non-cohesive (broadly sands and gravels), which is useful when thinking about the behaviour of the ground. One of these behavioural differences is how long it takes for the ultimate settlements to take place. In non-cohesive soils the settlements will take place as the loads are applied. Cohesive soils often take years to reach a final settlement and this time-dependent behaviour has to be investigated for a safe design to be achieved.

The observed collapse failure in soil shows how a shear plane develops in the soil which reaches the surface some distance from the foundation. Figure 11.14b shows two symmetrical shear planes. In reality only one needs to develop for failure, the foundation rotating with the one-sided slip. Comparing the two illustrations, pressure bulbs and shear planes, indicates that the volume of soil involved in collapse is not the same as the soil involved in settlement. Together the two diagrams show the extent of the investigation and the type of information required to predict the behaviour of the soil. The stress–strain (load–settlement) relationship is necessary to predict settlement and the shear strength needs to be known to check for collapse. These can be identified from laboratory tests on samples taken in the investigation or from in situ tests taken in the undisturbed ground.

These direct limits of depth and spread of the area of investigation for settlement and collapse may miss the third failure mode, the disturbance or overall instability from subsidence. This would cause the movement of a mass of soil that removes the support for the volume of soil stressed by the loads from the foundations. Subsidence may be geological in origin, with landslips or swallow holes, or may be man-made, with cut embankments or mining activity. This also has to be investigated, but the greater scope of area of investigation and the type of information required would probably need a desk study or topographical survey, although this may identify the need for a more widespread physical site investigation.

Another disturbance to the foundations not related to the direct stress in the soil is volume change due to changes in environmental conditions in the ground. Changes in moisture content in clay can significantly change the volume of the soil. In these 'shrinkable' clays, foundations must be taken below any risk of moisture change or the foundations may settle (or heave) independent of the stress in the soil from the building loads. Frost heave is another environmental change causing a volume change in the soil. Soils susceptible to movement under these conditions include silts, chalk, fine silty sands and some lean clay. Ground movement is discussed in greater detail in Chapter 12.

Other ground movements such as earthquakes generate stresses and changes in the soil properties, leading to new forces on the building, but these effects are beyond the scope of this text.

The major structural forms

It is now possible to return to the three major structural forms introduced in Chapter 3 and see the types of members and materials that can be used economically to achieve a variety of shapes and sizes of buildings for a variety of different uses.

Loadbearing wall structures

Structures based on loadbearing walls, by definition, use walls for support. In the past this has been the main structural form that has been developed for domestic-scale construction to larger buildings carrying high loads from relatively large spans, but this generated thick walls, increasing dead loads and limiting architectural form. The walls would have been based on masonry, probably stone or brick, or timber originally used to frame up whole walls and more recently to frame panels that can be used to create walls of varying lengths. Developments in materials such as concrete and steel have provided other options in skeletal frames and long-span roofs for these buildings but also other options for loadbearing solutions. The loadbearing solution is now perhaps most widely used where floor spans are limited to, say, 4 or 5 metres. This limits loads, giving relatively thin walls even for relatively high-rise structures. Where the building requires relatively small internal spaces on a regular pattern that repeats on all floors so walls can be continuous to the ground, loadbearing wall structures are an option. This is particularly true if the walls have sound or fire requirements, which tend to lead to construction that is good at carrying loads.

Modern materials have given stronger and more quality-controlled products that have allowed buildings up to 18 to 20 storeys to be constructed but only where there are many internal as well as external walls available to carry loads. Typical of these buildings would be hotels, student residences and flats. If internal layouts allow, low-rise commercial development with greater floor spans may be considered for a loadbearing solution.

Floors for commercial and high-rise buildings are likely to be concrete slabs, although, given limited internal spans in a single dwelling the timber joist (small beams at close centres) and boarding floor is probably the economic solution where sound and fire requirements are less. Roof structure is likely to be timber truss rafters (pin-jointed frame) on low-rise buildings.

If the cellular nature of the internal room layout provides walls in both directions, the resulting structure is naturally stable against wind forces. If, however, the loadbearing walls are predominantly in only one direction on plan, the building will be unstable in the other direction and some other stability members will have to be introduced. These may be from internal shafts such as stairwells or may be short returns, known as piers, at the back and front of the building on the ends of the wall to stiffen the building in the vulnerable direction.

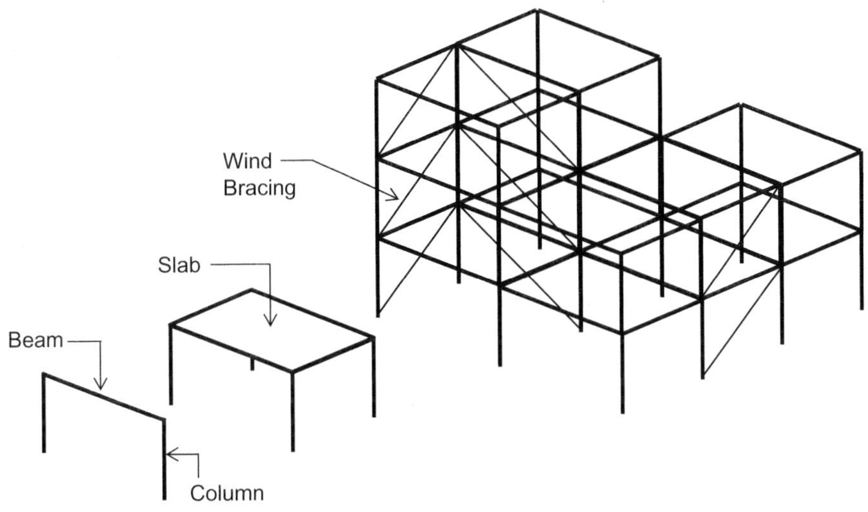

Figure 11.15 Basic members and arrangements of skeletal frames.

Construction where major loadbearing walls are used only in one direction is known as cross-wall or box frame construction and allows a variety of front and back facades to be chosen for the building.

Single-storey industrial building with long-span roofs can be constructed with loadbearing walls in both masonry and concrete. These, along with potential solutions to high-rise loadbearing structures, are discussed in Chapter 27.

Skeletal frame structures

The development of concrete and steel as structural materials allowed the design of effective and economic bending members, particularly beams and slabs that form two of the major structural elements of the skeletal frame. Steel has high strength in both compression and tension. However, steel is expensive and to develop economic bending sections it has to be formed into the classic 'I' cross section. Concrete, while having no significant structural tensile strength, has sufficient compressive strength to make a composite with steel to resist the tensile forces, forming reinforced concrete members. The success of reinforced concrete as a composite is based on more than the relative strength characteristics of steel and concrete. The strain, thermal expansion properties and cost are also complementary. The shrinkage of concrete on curing, plus some ribbing on high-yield reinforcement, provides the bond necessary for the two materials to act as a composite.

With the newly developed high-strength materials comes the ability to take vertical loads on small cross sections so walls could become columns. The basic form of the skeletal frame shown in Figure 11.15 could now be formed effectively to offer not only more clear internal space in buildings but also greater height without the greater thickness of materials that went with the limited strength of masonry. Floor slabs are the third major component of the skeletal frame. Almost universally constructed in reinforced concrete, even for structural steel frames, these floor slabs have been developed in a number of forms both in situ and precast. This gives a range of options that can accommodate spans between 3 and 12 metres, but their economy is determined by their choice as part of the frame layout as a whole.

It is not normally economic to have columns at much less than 6-metre or more than 12-metre centres. This is linked to the choice of slab. This would mean, for instance, the use of secondary beams for a 3-metre one-way spanning slab or up to a 12 × 12-metre column grid if a two-way spanning waffle slab were used. Slab developments have also led to reinforced concrete frames without beams where plate or flat slab floors have slabs transferring their loads directly to the columns. This generates high punching shear stresses around the columns. These floors with office loading can span 9 metres at around 350 mm deep in reinforced concrete and even greater spans (or shallower depths) if pre-stressing is employed. The economy in the flat slab is achieved in the simplicity of the formwork and falsework systems, which are a major proportion of the cost of reinforced concrete structures.

Wind stability members make up the other major structural member of the skeletal frame. Shear walls are probably the natural choice for in situ reinforced concrete frames. Compression or tension – tension bracing may be used with steel or precast concrete frames, although concrete or masonry cores are often an alternative, depending on the layout of the buildings.

The skeletal frame no longer requires external walls to carry dead and imposed loads from the building, only to transfer wind loads to the structure. This has allowed a wide variety of facade materials to be employed to give expression to a range of architectural forms.

Long-span roof structures

The need to roof over large areas with little interruption even from columns characterises a number of buildings that are commissioned. Initially developed for warehouses and factories, this form is now employed for retail and sports and leisure use. Predominantly used to cover ground-floor areas, they can also be used to form a roof over upper floors where the activities on lower floors can be accommodated in a skeletal frame structure in buildings such as airport terminals.

Although long-span roofs are often thought of as having a 35-metre span, or more, the simpler plane frame, and even some of the three-dimensional forms, can be economically used over much shorter spans. Although not universally accepted as a definition, it is useful to think of these forms as being used on short spans of up to around 15 metres and as medium span from 15 to 35 metres. This gives some clues as to which forms will be economic.

While, like skeletal structures, they are based on the frame (a bending structure supported on columns), they have been developed into a much wider variety of forms. Many are expensive and gain their value in their dramatic forms over medium to long spans and are often left exposed as part of the architectural design.

The simplest and generally least expensive form is the plane frame. There are now two common forms: the pin-jointed frame, such as trusses and girders supported on columns, and the portal or moment frame, where fixed joints distribute the bending moments into the columns, giving economic beam depths even over large spans. The triangular-shaped truss used to form a pitched roof is only normally used for short spans, but both girders and portals can be used for a long span of up to 100 metres if required, although smaller spans are more economic.

The spacing between the frames will be dependent on the ability to provide purlins and cladding rails to support the cladding (see Figure 11.7). On longer spans the individual frames become more expensive so setting them further apart can reduce the cost of the frames, but the purlins and cladding rails become more expensive. On short to medium spans, frame spacing of 4 to 8 metres with cold-rolled steel Z purlin and cladding rails can be used. On long spans, 8- to 12-metre spacing of frames with light truss purlin and cladding rails may become more economic.

Wind stability provision is less straightforward than for skeletal frames. Portal or moment frames are inherently stable against side wind, which will put additional bending forces in the frame (sway) that have to be analysed. Being plane frames and not connected together other than by purlins and cladding rails, they are unstable against end wind loading and require bracing in this direction. The long span creates large roof areas and the light roof cladding makes these frames susceptible to wind uplift.

Given that the roof sheeting remains connected to the structure, the frame may be subject to bending reversal. The top of the beam will go into tension and the bottom is subject to compressive forces. In these frames one failure may become lateral torsion buckling, as the bottom of the beam is unrestrained. In steel portals, for example, struts have to be specified to the bottom flange of the 'I' beam against this condition.

Pin-jointed trusses and girders, with pin joints to the top of the column, are vulnerable to wind in both side and end loading. The end-loading wind may cause the frames to overturn not only at the connection at the bottom of the column to the foundation but also at the connection of the truss to the top of the column. Bracing for the walls and at the eaves will be necessary.

As the height to eaves of these frames and/or the spacing apart becomes greater the length of the bracing member becomes longer. If single bracing is used, the increasing slenderness makes the buckling failure more likely and large-section members have to be used. If tension – tension bracing is used the increased length does not increase the chance of failure, making the two-member form of bracing economic.

The second broad type of long-span roof is the three-dimensional form. Either it takes advantage of the two-way spanning pin-jointed frame forming flat, two-layer grids such as space frames, or it gains the spanning advantage by curving (or folding). Curving can be achieved either in single-layer grids such as geodesic domes or from solid slab-like structures such as shells or folded plates. These forms are relatively stiff, but curving can be used in the more flexible tension structures based on suspended nets (cables) or fabric structures.

These three-dimensional forms can all be used over medium and long spans in a variety of construction options and materials, achieving a wide range of architectural forms, making the analysis of their choice complex, with a need to be well informed. It is not sensible to offer broad advice on form and spacing of these three-dimensional forms as it is with the skeletal frames and the plane frame solutions for these long-span roof structures.

All these forms of roof structure are discussed in more detail in Chapter 26.

Summary

1. To consider structural behaviour it is first necessary to understand the external loading conditions, primarily dead, imposed and wind loading, although they may include impact and moving loads. These loads will move or disturb the structure unless it is stabilised, after which the load will distort and potentially destroy (break) the members.
2. While all parts of construction are subject to at least some of these loads and therefore have to be considered for strength, damage resistance and adequate methods of fixing, there is for all buildings a recognisable structural system transferring these loads to the foundations. This system will comprise a series of structural members and supports jointed together to provide a load path through the building.
3. Five basic structural members can be identified: the strut, tie, beam, wall and slab. Each can be used for a range of loading conditions and support arrangements which will determine the distorted shape and will give clues to the distribution of internal forces that will be generated. These forces are compression, tension, bending, shear and torsion. Closer analysis of these forces shows them all to resolve into either compression or tension.

4. In addition to the distorted shape that arises in the direction of loads, lateral distortions will appear in slender members in compression at right angles to the direction of loads. Eccentric loads will also introduce new distortions, leading to new internal forces.
5. Curving spanning members can be used to advantage. Curving in the direction of the span can eliminate bending forces, leaving the member in compression (arching) or tension (sagging), although this leaves horizontal forces at the supports. Curving at right angles to the span increases the depth, giving a more effective bending section with less material.
6. The way members are connected also affects the distribution of internal forces. Pin joints, which allow rotation, do not transfer bending, while fixed or rigid joints do redistribute bending stresses into the support.
7. The pin joint can be used to build up spanning members into grid structures, such as trusses and girders, which become a framework of struts and ties, making large, relatively light, spanning frames for longer spans.
8. When members are arranged into building structures, it is possible to see the way in which wind loads are applied. It is likely that the structure visualised to carry the dead and imposed vertical loads will be unstable against wind loads, and other structural members may have to be added. The forces in these new members will have to be analysed and the forces in the original members reassessed.
9. With the whole structure stabilised and the pattern of external and internal forces established, the shape, size and materials for each member can be specified. The shape and size of the cross section of the member will determine the area over which the internal forces will act. This will give the stress that will be experienced by the material. Knowledge of the stress and strain characteristics of the material can be used to check the suggested shape and size for sufficient resistance to keep suspected failure within factors of safety.
10. All this analysis assumes that movements in the building do not cause a transfer of load from or to either structural or non-structural parts of the building.
11. The ground is part of the structural system of the building but cannot be defined and specified in the way other structural members have been considered. It has to be investigated, and the scope of the investigation is determined by the way stresses appear in soil causing settlement and collapse.
12. It is now possible to return to the three major structural forms – loadbearing walls, skeletal frames and long-span roof structure – to identify the economic arrangements, sizes and materials that may be chosen for a building structure.

12 Physical Behaviour Over Time

This chapter deals with the third major aspect of the physical behaviour of the construction: consideration of the analysis of deterioration, movement and wear. All these processes will happen during the life of the building and may reduce the reliability of the construction to continue its contribution to the performance of the building. Concerned with the changes that may take place with time, it considers the potential failures and the implications for maintenance and repair during the life of the building.

Reliability: renewal, maintenance, repair and disposal

The analysis of behaviour over time is about understanding possible changes that take place in the materials, components, joints and fixings that will affect the ability of the construction to perform. Changes may reduce the chance of performance the next time design conditions are experienced. The reliability of the construction is reducing. Its chance of failure under the next set of testing conditions becomes greater.

However, in most cases, construction is not just built and then demolished at the end of the life of the building. Its reliability may be dependent on interventions during its life to keep satisfactory performance. To evaluate a proposal against its reliability over time it is necessary to specify:

- How long it is intended to last before it is renewed
- What maintenance it can expect during its life
- What is the impact of failure and the ease of repair

To complete the specification it is then necessary to identify:

- How it can be dismantled
- What options there are for safe disposal

Without both the initial detailing and specification and a clear statement on these interventions in the life cycle of the construction, the proposal has not been fully evaluated.

Not all parts of the building have to last for the same length of time before renewal. Clients' and users' requirements change and, while not foreseeable in detail, they are likely to involve finishes and the division of space more than structure. It may be more economic to renew than to maintain, although this may not be desirable against sustainability criteria. It is necessary to establish as part of the analysis the client's commitment to renewal cycles and maintenance plans.

Another strategy for some parts of the building is to expect failure and renew or repair if and when this happens. For the client this has a cost burden for loss of use and in establishing an organisation of repair in the case of failure during the operation of the building.

These decisions on repair and maintenance concern all parts of the building from tap washers through to decorations and roof finishes to the life cycle of the building itself, including expansion and change of use. Clients will have a view on the management of the facilities they use for their everyday operations. Decisions on value for money and sustainability are not just concerned with initial costs but the whole life cycle of the construction.

The basis of choice is therefore to understand the changes that take place over time and then to consider the options for renewal, maintenance and repair to keep operational performance. Proposed construction will have to have an initial specification that should have foreseen the anticipated maintenance, with an analysis of the consequences of failure (safety, cost and disruption) and the ease of repair, with requirements for access that have to be incorporated into the design. If the client does not favour maintenance, and repairs cannot be made easily, then initial specification and detailing may have to be changed before the final choice is made.

Finally there is the consideration of dismantling and disposal, which may also include decommissioning. Whether considering renewing a part or the demolition of the whole building, the ease with which it can be taken apart and the impact of disposal should be anticipated in the original specification and details. When this is seen as an economic issue, it may be considered to be so far into the future that it can be ignored at the stage of initial design. However, with the increasing emphasis on sustainability with its focus on the quality of environment for future generations, dismantling and disposal can no longer be so easily dismissed at the design stage.

Basis of analysis

In the past the limited number of construction materials coupled with relatively low performance expectations allowed knowledge to build up over the years. Details were developed to protect parts from decay. Where protection was not possible, there was a greater acceptance of maintenance activities to keep the construction in good condition. The details that developed were often regional, if not local, to fully take into account the particular conditions for the locality. More than the detailing, the orientation and grouping of buildings took into account local exposure conditions, further protecting the vulnerable parts of the building from the worst ravages of the weather.

This description of the close links with experience, locality and building traditions no longer holds true for the majority of buildings erected today. This places greater emphasis on the designer's ability to undertake an analysis of behaviour of the construction with the passage of time.

There are three broad areas of analysis that have to be considered as significant to the reliability of building, each with a potential to lead to failure, invoking the need for maintenance, replacement or repair:

1. Environmental deterioration (durability)
2. Movement
3. Wear

There is a need to understand their direct effects in order to undertake an analysis of the proposed solution to identify ways in which the building may become less reliable, require effort to maintain and ultimately fail.

Soiling and cleaning

Before considering in detail the three broad areas of analysis identified above, all of which may lead to physical failure, the less threatening but disfiguring accumulation of dirt on both internal and external surfaces needs to be discussed. This soiling affects appearance and may be sufficient to justify cleaning during the life of the building. This has been recognised in Chapter 10, which discusses cleanable surfaces where conditions over and above soiling were identified to include the harbouring pathogens that may be harmful to health. Externally soiling is part of what is sometimes known as weathering, the other part being the environmental deterioration, as a visual change to the facade.

Like the more destructive aspects associated with durability of the materials, soiling is dependent on the environmental conditions. These determine both the type of dirt deposited and the natural cleaning action of rain. Perhaps the most recent period when soiling was a major problem was associated with the burning of coal to heat individual houses. This led to the Clean Air Act, which banned the burning of coals that produced excessive particulate matter. This did not stop gaseous pollutants such as sulphur and carbon emissions, but it did take away the dirt that was being deposited on buildings and causing bronchial diseases. Generally, even city air is now much cleaner, but some dirt is still present.

On a facade both deposition and natural cleaning can be quite localised and this can lead to characteristic patterns of staining. In splash zones, where water is splashed onto vertical surfaces from roofs, ledges or the ground, soiling can be heavy. Where water runs from ledges, such as the ends of window sills, deposition may be greater than on the general face of the facade. At the same time sheltered areas under ledges may not be washed by the rain. Whole walls either less exposed or sheltered by other buildings may accumulate dirt more than open walls even on the same building.

Some materials are more susceptible to soiling than others. This may be due to colour, where light shades are more vulnerable, or the surface provides for greater adhesion, making cleaning processes less effective. Some of the more traditional smooth renders have these characteristics. Detailing is also significant in that it can encourage both deposition and natural cleaning processes. Joints and overhanging details need particular attention. If excessive soiling occurs on a building, cleaning may have to be considered.

Windows, and possibly roof lights, have surfaces where cleaning always has to be considered. While it is possible to specify a self-cleaning glass, most buildings will rely on a cleaning regime undertaken at regular intervals during the life of the building. Modern systems are now available using brushes on long poles that deliver water that is processed so that it does not need to be squeegeed or leathered off. This gives the opportunity to clean windows from the ground in a safe and efficient manner. For taller buildings some form of anchorage for window cleaning access may need to be designed to connect directly to the structure, often on the roof. Access equipment can then be suspended to travel down the face of the building for both cleaning and maintenance or repair.

Durability of materials

Over time, changes take place in the properties and integrity of the materials of which components are made. This will alter their performance in the future. It should be acknowledged that some of these changes are beneficial in that they improve the durability of the materials. The weathered surface of stonework is a good example of this. This makes the material more durable. It is important to recognise this not only when specifying new building work but also when considering cleaning, maintaining or altering the building. However, many changes that take place reduce the durability of the material.

Durability of a material is dependent on the conditions to which it is subjected. Each material has specific agents of degradation or preservation. It is not possible to determine the durability of a material in a building without a clear definition of the conditions under which it will have to operate. Each material has associated agents and mechanisms that are known to degrade or preserve the materials. Either material must be chosen to survive in the expected conditions or, unless a new material is specified, some protection or preservation must be specified.

Agents and mechanisms

The mechanisms of change can be identified as taking place either within the material or at the surface. In both cases there will be a change to the internal structure of the material that is affected. This internal change can either affect the properties of the material or disrupt the internal integrity of the material, both of which result in a change in performance, most often for the worst. Changes in properties include embrittlement in plastics and fatigue in metals, while examples of internal disruption are wet rot in timber and sulphate attack in ordinary Portland cement-based materials. In each of these examples changes only occur in the presence of specific agents. Embrittlement is associated with ultra-violet light, fatigue with reversing stresses, wet rot with dampness and sulphate attack with sulphates and moisture.

These examples also serve to bring out another useful way of thinking about these agents of change. They are chemical (sulphate attack), biological (wet rot) or physical (fatigue). There is often a coincidence of conditions as in frost disruption where saturation and cold temperatures cause a change of state, not in the materials but in the saturation water, causing a physical internal disruption of the material normally just below the surface.

Study of the major materials used in building shows that water plays a part in many of the deterioration mechanisms, as does air with its ready supply of oxygen. These components are significant not only in many biological processes but also in chemical processes, for example oxidation.

This is not perhaps surprising, for the processes of nature are dependent on decay and regeneration. Many of the materials from which we make our buildings are taken from nature and they are, therefore, part of the decay–regeneration cycle. Understanding the mechanisms and agents in this cycle becomes the key to predicting the durability of many materials and diagnosing failure.

Many of the more recently adopted materials we use are not naturally occurring but have been refined or processed. The pathways and agents of change in these cases are often to revert to their original or lower-energy states. Many processed materials have proved to be stable with time, for example glass. However, there is a tendency in some processed materials to change to lower-energy states, which may not have the properties that we carefully designed into the processed materials. Many of these mechanisms also need agents, again often water and oxygen, with other chemical agents accelerating the process. The mechanism of oxidation in metals is a good example of this.

Oxidation is a surface process where the oxidised layer does not have the properties of the parent material. Oxidation of iron (rust) is recognised as reducing the life of a component. Not only does the oxidised material have a greater volume than the parent material, which increases the dimensions of the original component, but also the rust layer itself ruptures and allows the agents of oxidation to reach down to new parent material, continuing the cycle of degradation until all the material has rusted away. However, this is another example where not all changes reduce durability. Non-ferrous metals, such as aluminium and copper, oxidise, but this new surface, while of greater volume, changing the dimensions of the component, is not great enough to rupture the surface layer itself. This new oxidised layer, or patina, is stable at normal temperatures and therefore protects the underlying parent material by significantly slowing down the rate of degradation. The component's life is potentially increased.

The identification of agents associated with the mechanism of deterioration or preservation for each material is a key aspect of the analysis of the durability of a material. It involves the identification of the environment in which the material is to function.

Cycles and concentrations

Because the rate at which changes occur in materials varies with exposure to the agents of change, it is necessary to project an expected pattern of exposure. The pattern must identify the frequency with which changes in concentrations of agents, which are in contact with the materials, are expected to occur. Because many deterioration mechanisms have a number of agents involved, a pattern for each has to be established so the frequency of coincidence can be identified. In frost damage, not only is the coincidence of saturation and low temperatures

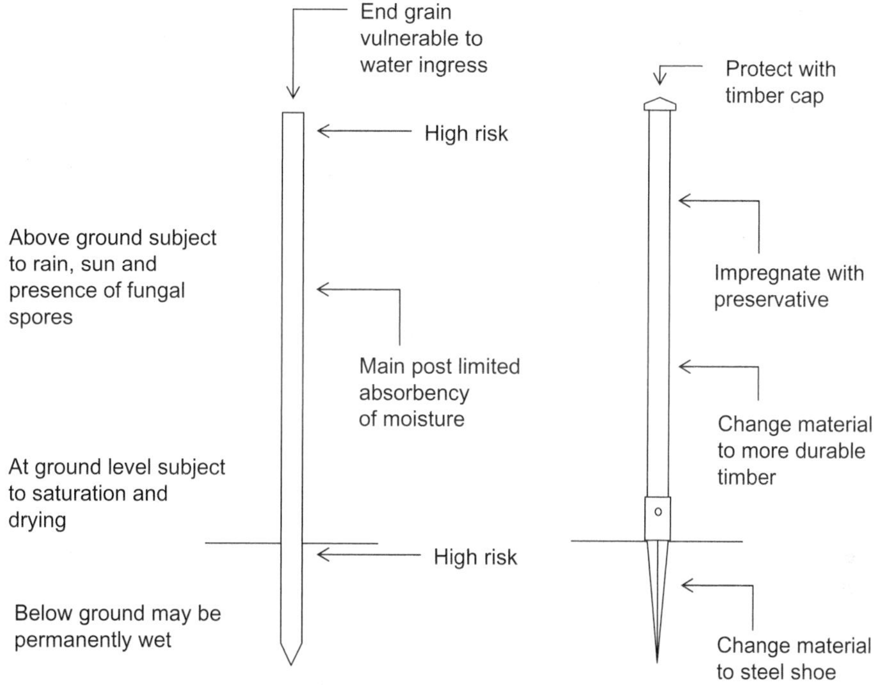

Figure 12.1 Timber fence post – deterioration and improved durability specification.

important but also the frequency of the freeze–thaw cycle is crucial to the rate of deterioration. If the material remains dry at low temperatures, there is no deterioration. If the material is saturated but remains frozen for long periods, there is little chance of deterioration. If the material is saturated and freezes at night and thaws during the day, there is a greater chance of deterioration.

It is very important to carry out this analysis of conditions very locally, in immediate contact with the material. It is very possible that even different parts of the same component are subject to very different patterns of exposure. Failure is often very localised, as with the wooden fence post rotting around ground level shown as a high-risk area in Figure 12.1. Here the patterns of concentrations of moisture, air and decay organisms are at their greatest and the cycles of change at their most frequent. This is also a good example to identify the preserving nature of some agents if their concentrations remain high but steady. The example here is water and timber. If water is constantly present, maintaining the timber in a saturated condition, it acts to preserve. The fence post buried deep in permanently wet ground does not decay. To continue the fence post example, the top of the post is also vulnerable as it is end grain that allows the water deeper into the timber, providing a pathway to high local concentrations of the decay agents.

This pattern of conditions allows an analysis of the rate of deterioration that takes place at the critical point in a component. Most materials in the building will have periods of decay and periods when little change is taking place. This may, however, in most cases be considered in the long run a gradual change. The changes taking place are making the component less reliable the next time the component has to operate at or near its design limits. The fence post breaks in a high wind but no worse than winds it has survived in the past.

Where deterioration patterns are identified as creating a high probability of failure (or the consequences of failure are catastrophic), consideration has to be given to changes in the suggested specification. It may be possible to protect the vulnerable parts from the agents of decay or provide some preserving process that limits the effects of the decay agents. It may be economic to change the specification of the material. In the case of the fence post the end grain at the top can be protected with a cap, the timber preserved particularly where it enters the ground or a metal shoe placed in the ground to receive the timber above the aggressive line of the ground. In this case the deterioration of the metal shoe would have to be analysed.

The maintenance expectations for this specification of the fence post may include the gradual reduction of the effect of the preservative, which would have to be replenished, applied by brush some time in the future. Maintaining and/or repairing of the cap at the top of the post would be cost-effective as it is easy to do. The ultimate failure may well now still be at the base of the post even with the metal shoe. If maintenance of the metal shoe is not contemplated, it may be that repair of the first few posts to fail could prolong the life of the fence, but ultimately the renewal of the whole fence would be the preferred option. Dismantling would not be problematic and neither would be the disposal of the timber as long as the preservative was not still active. It may be that at this stage the metal shoes may be cleaned, renovated and reused.

Movements in components

The term 'movement' here refers to changes not only in position but also, more generally, in size or shape of the components. These are changes in the dimensions of components. All buildings and their components will be subject to these movements, but, if correctly detailed, this need not lead to failure. Failure would be manifest in cracking or distortion. When small, these cracks or distortions may disturb finishes and allow water penetration but in the extreme they can threaten structural stability. Failure is, however, not inevitable. Movements of this kind can be accommodated and allowed for in the choice of materials and in the detailing.

The dimensional changes are often termed inherent deviations as they occur due to the inherent properties of the materials. This is in contrast to the induced deviations that are the result of the production process and need to be considered to ensure that the initial construction will fit together, as discussed in Chapter 4.

Many dimensional changes pose little threat to the building. If they can take place freely, without impacting on adjacent construction, they will have little effect. Movements that are of concern are generally differential, where one part is moving more than another to which it is connected, or restrained, where one part is moving in one direction and meets another that cannot move (or is indeed moving to meet the first part). However, some movements need be very small to have a large effect. The shrinkage of cement mortars in brickwork breaks the bond with the brick, causing fine capillary paths leading to the potential for damp in the building, as discussed in Chapter 10, where the most common solution is to introduce a cavity to break the damp path to the inside of the building.

Much of this movement can be accommodated in the normal detailing as long as care is taken in thinking through the probability that induced deviations will not eliminate the gaps and/or that the limit of flexibility of jointing materials that allow the movements to take place freely is not exceeded. These details then have to be built with care to ensure that the expected allowance is achieved in reality. Some movements are so significant to the integrity of the construction that specific movement joints have to be designed. These may be required where structures of different heights come together with different foundation settlements. Another example is the design of large panels of brickwork where environmental changes will cause internal stress in continuous construction.

The origins of these movements can be traced to one or more of the following:

- Initial equalisation immediately after installation
- The application of loads
- Changes in environmental conditions
- Volume change associated with deterioration
- Ground subsidence

Initial equalisation movements

Many materials undergo dimensional changes early in their life while the material is developing its final properties, often through changes in temperature or exchanging moisture. These movements are normally irreversible. In processing and manufacture, many materials have to be prepared to make them workable to produce the shapes and forms required in the component. For some components made off site, at least some of these equalisation changes in dimensions take place before they are delivered. Many components, however, have not completed this change when built into the building, and for work carried out in situ all the dimensional change takes place in the final position.

Materials most often built into the building while this change in dimension is still taking place include timber, concrete and clay products such as bricks. Timber components are manufactured at moisture contents above those that will eventually be achieved in the occupied building and therefore will continue to shrink after installation. The equalisation movements in concretes, plasters and brick also include changes in moisture but in different directions. Concretes and plasters need water for setting and workability and then dry out and, therefore, like timber, shrink. Bricks that are fired and therefore very dry at the end of the manufacturing process take in moisture and therefore expand after the initial cooling shrinkage has taken place.

The important factor is often the timing of the installation. Concrete blocks laid too soon after casting will continue to shrink, causing tensile forces, leading to cracking in the wall. While the majority of expansion in bricks takes place in the early weeks after firing, they continue to expand well into the initial life of the building. Some continuing expansion of brickwork should be anticipated. As brickwork normally has sufficient internal compressive strength this expansion will lead to the whole panel sliding on the damp-proof course (DPC) or cracking at short returns.

With in situ work, all the initial equalisation takes place in position. Drying shrinkage in concrete laid in large areas such as ground floors can lead to cracking due to internal tensile forces that build up during the equalisation phase.

Loading and environmental movements

These movements are dimensional changes dependent on the properties of the materials and conditions experienced by those materials during the life of the building. Movements caused by loads will depend on the stress–strain relationship of the materials and on the distribution and magnitude of the loads themselves. The two major environmental agents are temperature and moisture. Nearly all materials change dimension with change in temperature. The amount of movement that takes place in a material for 1 °C change in its temperature is its coefficient of linear expansion. Only permeable materials will be subject to moisture movement. Like temperature changes, the change has to take place within the material, not just in the external conditions. The amount of moisture movement that can be expected is normally quoted as a percentage of the original size.

The timing of movements caused by environmental and loading conditions can be seen to be associated with two phases that may overlap in time. The first phase takes place predominantly during construction and in the early stages of occupation, where the movements are usually not reversible. These are taking place at the same time as many of the equalisation processes and are initially associated with loading. Structural members are gradually loaded (dead and imposed) as the building is constructed and during the initial phase of occupation. This loading increases stresses that increase strains, causing the distortions discussed in Chapter 11.

In addition, the temperature and moisture conditions can be very different during construction depending predominantly on the time of year. This determines not only how quickly equalisation processes take place but also the size of components at the time when adjacent construction is installed.

Once the building is occupied, most of the loading is completed, but the internal conditions are now being established for the first time. This will move the temperature and moisture to working conditions, and during this period more dimensional changes will be taking place. Not all strains from the applied loads take place immediately. While settlements in non-cohesive soils will take place as the loads are applied, consolidation in clay soils may take years to complete, meaning that not all foundation settlement is experienced in the initial loading period. Some materials, including concrete and timber, exhibit creep. This is a continuing strain under constant sustained load. The rate of strain decreases with time but can manifest problems some years after the construction of the building. This period could cover the first few years of occupation.

The second phase in which movements may take place covers the rest of the life of the building, after the first few years of occupation. These movements are usually reversible. Most of the equalisation process will have taken place.

The effects of environmental changes in moisture and temperature on external materials become significant in this second phase. Exposed areas can experience wide variations in temperature. In temperate climates these are not so acute on a daily basis but are experienced in the annual cycle of the seasons. While air temperature varies from summer to winter, materials exposed to the radiant effects of the clear sky experience the greatest surface variation. Mid-day summer sun can raise the temperature of surfaces, particular if dark in colour, to temperatures of 60 °C. Conversely, clear night skies in winter will reduce temperatures of surfaces directly exposed to the sky to −20 °C, again particularly if dark in colour. This potential 80 °C difference will significantly alter the dimensions of many of the common materials used externally on buildings. Particularly vulnerable are continuous roof coverings, leading to the specification of white-coloured protection to limit the temperatures experienced by the roof covering itself.

External materials also experience wide changes in moisture, from prolonged periods of sunshine to extended periods of rain. Perhaps this is most often noticed in timber doors that 'stick' in wet weather. The materials affected, which have to be permeable to absorb the moisture, fall into two categories: rigid internal structures and elastic internal structures. These are broadly ceramics such as brick and concrete and organic materials, predominantly timber, respectively. Ceramics are subject to relatively small changes, while timber is subject to significantly larger changes. Timber shrinks and expands differently in a different direction from the grain, changes being greatest across the grain.

Interior components are subject to less variation in temperature and moisture unless the activity undertaken in the building creates variations in these environmental conditions and no specific measures such as ventilation are taken to limit the areas affected. It should not, however, be assumed that all internal components will remain stable. Windowsills are particularly vulnerable to sunshine through the glass, raising their temperature, and fittings associated with heating systems will be subject to wide changes in temperature. Pipes should be able to move freely where they pass through the construction or they may 'creak' as the movement takes place.

Movements due to temperature and moisture changes only take place if the changes occur within the material, not just at the surface. However, surface changes can create internal stress as the surface tries to change but is restrained by the mass of the material. This image of layers of construction expanding and contracting at different rates is particularly important if layers of different materials are to be bonded together to act as a composite. If thermal expansion properties are significantly different, changes in temperature from those at the time of bonding will create bending stresses in the materials and shear stresses along the bonded surfaces. This is also true for moisture changes. Materials bonded together that get wet can suffer similar stresses to those suggested by temperature changes if their moisture movement characteristics are significantly different.

Volume change associated with deterioration

In considering durability earlier in this chapter, some of the changes were in the properties of the materials, but some involved changes in volume. They were mainly associated with the chemical changes, such as sulphate attack or corrosion, but would also include frost disruption. These changes in volume create the potential to disrupt the component itself, but the overall dimensional changes have the potential to create additional stress in adjacent components with which they are in contact. An example of this is the attaching of stonework using iron fixings that rust.

Ground movement, settlement and subsidence

The ground is subject to two types of movement: settlement and subsidence. Settlement is the result of loading and moisture changes in the soil and therefore its analysis is the same as that for loading and environmental changes of any other part of the building.

In assessing the consequences of settlement it is important to remember that it is the overall flexibility of a building that determines the amount of settlement it can sustain without distress. Brick structures in cement mortar can take less differential settlement without cracking than brick in lime mortar. Frames with pin joints can take more differential settlement than continuous structures without a redistribution of stress that may cause failure.

To assess settlement, soils are often characterised as cohesive and non-cohesive, representing the clays and the granular soils respectively. This is a useful distinction when considering movements in the soil carrying the building. Under loading conditions, clay takes considerable time (sometimes years) to compress (the action is consolidation) and even then may be subject to creep. Granular soils compress (predominant action compaction) almost immediately the load is applied and do not change unless the loading is changed.

Clay soils undergo high volume changes with changes in moisture, not a characteristic of granular soils. Processes of desiccation in clays include seasonal drying of soil in the summer and removal of moisture by vegetation, particularly trees. These movements can be very large and very powerful. Not only will a drying and therefore shrinking clay cause additional settlement but also clay taking up moisture will heave with sufficient power to lift the building. Generally, foundations should be established sufficiently deep to be below any chance of these movements affecting the support of the building. Piles may need to be sleeved, as heaving clay will grip the pile and lift the whole pile with the building on top of it.

The freezing of water within the structure of the soil can also cause heave in some soils, typically silts, chalk, fine silty sands and some lean clay. Foundations should, therefore, be placed below the frost line in these soils (unless building on permafrost, a condition found in some cold regions of the world).

Increasingly buildings are being built on ground that is neither cohesive nor granular as defined by geologically laid down material. Landfill, be it from excavation, demolition or domestic waste, is creating new development sites. This material is highly variable and cannot be easily generalised. It may have been well laid with some measure of compaction or it may have been just tipped. The engineering properties of such materials have to be carefully investigated before foundation decisions are taken. Taking deep foundations through the fill, 'floating' the building on a raft or improving the ground to allow shallow foundation in the fill may represent the economic answers. This will depend not only on ground conditions but also on the building structure, taking into account the scale of development and the flexibility of the construction.

Subsidence causing a loss of support to the soil acting as the foundations can occur with geological changes or from human activity. Geological changes can be deep, such as earthquakes, which, while a real threat in some parts of the world, are not a design consideration in many others. Geological processes can also be surface actions, with slope stability being a particular threat to buildings. Natural slopes can become unstable in extreme weather conditions, but man-made slopes created as part of the development process can also become unstable, threatening the support to foundations. If slopes have to be steep, retaining walls may have to be built to protect the buildings at both the top and bottom of the slope.

Mining activity gives a particular subsidence problem. When some mine workings are abandoned, they may collapse, causing a lowering of the ground surface in a wave behind the collapsing galleries of the mine. Buildings that exist as the wave passes are vulnerable, but after it has passed new development is less threatened. Shafts and swallow holes (natural shafts which open up in limestone areas) can also pose a threat, so site investigation and a search of the geological and mining records are essential in areas known to be active.

Movement and detailing

Movements occur because of changes in shape, size or position of the components that may or may not in themselves create any damage. The possibility for damage often lies in the detailing, the way the parts are connected and fixed. As soon as parts are connected or touching then there is the potential for the movement or its effects to be transferred. If the detailing allows the movement to take place freely, only the component itself need be considered. If no such freedom can be identified in the detailing, the movement will be restrained and force will develop between the moving and restraining components. This force may move adjacent components or may induce stresses that distort parts that are still held in position; finally it may induce stresses sufficient to crack or break components or the fixing between components.

Estimating possible movements in a particular component is, therefore, only the start of any analysis. It becomes necessary to be clear on the details to identify free and restrained movements.

The first part of the analysis of details involves estimating the possible position of the parts when incorporated into the building. This cannot be assumed to be the position shown on the drawing. Each operation on site can only be achieved within certain dimensional limits associated with induced deviation. The exact position within these limits cannot be known at the time of specification. It can only be assumed that construction will be achieved somewhere within the extremes of these allowable tolerances.

Movements take place from this initial installed position. Equalisation changes that have not taken place at the time of installation will cause movement, loading as it is applied will cause movement, environmental changes may induce movements and later deterioration may create a change in dimensions, again causing movement. Not all of these will always be cumulative. Expansions may tend to cancel out shrinkage, as with concrete where operational thermal expansions are really greater than equalisation drying shrinkages. The shrinkage in concrete may, however, aggravate other details such as in the connection of a clay brick facade, which will tend to expand while the concrete frame will almost certainly shorten by both concrete curing and subsequent loading (including creep). Detailing for the accommodation of movement between structure and enclosure is discussed in more detail in Chapter 29.

If this analysis predicts failure, modification needs to be made to the specification or the detailing. Materials with different properties may have to be considered. Protection for the conditions creating the movements could be considered. However, the most appropriate option may be the redesign of joints or connections to accommodate the movement, allowing it to take place in a controlled way where the detailing can include gaps or materials capable of taking the movements without loss of performance.

Wear of components

The section on movement has concentrated on the changes in dimension, shape and position that occur in components when loading and environmental conditions change. Wear is also the result of movement, but in this case in parts that are designed to move, such as hinges or moveable partition tracks. Wear is not determined by the conditions to which the components are exposed but by the frequency of operation. It is only a small part of the analysis of most construction because the majority of building construction is concerned with passive performance.

Wear is more of a consideration in services where moving parts are often associated with motors and machines that are part of services systems. Here the possibility of loss of performance at some future date requires access to remove and replace parts or even replace the whole unit. This has to be considered in the design of the building as well as the design of the parts themselves.

Summary

1. During the life of the building, changes occur in the materials and components that affect the reliability of the construction to perform the next time it is tested under its design conditions.
2. The three broad changes that can bring about a loss of reliability and failure are environmental deterioration (durability), movements and wear.
3. Environmental agents causing deterioration are specific to each material, and therefore it is necessary to determine the concentration and frequency of these agents in the environments associated with the specific components of the building, often very localised to one vulnerable part of the detail.
4. Movements, or changes in shape, size or position of components, can be caused by initial equalisation processes following installation as well as the loading and environmental conditions of temperature and moisture changes that may take place during the life of the building. Movements in the ground then require a separate understanding associated with subsidence, which is in addition to the loading and moisture changes that cause settlement.
5. Wear takes place in parts designed to move and is, therefore, more significant in services design where provision has to be made for the replacement of parts or whole units.
6. Not all these changes lead to failure. Distress in the building leading to loss of performance will depend on the detailing limiting deterioration and accommodating movements.

13 Manufacture and Assembly

This chapter considers an approach to the analysis of the production process involving both manufacture and assembly. It suggests that it is not just the concern of the production manager but needs to be understood as an integral part of choosing an appropriate construction solution. Starting from the need to ensure performance, it is necessary to visualise the construction in pieces and then at its various stages of partial assembly. The work required to complete each stage has to be seen as a series of operations achieved through chosen methods that determine the mix of resources necessary. The resources of materials and labour are supported by two categories of production equipment: temporary works and plant and machinery. The methods chosen must aim for a safe, economic and timely completion to an appropriate quality to ensure performance. When the analysis shows that these can be achieved, the suggested solution can be said to be buildable. The chapter concludes with an exploration of the nature of in situ, traditional and prefabrication production options.

Realisation of performance

In considering the physical behaviour of construction in Chapters 9 to 12, the analysis has taken the completed building to be under operational conditions. The proposal represents a potential performance: only the final building can realise an actual performance. It must, therefore, be part of the evaluation to question whether it can be built at reasonable cost (economic and environmental), in a reasonable time and in a safe manner and achieve the performance potential.

The purpose of the analysis in previous chapters has been to specify and detail the components, their materials, shape and dimensions, and their relative positions in the completed construction. Any analysis of production needs this information in order to consider methods and resources that might bring these requirements to reality. This does not mean that it is impossible to think about production until the solution is finalised. Even if the solution is at an early stage, with only broad ideas of possible materials and dimensions, the early indications of production options are established. It is possible to recognise production implications as soon as a physical form is emerging. This may then influence the choice of construction as the solutions become refined.

Visualising stages and sequence

The analysis of the behaviour of the final construction, which has been considered in previous chapters, has required the ability to visualise the completed works. It has been necessary to be able to see the construction fully assembled and under working conditions. To consider either manufacturing or assembly processes it is necessary to be able to visualise the building in pieces, to be able to see the materials and components at various stages of manufacture, to see how each part might look when it arrives on site and how it has to be prepared for incorporation into the final assembly. This then leads to the ability to visualise how the building will look at each stage of construction, partially erected. This is necessary to analyse not only the methods and resources of production but also the behaviour of the partly assembled building to ensure it remains in a safe condition at each stage of production.

The production sequence has to be visualised as a series of stages, each being a recognisable safe condition. It is then necessary to identify the operations that have to be performed to move from one condition to the next. These operations can often be undertaken in a variety of ways, each with their own mix of resources that have to be of sufficient quality and be available in sufficient quantity. Being able to visualise an appropriate method to produce the building safely, economically and to achieve performance is the criterion in declaring the construction to be buildable.

The basis of the analysis of buildability is to be able to identify all the operations that are necessary to complete each stage of the construction. A very large range of operations is involved, from actual handling, preparing and assembling of the parts to technical and managerial control. Each of the operations has to be identified and a method chosen that will determine the resources necessary to achieve the next stage of production.

It is necessary to be able to see these stages, identify the operations and decide on methods before it is possible to check that the resources can be obtained to complete the works.

Manufacture and assembly

The overall process of production can be characterised as taking processed materials, forming them into components and then assembling them to create the building. Among the primary decisions to be taken in this process is where these activities should take place. In an industrialised society there are more options available as to where the components can be made and assembly processes undertaken.

In traditional construction the components are created and assembled by craft skills. This involves cutting and fitting on site, leading to site operations dominating the process. This is not a good description of some contemporary building. Components are made and even assembled off site. Components arrive on site as sub-assemblies, or prefabricates, made to size with no expectation of site adjustment. Not only does this mean that the production process now becomes more related to manufacturing but also site practice becomes focused on jointing and fixing processes to allow for adjustment and fit. In yet another process, in situ forming of components in their final position can be used, as, for example, in the reinforced concrete frame. Off-site manufacture may be adopted for the temporary works and reinforcement, while the site operations are focused on forming.

While some highly prefabricated systems have developed, building is still likely to involve a mixture of traditional, prefabricated and in situ processes. It is, therefore, important to be clear about the balance between manufacture and assembly, on and off site. This is then related to the availability of trained operatives, technical and managerial staff with the skills necessary to achieve the quality required within any constraints of time. The aspects of these production options are discussed in more detail at the end of this chapter.

Identifying stages

The process of production as a whole has been identified as taking processed materials, forming them into components and then assembling them to create the building. Production can be seen as a series of stages or, as they are known in production planning, activities, each defined by the creation of the next part of the building. Each one of these activities will take place in a sequence and be associated with a series of operations that suggest methods and sets of resources that will bring the construction to a natural point at which inspection may be made prior to starting the next stage.

The stages can be as large or as detailed as required, depending on the stage of thinking about the production process. Even at the design stage it may be necessary to think through some parts of the building in some detail, perhaps because of the innovative nature of the design.

Clearly, when the production process itself is being planned monthly and even daily programmes will be required.

The reasons for breaking the construction down into activities are to gain a sense of the methods and resources that may be employed, which will determine the cost, time and safety aspects, as well as judging the ability to ensure performance in, for instance, working within tolerances. It is difficult to lay down rules as to how big these stages should be for it does depend on how novel the solution being suggested is. In well-established solutions with well-known details, whole elements such as foundations, walls and roofs may be sufficient. If the solution is new (or at least unknown or even forgotten), care must be taken to think through often relatively small activities to be clear that the suggestion is buildable.

When these divisions become very fine, they become the domain of the individual operative. They become part of the craft or trade and can be sensibly left to the individual undertaking the operations. There may at first seem little need to understand production at this level. However, as detailing changes it may become difficult to achieve to an appropriate standard with current practice. An example of this has been suggested as the reason for some of the early failures of mastic when first introduced into construction. The specification called for surfaces to be 'clean'. Craft practice at the time gave meaning to the word 'clean' that was not appropriate. The methods used to complete the operation of 'clean surfaces' involved products and tools unable to achieve the standards required for performance of the mastic. Without the operative being made aware of the important aspects of the specification or detail to achieve performance, common practice may be used in an inappropriate way, which may jeopardise the ultimate performance. An appreciation of craft or trade practice is necessary to question the buildability of a suggested solution.

One other type of activity that may be considered in some detail right down to the operational level is one that is repeated many times in the assembly of the whole building. Small savings in these activities can bring high rewards in both cost and time without jeopardising performance if they are thought through at the time the solution is being evaluated. This type of analysis is important in evaluating details for highly prefabricated systems.

More generally, the depth of analysis required is associated with the risks involved. Risks are normally associated with the safety, technical and financial aspects of a project. These are normally at their greatest with novel construction or where site conditions are complex, such as working in or near buildings that have to remain in operation.

Safety is a prime reason why it is necessary to carry out a risk analysis. The danger to health and safety at any stage of construction can often be reduced by changes in materials and details. The reduction of health and safety risks should be a constant aim in the choice of a solution as well as in the analysis of the production operations themselves.

Technical risk is associated with the possibility that the building will, through inappropriate production methods, not perform when completed or will suffer a premature failure either during construction or in the early years of its occupation. This is most often associated with not being able to achieve the required specification of materials, quality of workmanship or dimensional accuracy to ensure performance either in that part of the building or in parts to be completed subsequently.

Although both safety and technical risks carry a financial penalty, some risks directly affect budget expectations. Financial risk of this type can arise in many ways, often not just associated with the direct costs being more than anticipated. An example of this is financial penalties that arise from not meeting occupation dates. Risks to meeting occupation dates can be reviewed by identifying activities that are on the critical path. This defines the stages that form part of a chain of activities that represent the minimum length of time of the project. Delays in these activities directly affect the date on which the building can be handed over to

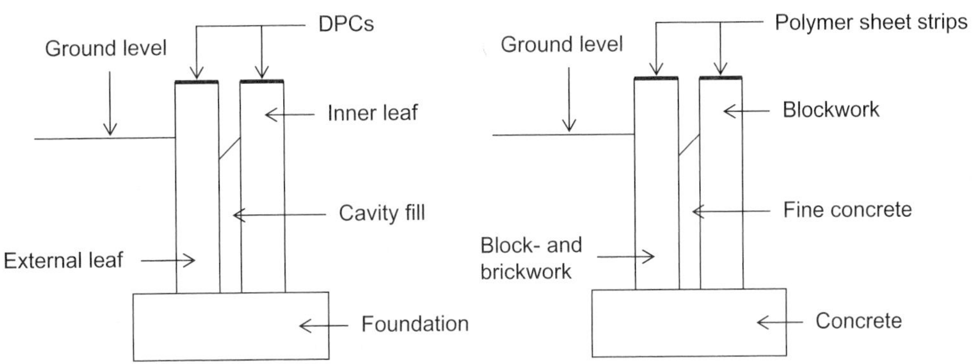

Figure 13.1 Strip foundations – parts and materials.

the client. Planning techniques such as critical path analysis can be used to identify critical activities. Knowing in some detail the methods to be employed on these activities will lower the risk of running over time and incurring financial penalties, without jeopardising the performance of the occupied building.

Analysis of operations, methods and resources

Once decisions concerning the level of analysis required have been taken and an appropriate division of activities and a sequence identified, the starting and finishing conditions of each activity can be visualised. It is then necessary to identify the operations required to complete that stage of the work. From an understanding of the operations, the methods and their associated resources can be chosen for the stage under consideration. However, the methods and resources may be influenced by those required for other stages, to give continuity. Repetition of operations and continuity of resources have an influence on productivity and the time–cost–quality relationship.

Figure 13.1 shows a cross section through the foundation to a house. A concrete strip footing supports the cavity masonry wall. While this is typically what would appear on the construction drawings, there are other activities or stages in the construction that have to be undertaken to achieve the design. These are normally associated with the production equipment discussed later in this chapter. In this example these would include the removal of the topsoil and the excavation and possible support of the trench. These activities plus those associated with the construction that is shown on the drawings are illustrated in Figure 13.2. This illustration shows the sequence as the activities are shown on a production planning Gantt or bar chart.

It is now possible to identify the operations associated with each stage in the construction of the foundation. This requires the introduction of a new vocabulary to visualise and describe the proposed construction. Activities or stages are described by reference to the parts of the construction to be created: foundations, wall, damp-proof course (DPC). Figure 13.3 suggests that operations are actions and therefore need active words (e.g. dig, lay, pour, transport).

Figure 13.3 also illustrates how often the activities involve a number of operations. Trench would involve digging but may also require temporary works support. Temporary works do not normally appear on construction drawings. Only an understanding of the production process allows the identification of the need for temporary structures and some ideas as to what methods may be appropriate and resources necessary to ensure safe and economic production. Forming the footings will involve the chain of integrated operations asso-

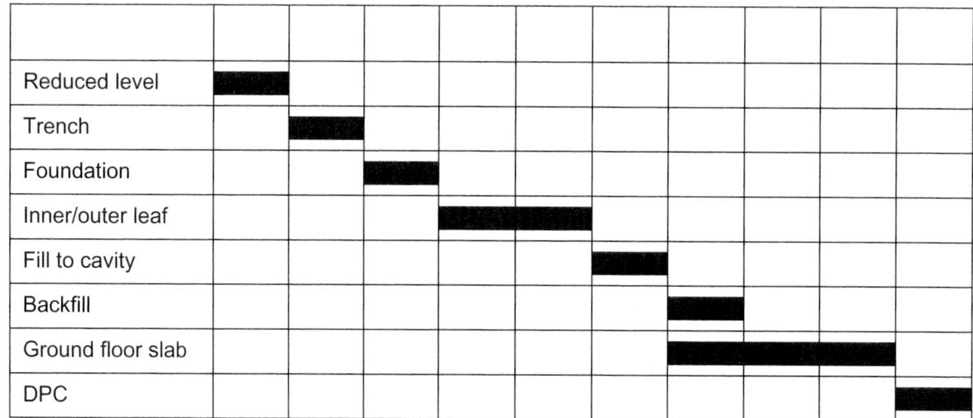

Figure 13.2 Strip foundations – stages and sequence of construction.

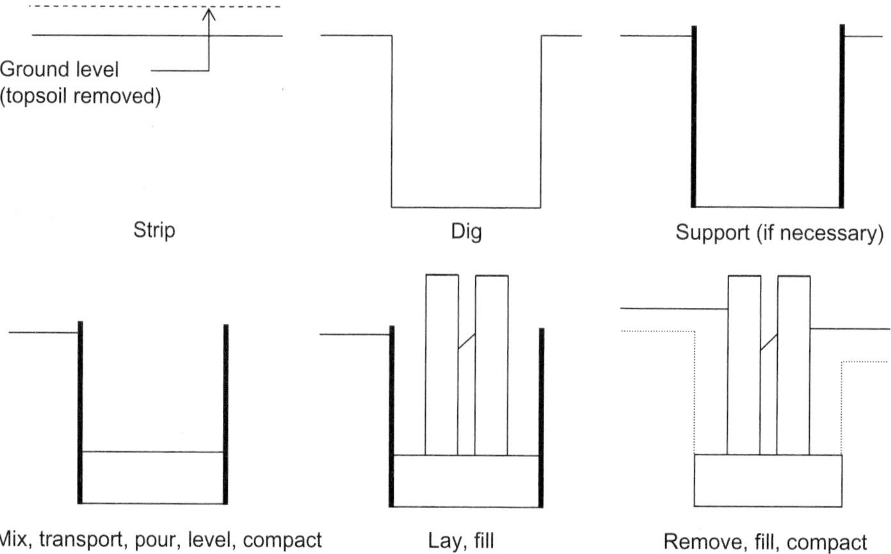

Figure 13.3 Strip foundations – operations associated with stages of construction.

ciated with concreting: mix, transport, pour, level and compact. All these have to be controlled to ensure that the quality of the concrete and the position of the concrete are within tolerances.

It is these operational actions that give clues to the alternative methods and the mix of resources that could be employed. For the excavation of the trench, hand methods may have to be employed in very confined spaces but are unlikely to be economic if access is available for machines to be used. If machines are used, the skills of the labour change to driver and banksman, whose job it is to guide the machine driver and ensure safe operations. Concreting is likely to be manual in the bottom of the relatively small trench, but options for the methods of transport will depend on the quantity, rate of delivery and site conditions. For small works, wheelbarrows may be appropriate, but again on open sites machines for transporting the concrete will probably be more economic. Laying the DPC is likely to be part of normal bricklaying operations when the external walls are built.

Returning to Figure 13.2, the Gantt or bar chart is used to show not only sequence but also time. It is the method and the quantity of resources that determine the time an activity will take. This is clearly most important in production planning for the organisation of the appropriate resources in the right place at the right time and in the right quantity. However, even at the design stage, clients are interested in time as well as budgets, so time often becomes a key aspect in the choice of a construction solution.

It is now possible to see the range of resources that may be used in the manufacture and assembly of buildings. The direct resources involved will include the specified components and materials, the trained and skilled operatives required to complete the work and the production equipment necessary to assist in the production process. Management and technical support will also be necessary with the requisite expertise and experience. Indirect resources associated with corporate organisation and capital will also be necessary but will not be considered in the analysis of manufacturing and assembly, which is the focus of this text.

Resources may include the need for temporary works not shown as materials and components on the drawings of the completed building. It is often necessary to provide a range of temporary works and safety measures in order to complete the work. In some cases these temporary measures can form a significant part of the costs and time associated with the completion of the activity. This is the case with in situ concrete construction. Temporary works have to be assembled and removed, and this involves a series of operations requiring resources. In a detailed analysis of the production process they may be seen as activities in their own right.

The possible methods and mix of resources are primarily determined by the choice of construction materials and detailing. Production cost and productivity have a baseline set by the construction chosen for the design. Careful choice of a production method and good management can only create efficiency gains. It is at the stage where construction solutions are chosen that basic cost and productivity are established.

Materials and labour

Fundamental to nearly all construction activity is the notion of labour operating on materials to bring them closer to their final state and position in the completed building. It is a characteristic of industrialised societies that the materials will pass through a number of processes, worked by a number of different people. Each of these people will have different skills and expertise and may work in a variety of places, from winning the materials to final placing and finishing in the building. This division of work on the materials is then reflected in the number of organisations that undertake this work and the levels of technical and managerial expertise that need to be applied. This in turn has implications for communication within the construction process, including those activities involved in the choice of construction.

The construction industry still relies on a number of traditional skills, such as carpentry and bricklaying, although there is now an increasing diversity of trades. These are normally more specialised, capable of a smaller range of operations, normally carried out within a well-defined stage of production associated with one part of the construction. In whatever way the skills are distributed among individuals, each operation will need to employ operatives with the required expertise and experience.

Health and safety

While labour remains at the heart of the operations for construction, there is a major concern for the health and safety of the individual. Industrial operations carry with them specific risks. The analysis of the operations will reveal the hazards to which individuals will be exposed and the harm these hazards may cause them. When the chance of an incident causing harm has been established, the risks to individuals can be evaluated. A hazard that could cause great harm or a fatality, even if it is not very likely, should be considered as a high risk. More commonplace incidents may be considered as high-risk even if they are unlikely to cause such severe harm. This type of analysis needs to be part of any evaluation carried out on a proposed production process. The evaluation of the risks may lead to changes in the choice of the construction specification and details as well as to the need for safety measures being incorporated into operational procedures.

Risk extends beyond the individual carrying out the operations. Any work may endanger others in the vicinity of the operation, either on site or in its immediate environment. This includes not only the immediate danger of the accident but also exposure to both short-term and long-term health risks. Most of these risks are covered by regulations, and operating a safe place of work is a requirement of the law. The law places responsibilities on various parties in the design and construction of buildings.

It is recognised that the most effective way to reduce risk is to eliminate by design. Once the design is established, the next approach is to change people's behaviour, and then to devise safe working procedures during the production process. Personal protection is considered the least desirable option. Although personal protection is still a major contributor to reducing accidents and long-term health problems in construction, design should aim to reduce the individual's reliance on personal clothing and equipment for his/her ultimate safety.

The specification of construction materials and details can help to reduce this risk. Avoiding the specification of materials with known health risks from fibres, solvents, etc. can make a direct contribution to health and safety. Where the risk is very high, as was the case with asbestos, substitutes will need to be found. With materials such as cement the risks are identifiable, but it is still specified, since alternatives are not easy to identify and protective measures, which are relatively simple, can be taken.

Detailing can help to determine where individuals have to be standing when completing fixings and the size and weight of parts they will have to handle. Loads to be lifted should be limited and working platforms should be stable and allow posture to be maintained. Working at height, particularly with loose parts, increases the danger of falling, requiring more measures against accidents. Hazards exist when gaining access and egress from the working place, particularly if this includes carrying materials or components. Risks can be highest when operations are being carried out in close proximity, particularly if one is above the other. Other high-risk operations include working in confined spaces and lone working.

While this is not intended to be a comprehensive list of safety issues, the few examples given above are meant to indicate that risk evaluation comes from an analysis of the operations dictated by the choice of construction specification and details. Methods chosen have to incorporate safety measures to create a safe working procedure. Real contributions to health and safety can be achieved in the choice of construction solutions.

Production equipment

Most building operations are still completed by an individual working on materials or components by hand, now probably assisted by powered tools. The worker will, however, almost certainly be supported by a range of production equipment. The two major categories of this equipment are temporary works and plant and machinery. These have always been employed by constructors to make the work possible, but now also to make it safer, easier and to increase productivity. Many construction solutions are now only possible because machinery and support systems are available.

Typical examples of temporary works and plant and machinery are given in Table 13.1. This production equipment does not form part

Table 13.1 Production equipment.

Temporary works	Plant and machinery
Structural support • Partial erected construction • Forming • Earthwork support • Existing building support • Working platforms Site operations • Access and egress • Environmental enclosures • Security • Health and safety • Environmental protection	Materials handling Excavation and earthworks Concreting operations Manual operations Power plant

of the permanent construction. It is normally dismantled and removed once the permanent works can be used safely. It does not normally appear on any drawings or in any specification for the construction itself. It is, however, implied in the choice of construction and detailing. Changing the construction changes the need for temporary works and plant and machinery.

In most cases the economy of these production equipment systems lies in their ability to be used a number of times in different locations, often assisting in the construction of a number of different buildings. This means that any analysis of their useful life includes an analysis of the ease with which they can be dismantled and re-erected into similar, but often not identical, forms. This process of removal and reforming becomes a major cost in the life of many items of equipment. This may be as straightforward as transporting an item of plant between sites or the complete dismantling and re-erection including retesting of falsework systems or a tower crane. This requirement of reuse has a profound effect on the approach that would be taken in their design, both in the standardisation of components of systems and, particularly, in the methods of jointing the components. Joints will be designed to be quick in both erection and dismantling. This can be seen in many temporary support systems such as scaffolding. It is these characteristics that determine the economy of a proposed construction specification or detail.

Production equipment has to be reliable. This is a relatively straightforward idea of breakdown for plant and machinery, but more generally the association of risk becomes linked to safety rather than to loss of performance. In addition to aspects of durability, the use of production equipment means it is more prone to damage, misuse and unauthorised alteration. This potential for unsafe equipment creates the need for regular examination and even retesting at regular intervals.

While both temporary works and plant and equipment have many characteristics in common, they have their origins in two distinct functions associated with the production process. Temporary works can be identified as fulfilling a primary function in that without them labour by itself could not safely undertake the activities dictated by the construction solution chosen. Plant and machinery, however, have their origins in assisting in operations that were once carried out by hand. In this process of mechanisation, machines replace some labour (often associated with heavy, dirty or unsafe operations), although they normally generate new jobs, often of higher skill. Lifting equipment replaces large numbers of operatives on site with drivers, banksmen, mechanics, etc. Much of what is now built could not be achieved by hand, as solutions are adopted on the understanding that certain types of plant and machinery are available.

Increasingly, there are items of production equipment where the distinction between temporary works and plant and machinery is becoming blurred. The emergence of the MEWP (mobile elevated working platforms) is perhaps the most striking example. While clearly appearing to be part of the process of mechanisation (a vehicle able to be powered to the required location), what has been mechanised (holding labour safely while they complete the operations) would have been previously provided as temporary works.

The distinction does, however, hold good for a large proportion of production equipment and can therefore be considered separately, even if it could be argued that some equipment could be in either category.

Temporary works

Temporary works generally have either a structural support or a site operational role, including environmental enclosures and security, as indicated in Table 13.1. In all of these are elements of health and safety, and these will influence the design, detailing and timing of the introduction and removal of these works.

First, there are those systems that provide structural support until the permanent works can be used safely. Some forms of construction such as arches are constructed from components that only function as a whole when they are complete. This illustrates a fundamental question that has to be asked about any construction solution: 'Will the components be self-supporting as the assembly progresses?' If they are not, there will be a need for some temporary works to support the partially erected structure. In the case of arches this is known as centring.

If the components are made from a material such as concrete that has no initial form, it will need a mould. Concrete can be formed on site in the final position, where it is known as in situ. It can also be formed in a mould, often not even on the site, and then transported and fixed in its final position as a component; this is known as precast. For both in situ and precast a basic mould known as formwork will be required. If in situ, the mould will have to be held in the position required in the completed building. This support is known as falsework. Since the materials from which the moulds are made (timber, metal or even glass fibre) are very different from the material to be moulded, different trades may be involved. Thinking through the mould and support requirement for this type of construction at the time of design, and detailing for simplification and repetitions, can have a significant effect on the cost of the final construction. This has been seen in the design of in situ reinforced concrete slabs in the development from the beam and slab to flat slab designs, where the simplification of the formwork is the major source of cost saving.

Another major area of support that may be required is that of excavations. Because of the great variety of ground conditions and configurations of permanent works below ground, this is a major area of consideration. Where excavations are deep and/or near existing structures, earthwork support often involves a full engineering design. The need for earthwork support has implications for both cost and construction time. The choice is often left to the constructor, where considerations of sequence and difficulties of constructing permanent works around temporary structures have to be weighed against the risk of ground movements during construction. However, the development of solutions such as diaphragm walling and contiguous piling, allowing the permanent works to be installed without traditional earthwork support methods, has made a major contribution to the time, cost and safety of the construction of basements, particularly in city-centre developments. These are discussed in more detail in Chapter 28.

When works above ground are adjacent to an existing building, or part of an existing building has to be retained, or alterations are being carried out, support offered by the original building may be removed. While the permanent works will have to reinstate this support, temporary support may be needed during the production process. This is normally known as shoring.

The temporary works identified above have all been to support part of the construction or the ground until the permanent works can safely take the loads themselves. The final group of temporary support systems with a structural role is working platforms. These do not support the construction itself but support the labour and materials associated with the operations as they are carried out. They are required until the operations of assembly are complete. Working platforms are most often scaffolds and have to support the loads of the components and materials as well as the individuals themselves. Because of the direct use by individuals in order to carry out operations, health and safety becomes a major feature of the detailing of the scaffolding. It is important to consider not only the posture and ergonomics associated with carrying out the operations but also the possibilities of accidents.

Statistics show accidents from falling as a major source of injury and that most of these occur when work is being carried out at height, not only on temporary platforms but also from the permanent construction. Dangers from falling include both individuals themselves falling and materials or tools being used during the operations falling and injuring those below. This makes edge protection of particular significance. On temporary platforms such as scaffolds these have to be incorporated in the design as handrails and toe boards. If the permanent works form the working platform, specific temporary edge protection will be required.

Detailing can often dictate where the operative has to stand and what operations have to be completed on or near to the edge of the building. For example, fixings for cladding can sometimes be arranged to be on the top of the slab or on the face of the edge of the slab. Fixings can involve drilling or the use of cast in fixing channels. While these have implications for tolerances and achieving fit, they also determine the options associated with the operations to be carried out at the edge of the building.

Working platforms can be associated with mobile equipment (MEWP) more easily recognised as plant. These are of particular use when operations have to be carried out in a series of relatively isolated positions, at height, above a clear area, with an adequate surface for the equipment to move around safely.

Many temporary structures do not have an essentially structural role. They are more generally associated with the organisation of the overall site. The influence of the construction chosen is still very strong. Many systems will be associated with temporary access for the movement of materials, components and individuals to the place where the operations are to be carried out. Access (getting to) and egress (getting from) the place of work cover both horizontal and vertical movements. The type of construction will determine the range of loads and number of movements.

The analysis of these temporary works is at the heart of site organisation. Planning routes for the transfer of materials, areas of storage and methods of handling have to be considered with aspects of temporary access. The loads to be carried, the routes to be taken and the physical size of access requirements determine these routes. Cranes carrying materials through the air need only landing areas, while hoists require support from either permanent works or temporary scaffold structures. Vehicular access not only will need roads as well as hard standing for unloading but also may need environmental control equipment to limit the spread of mud. Separating out pedestrian access from vehicular routes contributes to safety.

The link here to plant and machinery is very strong. The examples above of cranes and hoists for materials handling indicate how levels of mechanisation determine the need for temporary access and egress structures.

While the idea of individuals having to move both to and from the working area is fairly straightforward, it may at first be assumed that components and materials only have to be taken to the place of assembly. All operations create some level of waste. This may come from preparing materials, shaping components, cutting to fit or just from the packaging in which the components were delivered. Some estimates of the amount of this waste and the form it will take have to be taken into account when planning disposal.

Temporary construction may have to be provided for environmental enclosure. Nearly all sites will require temporary office and storage buildings, the extent of which will be proportional to the value of the works. It is possible on some sites to prepare materials or even complete sub-assemblies in areas on site, but not in the final position in the building. These operations may be covered, creating workshops or small manufacturing areas. It may even be beneficial to provide cover for the actual assembly process itself if adverse weather is expected, although many assembly operations would be carried out under open site conditions.

Security works may be necessary at the boundary of the site not only to protect the works but also to limit access of the public to the potential dangers of site operations. Further levels of security may be required for storage compounds to protect materials and components from theft and vandalism. Some items, such as bottled gases, pose particular safety hazards and these may carry a legal requirement for special secure storage conditions.

Although the temporary works so far identified will achieve the majority of health and safety provision, some operations may need specific provision. Particular areas of danger such as edges of excavations, overhead power cables or holes in slabs will also need protection for the unwary.

Many site operations may threaten the environment. This may include the soil, air, watercourses or the local wildlife balance. While this is best avoided by devising procedures that do not affect the environment, it may be necessary to provide specialist disposal and/or temporary screening or settling arrangements.

It has been suggested that some temporary works are necessary as support structures and some because of the needs of the overall site operations. In all cases they are influenced by the choice of construction. Analysing how a proposed solution has to be manufactured and assembled includes identifying the temporary works that the suggestion will require.

Plant and machinery

The distinction has already been made that, while temporary works are fundamentally necessary to assemble buildings, plant and machinery have their origins in the mechanisation of manual operations. However, many buildings could not now be built without machines as the choice of construction is made in the knowledge that the plant and machinery exist. An understanding of the levels of mechanisation that are available has to be part of the analysis of the proposed solutions.

The most successful set of manual operations to be mechanised is materials handling. The size and weight of loads that can be lifted and carried have a direct influence on both the size of components and the potential size of building as a whole. The range and scope of materials handling plant are very wide. They vary not only in the size of loads they can carry but also in how versatile many have become, being able to move a variety of materials and components around the site and into position.

Perhaps the most obvious distinction is that between horizontal and vertical movement. Dumpers, trucks and lorries move horizontally; hoists and winches vertically. Cranes, forklifts and pumps and even helicopters can achieve both horizontal and vertical movement. All of this plant has an operational envelope defining the limits of movement in the horizontal and vertical planes that has to be matched to the site access and the form of construction chosen.

Excavation and earth moving is a specific site activity that has developed a range of plant for specialised materials handling. The action of digging involves additional forces over and above the action of picking and carrying the load. Machines range from the versatile slew machines with a range of buckets to specialist trenching machines. Pumps are used to mechanise the manual operation of bailing out flooded excavations. There is a range of lorry and dumper transport options, again dependent on the size of operations.

Another site activity that involves a range of operations that have been mechanised is concreting. While most of the mechanisation is materials handling, other operations are associated with preparation and final assembly. For all but the smallest amounts of concrete mixing will be mechanised, ranging from small hand-fed mixers to fully automated batching plant. Moving the mixed concrete to the formwork can use a variety of materials handling plant, depending on site conditions and the stage of construction. Once in the formwork the compacting and even the spreading and levelling operations may be mechanised. One concreting activity that has become even more mechanised is the laying of large areas of concrete ground-supported floors. Not only is the plant highly specialised in placing, spreading, levelling and finishing, but also it incorporates some elements of automation.

Many operations remain essentially hand operations but are assisted by hand and even bench tools. Cutting, drilling, screwing and spraying are just some of the hand operations that are now commonly undertaken with power tools. Many of the smaller versions of these tools can be battery operated, but for larger, more powerful applications a power supply has to be available. On-site electricity and compressed air are the two main sources of power supply for small tools. This may lead to the use of generators or compressors as on-site power plants.

Most plant is either hand-held (some simple machines such as winches are even hand-operated) or controlled by a driver. There have been some attempts to automate the process with driverless plant using computer control. While this has found limited application on site, it is increasingly used in factory production. The use of computer-controlled technology in manufacture is leading to systems that link electronically produced design information with the manufacturing facility to automate the fabrication process.

Understanding the process and possibilities of mechanisation identifies the relationship of the construction solution to the production process. Developments and availability of plant and equipment have an effect on the possibilities for the development of the construction solution. In some cases the two may have to be developed together. In early civil engineering works the engineer would have designed machines for production as well as the civil engineering works themselves. Most current common construction is only possible because of the availability of site equipment and manufacturing techniques. Most common approaches to the delivery of components and materials are only possible because of the availability of materials handling plant. This again demonstrates why production analysis becomes a necessary part of the evaluation of a suggested construction solution.

Production options

The production process can be characterised as taking processed materials, forming them into components and assembling them to form buildings. This implies a sequence of operations each with an associated method that defines the resources required to complete the work. This series of activities associated with processing, forming and assembling is greatly influenced by the detailing and specification chosen to realise the design.

To carry out an analysis of this production process the construction has to be visualised in pieces. It is necessary to be able to envisage each part at each stage of its manufacture and assembly. Visualising the components and the building partially formed at each stage of production and the operations to be performed to bring it closer to the finished building is as important as being able to see the building completed and responding to operational conditions.

Very early in the design stages of a building the decisions that are taken influence the nature of production activities, not only the chain of events necessary to produce the building but also where these events may take place. There are broadly three places where the actual work of processing, forming and assembly can take place:

- In position
- On site
- Off site

It is clear that the final stage of assembly must take place in position but there is a question as to how much of the processing, forming and assembly can be completed either adjacent to the emerging building on site or even off site under factory conditions.

This choice of where operations take place is often closely related to the choice of materials and the trades and processes associated with their manufacturing and assembly. For a traditional material such as timber the skills and know-how have been developed to work the material both on site and in the workshop. Initially this may have been associated with carpentry and joinery, but modern manufacturing technologies have allowed development not only in the scale of forming and assembly in the workshop but also in the processing of timber to form boards and laminates that have greatly expanded the possibilities to employ timber in a wider range of building applications and design concepts.

The possibilities to employ steel in building came with the industrial scale of processing, and this remains true today. The forming of steel is also an industrial and workshop activity with most steel items being delivered to site in component form ready for assembly. While it is possible to see some forming, cutting and drilling take place on site, with wire, bar or sheet materials this is less appropriate with heavier sections. The need for such site operations may well warrant establishing a temporary workshop to complete the operations.

Concrete as a material that needs forming in a mould has developed in two major forms: in situ and precast. The distinction is one of place of forming. For in situ work the moulds, known as formwork, are supported in final, assembly position by falsework, so forming and assembly take place in one activity. In precast the concrete is formed in moulds (again known as formwork) normally at ground level that then requires the cured component to be transported and assembled. The site process now focuses on the jointing and fixings, where tolerances have to be accommodated and fit achieved. Where moulding should take place will be dependent on factors such as size of component, technical aspects of the materials and specification, number and type of components involved as well as lifting and transport implications. Moulding may be completed on site to minimise transport but will need some workshop type of operation to be established in, for example, tilt-up construction (see Chapter 27). It may be complete in a factory dedicated to the production of these components and then transported to the site.

These manufacturing and assembly options are determined very early in the design process, not only in the choice of materials but also in the detailing. Starting with the choice of dimensions and the repetitive nature of components, this can be extended to the detailing of joints and fixings to determine the boundaries of assemblies, opening up options for prefabrication.

Prefabrication is the assembly of a number of components, possibly of different materials, in factory or workshop conditions to be lifted into position at a later date. Again, this can be achieved in on-site temporary assembly shops best placed close enough for direct lifting into position but can be produced in factories away from the site and transported in their preassembled condition.

This simple analysis of materials and their traditions in processing, forming and assembly and the discussion of where each of these three key processes takes place give a picture of three broad types of production options:

- In situ
- Traditional (incorporating factory-produced components)
- Prefabrication and system building

While there are distinctive features of each of these options, it is important to be aware that they often exist on a continuum where it is difficult to identify the move from one approach to the next. It is also possible for different elements of the building to use different approaches. If this is proposed, the inherent deviations and tolerances in each of these approaches need to be resolved in the junctions between different approaches.

In situ processes take essentially formless materials such as soil or concrete and create components in position. The whole process of forming size and shape takes place in position. Only the preparation of the material takes place prior to the forming of the components. The skill and know-how associated with the temporary works often plays a major part in the success of this type of production as this determines not only shape but also dimensions including achieving tolerances and fit.

Traditional construction is based on operatives who work on site where the individual has the skill, often in working with one material, to shape and/or fit the components and then fix and joint the components into position. Originally the skills of these operatives would include the ability to carry out many of the shaping and forming processes for the components on site. Increasingly, however, the components are made off site and, while traditional trades are still employed, the nature of their work on site is to fix and joint the components. Components are not normally designed to be cut on site. In fact any need to cut and fit the components would be considered a failure of control of tolerances and fit. Often these components may be built in, gaining fit by cutting the surrounding, more traditional construction. Typical of this change would be the work of carpenters with the use of gang-nailed roof trusses replacing the timbers of the cut roof. However, it is clear that with more components being used there is less opportunity to build in, so fit has to be accommodated at the joint or some cutting to fit designed into the component.

The use of components in traditional construction has led to the development of what are often seen as specialist systems, normally erected by operatives who do not have the wide range of versatile skills of the traditional trades. Specialist operatives have skills associated with a small range of standard sections and components. There is often still a need to cut and fit some of the standard sections but these are often then connected or supplemented with a range of components. This can be seen in many of the internal finishing systems such as metal stud partitioning (dry lining) or suspended ceilings.

The development of manufactured components has opened up the opportunity to design significant parts or even whole buildings from standardised components. Often known as system building, components are manufactured in a limited range of sizes to a dimensional framework with standardised jointing and fixings for site assembly. In these systems, cutting and fitting on site will be minimised in the design and should be eliminated for the pre-sized components on site. Fit now has to be accommodated in the jointing, often with adjustments being designed into the fixings. This represents a change from the traditional site processes even if they do include the use of many components.

Many components delivered to site will already be made up of components assembled at the factory. Many parts may be fixed together before being delivered to site, for example the door hung in the frame with lock or latch fitted in door sets. 'Prefabrication' is the term given to an assembly of components that are put together in a factory or assembled on site prior to being lifted into position in the building. The amount of pre-assembly involved can vary. The level of prefabrication can range from the production of a basic structure onto which enclosure, finishes and services can be attached on site to the prefabrication of whole sections of a building complete with cladding and services to be lifted into position on site. These options are discussed in more detail below

While the possibility of on-site components manufacture and prefabrication still exists, the current levels of infrastructure both in manufacturing capacity and transport means that prefabrication can now be achieved in factory conditions remote from the site. Initially components, systems and prefabricates had to standardise shape and dimensions for production efficiency and to lower cost, but increasingly flexible manufacturing methods mean that even bespoke sizes and shapes can be produced at little extra cost for each project. An example of this is the production of timber wall frame panels for single one-off house designs.

Time, cost and quality

The need to consider a particular production option for a project can be associated with resource availability, often either materials or skill shortage, to achieve construction against one of the following criteria.

- Time
- Cost
- Quality

It is a mistake to think that there is direct relationship between some of these criteria and the production options. The production options for any one project will depend on the context and prevailing resource base, including the design skills currently available. It may be better to adopt different basic production approaches for different parts of the building. However, as pointed out above, this needs to be done with care to match deviations both induced and inherent, to ensure manufacturing fit and accommodate long-term movement respectively. It puts a specific focus on the interfaces or junctions: the actual joints and fixings between elements of the construction with different production approaches.

Time may be a consideration for an individual project but it may also be a concern for construction of a type of building in society such as housing where the capacity of the industry as a whole is in question. When demand for housing is outstripping supply because of the industry's lack of capacity to produce in a traditional method due to limited resources, prefabrication may be considered. This was the situation in the UK following the Second World War that, combined with an excess of factory capacity released from the war effort, led to the prefabricated systems such as the Woolaway and Airey precast concrete houses being developed at that time.

When considering time, it is important not to judge the speed of a project based on the time on site. This may lead to the impression that prefabrication is the choice for speed. The time of a project for the client (who may have to pay a premium for speed) is from commissioning the design to hand over for occupation. This includes all the design time, the pre-production lead times and the factory production time, as well as the site operations. If a system, fully thought through with all the components, joints and fixings standardised ready for ordering into a variety of building configurations is available, some of this time can be reduced. This assumes that the scale and context of the building is suited to the pre-designed system. If the design is to be specified and detailed for bespoke prefabrication, the design, procurement and manufacturing time needs to be anticipated as well as the time on site. For many traditional designs the materials and expertise are widely available and can often be procured and mobilised relatively quickly, even though the time on site may be longer but still giving the client an earliest occupation date.

Cost is more than the inherent labour and materials in specification and detailing. Traditional building needs relatively little capital investment. In the past the development of skills through apprenticeships and training needed investment as did the owning of plant and machinery. With the development of components, and particularly component systems, what may still be seen as traditional trade-based solutions reduce the need for investment in training, but do not eliminate it. With the widespread availability of components, plant and machinery, traditional production method costs are mainly incurred during the production process itself. There are, however, still implications for training, knowledge and skills for both operative and management in all production options that are no less significant in traditional construction than in any other.

For the production of many components and prefabricated units there is a capital investment in factory and manufacturing capacity. This achieves economy only if steady demand exists to maintain manufacture at a sufficiently high level for the capital to be recouped in the cost of each item sold. While this can be achieved in many components such as lintels for cavity brickwork, which are specified in a range of types and scale of building, it can make more bespoke and complex prefabricates expensive. The need for a demand for a limited range of standardised components is less true in the more flexible manufacturing plants but the machinery to give this flexibility, often based on computer numeric control (CNC), is expensive but can accommodate changes in the detailing and specification, particularly of dimensions, very easily. This gives the opportunity to provide one-off designs for individual clients.

Capital costs can be reduced if the prefabricates are detailed as traditional construction. This may allow a flow line approach with traditional trades working in a workshop environment with inherent improvements in working conditions, including health and safety, with the potential for reduced time and increased quality.

Quality is the third reason to consider carefully the production options, as quality, like time, can be a justification for increased cost if it is of value to the client. All buildings need to be formed and assembled with care and levels of skill and knowledge and quality of materials appropriate to the design. Each production option, with the material chosen, has the capacity for quality both in finish and dimensional accuracy, important in realising the design concept. There will be associated with every design a level of dimensional and workmanship quality that if not achieved will not realise the design intention. Vernacular or modernist in design, there is an implied quality that, if not delivered in the production process, will confound the design in its execution. It becomes part of the analysis of a technical solution to check if the production process implied in the specification and detailing can deliver the quality required.

Sustainability and environment

Part of the sustainability analysis introduced in Chapter 15 will be associated with production options. As the buildings themselves produce less carbon in operation the CO_2 release associated with the materials in embodied energy and carbon becomes more significant, and this includes the production process. Embodied carbon in materials is mainly associated with processing the material and then in the forming of the components and the assembly process. In the full cradle-to-grave analysis all these stages have to be taken into account, including dismantling and disposal options. Figures for embodied energy and carbon materials are normally only quoted from cradle-to-gate normally meaning the site gate. Increasingly energy and carbon figures are associated with materials formed into component to help to refine the analysis, but carbon released is also a concern of transport and assembly processes. This leads to notions of local supply of materials and components but this has to be extended to the full journey of the materials and components if prefabrication even from local suppliers is to be considered. In situ and the other trade-based processes require the delivery of materials and components to site and then distribution around the site. On-site (or even local) fabrication may prove effective in limiting carbon. Off-site fabrication process may require more transport and more machinery but may mean less waste.

Waste is another key issue for the environment. In addition to concerns for embodied energy or carbon material, processing will produce waste not only as solids but also often as liquid and gases that can pollute water and the air. This type of waste is normally less in the forming and assembly phases and more associated with materials processing. However, many factory and site processes produce waste, and without both design and management controls the waste may not be reduced or recycled. It is a challenge for site operations to reduce waste and to have procedures in place to separate and stream waste disposal to reduce landfill and promote recycling.

Demolition operations have a specific waste problem which has led to the development of many technologies and procedures to deal with waste streams. Brick and concrete can be crushed on site to produce fills, while steel can be collected for recycling. Timber can be selected for reuse or sent to be processed into timber-based composites or for fuel. It can now be a conscious design decision to use materials that can be reused or recycled at the end of the building's life. This can be further aided by considering dismantling in the detailing and choice of fixings. Ease of disassembly can greatly improve the chance of reuse of a component at the end of the building's life.

Many factory operations can manage waste successfully as they have a predictable supply of waste that can systematically be reduced or put to another purpose. In the production of timber components waste can be processed into boards or larger structural sections, processed for insulation or even burnt to provide energy and heat for the factory.

Both the choice of materials and the production process have implication for energy, carbon, waste, transport and the opportunity to procure from sustainable sources. To these can be added issues of water used in the processing of materials and fair trade in the overall sustainability argument. Both the specification and procurement processes can help in responsible sourcing to meet sustainability objectives.

This means that there becomes a need to evaluate the materials and components not just one with another but from alternative sources of supply. Rating and certification schemes can help in this evaluation but the schemes vary in their complexity, the factors that they take into account and the amount of explicit information they give. Many may be useful, for example the scheme to certify that timber has been obtained from sustainable or renewable sources. However, for a major supply contract involving a significant proportion of the works or where particular pollution, transport or scarce or over-exploited resources are involved, individual evaluations based on the above analysis may be necessary.

Production options – design decisions

The possibilities for production options are significantly affected by the choice of technical solutions at the design stage. Each of the technical choices made at the design stage comes with a set of production options. The production options are often very limited, being closely associated with the development of the technical solution itself.

Well-established (robust) detailing implies a set of resources and methods that have been developed to make those details work. These are normally associated with in situ or on-site trade operations and will achieve required performance levels only as long as they are produced in an appropriate way, often a way that has been established over time along with the refining of the details themselves. New, untried detailing may imply a change in the production process and any desired change in the production process will imply a re-evaluation of the detailing.

The need to consider new technical options to achieve new performance requirements based on the demands of society and of clients is constantly changing. This is in addition to the pressure from the industry to achieve economies and efficiencies in production and this can often best be achieved by modifying details and specification.

All designs have to ensure that an appropriate production process is viable to realise the building. Some may even include exploiting a particular production option to express a design concept. It is necessary to decide on a basic production approach to be integrated into the choice of the technical solutions and even the design concept itself. This type of analysis is particularly significant if prefabrication is seen as a way forward in realising a particular design.

Prefabrication and system building

Prefabrication and system building both look to produce factory-made items that need no cutting or fitting on site and need only to be made in a limited number of coordinated sizes to control costs. Therefore, the first decision to be made for systems building and/or prefabrication is the identification of a dimensional framework. Dimensional frameworks and repetition bring economies and efficiencies to site-based trades. However, in prefabrication a dimensional framework is necessary to limit the number of pieces to be manufactured yet give maximum freedom to the design of space for the efficient use of the building. Often identified as a module size, it is chosen to determine the size of prefabricates to give maximum flexibility in design. The most used module is 300 mm leading to larger prefabricated units being based on 600 mm, 1200 mm or even larger units. This not only gives economic division of internal space but has to provide units that are easy to handle. As indicated above the introduction of flexible manufacturing methods has reduced the need for this standardisation but some dimensional discipline will lead to economies in the detailing and production process.

In addition to the choice of a dimensional framework, decisions have to be taken on the extent of prefabrication for off- and on-site processes. High levels of prefabrication involving finishes and services off site reduce the time on site and limit the skill base required to erect the building. However, it increases the chance of damage and can make joints more problematic as they may become visible between prefabricated sections. While it is a principle of prefabrication that cutting and fitting on site should not be necessary, the greater the level of assembly achieved off site makes this even more imperative as any modification will almost certainly affect appearance as well as potentially decrease performance.

The simplest level of prefabrication provides only a basic structure on which the other elements can be fixed. While this can be the loadbearing elements of the building carrying the dead and imposed loads, it can be a sub-system of panels or pods that will carry the enclosure elements such as insulation and finishes. For example, precast concrete panels can be employed to form a loadbearing structure or cast as cladding panels to stand or hang on a building frame. Timber- (or steel-) framed panels can also be used as loadbearing or infill construction. These framed panels are normally delivered to site as empty panels with boarding on one side ready for site processes of insulation and services carcassing prior to boarding the internal surface and cladding the outside, all as site-based trades. However, panels may be delivered with both sides boarded ready for filling with blown insulation ready for finishes to be applied on site.

Another approach is to develop the system of building components made off site and delivered to be assembled into the building but with standardised joint and fixings. In many of these systems the components can only be used with other components in the system. These can vary in scope from single-element systems such as building frames to systems that create whole buildings. The dimensional ranges of the components and the detailing of the joints and fixings become particular to the system. These may be developed and procured through commercial companies or through consortia of clients who have demands for similar buildings. Where buildings systems are developed, they may be limited to certain phases of the building, for example to watertight providing a dry envelope, or may include all components including internal partitions and fittings. An example of the consortia approach is Scape System Build Ltd. set up as a trading company for the Consortium of Local Authorities Special Programme (CLASP) to provide for their school-building programmes.

Rather than try to develop a system that can be applied to many buildings it is possible to design a bespoke system of components for a single project. This system approach combines the off-site manufacture of components with on-site assembly. The on-site assembly is now not dependent on traditional trade processes, as it should eliminate cutting and shaping and provide simplified jointing that, while it may be carried out by operatives with a traditional trade, becomes an assembly process. This will limit the number of site operations, reduce the range of tools required and give repetition that, when learnt by site operatives, should speed up site production. This speed on site can also be claimed by the commercial and consortia systems. However, they can benefit from the design and detailing investment over many buildings as well as economies from component manufacturers that have some expectation of repeating orders from a number of projects. The bespoke system may have an increased design time and require one-off negotiations with manufacturers, but this may be deemed worthwhile for the nature of the building or the design and build quality.

Another level of prefabrication is to complete a significant amount of the assembly of enclosure, finishes or services components on the basic structural unit in a factory before delivery to site. With more complete units being delivered to site, erection time can be minimised, particularly if joint design is focused on achieving fit and simplifying site operations. It may put more emphasis on tolerances and fit and leaves the construction more vulnerable to damage.

As these issues become more of a concern it leads to the notion of volumetric prefabrication. This involves replacing panels and flat pack deliveries with whole rooms or sections of building, including floors and walls as pods. While the outside of the pod may need to be clad later, it leaves the inside less vulnerable and therefore finishes, services and even fixtures and fittings can be completed in the factory within the pod.

Potential for prefabrication can be identified if the building programme is examined to identify either part of the building where early completion is helpful or lengthy activities on the critical path, particularly around congested areas, take time to complete. Two such areas are staircases and highly serviced areas such as bathrooms and kitchens. These areas do, of course, have to be repeated a number of times to make the setting up of factory prefabrication worthwhile. This has made hotels and student residences particularly amenable to this level of prefabrication.

In the past, systems and prefabrication have been associated with building types such as houses, schools and hotels at a time when demand for these types of building has been high. It has also been successfully applied to high-quality, one-off development. It has also proved helpful to consider elements or parts of the building to be prefabricated to speed up overall project times. It is necessary to evaluate each building in its design and resource context to determine the merits of a particular production option. However, it is also necessary to be clear which production option is being implied by the design to ensure coherence in the detailing. The junction where different production options meet needs to be designed with particular care.

Knowledge and expertise for the analysis of the process

Carrying out this analysis of manufacturing and assembly requires knowledge and expertise gained from practice.

At the most fundamental level this knowledge is derived from craft and operative practice with modern methods, now including considerable use of plant and machinery from hand tools to high levels of automation. The basis of the understanding of the ways in which basic operations can be achieved comes from the know-how of those who work the materials and components and complete the assembly tasks. These are the people who know what can be achieved within the deviations demanded by the design and the time this will take. They will also have a view on temporary works and safe working procedures.

With modern specifications of materials this know-how has to be supported by technical knowledge. When new materials are introduced, the key aspects of the operations that control their properties, and therefore performance, may not be immediately apparent from previous know-how. This has to be worked out through the technical knowledge, although with time this may become part of common practice. In some cases the technical support becomes part of common practice. An example of this is the determination of concrete strength to allow the decision to remove the temporary works.

This is then overlaid with knowledge and expertise gained by those who organise and manage the production process. They have the ability to take an overview of the whole production process. They have to see all the stages or activities that have to be carried out. Management will gain a view on sequence, temporary works, factory or site organisation and safe working practice. They have a perspective concerning overall productivity, cost and time. They then need to be informed by, and understand, the operative and technical knowledge to ensure it is integrated into the plans and programmes that they provide.

In the same way that the choice of the construction is guided by an overall design but has to be solved in the choice of specification and details, so production has to be guided by overall programmes but has to be solved in choice of methods and resources. These require a different knowledge base and approach to the analysis, but it does not mean that these two creative acts should be carried out independently. Overall productivity is governed as much by the design as the efficiency of the production operations. Overall performance is governed as much by methods of production as by the materials and details of construction.

Summary

1. The analysis of the production aspects of manufacture and assembly is necessary as part of the evaluation of a suggested form of construction to ensure that the potential performance can be realised in the final working building.
2. The analysis is based on visualising the building in pieces and at the various stages of assembly in its partially erected condition.
3. How many stages it is necessary to divide the building process into will depend on the novelty of the construction and in particular the perceived technical risks. For productivity, repetitive operations may also be seen as worthy of a high level of analysis.
4. Once stages are determined, the analysis continues through an identification of the operations required to prepare and assemble the works and the choice of methods to be used, which determine the resources necessary.
5. The primary resources for construction work remain materials and labour supported by temporary works and appropriate levels of mechanisation.
6. Because labour remains a key resource in production, health and safety is a major consideration in choosing methods and resources, both of which can be influenced by the choice of construction materials and details.
7. Performance, quality, cost and time are determined by the sequence and methods; therefore the analysis of production must be an integral part of the choice of the construction materials and details.
8. Broad production options characterised as in situ, traditional and prefabrication are determined by the design specification and detailing.
9. The knowledge to carry out an analysis comes primarily from the know-how developed from practice. However, this is increasingly being supplemented by both technical and operational expertise.

14 Cost

This chapter explores the issues concerning the analysis of the cost of a proposed construction solution. The ideas of cost–price chains and the need to identify the costs in a clear time frame within an economic and fiscal context demonstrate the necessity to be clear about whose cost it is and what the cost should include. There is no single definition of the cost of a solution. The most usual interpretation of the cost is the actual construction cost to the client. However, this may be misleading, and design, running and even disposal cost may be a better basis for choice. This depends on to whom the cost applies and the value he/she puts on current and future expenditure.

Cost and value

It is clear that the cost of a solution is important if for no other reason than that ultimately a solution which is outside the economic reach of the client will not be built. It is also true that clients are unlikely to want to spend more than they need to obtain what they want. Two solutions of similar performance are likely to be chosen on price. What clients want and what they can afford may be two very different things, which may conflict. This will have to be resolved in the evaluation of a proposed solution. All of these need a clear basis for the analysis of costs.

What clients want, and the proportion of their money they are prepared to spend on it, is related to the worth or value they place on the service they get from the building. This sense of value or worth that clients are likely to place on different features of the building has to be established at the briefing stage, along with the more straightforward idea of the budget that will be available. Judgements on value will be made in different ways and these will influence the way in which cost should be analysed.

Perhaps the simplest notion of value comes from the provision of what may be termed the basic performances, such as weatherproofing and fire performance. Here the value is in a level of performance that the client would probably suggest should be achieved at least cost. However, clients may be prepared to pay a great deal for the appearance of the building, for its value may be in the symbol or image it represents of the individual or the corporation. This is evident in the premium that clients are prepared to pay for tall buildings located in city centres. Other clients may value the quiet that can be achieved with acoustic measures (assuming location has been fixed) or the quality of light from additional money spent on lighting. These judgements on value are predominantly issues of design that need to be translated into solutions, and here costs will be included which will have to be borne by the client. These design issues are often difficult to justify in cost terms as they can, in basic performance terms, be seen as useless. However, useless should not be confused with worthless, as it is in its value to the purchaser that the cost is justified.

There is another notion of value: one associated with paying more now for lower cost in the future. This is the argument for the value of thermal insulation, with its associated payback periods of lower heating bills. However, some clients may not be prepared to pay for insulation over and above the statutory requirements, for they may wish to invest their current resources in other assets and be willing to fund future costs from future income.

Cost and price

The cost incurred in adopting a particular solution for the construction of a building is computed from the price paid for the various resources employed for the goods and services in order to complete the work. Costs are created from a chain of prices charged which create the costs of each stage in the process. Hence the cost to the client for the construction of a wall is not the cost to the constructor. It is the price the client agrees to pay the constructor. The cost to the constructor is the price the constructor agrees to pay suppliers of goods and services needed to complete the works.

If, therefore, it is required to analyse the cost of a solution, the question that has to be asked first is 'To whom?' The usual answer to this would be the client, for it is the client that has the resources to invest, and the expectation of a return on the investment.

The scope of much analysis of cost that is carried out for the client is limited to the cost for the construction itself, the price paid to the constructor for the final assembly. This is often represented by the tender price. In the professional and industrial context there is considerable thought put into this concept of cost, for its commercial significance is high and hence data are available for its analysis. It is, therefore, tempting to use this as the basis on which a suggested solution for the construction should be evaluated for cost. This is, however, the client's cost, representing the price paid for the services from the constructor, and any evaluation will be based on historical tender data. This may not be the most appropriate basis on which to analyse the cost of a solution.

The two major limitations in applying this basis of analysis are, first, that the cost data are historical, making interpretation of the figures for new or novel solutions difficult, and, second, it does not take into account the design or running costs, let alone the disposal and redevelopment implications.

Cost and time frames

The construction cost forms only part of the outlay the client will have to make. Even staying within the same time frame of cost incurred after land acquisition and prior to occupation of the building, there are costs associated with designing the works. Generally, where construction solutions are well established, the design costs will be lower than for novel designs. If experience and the gathering of existing knowledge still prove inadequate to determine the performance with an appropriate level of confidence, research and development costs may be incurred, although these are rare in construction generally.

The time frame of land acquisition to occupation also represents for the client a period of expenditure prior to gaining an income from the building. Designers and constructors will require payment in stages as the design and construction work proceed. The longer it takes to bring a building into occupation, the higher this cost. If the time between initial investment in the land and occupation can be influenced by the selection of a construction, it could be argued that this is a cost associated with that choice. Hence it could be seen as part of the analysis of its cost.

Different solutions have different costs after occupation of the building. Deterioration of materials suggested for alternative solutions will take place at different rates, requiring different renewal cycles to maintenance requirements. The reliability of components will vary, leading to different replacement costs. The detailing will determine the ease with which renewal and replacement can be achieved. This not only affects the direct cost of maintenance but also may carry a cost of disruption to the occupants while the work is carried out.

The overall building will have associated running costs directly affected by the construction choice. The example of the inclusion of insulation has already been mentioned and represents a part of the energy-use cost of a building. Another example is associated with the choice of surfaces, where the cleaning of a commercial building over its life may be many times that of the original construction costs. If cleaning could be achieved with less staff, albeit that the initial cost of the surface was more expensive, the overall cost might be less.

Finally, different solutions will have different disposal costs. This is not just the cost of demolition but also, increasingly, the cost of safe transport and deposit of waste from the demolition of the building. However, some materials that are waste to one project may have value to another. Recycling of architectural salvage has long had a commercial value, but increasingly building waste can also provide an income when the building is to be demolished.

There are cost implications associated with the land that has to be purchased by the client. The availability of certain types of construction solutions will bring land into economic use. This is particularly true in the case of foundation and geotechnical ground improvement processes that can make poor land economic or allow usable space below ground. This can also be the case if lightweight superstructure solutions can lower the cost of foundation solutions.

Increasingly the development of brownfield sites where previous development has left land contaminated is becoming more viable as technical solutions to cleaning or capping sites are developed. This is also true if projected to the end of the life of the building in order to consider the redevelopment costs. While it is unlikely that new development will be allowed to contaminate land in the way past users have, certain foundation solutions such as piles may be difficult to remove. This will make the value of the land less for redevelopment at the end of the life of the building.

The analysis of the cost of a solution has to be qualified by the question of not only whose costs are to be analysed but also over what time frame does this individual want the costs evaluated. It is unlikely that the full spectrum will be taken into account, as the client has to balance the use of current capital against future income. However, increasingly a fuller cost evaluation, often defined as lifecycle costing, is being undertaken and is affecting the choice of the construction.

Cost and cost data

It has already been identified that while data concerning construction costs are readily available, they are predominantly based on historical prices for which constructors were willing to undertake the work once it was fully specified. The costs are based on measurements of parts of designed buildings and therefore rely on new work being similar to previous construction. There are reliable processes for estimating the cost of this measured work, but this is linked to construction costs of items measured in a standard way.

Extending the requirements of a cost analysis beyond that of construction costs into extended time frames has the difficulty of establishing reliable cost data. While design costs can be estimated, it may be difficult to extract costs for designing a specific part of the building. Feedback on the cost of running a building may be limited, particularly for costs expected to arise some decades into the future. The lives of buildings are long and ownership often not continuous, so gathering these data is difficult. Even if these data were available there would still be the need to project this into the future, where resource availability and economic conditions would be difficult to predict.

A more fundamental basis of analysis

Costs arise from the deployment of resources. The price that has to be paid for resources depends on economic (and fiscal) conditions.

For example, resources to be deployed for manufacture and assembly can be identified from an analysis of the operations required and the methods that could be used to complete those operations. This is the analysis undertaken to identify the buildability of a solution identified in Chapter 13. It involves taking materials and components and operating on them with appropriate labour and equipment in a safe manner. This will give not only the quantity of resources required but also the time over which the labour and equipment will be employed. To these direct operational costs one must add any management and technical support and a proportion of the organisation costs, for these are also deploying resources.

In a market economy, price is primarily influenced by supply and demand; with the tendency to surplus, price falls, whereas with scarcity, price rises. In a competitive market the bid price will be based on a judgement of what others may be asking for the equivalent goods or services. The fiscal conditions influencing costs are either direct taxation or inducements such as employment schemes or regional development grants.

The further into the future these predictions have to be made, the greater the chance of errors in the cost analysis. Running costs in the first few years can be identified with some certainty, but the long-term renewal and replacement costs are often some years or decades ahead. Alternative resources may become available (or existing resources become unavailable), which may make an analysis of the expected operations and methods inappropriate. Changes in demand and supply or in the presence of new alternatives increasing competition will distort price differentials. New taxation will change prices, as happened in the UK with the introduction of VAT on maintenance work as well as new-build. At the end of the building's life, demolition may be made very expensive with the discovery of new hazards, as happened with asbestos, or made cheaper by the introduction of recycling methods, as with concrete.

All this means that estimating costs in extended time frames can only be carried out with less confidence. The figures will be subject to possible errors and risks that need to be analysed before choices are made based on these extended time frames.

Building up the price

While this chapter is about evaluating cost, it has been established that costs are built up from a cost–price chain where economic (and fiscal) factors influence price. It is therefore necessary to understand the generation of prices.

Price can be fixed in one of two basic ways: what the market will stand or the application of rational margins. In practice there are likely to be aspects of both in the setting of a price, particularly in bidding in a competitive situation. In both cases, knowledge of the expected costs is important, plus some estimate of how certain those costs will be. This allows some judgement on margins. The margin is the excess of price over cost that an organisation needs to survive and grow. The need for margins and the role of profit is beyond the scope of this text. Even if rational margins are applied to the costs, there is still the question of what the market will stand. If this is below costs, it may not be sensible to bid.

While, as has been indicated, a full analysis of pricing is beyond the scope of this text, there is one more aspect that could be usefully highlighted at this stage. The relationship between cost and price should also reflect the risk. This may represent the technical risk of non-performance of the completed works and the subsequent costs this will incur. It may represent a financial risk associated with delay and non-performance of the contract. For these and other risks it may be possible to identify insurance that is at least a known cost that can be included. If, however, the risk is to be carried, some judgement on what price makes the risk worthwhile has to be made. Generally it should be possible to get a lower price if you are willing to take the risk. Who in the cost–price chain takes what risk is dependent on the contract entered into. Designs based on manufacturers' components and calculations transfer risk to the manufacturer. Developments such as serial tendering and partnering have the effect of limiting or even sharing the risk through shared understandings and information between the contracting parties. This is likely to affect the construction solutions seeking lower costs for all parties involved.

Summary

1. It is necessary to be clear what the client sees as value for money in establishing design and performance criteria, following which it is reasonable to seek economic solutions.
2. Cost is not a single figure but depends on the answer to whose cost it is and what time frame the cost is to include. Cost could include construction, design, running and disposal costs, which can be further refined by looking at investment cost and fiscal advantages.
3. In accounting for costs that will be incurred in the near future, price information may be available. Allowing for expenditure in the future requires knowledge of future technical developments and the economic and fiscal context and is therefore less easy to predict.
4. Price may be based on market forces and/or rational margins but will also reflect the risk, technical and financial, that is to be taken in designing, constructing and operating the chosen construction solution.

15 Sustainability – Social Concern

This final chapter in Part 1 considers sustainability as a new social concern requiring a technical response. Specific technical knowledge concerning sustainable development is given in previous chapters where it needs to be embedded in the normal process of choice. Sustainability is here set in the context of social demands requiring solutions outside the current knowledge base and the need to understand the care required in making choices until this knowledge is established.

Two major forces

There are two major determinants of the construction solutions that we adopt. One is the law of nature and the other is the social context within which we operate. These two great forces condition the technology we develop and apply. This is the context in which we build. We strive to understand the natural resources that provide our materials and the forces of nature that act on the final construction. This is the knowledge we used in the analysis of the physical behaviour. Our knowledge in this area is now well established and, while it will improve, it is unlikely to change fundamentally. We then need to understand the cultural, economic and political conditions that exist and that these will change with time. These shape the designs and determine the resources that are available for construction. They affect the commercial and ethical limits that govern the organisation of the industry. This influences the skills available and methods of operation that make solutions buildable and become the major determinant of the cost of construction. These changes may well require new technical knowledge with the need to return to our fundamental understanding of systems and systems behaviour.

Any extended discussion on the nature and impact of cultural beliefs and values and an ethical framework is beyond the scope of this text. It is most often so embedded in a society that it is covered in what would be commonly accepted to be a 'reasonable solution'. However, when society is questioning its own values and ethical framework it is inevitable that these will be reflected in the questioning that has to be undertaken about what has to be 'reasonably' achieved and what is a 'reasonable' solution. This leads to changes that have technical consequences. Technology has to respond to shifts in values to be accepted by society in its own continual process of evolution and change.

As those involved in construction are usually part of the society they serve or are at least culturally linked to that society, most of these influences are inherent in their decision-making. This was not true when, in the second half of the twentieth century, the developed countries took their technology to the developing countries with the best of intentions, yet errors were made. It gave rise to the term 'appropriate technology' to remind those engaged with other cultures of the significance of taking their social context into account when choosing a technical solution. We should always strive to provide appropriate technology.

At the turn of the twentieth century the new social concern for our impact on the natural environment has emerged and is having a profound effect on buildings and the way we build. It is making us question what is appropriate technology for the twenty-first century, which implies changes beyond our current knowledge. This concern for the environment has been integrated into the idea of our ability to provide for the future without any compromise to the quality of life we have now or to limit the hopes of future generations. This overarching concern for the future has been encapsulated in the concept of sustainability.

Sustainability

All generations have worked to improve the quality of life for their children. As societies develop this has been dependent on an economic process, the creation of wealth, and then in its distribution that determines social equity, providing the rewards of economic development to individuals. Linked to political beliefs associated with values and ethics and the best ways to stimulate economic growth and social equity, this has determined the demand for buildings and provided the means to produce them. In recent years we have made such progress in economic and social terms that development is now at an unprecedented scale.

This has brought the two great forces – the natural environment and society – together. Development is now changing the environment to such an extent that we have to face the possibility that we will not have the resources to maintain our current rate of development or, even worse, may make the environment so hostile that we will not be able to provide technological solutions to maintain and improve the quality of life. It is now widely believed that the choices available and the quality of life for the next generation will be harmed if we do not bring together concerns for the natural environment as well as our concerns to maintain stable economic and social conditions.

This movement to bring concerns for the natural environment into our analysis and make it part of our process of choice has been termed sustainable development, normally just called sustainability. It sees the stewardship of the environment as equally important as the management of the economy and ensuring equity among the members of society. This accounting for the cost of development in environmental, economic and equity terms is known as the triple bottom line (or the 3 *e*'s).

Sustainable development is not just about buildings, but buildings do have a significant impact on the environment as well as on the economy and within society and will therefore have to take the new concerns into account. In the same way as the choice of construction is mediated by economic and social conditions so it will need to take into account its impact on the environment.

Established influences and new concerns

Many of the economic and social influences seen as important in sustainability are well established and have been shaping our designs and the construction solution for many years. These would include the performance expectation of clients and the levels of infrastructure developed within a society for industrial activity. These are established but often changing and evolving. However, our concerns for the environment are new. These concerns are now evident in intellectual, political and social activity. These will need to be tested and given reality in the technological solutions to be put into practice.

This has generated the need to identify the important issues and look for knowledge and understanding sufficient to suggest 'safe' solutions. It may be that designers cannot analyse the behaviour, constructors fail to manage the critical aspects of quality or that the operatives have little experience or misinterpret the experience they have. The technological knowledge resides with all those involved. This text is prepared for the analysis and evaluation of choice and suggests that successful construction will require the existence of complementary knowledge and the know-how of all involved. It becomes part of the analysis of a solution to ensure that this complementary expertise exists before choice is made if performance is to be realised. This highlights the need for education, training, and research and development to be a part of the culture of both society and the industry and professions that contribute to construction.

It is often difficult to differentiate between the established and truly new issues. It could be argued, for example, that the most recent changes in health and safety requirements to include design as well as production do not represent the fundamental change in our view of the value of life but reflect our improving economic standards, at least part of which should be spent on greater safety standards. The more fundamental change in values concerning health and safety took place with the public health reform, which also had a profound effect on construction solutions, changing not only the detailing of buildings but also the site operational arrangements with provision for inspection and enforcement. It was part of the general movement towards equity to improve social conditions and ensure greater sharing of the new economic wealth being generated from the Industrial Revolution. Health and safety is now part of any analysis for buildability and the operation of a building. New requirements are absorbed and acted on within the broad base of knowledge and experience available to the designer, constructor and operator of the building.

Another example of a social concern that is having an influence on construction is inclusion and in particular the need to take into account the needs of people with disabilities in our development of the built environment. This is essentially just an extension of our understanding of designing for people and is unlikely to have the far-reaching effect on construction that sustainability will have, but it will have an impact on the way we build. While it is not always easy to modify existing buildings, all the indications are that in new buildings we should be able to include everybody by modifying the solutions we already employ. Perhaps the simplest example is wheelchair access at external doors. The need for level thresholds will change the relationship of internal floor levels and external ground levels unless ramps are used. By bringing the internal and external levels together, the potential for water to enter under storm conditions, particularly in flood areas, is increased. This will lead to new detailing around thresholds involving linear drainage gratings and possible flood defence measures designed into new buildings. There will be consequences for the levels of damp-proof courses (DPCs) and the detailing of ventilations to suspended ground floors, but these are not new requirements, just modifications of well-understood construction.

The essence of an emerging issue such as concerns for the environment is a lack of a knowledge base and experience. There will almost certainly be no direct technological understanding of these issues among those engaged in the design and realisation of the building. This lack of understanding will range across individuals who hold the complementary knowledge, the total technology. These represent all those involved, from researchers to craftsmen. Indeed, it is difficult to identify who may have the most relevant knowledge to both think through and put into practice the first attempts at solutions.

An example of this was the drive for energy conservation following the oil crisis of the early 1970s. It was essentially an economically driven crisis that meant that society as a whole recognised, or at least did not question, the need to act. Social acceptance for the principle meant that political action was ensured, which led to demands for a very rapid response from the technology. One response was to increase performance levels from insulation. In many ways the demand for this response was beyond the ability of the technology to develop and provide solutions that would maintain all of its performance expectations. While increased resistance to the passage of heat might have been achieved, the detailing and the site operations were not well enough understood to provide 'safe' solutions. With the increased insulation many cases of increased dampness were recorded, through either rain penetration or condensation, with the new materials and detailing. There was a breakdown in perhaps the most fundamental requirements of a dry environment. It is helpful for sustainability, where reducing energy use and carbon dioxide pollution play a major part, that these early difficulties with insulation are now well understood and we now have many 'safe' solutions to increasing insulation values.

The concept of sustainable development, as defined at the beginning of the twenty-first century, introduces environmental impact into decisions on construction detailing and specification. The analysis brings into question the origin of the materials, their manufacture, assembly and disposal, with the contribution to pollution and energy use as well as direct issues of resource depletion and changes in ecosystems. Sustainability requires the analysis of the operation of the building and its impact on the environment in pollution and energy terms over the life cycle of the building. It questions the source of energy, for if energy could be made available economically with little or no environmental impact then it might be used as part of sustainable development. Further, it demands this analysis be undertaken against global and regional as well as local impacts on the environment.

There becomes a need to act in conjunction with other agencies who are considering sustainability at other levels, such as settlement development, transport, etc. Sustainability also looks at the impacts over extended periods, with its underlying objective of leaving resources for following generations.

Responding to the sustainability agenda

If the concern for the environment has the status of a 'new concern', it is important that in providing a response we are asking the right questions. Sustainability is a very wide movement exercising the thoughts of many sectors of society. It introduces the idea that ensuring the quality of life for current and future generations must take into account the environmental as well as the economic and social systems. The focus of the new concern is environmental impact, and even the simplest analysis shows that building and buildings in occupation have a significant environmental impact. It is also clear that, while not directly associated with the technological solutions adopted for the buildings themselves, building development and urban design have implications for travel and journey times that will influence another sector having a high impact on the environment, that of transport. It is the nature of these problems, their interactions and interdependencies, that play a significant part in the analysis.

The concerns of this text are building and buildings in occupation; however, it is necessary to be mindful of the environmental challenges for all aspects of human activity so the contribution made by construction can be effective. The concerns include global warming and climate change, eco-systems and biodiversity, resources and waste management. There are two distinct issues that can be identified as common threads.

- Pollution
- Resource depletion

The first, pollution, has its impact associated with changes to the environment. Its effect may be direct in poisoning the air, water or land and thus influencing, among other things, eco-systems, biodiversity and the food chain, but it also has indirect effects. One of the major concerns is global warming. These changes happen naturally over geological time as constituents in the atmosphere change. Our burning of fossil fuels releases carbon dioxide that we now recognise as a greenhouse gas. When it accumulates in the atmosphere, it permits the short-wave radiation from the sun to heat the earth but limits the long-wave radiation from the warmed earth back into space, acting the way glass does in a greenhouse. The concern is that this will change our climate at a rate that the human population will find difficult to respond to. As this process of climate change will influence our capacity to maintain a sustainable quality of life we characterise CO_2 as pollution, although there are other greenhouse gasses such as HCFCs and nitrous oxide (NOX) which are produced in the operation of the building.

The second issue, resource depletion, affects the continued availability of the resources with which we construct and maintain our quality of life. Some resources such as timber and water are identified as renewable, but this still has to be managed as demand can still outstrip supply. They are part of natural cycles that, if not respected, will create conditions that will make supporting our aspirations for quality of life unobtainable. Some resources are acknowledged as finite (either physically or economically) and some of these could run out (or be priced out) if we continue to use them with the same rate of increase we have over the past decades. Reducing the availability of resources and maintaining a continuity of supply may have a significant impact on economic and social systems before they have irrevocable environmental impact, but this is all part of the concerns for sustainability.

Among the finite resources are fossil fuels, from which we currently obtain most of our energy. This makes energy a key focus for the design and construction of buildings. However, energy is not the problem; indeed, a continuing source of energy is vital to a sustainable future at even current levels of quality of life for a growing population. It is pollution (CO_2) and depletion (fossil fuels) that are the issues; energy saving, however, turns out to be a key strategy in achieving sustainability objectives.

It is important to recognise that all building has always had an impact on the environment and resource availability. In the past the impact has often been local and it has been within the capacity of the environment to recover with time. When we judge recovery, we are generally using the criterion that change does not significantly threaten human life or the quality of life both now and in the future. We are now facing the probability that recovery is not possible and that we may not have the technological capacity to deal with the range of conditions to maintain the well-being of the human population.

The task is therefore not to have no impact on the environment but to limit the impact to maintain convivial conditions. Opinion is divided on how close we are to irrevocable change, but there is now widespread agreement that the impact has moved from local and regional effects to global influences and that we must reduce the adverse impact we are currently having on the environment. The aim must be for low environmental impact construction.

It is now recognised that we must look for a convergent solution. For example, it is generally accepted that current levels of our use of energy and resources cannot be replaced by managed renewable sources in the timescale required. We need to reduce demand for energy while increasing the used of renewable sources; but, at least for the near future, we will still have some need for fossil fuels.

There is no agreement on by how much we must reduce pollution and resource use and we know that there are many pollutants and limited resources. There is currently a focus on energy use and energy generation to reduce both CO_2 production and slow down the depletion of oil and gas. It is one of the significant areas of analysis if buildings are going to reduce their impact on the environment, but it is not the only concern.

There is a need for consumers to rethink their behaviour. This is perhaps most evident in the exaltations to 'reduce, reuse, recycle'. As constructors of buildings we are consumers of materials as well as energy and other resources such as water and, like the users of buildings, we produce waste. Buildings have to be designed not only to have a low environmental impact in operation but should also aim to lower the environmental impact of the construction process itself. In fact, the two must be considered together. Concrete can be characterised as environmentally unfriendly because of the energy used to produce the cement. However, if the cement is sourced from factories using renewable energy, and specified with cement replacements and then it is used in the design for its thermal mass to save energy throughout the life of the building, it may provide a solution with the least overall environmental impact.

Limiting environmental impact

It is necessary to see the building as polluter and resource consumer as well as contributor to economic and social activity. Like the building's economic and social contribution, it is also necessary to see the building's environmental impact through its whole life cycle: construction, use and disposal (redevelopment).

The process of selecting construction solutions already takes many factors into account, from physical behaviour and appearance to production and cost considerations. Looking for low environmental impact is just one more factor in this process of choice. By considering the creation of the internal environment through heating, lighting and ventilation, low-energy solutions can be identified. Looking at processes such as water consumption and waste systems, the impact on the environment can be reduced. Making choices about components and materials while considering their origins, embodied energy and waste implications leads to less environmental impact on the building.

It becomes helpful to have a framework into which the choices about building design and detailing can fit into the concerns of sustainability. One such approach has been developed by the joint initiative of BioRegional and the World Wildlife Fund (WWF): One Planet Living (or OPL). Global challenges are identified from which they have derived what they term the OPL principles. From these a series of goals and strategies are set that can be adopted by communities. These OPL principles give a set of headings that can be used to prompt more specific goals to be achieved and strategies that can be adopted in the design of buildings and the choice of technologies. The 10 principles are given in Box 15.1. It is clear that some of the areas identifying action are more applicable to building and building development than others, but it is possible to see connections to most and therefore to identify actions that can be taken throughout the cycle of building development. These will then influence the specific activity of selecting technologies and preparing specifications and details.

Box 15.1 One Planet Living principles.

Zero carbon
Zero waste
Sustainable transport
Local and sustainable materials
Local and sustainable food
Sustainable water
Natural habitats and wildlife
Culture and heritage
Equity and fair trade
Health and happiness

The emphasis in OPL is on communities, so it becomes necessary to ask how buildings contribute to communities. It is then possible to apply the principles by adopting appropriate strategies in the design and choice of technologies employed in creating, operating and ultimately disposing of the building.

Frameworks such as OPL show that there are many ways in which a building can contribute to reducing its environmental impact and support sustainable development. With the introduction of regulations and governmental pledges to meet targets comes the need to measure a building's contribution to sustainability. This leads to the development of environmental assessment rating systems. One example of this is the Code for Sustainable Homes (2007), which awards credits for various aspects of the design of new homes in England that, when weighted, gives a percentage points score to award a building a code level rating. The weighting gives most points for savings in energy/CO_2 as this is the major governmental concern. Government environmental assessment rating schemes are instruments of policy and are based on an assessment of impacts that need to be made against national criteria. The context of a specific development may make local needs such as flood risk or specific site ecology of greater importance than the credit awarded in the rating system.

Buildings as systems

This book has suggested that seeing buildings as a system of a series of flows and transfers is fundamental to understanding the dynamic behaviour of the building. This again proves useful in evaluating building design against environmental impact criteria. By looking to minimise inputs, make best use of resources while they are in the building and then minimise outputs, many of the aims of low environmental impact can be achieved. This can perhaps be best seen in what has become known as passive design. If flows of heat, light and clean air can be kept in balance to maintain the required internal conditions through the passive and active fabric (see Chapter 10) rather than the active building services, the building will have a lower environmental impact. Active services use energy and therefore designs that can maintain the balance of flows passively, making a significant contribution to energy/CO_2 reduction. The main users of energy are heating, cooling, lighting and ventilation, so these become the focus of passive design strategies and low-CO_2 technologies.

Passive environmental design engages not only the fabric for gains and losses but uses the spaces in the building to introduce natural light and capture heat that can be moved around the building by natural forces. The aim should be to minimise the need for active systems that use energy. Active systems that are still needed can then have efficient components and be controlled via building management systems (BMS) and the energy still required then can be supplied from renewable energy sources to further reduce CO_2 emissions.

These three aspects of passive design, energy management and renewable generation all need to be incorporated at the concept design stage. It is less satisfactory to add them later on in the design of a new building. However, efforts to improve the energy efficiency of existing buildings can make a significant contribution to reducing CO_2 nationally.

Passive design for thermal environments

In the UK seasonal variation is such that there is almost certainly a need to heat in the winter and possibly cool in the summer. This will be influenced by internal heat gains that will contribute to heating in the winter but increase the need for cooling in the summer. In practice this often means that for housing the major concern is winter heating, but for commercial and industrial buildings there is a significant number of days in the summer when cooling will be required. It is therefore necessary to have passive designs that can operate during both winter and summer conditions. Although this analysis is for comfortable thermal conditions in the building, it will be seen that the requirement for ventilation for clean air has to be considered at the same time and will become part of the design.

Figure 15.1 shows the flows that need to be considered to maintain the thermal environment by focusing on the analysis of the gains and losses from the spaces in the building. The design needs to ensure that there is a balance at the desired internal conditions. Heat gains come from solar gain and internal sources such as the occupants and the operational equipment as well as building services such as

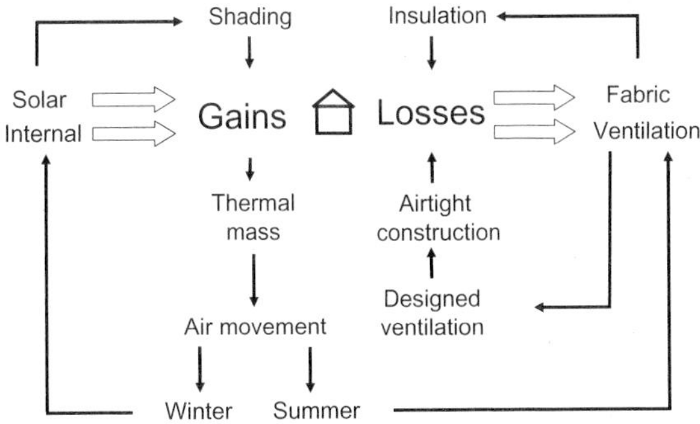

Figure 15.1 Thermal gains and losses.

lighting (although a successful passive design would minimise gains from these services). It is clear that these are useful in the winter but unhelpful in the summer condition. Passive losses occur through the fabric of the building and ventilation. Fabric losses can be limited by insulation, and this makes a significant contribution to limiting losses under winter conditions so the need for high levels of insulation is now well accepted for all buildings.

Some ventilation is necessary to provide clean and wholesome air and control humidity levels but can account for significant heat losses in winter, so ventilation needs to be limited to that required for healthy living. In summer, however, ventilation is used to help provide the correct thermal environment with the introduction of cooler air and increased air speed to help evaporation to cool skin. Ventilation in summer needs to be high to increase losses to maintain the balance between gains and losses of heat. To satisfy the needs of both summer and winter thermal conditions and wholesome air environments, there is a need to build airtight buildings and then design sufficient and controllable ventilation to provide good air in winter and direct cooling air in summer. In passive design this ventilation should be achieved using natural forces provided by the wind and thermal differences within the building. Rooms may be ventilated naturally by single-side ventilation, but if the internal layout and building shape and dimensions allow then cross ventilation and the stack effect can be utilised, as indicated in Figure 15.2. These will be considered in more detail below.

There are relatively few days in the year when there is a balance between these passive gains and losses to give a satisfactory thermal environment in all rooms. In the recent past, active services would have been designed to bring the conditions back within desirable limits. Heating in the winter and cooling, sometimes with air conditioning, in the summer. However, two other passive mechanisms, thermal mass and shading, can be brought into play to hold the balance between gains and losses.

The first mechanism is to exploit the thermal mass of the fabric. The fabric is used to store heat during times of high gains and release it when losses are possible. This proves useful over the 24-hour cycle where daytime gains are high and night-time losses can redress the balance. This time shifting of gains and losses has to be controlled and will require the heat to be moved around the building. The most effective passive mechanism to move heat is to use the air in the building. This requires predictable and controlled air movement through the building. This may be different from the ventilation process associated with clean and tempered air. Indeed, it is necessary to have different air movement patterns for summer and winter.

In summer the movement is to provide ventilation as a loss during the night, known as night purging; while in winter the heat has to be treated as an internal gain with the heat being distributed inside the building, while ventilation is running at a minimum for wholesome air. The forces that move the air in summer are likely to be those being used for daytime ventilation. Single-sided, cross-flow and stack effects can be used, but the air has to be directed towards the thermal mass being used for storage not over the occupants to increase cooling. The position of the thermal mass is important for both effective storage and transfer during purging. The most usual part of the structure is the underside of the slab. This approach is shown in Figure 15.3. Hot air rises in the day and the thermal mass in the slab will start to absorb the heat. In addition to absorbing the heat, the high thermal capacity will mean that the soffit of the slab will take time to rise in temperature. All the time the surface temperature is below air temperature, the occupants will experience cooling as they radiate heat to the slab. When the air outside cools sufficiently to receive the warm air in the building, windows have to open automatically to start the night purge, drawing in the cool night air that passes over the slab soffit and then ventilating back out into the night air. This 24-hour cycle utilises about the first 40 mm of the slab as effective thermal mass. However, the whole thermal

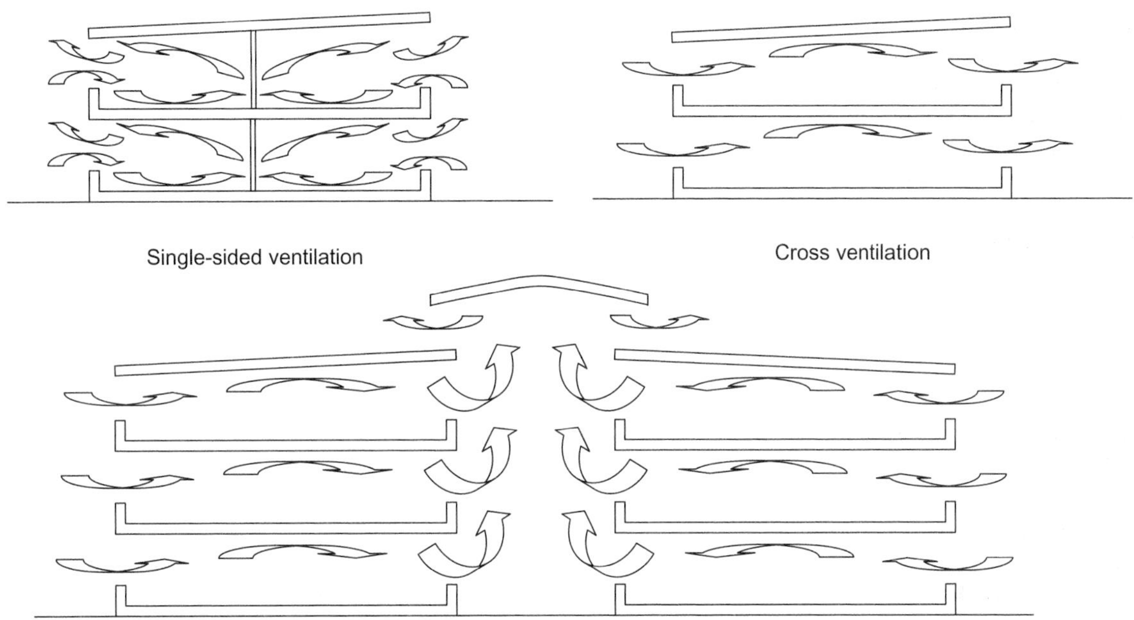

Figure 15.2 Side, cross and stack ventilation.

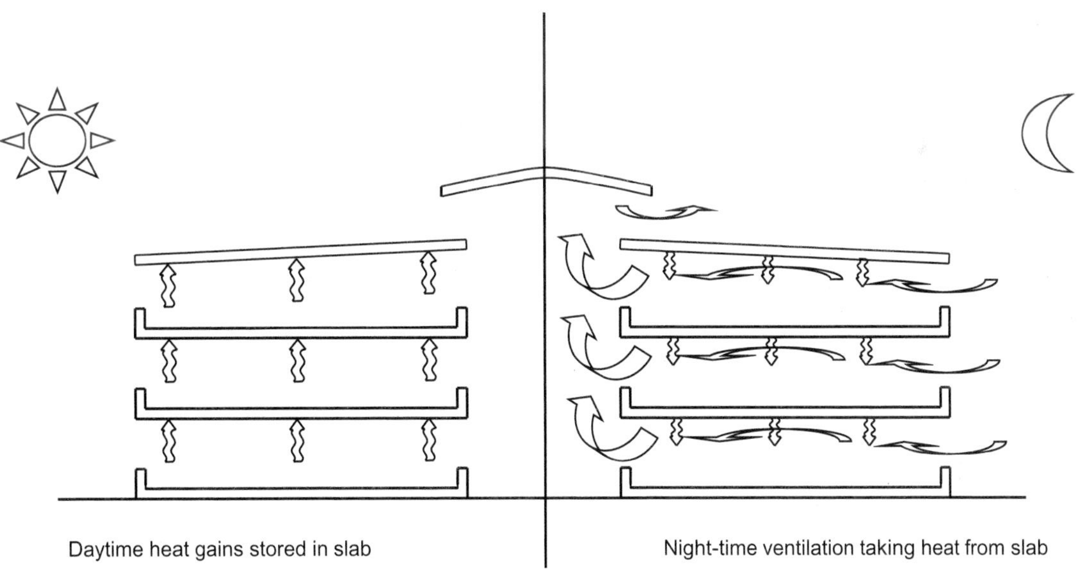

Figure 15.3 Summer night-purging (commercial building).

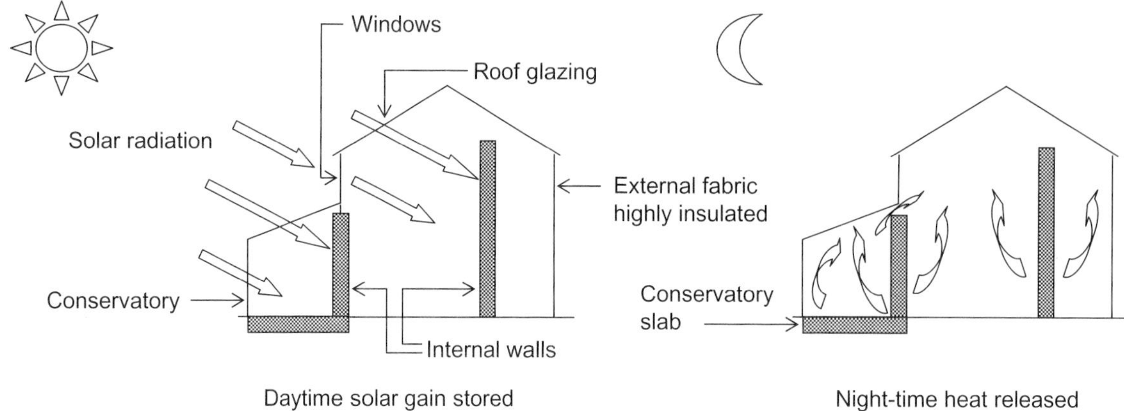

Figure 15.4 Winter heat gains (domestic building).

capacity of the slab is utilised indirectly as any thermal mass will absorb heat during the day and will give it up to cooler air at night. Any thermal mass will assist the gains–losses balance, and this is particularly useful under winter conditions.

In winter the night-time air movement must remain inside the building and so cannot take advantage of the external pressure driving the cross-flow and stack effects. The only forces come from thermal gradients and perhaps the small amount of ventilation for wholesome air. The other difference between summer and winter is that in summer it is unwanted heat in the air which is taken into the thermal mass while in winter it is solar and internal gains that are to be captured. This heat can then be used to limit the fall of the internal air temperature at night when no significant gains are available.

Daytime internal gains will be collected by any thermal mass and redistributed during the night, even if only back into the room in which it was collected. Rooms that continue to have internal gains through the night may need some provision to vent the air from the room to other parts of the building. For the maximum effect from solar gains in winter, the thermal mass has to be internal and in direct line with the sun. This means facing south and behind glass. This condition will create significant summer difficulties in most commercial buildings but can be useful for house design. Both heavy walls and floors can be used to capture heat from conservatories, windows and roof glazing and help keep the house warm during winter nights when losses are at their greatest. This approach is shown in Figure 15.4. There is now a potential for rooms such as conservatories to overheat in the summer when it may be difficult to keep this heat from being transferred into the house during the day.

The dilemma between using solar gains for winter conditions and the overheating problems this may cause in the summer leads to the second passive mechanism: shading. Solar gains are greatest from the south but even facades orientated more to the east and west will have direct solar gains for part of the day. Any elevations facing south will have the potential to maximise gains through glazed areas in the winter but will have to be protected from excessive gains in the summer. The mechanism used for this protection is shading. While it is possible to use internal shading such as blinds, this has limited effect as they will be warmed by the sun and this heat will then be inside the building and become a contributing gain. External shading in the form of shutters, brise soleils or secondary louvre walls, or even trees is more effective. Brise soleils can be arranged to cut off the sun when high in the sky in summer but allow the low winter sun to enter to make the most of potential gains. This approach is shown in Figure 15.5.

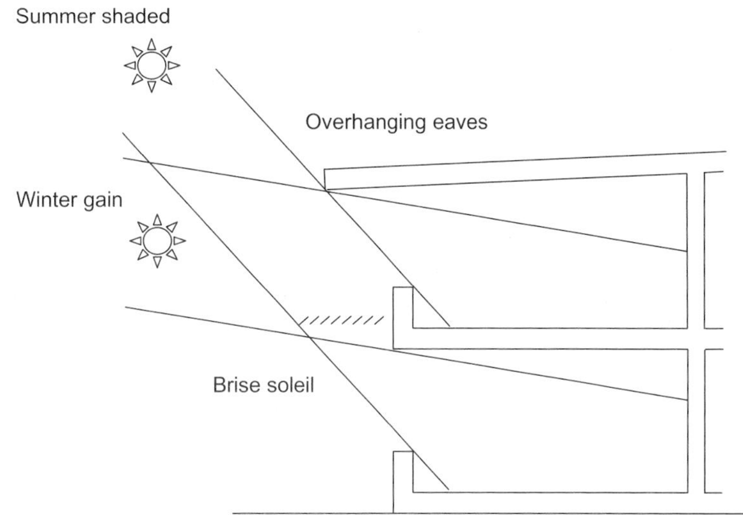

Figure 15.5 Winter gains, summer shading.

Deciduous trees can similarly shade in summer when in full leaf but let through winter sun when the leaves have been lost.

Building form and internal layout

If the flows and transfers identified in the ideas of passive design are to be exploited, the need to use natural ventilation will have a significant influence on the form and layout of the building. As natural ventilation is using only small forces to move the air, the size and proportions of the spaces and sizing of the openings need careful attention to ensure that sufficient air flow is achieved. Perhaps the most significant dimension is the width of the building between opposing ventilating facades. The pressure differences are created by wind flowing around the building giving positive and negative pressures that can drive either single-sided or cross ventilation. If cross ventilation is to be employed, the width should not exceed around 4.5 times the floor-to-ceiling height. For single-sided ventilation the depth of the room from the windows cannot be more than around 2.5 times the floor-to-ceiling height. Floor-to-ceiling heights are unlikely to exceed 3 metres, giving a limit to the building width of around 13.5 metres for open-plan layouts with cross ventilation and perhaps 16 metres if there is a central corridor for single-sided ventilation. Both cross and single-sided ventilation need sufficient areas of opening window, and for single-sided ventilation openings at low level for inlet and high level for exhaust can enhance ventilation rates with a local stack effect.

The stack effect is a vertical movement of the air that can also be used to move air through the building from openings on each floor level, normally to openings at or above the roof level. The stack can be driven by buoyant warm air rising through the building and out at the top. This can work even in the absence of wind on still summer days. However, in peak summer conditions, if the passive design works well and shading and thermal mass maintain low internal temperatures, the stack may not work well during the day until the building has started to overheat. Buoyant thermal effects will, however, work well at night for effective night purging.

The stack effect is also driven by the wind-induced pressure differences that create the cross and single-sided ventilation. On all but still days this is likely to be the predominant force moving the air in the stack, particularly if used in conjunction with cross ventilation. The top of the stack has to be high enough to ensure a lower pressure at the top of the stack than at the window level on the highest ventilated floor or a reverse flow will occur at this floor level. It is also possible to use the wind direction to introduce fresh air at the top of the stack. This requires some mechanism to ensure that the inlet section of the vent is facing into the wind. The stack can be created from vertical open spaces such as atria. If stack shafts are to be used, there needs to be around 1 m^2 of area for each 6-metre section of building to serve three storeys. Smaller stacks can be used but they may need fan assistance.

Natural lighting as passive system

Artificial lighting also uses a significant amount of energy, and therefore designs that maximise the use of natural light make a significant contribution to passive design. The analysis of passive thermal design utilising natural ventilation led to narrow plan buildings that also give benefits in using natural light, as do atria used for stack ventilation. To give required light levels towards the centre of even these narrow plan buildings there will need to be more glazing than is required for the openings necessary to ventilate. The penetration of light into a room is dependent on the height of the head of the window and the width of the room. It is also dependent on the reflective values of the surfaces in the back half of the room. For example a room 3 metres wide with a window head height of around 2.5 metres and a surface reflective index of 0.5 will only be possible to light naturally up to around 5.5 metres from the window, however big the window. However, a room with the same window head height but 10 metres wide with surface reflectance of 0.6 could be lit to a depth of around 7.5 metres, matching the room depth for maximum natural ventilation. These room dimensions still only show a potential for natural lighting. The area of glazing still has to be designed to achieve this potential.

This arrangement of glazing for adequate natural lighting may aggravate thermal gains. As with thermal gains, orientation is a significant factor in lighting. North light is consistent while light from the south has the potential for high light levels that can produce glare. Light from the south may need shading, particularly in summer. Any external brise soleil used to limit thermal gain would provide shading in the summer but not winter, when internal blinds may still be necessary.

Buildings clustered around atria or courtyards for natural ventilation can also use the atria or courtyard to bring light into the building, as long as they are wide enough to ensure adequate light levels in the absence of a direct view of the sky. Although natural lighting has to be considered separately to the thermal design and may need more glazing and/or may limit building widths, the general building form and orientation for both thermal and lighting requirements are compatible.

These design considerations for both ventilation and lighting make deep plan building difficult to design passively unless the internal layout can create paths for the air movement for natural ventilation and lighting. However, narrow plan buildings can be grouped to provide streets or courtyards with either covered atria or open spaces that allow for facade ventilation and lighting to both sides of the internal spaces. These arrangements are shown in Figure 15.6.

Mixed systems

Even with passive design it may be necessary to provide active systems to deal with conditions towards the extremes of the design conditions to keep the gains and losses in balance. This may be particularly true on difficult sites where space is limited and there is little choice in the configuration and internal layout of the building. Some form of heating will probably be necessary for both space-heating in winter and hot water all year round. With high internal heat gains, it may be necessary to have fan-assisted stack ventilation or even some areas mechanically ventilated or cooled to increase losses in summer. Heat recovery from exhaust ventilation air may prove to be energy efficient if it can recover more energy than

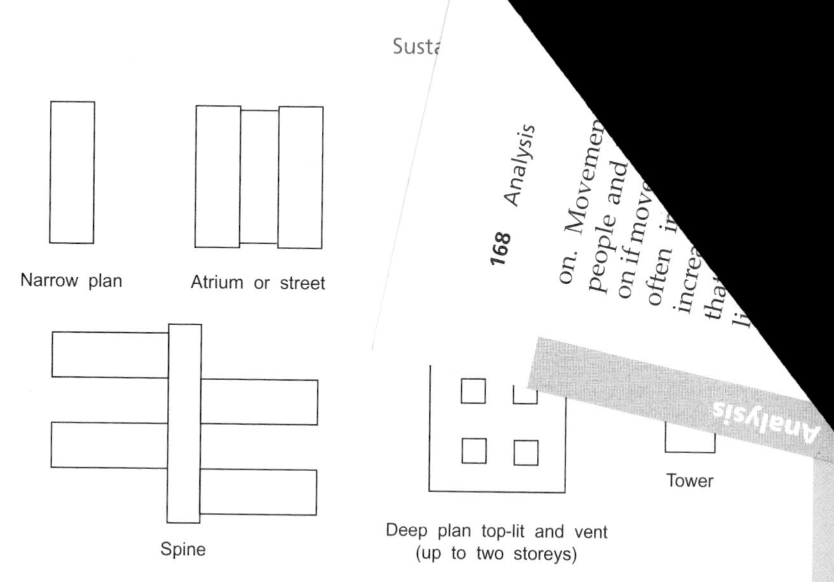

Figure 15.6 Building layouts capable of using natural ventilation and lighting.

it consumes to run. Areas that are difficult to ventilate naturally may be used for toilets, where the need for high levels of ventilation for wholesome air might have meant that mechanical ventilation would have been necessary in any case. Artificial lighting will be required for night-time operations and to supplement daytime lighting levels, but these areas should be kept perhaps for low-occupation activities, such as storage.

Reducing CO_2 even further

Passive systems aim to create and maintain environmental conditions using natural forces, although some use of active fabric will need small amounts of acquired energy. The services systems plus the operational equipment such as computers will still need energy. Beyond passive design it is possible to reduce CO_2 emissions of these active systems still further. The two main ways of achieving this would be:

- Choice of efficient equipment
- Management and information systems

There will be equipment specified within the building that will need energy. For many items of equipment, energy-efficient versions can be specified. Energy-efficient light bulbs and condensing heating boilers are, at the beginning of the twenty-first century, becoming the norm, with their more energy-thirsty predecessors being phased out by the government. Many appliances are now rated so choice can be made against energy-efficiency criteria.

Building management systems (BMS) can make a significant impact on the energy use of a building. Full BMS are linked through computer-based systems and normally associated with commercial buildings, although such systems for homes have been devised. The management part of the system is about control. It makes use of sensors to provide automatic control. As was indicated above, automatic control will be necessary if natural ventilation and night purging are employed in passive systems with active fabric. The opening of windows will need to be on a temperature and/or time base to ensure the air flows at appropriate times. Motorised vertical shading that follows the sun would also need automatic control.

BMS can also be used to control the active services to reduce energy waste. The simplest sensor systems do not need computer control and can be directly linked to switching the energy using appliances. Thermostats can be used to monitor temperature and turn heating systems on and off while movement, sound or light sensors can be used to switch lights off and t sensors detect an absence of therefore turn lights off (and back ement is detected), while light sensors, the light fittings themselves, detect sing levels of light and turn off any lights are switched on. The most effective use of ght sensors is to have lights wired in banks parallel to the windows so changes in natural lighting levels that normally occur across the depth of the room, away from the windows, can be compensated for with the artificial lighting. This will maintain lighting levels without the need for the occupants to respond to the changes in light. The occupants are less likely to switch lights off as natural lighting increases, although they will switch lights on as lighting levels drop. None of these automatic controls remove the need for switches that the occupants can use to override the automatic systems.

The BMS can also incorporate the security and fire detection systems that also use sensors, but in this case the system will switch on alarms rather than control energy use.

The information function of the BMS should be designed to give the user (often more specifically the building manager) both instant and time-based energy use information for both heat and power usage. This should be related to specific areas of the building where occupant behaviour can be investigated and informed in a way that, hopefully, will modify behaviour to reduce energy use in the building. This will require the building operator to inform and train the occupants in how best to use the systems to maximise energy efficiency. This may include everyday behaviour such as switching off equipment overnight but may also draw attention to the way the building is being used, for example the introduction of half-height partitions and items on the walls that restrict the penetration of daylight. These modifications will introduce obstructions and reduce the reflectivity of the walls that were decided at the design stage to ensure adequate levels of natural lighting.

Energy-efficient appliances and simple sensor systems such as thermostats and light sensors can be installed without computer control, but there is then a need to inform the users of how to make the best use of these systems. Here the information should be available through a user guide. This should inform the occupants of how the systems should be operated and used to their full advantage. User guides are particularly significant in homes where there is unlikely to be a building manager to inform and train staff. The occupants need the information specific to the installation in the dwelling and need to take full responsibility for maximising energy efficiency. Smart meters will help provide this information.

Renewable energy sources

Even with passive systems and the energy efficiency achieved in the choice and management of equipment, energy will still be needed in most buildings. In seeking the convergent solution to energy needs the use of passive design and system efficiency reduce energy demands but there is still an opportunity to reduce CO_2 by considering supply from low- or even no-carbon technologies. While this policy of renewable energy sources can be pursued at the national level, it is also possible to look to technologies for individual buildings or small community schemes. These are often known as micro-generation technologies.

There are now a number of micro-generation technologies that capture the energy from the sun, either directly or from wind or water movements. This energy is carbon-free in generation but needs energy to make and maintain the generation equipment. The equipment falls into two categories: capture of direct heat or use to generate electricity. For the individual building the capture of heat in solar panels, often mounted on the roof facing south, can provide much of the hot water for washing and cleaning but is not sufficient to exploit for the heating system. Electricity generation for an individual building is limited in the UK, although photovoltaic systems could be considered if a sufficiently large array could be installed. These systems also become more viable if the electricity can be stored in batteries or fed back into the national grid when generation exceeds the demand from the building. Wind for the individual building, particularly in urban areas, is unlikely to be efficient unless the development is large enough to install a large independent mast-mounted turbine. Proximity to a fast-running stream offers the possibility for hydroelectricity, but this is relatively rare on development sites. Both wind and water will be used for national generation schemes for distribution via the grid.

Beyond the carbon-free sources there are systems that can be considered carbon-neutral. These generate heat and power from plant-based biofuels that can be considered carbon-neutral in so far as the carbon that is released is captured in the next generation of fuel plants currently being grown. For these to be truly neutral and renewable there must be a growing programme of supply to match demand. Unfortunately this is limited as the plant-based biofuels are competing for land food growth and natural habitats for biodiversity. Leaving this to the market to seek a balance on price may not achieve the sustainability goal of limiting environmental impact or social equity.

Beyond carbon-free and carbon-neutral, carbon-efficient systems such as heat pumps can be used to draw heat from the air or the ground to provide the space heating and possible hot water for the building. These systems do use power to transfer the heat, but the heat itself comes from the sun that warms the air and the ground and is, therefore, carbon-free. Even air at below 0°C has heat in it, so air systems work even in winter conditions. Air pump systems normally move the heat from outside air directly to inside air so either are room-based systems or need internal air movement if units are to heat more than one space, clearly best supplied by natural air movements within the building. Ground pump systems draw heat from depths where the ground is naturally less variable in temperature than the air. These need not be too deep as surface systems laid in trenches around 1-metre deep can be effective but do need a loop of around 200 metres for an average house. Where this area of ground is not available, systems can be installed in vertical bore holes.

Ground source heat pumps are suited to exchanging heat to the water of a central heating system. However, the potential flow temperatures are low and therefore the use of radiators would not be efficient requiring over-sized radiators. Low flow temperatures are better suited to under-floor heating, where the large area of rising heat provides a comfortable environment that, in conjunction with good insulation below any ground floors and the high thermal mass of the floor, can give good internal temperature control even with large changes in external temperatures.

Another carbon-efficient approach is to use combined heat and power units that obtain the heat for the space heating and hot water from units that generate electricity. This technology can be used to gain efficiencies in burning oil and gas or with biofuels. The use of biofuels is being based on power units such as the Stirling Engine, which, while the principle has been established for many years, is proving effective for these innovative applications. These units are available for individual dwellings, but larger units for commercial buildings or small communities help to make the most use of both the heat and the power that is produced simultaneously.

There is increasing interest in new fuel sources that reduce the emission of CO_2, limit the rate of depletion of oil and gas and/or help maintain the security of supply. Biofuels are being developed from animal waste (biodigesters) that produce methane to be burnt as a gas, normally to produce electricity to be supplied to the national grid although they may be used for small community heating from combined heat and power units in rural areas where a supply of animal waste is close by. This has the additional benefit in that, while it releases CO_2, it stops the release of the methane that is a more potent greenhouse gas than carbon dioxide. Fuel cells that are powered by hydrogen are also being developed for heating units.

It is not clear at the beginning of the twenty-first century what fuels may be available for the energy requirements of buildings in the future.

However, the precautionary principle to seek convergent solutions that both seek to reduce energy demands and look to reduce CO_2 and other pollutants from renewable secure sources should be significant criteria in the development of new and the upgrading of existing buildings.

Water and waste

The building as resource consumer and polluter goes beyond energy and global warming. Another natural cycle which building development affects is the water cycle. Vital for health, the availability of a good source of drinking water has always been a significant factor in deciding where developments have taken place. With modern supply and distribution systems this direct link between source and development has been broken. Water has provided support for a growing population but, while renewable, cannot have a guaranteed continuous supply if the natural cycle is not understood and respected. It could become another factor that would make the future unsustainable for the next generation.

There are two strains on the supply and distribution system. The first is the use of drinking-quality water for all water use during the life of the building. The need to supply this high-quality water in sufficient quantity for all the uses is becoming unsustainable. The second strain is our disposal of surface water. Our approach has been to take the rain that falls on developed areas and pipe it away to water courses for disposal. This puts the rain back into the water cycle but significantly increases the natural rates of entry of rain into rivers and bypasses the natural evaporation processes and infiltration rates to aquifers, all of which are the main sources of our quality water. These piped services also increase the times to concentrations of water at certain points in the water courses following heavy rain, thereby increasing the risk of flooding for some specific areas.

As well as surface water, development creates foul water that has to be piped away for treatment if significant local pollution is to be avoided. Expanding development demands greater infrastructure and again leads to water bypassing the natural water cycle, with treated water being returned directly into water courses. In addition there is a growing acceptance that some surface water, particularly from roads, is polluted and, while not requiring full treatment, need to be disposed of with care if habitats are not to be polluted or heavy-metal pollutants allowed to enter the water cycle.

These issues of the quantity of drinking water supply and maintaining natural water sources along with concerns for flooding have produced a number of options, most of which address a number of these concerns as they help restore the natural rhythms of the water cycle.

The simplest systems just reduce water use in the building with appliances such as dual-flush cisterns for WCs and spray taps for hand washing. The next step is to recognise the variety of water qualities for different uses in the building. While water of the highest quality will be required for drinking and food preparation, water for washing and cleaning need not be of such high quality and water for flushing toilets and external use has even lower quality demands. All water has some minimum requirements for quality, often in colour and particles to be acceptable for use in buildings and must have biological (pathogen) and chemical limits. However, within these limits, recycling and alternative sources of supply are possible.

Grey water systems recycle water that has been used for washing and cleaning to flush the toilets. Some filtering and disinfecting will be necessary but this system reduces both the need for supply and the volume of foul water requiring disposal. Rainwater harvesting provides an alternative supply of water for flushing toilets and external use (watering gardens and washing cars) by collecting rainwater for use in the building. This reduces not only the need for supply but also the surface water flow rates during periods of heavy rainfall. Both these systems require water to be stored to help ensure supply, but there are likely to be times when demand outstrips this supply, so dual systems are required particularly to flush toilets when the supply of grey or rainwater is not available. See Chapter 21 for more details on grey water and rainwater harvesting systems.

The volume of storage is likely to be large, particularly for rainwater harvesting, where timing between supply and demand is less well matched. Storage provided by tanks will be on or, more likely, in the ground to save heavy loadings on the structure. This will require some pumping, perhaps to a smaller storage tank at roof level, to supply water to appliances on all floors of the building. In addition, when supply exceeds demand, overflow from the tanks must be provided. For grey water this will be into the foul sewer system but for rainwater harvesting this could be to a soakaway.

The reason for the use of the soakaway (rather than a piped drain) is part of the thinking about creating surface water disposal systems that better match the natural run-off characteristics of the development site. These are collectively known as sustainable urban drainage systems (SUDS). The idea is to hold water in catchments to allow infiltration and evapotranspiration (where vapour is released into the air by evaporation from open water and from the ground through plant transpiration) between periods of rain. These systems not only reduce the speed of run-off and help control flooding but also replace water into the aquifers, helping to safeguard water supplies.

SUDS employ many techniques to mimic the site's natural drainage pattern, including green roofs that delay water run-off, permeable hard surfaces that allow surface water to drain through car parks and roads, and landscaping and ponds to hold water and improve amenity. Each site will be able to help but SUDS are another example where community schemes over larger areas can be more effective.

Like all community schemes, they need management and maintenance. In the absence of public authority adoption of these schemes, other management, possibly community or commercial, has to be put in place for the scheme to continue to be effective throughout its life cycle.

Solid waste

In the same way as water was once seen as being just one quality to be used throughout the building, solid wastes once used to be seen as all the same to be taken away for disposal, normally in landfill or by incineration. Considering solid waste from a building as all the same leads to two difficulties in achieving sustainability objectives. These are the two common concerns of the potential for pollution and resource depletion.

There is limited land available for disposal sites and the waste will decompose, with potential leaching of pollutants into the ground water and watercourses plus the production of methane, the greenhouse gas, into the air. It is possible to select landfill sites to limit harmful leaching, to collect the methane to be burnt for energy production and return the land to support wildlife or development once decomposition is complete, but these options are limited in so far as they are expensive, take time and may not be in convenient locations. Incineration creates potential air pollution and still requires disposal of ash and, again, will face the difficulties of suitable location. Technical solutions are available but, as with responsible landfill, they come at a cost.

Many items identified as no longer required by the building user or operators are made of materials that have the potential to be recycled. However, the processing of materials from waste is very specific to each material, and this leads to the need to see waste as separated waste-streams from the building. Identifying and separating these different waste-streams is best achieved at the point where there is the decision to throw it away. This identification and separation is only of value if collection, distribution and reprocessing are available for the potential of recycling to be realised. In the UK these facilities are well developed and therefore designing buildings to encourage users to separate waste for collection becomes

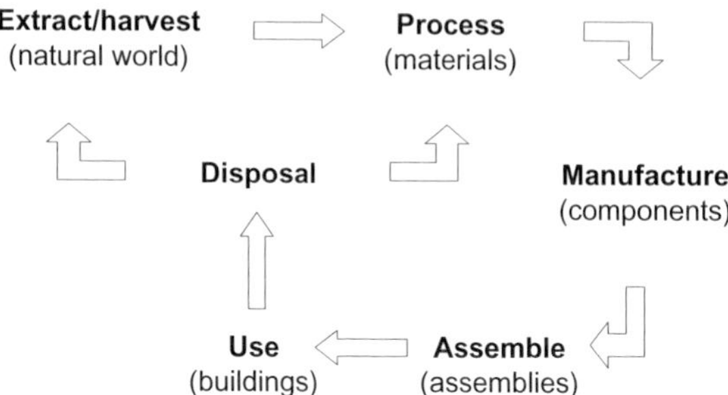

Figure 15.7 Materials lifecycle.

part of sustainable development. Organic waste can be used for compost, while paper, glass, metals and plastic can be reprocessed, saving on energy and avoiding depletion of the original feedstock still further. The list of recyclable material is growing but it is still better to reduce and reuse before recycling, but this is a matter for us all in our purchasing and lifestyle choices. However, there is waste associated with the construction of the building itself, and this is a matter for design and the choice of technology in the specification and detailing of materials for the construction.

Materials choice and detailing

As the design of buildings reduces energy and CO_2 emissions and promotes the management of water and waste during the operation of the building, the significance of the construction of the building itself as polluter and consumer of resources becomes greater. This is directly related to the choice of materials but is also influenced by the choice of detailing and the production options involved in taking the processed material and making it into components to be assembled into buildings.

Evaluating the environmental credentials of materials has to take into account the lifecycle of the material from what is often termed cradle-to-grave analysis. The stages in the lifecycle are shown in Figure 15.7. While we refer to the material's lifecycle, it is only in the stages of processing that it can be thought of purely as a material in terms of its environmental impact. When the material is passed to the manufacturing stage, it is formed into components and now a variety of manufacturing processes may be involved, each having its own environmental impact. Simply comparing in situ with precast concrete, with their different specification requirements, moulding and transport arrangements, shows that being concrete does not mean that both have the same environmental consequences.

This has led to what is often known as gate-to-gate analysis. This sees any part of the building as being traceable through a series of processes back to its source and on to its disposal, where each stage has a different environmental impact that is cumulative through the life of the material as it is used in that building. The gates referred to are the gates to the sites where work is carried out on the material to bring it to its component form and then to be assembled into the building to be used and then disposed of. Between each gate there will be transport requirements that have an environmental impact. Local materials with less processing will now be shown to have a lower environmental impact. It does, however, mean that each component in the building has to carry an individual analysis. At the time of choosing the specification and details, the impact of the component in use and its disposal will be in the future. For this reason comparative figures are often cradle to site gate to take into account the manufacturing differences as well as the extraction and processing impacts.

This may still be misleading. Materials are now being specified as part of energy-saving strategies and therefore, it could be argued, have environmental benefits in use. This is the case with concrete being used as thermal mass to reduce carbon release during the operation of the building. It also does not take into account recycling potential to conserve resources and reduce pollution.

However many stages are taken into account, the concerns at each stage remain the same. For each process (inside the gates) and for each method of transport (between the gates) there needs to be an analysis of the process resources to identify the effects of pollution and resource depletion.

The focus on carbon as a pollutant has led to the analysis of embodied energy, or more specifically embodied carbon, in materials forming components, transport and even in the assembly process itself. As with the operation of buildings, issues of water use and waste in each process can be addressed to reduce, reuse and recycle as part of the analysis of the process and manufacturing options. Extraction and materials processing often involve large quantities not only of energy but also of water and can produce some highly toxic by-products that are not evident once they are formed into components and incorporated into the building. These issues can be addressed in the choice of materials and detailing but have to be followed through in the responsible sourcing and procurement of the materials and components. Managed sources and local materials contribute towards a lower environmental impact of each component in the building under consideration.

Planning and design concepts

Achieving sustainability objectives needs to be incorporated into the very form of the building at the concept design stage. All choices about building form, appearance and internal layout that are considered at the concept design stage have implications for technical solutions; achieving sustainability objectives is no different. Aspects of analysis that need consideration at the early design stage have already been identified in the sections above. The use of thermal mass, tracing air movements for natural ventilation and ensuring the quality of natural lighting have, amongst other considerations, been discussed. Site landscaping and the potential for trees to provide shading or being part of a SUDS design have also been highlighted.

Another idea that has recurred in the above sections is that of community schemes and the potential for development involving many buildings at the master planning or even urban planning level to take advantage of technologies that support the development of sustainable communities. Local heat and power plant and urban drainage schemes again have been identified as examples.

Community schemes need to be managed and maintained. Responsibility for the effective use and maintenance of the scheme and for administering any charges has to be established at the design stage. This again has been referred to above, suggesting that if local authorities do not adopt the facilities, as they might have in the past, then some community or commercial arrangements need to be made and incorporated into the contracts for the operation and/or sale of the buildings.

Other wider connections to sustainability, possibly incorporating the OPL principles, can be considered at this level of planning and design. Perhaps the most direct is the link to transport and how development can shape the need for transport and in particular the use of the car. Although a full discussion of these issues is beyond this text, the provision for parking and the treatment of access for all forms of transport, including walking and cycling, will play a significant part in the wider objectives of achieving sustainable communities. This may be achieved with mixed development, the provision of local services such as local shops, health care, allotments or even in the emergence of new types of buildings such as life/work units to reduce commuter travel and reduce the need for personal transport.

Emerging technologies

All the above considerations to reduce environmental impact will give rise to the need for new technical solutions. Many may be existing technologies just applied in new ways but some may give rise to new technologies and ways of construction that will need to be evaluated not just for their sustainability credentials but also for their performance, cost and production potential.

Society will expect technology to respond and play its part and possibly be unforgiving of failure. However, much of this new analysis will involve knowledge of the process of production and assembly and of the materials themselves. In terms of this book the issues of sustainability have been introduced as part of the evaluation of the design, the physical behaviour and manufacture and assembly of a building. As knowledge and experience grow, they will become a natural part of the choice of construction. Treating them as special considerations runs the risk of the errors made with insulation detailing following the oil crisis of the early 1970s.

Of the two great forces suggested at the beginning of this chapter it is the social forces that change. Sustainable development is the new concern of the early years of the twenty-first century. As we learn more about the issues and devise solutions that prove themselves in practice and provide precedent, the need for environmental concerns will become an established response in our designs and production procedures. Sustainability is just the latest in the history of the need for technology to respond to new social concerns. It is unlikely to be the last.

Summary

1. There are two major forces that condition a construction solution: the laws of nature and the society in which construction is practised.
2. At any time in a society there may be deep changes in its values and beliefs, which have to be taken into account in a technical solution for it to be seen as a reasonable option. At the beginning of the twenty-first century this includes embracing environmental concerns in our understanding of sustainable development.
3. In developing a response to the sustainability agenda the two key areas to consider are pollution and resource depletion.
4. There is an immediate focus on global warming and energy consumption as our current energy production relies on burning fossil fuels that produce the pollutant CO_2. There are concerns over their depletion and the economic and social impact this may have. Energy saving is a strategy not an objective.
5. Sustainability is not exclusively about buildings. However, as polluters and resource consumers in both their operation and in their construction, choices about how we build have to be integrated into society's responses as they have a significant impact on the environment and on our social and economic well-being.
6. Passive design principles can make a significant contribution to achieving buildings with lower energy needs, particularly in meeting required thermal, ventilation and lighting performance.
7. These concerns need to be considered at the very early stages in the design, as internal layout, size, space and orientation are important to passive design.
8. The choice of equipment and building management also contribute to reducing environmental impact.

9. Water usage and waste, both liquid and solid, is another area for consideration in contributing to sustainability objectives.
10. When making the choice of construction itself, the detailing and the specification of materials, particularly in sourcing and considering embodied energy or carbon in the materials and components, is another important area for consideration.
11. These more fundamental shifts in values often demand changes for which knowledge is uncertain, making the analysis of physical behaviour of the construction even more significant to ensure 'safe' solutions.

Part 2
Choice – House Construction

16 Applying the Framework to Housing

This opening chapter for the second part of the book applies the process of choice determined by the earlier chapters. It then outlines the basis of a case study that will be used to introduce how this process has led to the common forms of house construction in the UK at the beginning of the twenty-first century.

The need for an integrated approach

The previous chapters suggest that proposed construction solutions have to be evaluated through a number of areas of analysis. There is a danger that this leads to the view that each area of analysis can be considered independently and they can be carried out in a given order to make the final choice. This view is not correct and it is hoped that this text has not given this impression. The first few chapters of the book are designed to give a holistic feel to the process of choice before the detail of analysis of construction is introduced.

In reality each project has its own balance of priorities set by the client and its own context that determine the appropriate approach. The emphasis is on evaluation as a holistic process, not satisfactorily complete until all the areas of analysis show acceptable predictions. While each analysis is largely independent, it is always being undertaken in a growing understanding of the outcomes of the evaluation as a whole. Early proposals need to be provisional until the outcomes of each analysis have become clear. This leads to the need to have techniques to establish approximate solutions for each analysis in order to see whether an overall solution is possible, as well as detailed design information to make the final choice.

This text is based on a framework, not a sequence. The process has been thought of as dynamic, open and iterative, moving towards a solution but without prescribing the route. The choice has to be made within a framework that indicates the range and scope of issues that have to be taken into account, a framework that can be used to establish the possible complementary, possibly competing, factors that will determine the success of the construction in technical and social terms.

It is difficult to reproduce this process in the pages of a book. It is, by its very nature, one that has to be experienced to fully understand how what sounds like a chaotic process is controlled and rational, and yet creative choices are made in practice.

This text has set out to explain a basis for the technical and rational aspects of the process and maintains that it is only through this understanding that the creative and innovative can be put into practice.

In order to continue to apply this approach the chapters that follow are written from the perspective of a case study. The case is the choice of construction for houses for the UK at the beginning of the twenty-first century. This chapter introduces the context in which these choices are being made, the nature of the environments, physical and social, and identifies the aspects of design, appearance and the resource base that will condition the choices to be made.

The basis of the case study

The study is focused on the house, a single-occupation dwelling with its own ground floor and roof, and therefore does not include flats, although the house could be detached, semi-detached or terraced. This description could include specialist accommodation such as sheltered housing, so the study will be further limited to family occupation. Given the time and place specified for the case study, the notion of a family, its members and their relationship is changing. Because the house is still the aspiration for many of these families, this text assumes that these changes will not significantly affect the construction of what these families are seeking, although there may be changes in the performance requirements with new patterns of use within the house.

While this description of a house will allow a broad understanding of the needs of the user, it says nothing about the resource base or industrial structures to both design and build houses. This case study will focus on the decision-making of the speculative, volume house builders as they construct the majority of houses built in the UK at the beginning of the twenty-first century. The methods of building they adopt will make resources available and therefore even the more bespoke individual development will tend to choose the same technological solutions. The more innovative designs may need to develop other technical solutions. These designs may have to adopt an approach more like that suggested in Part 3 of this book for commercial buildings, where solutions are more diverse.

The two ideas used so far to define this case study – family and industry (demand and supply) – have set much of the scope of choice. The depth of understanding that is necessary, of each of these ideas, should not be underestimated to appreciate why certain construction solutions will be chosen.

This was the thrust of some of the earlier chapters of this book, where it was suggested that it is necessary to understand something of the physical and social environments within which choices are made. This will include an appreciation of house design and the resource base available to the house-building industry.

Physical and social environment

The case study is set in an advanced industrial society on a highly populated island in a temperate climate. The social values include the distribution of private wealth with equity and respect for all members of society, with a growing awareness of environmental impact. This gives a high level of demand for housing, with limited land and high prices on the open market but a need for social housing.

These few facts start to create a picture of the context in which the choices will be made. Many of the details will emerge in the chapters that follow, but even this short description can lead to some more observations. Advanced industrial societies often have a great deal of legislation and regulation. We can take advantage of factory production and transport of component and prefabricated sections of buildings from almost anywhere in the world. We are, however, having a huge impact on the environment. The need for housing on our highly populated island means that we are increasingly building on previously developed land, brown-field sites, and have a problem in dealing with our waste. Our concerns for equity and social justice, with a fair standard of living for all, set relatively high performance levels for construction. The temperate climate defines our weathering problems and indicates that water penetration and winter heating are major concerns for house construction. There are now concerns that the levels of human activity we wish to sustain are causing environmental changes that threaten our ability to maintain this progress. Environmental concerns must take their place with economic and social considerations to ensure sustainable development.

There is limited value in this text in exploring these ideas in detail, as this is not the purpose of this book. The more significant aspects have been introduced in the previous chapters and the reader should become increasingly aware of the importance of these factors as the following chapters develop the basis of choice for the construction of houses.

The resource base

With high levels of house-building activity the supply of materials and factory-produced components is good, but there is a shortage of site labour and the traditional trade skills in particular. To meet demand considerable attention has been paid to specifying components and prefabrication, including modular and even volumetric systems to limit site operations (although these will not be discussed in detail in the chapters that follow). Machinery is readily available for materials handling. The introduction of cordless hand tools has influenced fixing and jointing processes, often now the focus of the site assembly processes. Money for housing is readily available where risk is low, although the resources for social housing are more limited and highly regulated.

Design and appearance

The basic design and appearance of the house is well established. The pattern of living is sufficiently common for the major rooms of living room, bedroom, kitchen and bathroom to be easily identifiable, with some circulation space including hallways and staircases. While there is increasing variety in the number of rooms in a house and some creative internal layouts can be observed, the size of the rooms has generally been reducing, particularly in the volume housing market where the whole building footprint and plot size is reduced. This has led in some cases to habitable space extending into the roof.

Externally the brick house is still a dominant form. Rendered finishes may be used, although they still maintain the sense of masonry construction. Roofs are pitched with tiles or slates to give the characteristic shape to either bungalows or two- and three-storey houses. There is then a variety of window forms, including bays and oriels often used to gain variety in housing estate design. It is not possible to generalise on the designs for the individual client, where expression in materials such as glass and timber may be featured along with a willingness to include aspects associated with sustainable development.

Internally finishes are plain but need to accommodate a variety of building services. Energy saving has featured in the regulations over recent years and has had a large influence on detailing, and this is likely to continue with increased environmental concerns. In the single dwelling fire resistance is only nominal within the dwelling, but walls between dwellings (party walls) and the roof finishes that are continuous between dwellings are subject to more stringent regulation. New regulations on soundproofing and air infiltration with the possibility of the need for performance testing and the introduction of robust detailing are also changing the construction of and specification for houses.

Environmental concerns are affecting the balance between passive and active technologies as well as the choice of the components, materials and detailing itself.

Common form

The conditions established for the case study have given rise to the common form introduced in Chapter 3. This will be developed in greater detail in the chapters that follow. With the overall form established it is possible to take an elemental approach to floors, roofs, walls, foundations and services. Each element contributes a major group of functions and is easily recognised by the basic construction solutions. However, the question arises as to what order they should be considered in. It has already been established that no one element can be fully selected without some reference to the probable form of the others. In this case the elements that can be discussed knowing least detail of the other elements are the floors and the roof. These transfer loads to the walls, and so it is necessary to discuss these before walls can be fully investigated. It is then possible to complete the structure by looking at foundations where the loads from a house are relatively low and the site conditions are unlikely to dominate the choices for the superstructure. Finishes will be discussed with each element to cover all the analysis of the passive construction. This leaves the active services, although aspects of them will have inevitably been introduced by their need to be integrated with the other elements.

Summary

1. The framework identifies the areas to be considered but does not suggest the sequence in which the areas of analysis should be undertaken. The nature of each project will determine the key choices, with no single choice being confirmed until all areas have been considered, at least in outline.
2. The case study approach to be used in the next five chapters concerns the choice of construction for a house in the UK to be built at the beginning of the twenty-first century.
3. The physical and social environment in which this case study is being considered is represented by the needs of the family in an advanced industrial society, on a highly populated island, in a temperate climate, in a society that strives for social equity and has a growing environmental awareness.
4. The advanced industrial society gives high labour costs and shortages but provides the infrastructure for factory production and prefabrication as an alternative to traditional craft-based construction.
5. Given the design expectations of the users of housing, a well-defined format for the house has emerged for the speculative and social housing market, although much of the detailing within this format will be evolving and must be confirmed for each project.

17 Floors

This chapter introduces both upper and ground floors for the house construction that could be adopted in the case study outlined in the previous chapter. It follows the analysis of the choice of construction detail and specification, suggesting alternatives and giving some typical details to support and illustrate the analysis.

Upper floors

It is sensible to consider upper floors and ground floors separately. They have a different set of functions to fulfil; the structural support options for upper floors where the spans are defined by room layout do not apply to ground floors.

Upper floors are structures spanning over the rooms below and providing a stable clean surface for the rooms on the upper stories. Being internal, they have no weathering function and are unlikely to be involved in resisting the passage of heat or in providing light or ventilation. While they will not contribute to these passive environmental functions, many of the services distribution systems may need to be integrated into the space or zone occupied by the floor.

Within a dwelling, security is unlikely to be a performance requirement for the floor. However, with changing patterns of living and the expectations of individual members of the family, privacy, particularly sound transmission, may be an issue. This may not be satisfied by just a structure with the addition of the finishes. Where floors are between dwellings in flats or apartments, they will have greater fire and sound requirements, which will have a significant effect on the choice of construction.

General forms for upper floors

There are two general forms that can provide an economic solution to this set of performance requirements: joisted floors and slab floors. Joisted floors are based on small beams (joists) set a limited distance apart with an additional component, the floor board, to form the continuous surface required to support a finish for the room above. The floor would normally also have to provide support for a ceiling, probably of plasterboard, for the room below. A slab is a solid (typically concrete) continuous structure that, while normally spanning only in one direction like the joists (one-way spanning), can be designed to transfer the loads on all four edges (two-way spanning).

In housing, floor spans would not normally be above 5 metres and that makes the timber-joisted structure economic, although greater spans are possible in timber if the shape of the joist is changed, as discussed later in this chapter. For longer spans, but particularly for the increased performance demands of sound and fire required between dwellings, the concrete slab may prove a better choice. Slabs in concrete can be formed in situ but are more likely to be precast, either planks or a form of slab known as a beam and block construction. The beam and block floor has been adapted for suspended ground floors (discussed later in this chapter).

Timber-joisted upper floors

For the upper floors of the house in this case study, a timber-joisted floor will be chosen. The basic arrangement and variables are shown in Figure 17.1. The floor structure illustrated shows a traditional arrangement of rectangular section timber joists at regular centres with a boarding to provide the floor surface. Two of the basic variables have therefore been established: the material and shape of the joist. While the shape is unambiguous and only needs defining by size, the quality of the material needs further specification. Just specifying the joist as 'timber' is not sufficient for a structural (or, indeed, any other) application. Some way of ensuring its strength in order to determine a safe and economic size is necessary.

Timber is a variable material, and one way it varies is in its strength. Even the basic distinction of hard and soft wood is not a good guide to strength. The species of the tree will give some indications, but variation still exists. A safe design would have to assume that the weakest timber produced by that species might be delivered to site and used in the construction. Timbers for structural purposes are stress-graded, where they are given a designation that guarantees the strength within a limited band, so the size can be chosen with some confidence of the strength of timber that will be used in the construction.

The size of the joist now depends on the centres and the ratio of the breadth to the depth of rectangular section. Larger centres mean fewer joists but of larger sections and thicker boards as the centres of the joists determine the span of the boarding. In practice, centres between 400 and 600 mm are economic. This leaves the ratio of breadth to depth of the joist. The depth has a large influence on bending strength and limiting deflections. Breadth influences stability associated with lateral torsion buckling and to some extent how live, or

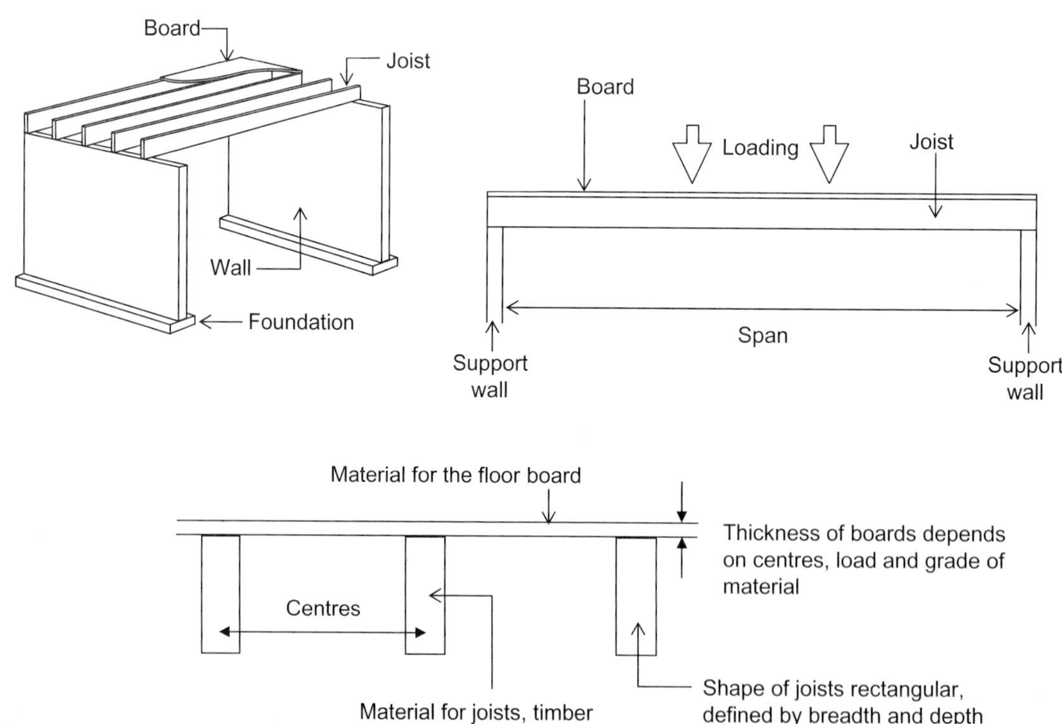

Figure 17.1 Timber-joisted floor – arrangement and variables.

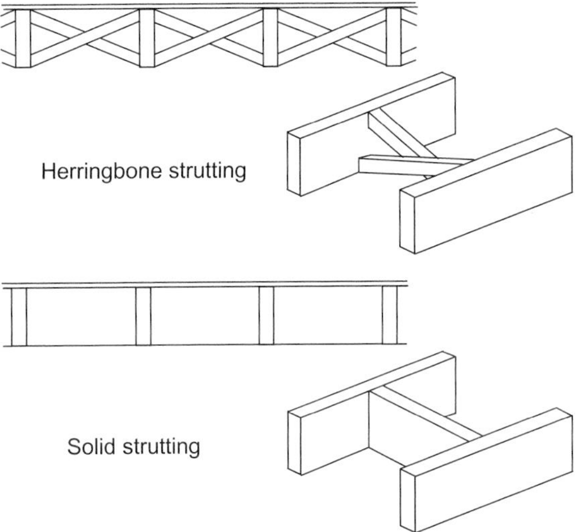

Figure 17.2 Timber-joisted floor – alternative forms of strutting.

dynamic, the floor feels. There is little option but to invest in depth for bending strength and deflection, but there is an option to help improve stability and the overall stiffness of the floor rather than just making the joist thicker.

Figure 17.2 illustrates two forms of strutting, but both serve the same purpose. They stabilise the compressive zone from the bending forces in the top of one joist using the stable tensile zone at the bottom of the adjacent joists. This allows slimmer joists than would otherwise be necessary for stability and stiffness. The amount of strutting will depend on the breadth/depth ratio of the timbers selected for the joists. Joists are most likely to be selected from sections between around 38 × 100 mm and 75 × 225 mm, the larger sections becoming expensive and possibly hard to obtain. With these joists, strutting is unlikely to be required up to spans of 2500 mm, with one line of strutting being sufficient at the centre of the span up to 4500 mm and then two rows at the third points above this span. It is unlikely that the timber-joisted floor will be used much above this span, so more than two rows of strutting are unlikely to be required.

As has already been identified, the selection of the joist is not unrelated to the choice of boarding. The board span is related to the centres of the joist, but first the material for the boarding must be chosen. Traditionally timber was the choice, but now it is more likely to be a composite sheet material such as chipboard. In the case of chipboard, defining strength quality is established by specifying flooring grade, but some issues of durability may have to be addressed if the chipboard is likely to become damp. The integrity of a composite material such as chipboard is influenced by the binder or glue used, and some may be susceptible to damp. Where damp resistance is required in areas such as bathrooms and kitchens, it must be part of the specification. Whatever the durability specification, the thickness of the board with joists at 400 or 450 mm centres, in a flooring-grade chipboard, will have to be 18 mm thick and for 600 mm centres will have to be 22 mm thick. Alternative board materials such as plywood or oriented strand board (OSB) may be used. These will be required to be of similar thickness to chipboard and the specification of grade of material will have to take strength and durability into account.

Joints and fixings

Now the materials and the basic size and shape of the components have been chosen, and their spatial relationship defined, there is now a need to be clear about the joints and fixings to be employed. These will occur between:

- The sheets of the boarding
- The boarding and the joists
- The joists and the strutting
- The joists and the support

The two long edges of the chipboard sheets are tongue and grooved and glued together as these edges will not be fully supported, as they will be laid across the joists. The other joint, on the ends of the boards, should rest on a joist to give full support. The boards are then nailed or screwed to a specified pattern that, with the glued jointing, not only holds the boards in place but also eliminates squeaking when walking on the floor. If nails are used, they

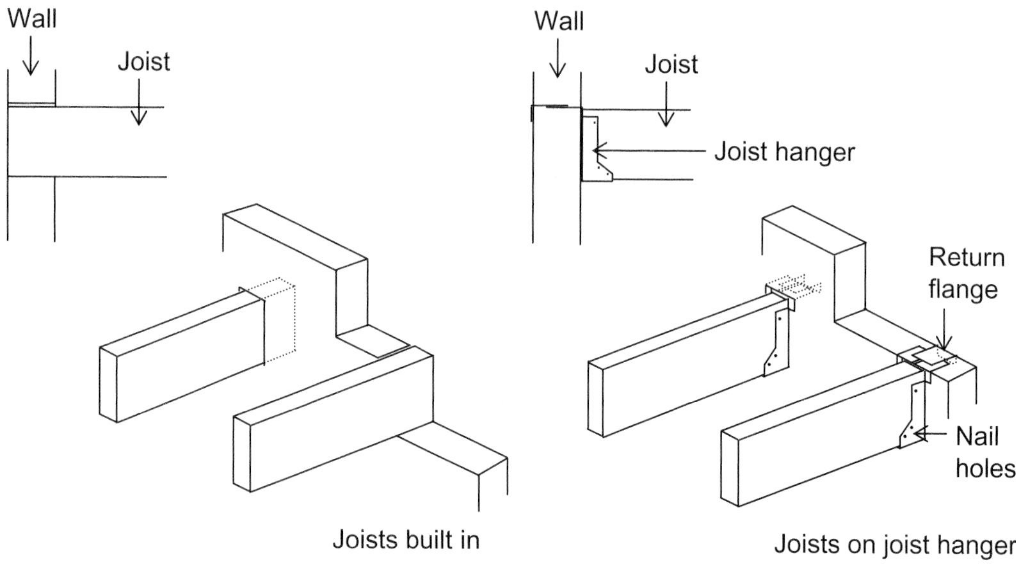

Figure 17.3 Timber-joisted floor – alternative forms of joist support.

should be at least 3 mm ring shank nails for improved grip in the board with a length 2.5 times the thickness of the board. To allow for movement, boards should not be cut hard up against the walls, but have a gap of 12 mm to the wall for an average room size and 3 mm under the skirting to allow for unimpeded expansion. Strutting is also either nailed or screwed to the joists.

The junction between floor and wall requires the transfer of load from the joists to the wall. The joint has to provide sufficient bearing so as not to cause excessive pressure (stress) at the point of transfer, which can lead to a crushing failure in either the end of the joist or the walling material. There is no particular requirement for the joint's structural action, but a simple pin joint would make the behaviour of the joist and wall more straightforward. The two options given in Figure 17.3 show, first, the simple building in of the end of the joist to give a full 100 mm bearing onto the wall and, second, the use of a joist hanger made of galvanised steel that can be specified with a return flange to provide additional stability to the wall. Both these simple, pragmatic connections provide a pin-jointed action so that the joist remains, in structural terms, simply supported.

Holes and openings

As with most elements of construction there is a requirement for openings, holes and voids to be formed. Small holes to pass services vertically through the floor can be accommodated between joists and through the boarding. There are few limits as to where or how big these holes can be so long as the edge of any cut (as opposed to drilled) board is supported on noggins, cross timbers fixed between the joists, at the edge of larger voids. For staircases or chimneys the opening is greater than the space between two joists, so the structure will require trimming. The arrangement for trimming and the use of hangers to make the joints are shown in Figure 17.4.

To run pipes and wires for the services systems horizontally the space between the joists is available. However, if services have to run across the joists some portion of the joist will have to be removed for the pipe or wire to pass through. Care is necessary not to affect the

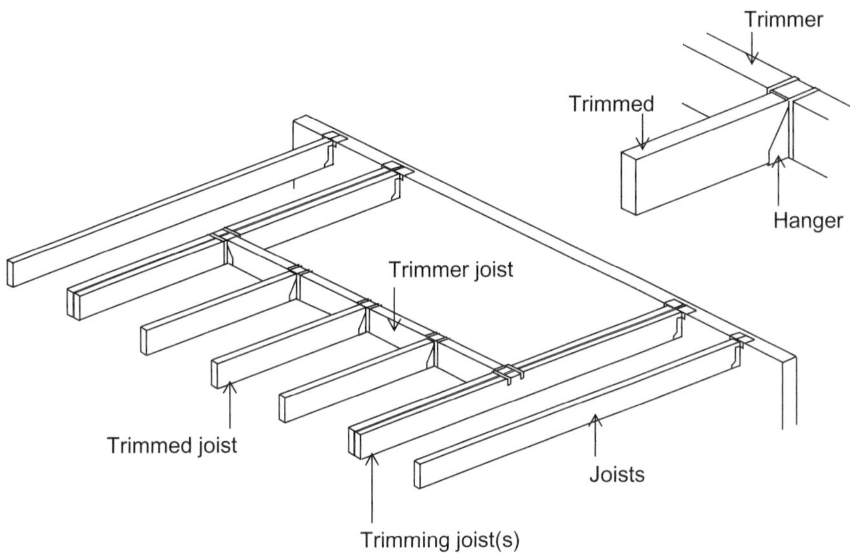

Figure 17.4 Timber-joisted floor – arrangements for trimming openings.

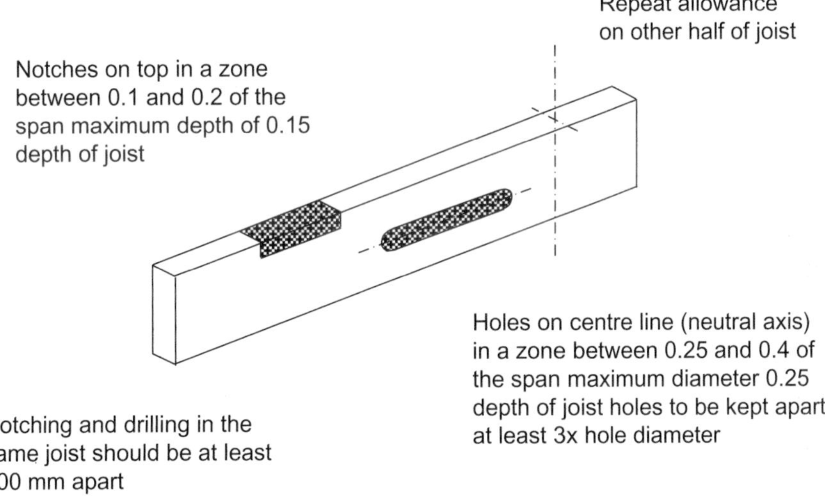

Figure 17.5 Timber-joisted floor – allowable areas for removal of timber for service.

structural integrity of the joist. The joist is a beam, and the way it is connected to the wall makes it simply supported, so the greatest bending stresses are in the top and bottom of the beam at the centre of the span and the associated shear is greatest along the neutral axis towards the supports (see Chapter 11). Removal of timber at these points is not advisable. Some guidance as to where holes and notches can be made is given in Figure 17.5. As bending stress reduces towards the supports, notches can be removed from the top (or bottom but not both). Towards the centre of the span reduced shear makes holes on the neutral axis possible. Across the centre of the span the design will have assumed a full section to resist the bending

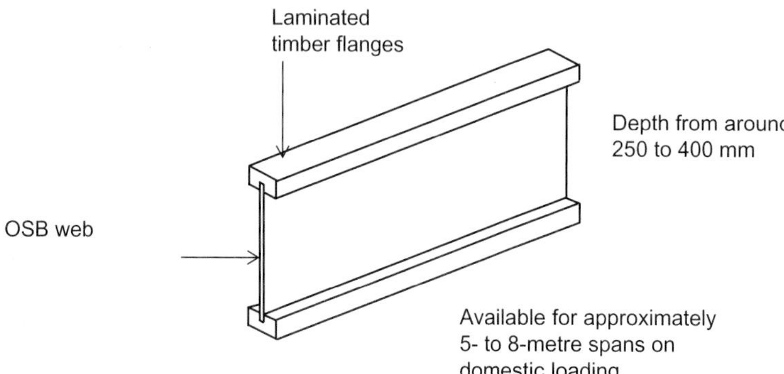

Figure 17.6 Timber-joisted floor – 'I' section timber joists.

moments and, while the material near the neutral axis is stressed less than that at the top and bottom, it is contributing to the overall bending capacity. Any reduction in this area may compromise the load-carrying capacity of the joist.

Return to original suggestion

This analysis began with the cross-sectional shape of the joist being taken as a rectangle. If solid rectangular timber sections are selected, the decisions on detailing and specification are all very logical, but, like all initial decisions, it limits the subsequent possibilities, in this case particularly limiting the economic span to around 5 metres. If greater spans were required without changing the basic material, returning to the shape of the cross section might prove useful.

A couple of other useful cross-sectional shapes for beams could be adopted: the 'I' section and the box section. Both provide solid material at the top and bottom, where bending stresses are greatest, give depth to limit deflection, while removing solid material to provide a more lightweight structure with less material. They do, however, have less material along the neutral axis and therefore have limited shear resistance, but on relatively lightly loaded longer spans this is unlikely to be the critical failure. Overall stability and dynamic behaviour is also likely to be less easy to achieve than with a solid timber section, so this will also have to be revisited.

Both of these alternative cross sections have been developed, but the 'I' section has proved to be more commercial. These are fabricated sections based on a composite timber board web and timber sections, often laminated to form the flanges, as shown in Figure 17.6. The use of composite and laminated timber products reduces movements such as shrinkage, bowing, twisting and splitting associated with solid timber sections, but as the basic material has not changed the majority of detailing and fixings can remain the same. Boards can be fixed as before and joist hangers are options for the connection to the wall. If these sections are used as trimmer joists or doubled up as trimming joists or partitions support, blocking at the web will be necessary where the loads are transferred and at intervals to stiffen the pair of joists. With the 'I' section the web contributes less in bending stress but has to take all the shear stress. This means that notches for services are not sensible at any point along the flanges, but there are more options for holes, particularly at the centre of the span and even near the supports if additional shear stiffening is provided.

The 'I' section beams are manufactured under factory conditions and would be ordered to length for each project. On site they are less stable than the solid section timber joists when laid out before the boards are fixed, so temporary strutting is necessary for the safety of the boarding operation.

Lifecycle considerations

Reflecting on the proposal so far, the issue of behaviour over time has only been considered for the possibility of dampness in the chipboard. Some more thought needs to be given to the main components so far suggested. Floor structures like this would be expected to last for the life of the building without any major maintenance. Having used the space for services distributions systems, however, these may need access for renewal or change to the pipes and wires. It may be better to consider the layout of the services and provide access hatches in the boarding, as the continuous nature of the glued joints makes subsequent partial access difficult.

Timber will deteriorate if the moisture content rises above 20% or conditions exist for a major infestation. This is unlikely for the bulk of the timber. However, if the timbers are built into the wall for support there is exposed end grain into the cavity. This will probably be protected by the insulation, keeping it both warm and dry, but if detailing suggests that this is not the case it may be better to apply preservative to the end grain on site, essential if the whole timber is specified with a preservative. If the timber is supported on hangers, the timber is likely to be cut to length on site, so end grain protection may still be necessary if the whole joist is impregnated with a preservative. The hanger will be galvanised and should, therefore, not be vulnerable as long as the galvanising is not damaged in site handling and installation.

Dismantling at the end of the lifecycle should not be difficult and the joists may be reused if the timber is still in good condition or possibly sent for conversion to new composite boards or some other recycling options. Given the gluing and nailing of the boarding sheets, removal may be difficult so disposal may not include reuse, always a possibility with old timber floorboards. Hangers can be recycled or cleaned for reuse at the end of the life of the house, depending on their condition.

Finishes and other elements

While some idea of a suitable ceiling finish using plasterboard was in the initial suggestion, it is necessary to return to the question of finishes to see whether the proposed structure can provide the background for both floor and ceiling finishes.

It is necessary for the background or substrate for the finishes to provide:

- Sufficient stable support
- Suitable fixing opportunities
- Compatible flatness tolerances
- Matching for future movement
- Avoiding deterioration mechanisms

For the floor the most likely finish is carpet. This needs full support and will follow the tolerances of the background. The boarding can provide these, but the surface may be rather hard (lack of resilience), so an underlay for the carpet will have to be specified. It is possible to use an adhesive to stick carpet down and this would be possible on the composite boards, but a better solution is provided by an edge fixing nailed to the boarding, which will overcome any relative movements. There should be no physical or chemical interaction between the carpet underlay and the boarding.

For the ceiling the underside of the joists does not provide continuous support but can provide good fixings at reasonable centres. The initial suggestion of the plasterboard will be compatible with the solution suggested so long as it is thick enough not to sag between joists. Plasterboard provides a good surface to paint or apply an additional finish. With the board fixed firmly to the joists initial movements may develop cracks at the junction of the ceiling and the wall where a coving may be used to cover and hide any cracking. All this can be achieved with a 12.5 mm board screwed or nailed to the joists.

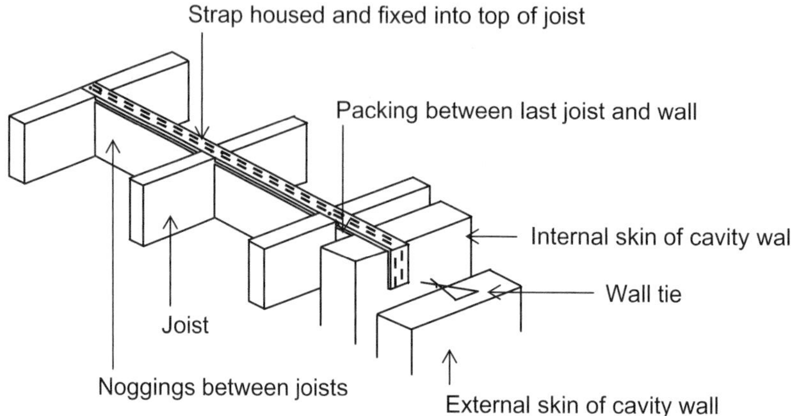

Figure 17.7 Timber-joisted floor – strapping of external wall to floor.

While the integration of services has been considered above in the section on holes and openings, it is necessary to see whether any other elements of construction might interact with the floor. In Chapter 19 the need for restraint for the external wall will be discussed and the floor can provide this. The hangers shown in Figure 17.3 can offer restraint where the joists transfer loads to the walls (note the need to nail the end of the joist to the hanger to ensure the joist provides restraint to the wall). Where the joists run parallel to the wall, straps can be provided connecting the inner skin of the wall to the first three joists, effectively holding the wall to the whole floor given sufficient fixing of the boarding to all the other joists. This detail, given in Figure 17.7, should be provided at no more than 2000 mm centres.

Internal walls that are not taking loads from the floor itself have to be built up to the underside of the ceiling. If these walls are made of blockwork, they are stable without any fixings at the top of the wall as long as the ceiling is not too high, which is unlikely in a modern house. However, if the partitions are lightweight studwork, they will need fixing to the underside of the floor structure to provide stability to the head of the partition. If the partition is running at right angles to the joists, fixings are available at each joist. If a partition is running parallel with the joists then, unless it happens to be on the line of a joist, no fixings are available. In this situation it is necessary to provide noggins in the floor, fixed between the joists at between 400 and 600 mm centres, specifically to support the head of the partition. These have to be shown on the working drawing for the floor, as, like the strapping for the external walls, they have to be fixed by the carpenters as part of the floor construction.

If lightweight partitions are built from the floor, they can be fixed to the boarding. Block partitions at right angles to the joists will spread the load across the floor, but where a heavy partition is required in the direction of the joists it may be necessary to provide a double joist immediately below the partitions.

Sound transmission loss

Changes in patterns of occupation show that bedrooms are increasingly being used during the daytime when other rooms below are also in use. This is changing the performance expectations of the floor, particularly for airborne sound reduction. The floor construction so far described might have been satisfactory in the past when bedrooms were used only at night but would fail to meet this new performance expectation. Performance levels for transmission loss (sound reduction index) of R_w of 40 dB are now expected from floors within

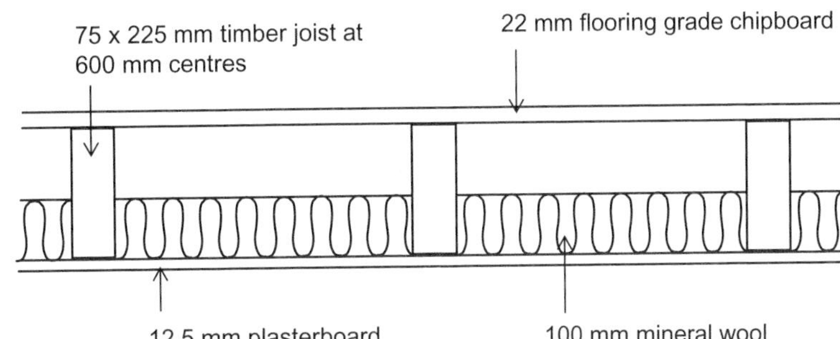

Figure 17.8 Timber-joisted floor – typical domestic floor construction for around 4.5 metre span.

dwellings. It is possible to achieve this performance by introducing a 100 mm thick sound-absorbent material in the cavity between the joists resting on the ceiling board. This, along with the other aspects of specification, is shown in Figure 17.8.

The inclusion of sound-insulating material will influence the space available for any pipes or cables being distributed in the floor cavity. It may now be difficult to use holes along the neutral axis, so services distribution will have to be across notches only allowable towards the supports. If electrical cables are laid under the quilting, its thermal insulating properties may cause overheating and will have to be taken into account in the sizing of the wires (see Chapter 21).

Ground floors

The choice of the construction for upper floors within a single dwelling discussed above was concerned primarily with establishing a structure capable of providing support for the imposed loads, stability for adjoining walls and a background for finishes to provide clean and stable surfaces with an adequate level of sound reduction. It is unlikely to have any other major function that determines the specification and detailing. This is not the case with the ground floor. Figure 17.9 shows diagrammatically the conditions to be expected below the floor and the conditions required above the floor. This indicates a number of different functions, including damp-proofing and resistance to the passage of heat. The conditions below the floor are potentially more aggressive to the deterioration of materials, and the risk of movement is greater. All these have to be taken into account in the choice of any proposals for the construction of the ground floor.

While the upper floor has to span over the rooms below, this is not the case for the ground floor. The ground conditions may provide sufficient strength and be stable enough to place the floor directly on the ground. This surface of stable ground has to be reasonably close to the desired ground-floor level. If achieving stable ground requires excessive excavation or fill, a suspended floor may be a more economic option.

If a suspended floor is deemed the economic choice, the construction for the structure is likely to be similar to upper floors, but it is now possible to build walls to support the floor at any desired position below the structure. More choice is available to select the span, giving an opportunity for economy. However, the additional functions and the more aggressive environment will make more demands on the detailing and the selection of materials.

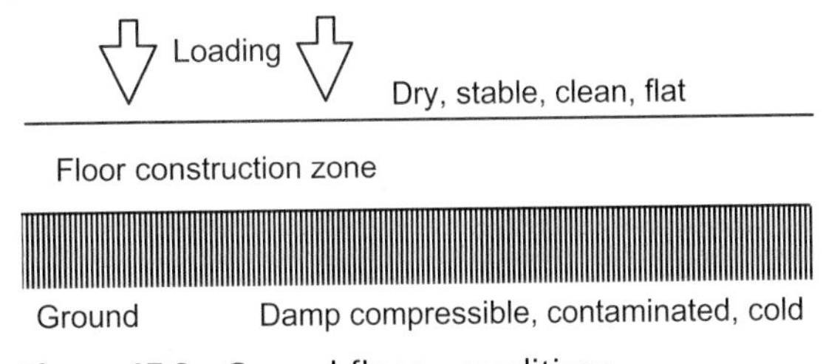

Figure 17.9 Ground floor – conditions.

Suspended ground floors

In traditional detailing, the floor would be a timber-joisted structure with similar details to the upper-floor construction. In a ground floor it is possible to gain some economy in the size of the joist by providing intermediate support on sleeper or honeycomb walls built on their own foundation with a damp-proof course (DPC) under a timber wall plate laid to receive the joists on top of the wall. This is shown in Figure 17.10. On top of the joists a board material would be provided, much the same as an upper floor. Under-floor ventilation and the sealing of the ground surface under the floor are important aspects of detailing for the timber solution. There is potential for damp air to build up under the floor, increasing the moisture content of the timber and leaving it susceptible to fungal attack. This can be avoided by a good flow of fresh air below the floor and a ground-covering layer below the floor. Although the ventilation is required to ensure the durability of the structure and not directly to provide clean air for the occupants, the mechanisms of ventilation discussed in Chapter 10 are being exploited in this detailing. They rely on natural ventilation gaining cross flow driven by the wind. This requires openings in the external walls on opposite sides of the building with external walls and an uninterrupted path for the air from one side to the other. Air bricks in the external skin, a duct across the cavity and then the holes in the sleeper walls provide this path for ventilation. The space between the ground cover layer and the underside of the joist has to be at least 150 mm and at least 75 mm to the underside of the wall plates on the sleeper walls. Air bricks have to provide an opening at least 1500 mm^2 per metre run of wall and be designed to prevent an infestation of rodents, as the space below a floor would be very attractive to animal life. These ventilation arrangements are also shown in Figure 17.10.

An alternative solution, a concrete beam and block floor based on joist-like beams with the surface formed with concrete blocks laid between them, has been developed. The general

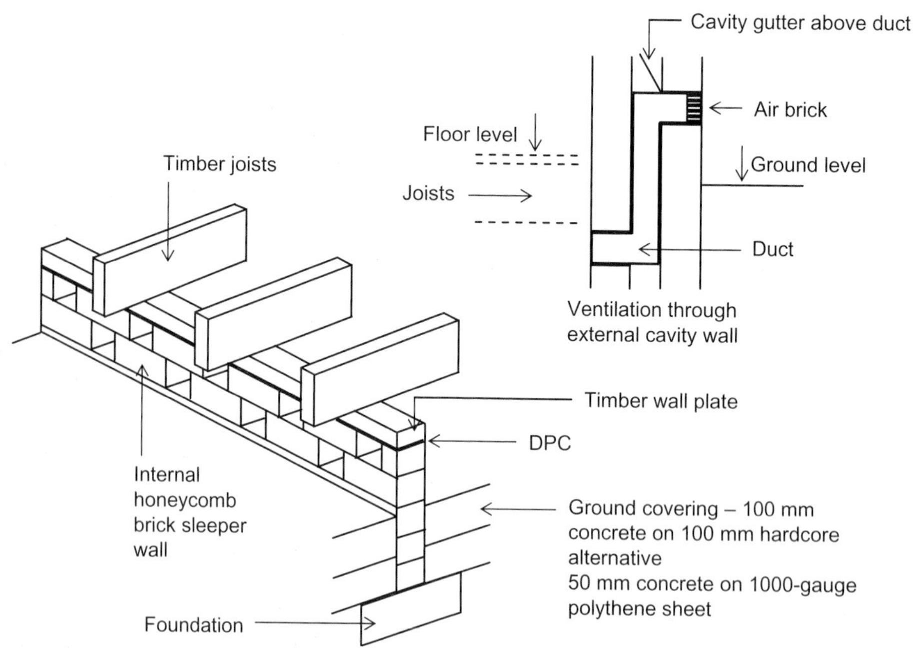

Figure 17.10 Suspended timber ground floor.

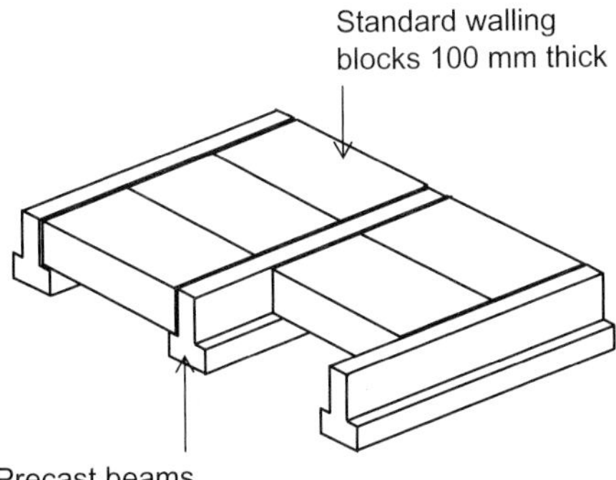

Figure 17.11 Beam and block ground-floor components.

form, shown in Figure 17.11, has been taken from upper-floor systems developed for commercial building, where spans and loadings are greater. However, a system of components suitable for the smaller scale of the house floor has proved economic.

The durability risk is reduced with the use of concrete instead of timber and therefore a ground cover layer is not necessary. However, under-floor ventilation detailing similar to the timber solution is still required. New recommendations are emerging, concerning areas where flooding is likely, which suggest that a means of inspecting and clearing out the under-floor space should be considered.

Where there is a risk of gas below the floor, it may also be necessary to provide a gas barrier on top of the floor structure protected with a screed. In these circumstances it will also be necessary to protect the wall cavities with gas-proof DPC cavity trays. The nature of these membranes and other provision for gas control are discussed in the section on ground-supported slab floors under 'Environmental control and lifecycle' below.

Resisting the passage of heat

The analysis so far of both the timber and concrete has shown that they have the potential to be economic structures, can control damp and the specification can be drawn up to ensure durability. If the walls are on sound foundations, the movement risk is now very low. However, they are not good at resisting the passage of heat, and the concrete beam and block will not have the required accuracy to apply a finish directly to the structure.

It will be necessary to introduce thermal insulation to resist the passage of heat. There are a range of materials that can be considered as good insulators, but their choice will not just rest on the best insulation value. It is, for instance, necessary to understand the form that insulating materials take: they can be loose fills, quilt, bats or boards. These vary in the stiffness of the layer they form, its compressibility and susceptibility to damage. These will determine what support is needed, the required fixings and aspects of the installation process.

Another consideration in the choice of material arises from environmental concerns. While one of the perceived needs for high insulation values is to reduce energy use and contribute to sustainable development, the materials chosen will have an impact on the environment during manufacture and right up to their delivery to site and installation procedures. The analysis of sustainable construction must account for energy at all stages of the lifecycle as well as pollution, waste and any direct effects on habitat and ecosystems throughout the life of the material. The rate of resource depletion and the consideration of renewable or recycled resources as well as the ultimate disposal as waste or the potential to recycle have to be investigated and understood. This may lead to the choice of a material with lower insulation potential that can be provided as a thicker layer to achieve the same insulating effect.

Yet another consideration when analysing proposals to insulate is the position of the insulating layer. While the overall resistance to the passage of heat (the U value) is not influenced by the position, the temperature gradient across the construction is. This is explored in Chapter 10, where the need to consider thermal capacity, condensation risk and cold bridging is also explained.

Practical considerations would lead to an initial suggestion that the insulation layer

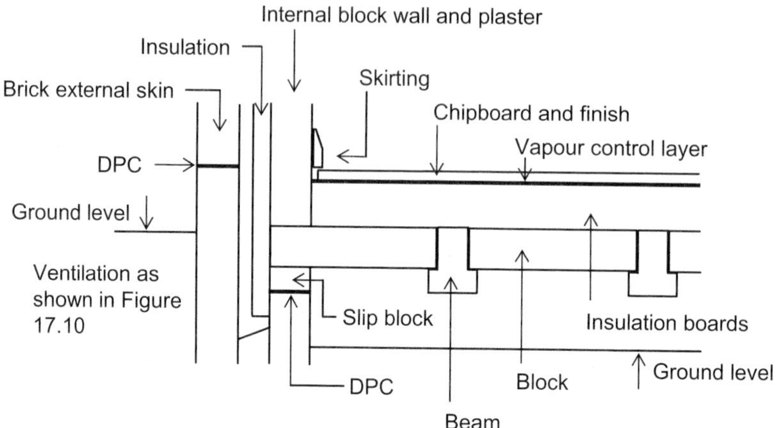

Figure 17.12 Beam and block ground floor – typical details.

should be positioned on top of the floor, as indicated in Figure 17.12. It can be installed when the house is weather-tight on a stable, cleaned surface. However, a condensation risk now exists across the floor structure. Condensation on the underside of the structure will be limited by any cross ventilation, but detailing a vapour control layer on top of the insulation will be necessary to minimise the risk of condensation below the insulation.

The insulation on top of the floor will be subject to the imposed loads of the occupants. It will have to have some strength and a board material may be the only option unless a cross-battening solution is considered, but this will introduce cold bridges, requiring thicker insulation to achieve the same U value. Even with a board material, impact or point loading will damage the insulation, and the surface will be neither stable nor clean and will not provide the fixings for the finish such as carpet. It will be necessary to 'float' (unfixed) a floor boarding material over the insulation on a vapour control layer. An example of this could be the flooring grade chipboard or OSB3 boarding with tongue and groove, glued joints on all four sides and a suitable movement gap at the edge under the skirting as the upper floor. The finish can now be fixed in the same way as on the upper floors.

If the insulation is to be placed below the structure, it will have to be suspended in some way, and if placed between the joists in the timber solution it will have to be thicker as the joists will form cold bridges, requiring more insulation to gain the overall U value. With insulation in this position the condensation risk is now reduced and other forms of insulation such as quilts or loose fills can be considered, but opportunities for edge detailing to reduce cold bridging may be limited if the structure is transferring its load to the external walls.

Having identified the position and type of material, it is necessary to specify a thickness. This will depend on the target U value and the conductivity of the insulating material, but the transmittance of ground floors is complicated by the way the heat is lost through the ground. Towards the perimeter of the building the heat will be lost back to the surface, while at the centre this effect is considerably reduced. The overall heat loss is dependent on the ratio of the perimeter to the area of the floor. The calculations for this are complex, but tabulated information is available to determine the required thickness.

Finishes

While this completes the analysis for the timber floor with all aspects evaluated, it leaves one question about the concrete beam and block solution: the required accuracy to apply a finish directly to the structure. If the insulation has been placed on top of the structure and a floor boarding floated on top of the insulation, this requirement is satisfied. An alternative to a board finish is a 50 mm thick sand and cement screed. If a screed is used, the carpet may have to be fixed with an adhesive, as nailing into a screed is not satisfactory.

Ground-supported slab floors

As the name of this solution suggests, the structural behaviour of this floor is not in bending but compression as imposed loads are transmitted directly to the ground through the slab. This simple view of all compressive forces assumes that the conditions for three other potential failure modes are not present: (1) that heavy impact or point loads will not punch through the slab; this is a shear failure but is very unlikely in a house and can be discounted; (2) that if the chosen material is subject to shrinkage then the tensile forces this creates can be resisted without cracking the slab. The most likely material for the slab is in situ concrete, which does shrink on curing. However, the size of the floor for a house is very unlikely to mean that the tensile forces will build up to create cracks of sufficient size to be a problem. This is, however, the case for large commercial floors where these tensile forces determine much of the detailing; and (3) that if the compressibility of the ground is not equal under the entire slab this uneven support may induce a bending action as the stiff slab bridges across soft areas of ground. This condition may exist on some sites and does have to be checked. This lack of strength or variable compressibility may come from natural variations in the ground, but these are likely to be small over the area of a house if the ground is of a reasonable bearing capacity. If the soil is weak, or the floor is being built over ground filled at some earlier date or it is on shrinkable clay, the possibility of this mechanism becomes much higher and needs to be taken into account. Another cause of this uneven long-term support can arise where fill (hardcore) is required to make up levels during construction. The process involves compacting in layers and, if this is not controlled, settlement of this fill will increase the chance of the slab cracking. If any deep fill is required to make up levels, a suspended floor may be a more economic solution.

As has been indicated the structural slab will be constructed in concrete. If the ground is stable, there is no need for this to be reinforced, as the compressive strength of the concrete will be sufficient to resist the forces involved. If the ground is suspected of being weak or variable under the slab, reinforcement can be introduced to ensure loads are distributed and any differential settlements do not distress the slab.

Reinforced or not, if the slab of concrete is built directly on the ground it will risk some cracking due to the relative stiffness of each material. For this reason the slab needs to be laid on an intermediate layer, stronger and stiffer than the ground but less rigid than the concrete. This material, often referred to as hardcore, is a fill of graded inert loose material such as quarry waste or crushed recycled construction waste compacted without any cementation binder. For a house on reasonable ground this will only need to be 100 mm thick and can be compacted in one layer. On top of this a floor slab between 100 and 150 mm thick will be sufficient for house loadings.

Environmental control and lifecycle

Having established a satisfactory structure, it is necessary to return to the information identified in Figure 17.9 to start the analysis of the environmental control elements of the floor. The construction of a concrete slab on hardcore will not stop damp rising through the construction to the floor surface and will provide little thermal insulation. It is not easy to change the specification of the hardcore to gain much advantage in its thermal properties, and while lightweight concrete for the slab would have sufficient strength and improved insulation it would prove more expensive.

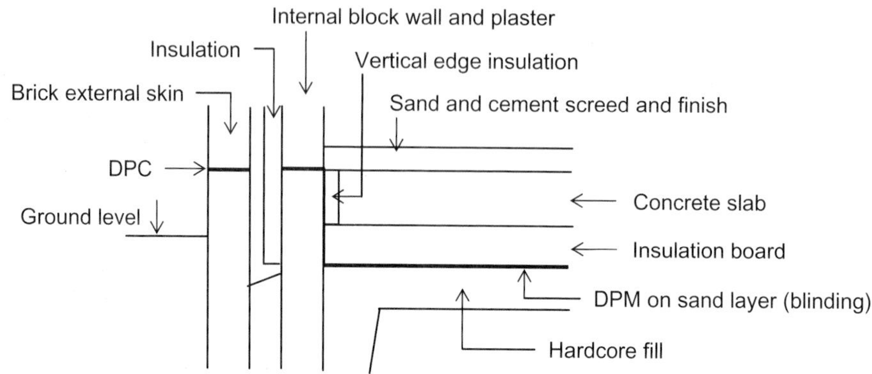

Figure 17.13 Ground-supported slab floor – typical details.

Damp resistance and insulation plus provision for a background for the finish will all require additional layers of materials with the right properties and of sufficient thickness to achieve the performance. Figure 17.13 shows one of the alternatives for these layers. It shows damp-proofing and the insulation between the hardcore and the slab and a sand and cement screed on top of the slab to provide the surface for the floor finish. An alternative is to place the insulation on top of the slab with a floated chipboard floor, as shown in Figure 17.12, over the beam and block floor. The damp-proof membrane (DPM) can then be either under the insulation or under the slab with a sand layer (blinding) over the hardcore to protect the DPM from damage. Whatever the position of the DPM, it is most likely to be specified as a 1200-gauge polythene sheet.

The polythene membrane can also provide a barrier for radon gas protection so long as the correct grade is specified. If a gas barrier for methane and/or carbon dioxide is required, a multi-layered sheet including an aluminium foil and a layer to improve tear resistance has to be specified. These membranes will also act as the DPM. For gas barriers, joints should be sealed with double-sided tape and the cavity should be sealed with a gas-proof DPC tray. Particular care to seal the membrane at service entry points is also required and is discussed below. The gas membrane represents a low-permeability barrier that may be used in conjunction with a high-permeability granular layer below the slab (in place of the hardcore) from which the gas can be removed in a controlled manner with passive ventilation.

The basis of choice of materials for insulation is similar to that explained for the suspended ground floor. Insulation placed on the slab will need to have a floating floor board or screed to provide the clean flat surface for the finish. Insulating the edge of the slab against the external wall to eliminate cold bridge is important. This heat loss path through the edge is almost directly to the outside air, representing a high proportion of the potential heat loss through the floor. As with the suspended floor the thickness of insulation required will depend on the ratio of the area of the floor to the perimeter of floor against external walls.

While the conditions below ground are the same as below a suspended floor, the materials that are being used are less susceptible to deterioration as the aggressive generators of decay for the hardcore and concrete are not automatically present. However, some care has to be taken in checking for harmful conditions. Given that the temperature of the soil under the house is unlikely to go below freezing, then, even if the DPM is above the slab, frost attack is not a problem. The presence of chemicals is probably the greatest risk. The most aggressive to be found naturally in the soil are sulphates, which react with the cement and disrupt the concrete unless a sulphate-resisting cement has been specified or the permeability of the concrete is very low. This is not likely to be the case with the relatively low grade of structural concrete necessary for this type of slab. With the increased use of brown-field sites some chemicals that are aggressive to hardcore or concrete may be present from previous occupations. For housing development, soil remediation will have to be undertaken to make the site safe for habitation and therefore many chemicals will be removed by pre-treatment or removal of the soil.

Another source of chemical attack may be brought in with the materials themselves. In talking about the fill layer it was indicated that the material should be inert. One of the major problems with hardcore from traditional demolition waste is that other materials such as plaster remain. With the advent of procedures to separate demolition materials for recycling and then some processing, including crushing, this recycled material is now much safer and can be specified with the confidence of using a controlled material. The other source of chemical attack comes from within the concrete itself, with the use of reactive aggregates in the concrete leading to alkyl silica reactions.

Services and other elements

Having considered the basic form of the floor, be it suspended or ground supported, it is necessary to think about the way other elements of the construction will interact with the chosen solution. Internal partitions that do not have their own foundation will be built from the floor. It may be necessary to strengthen the floor where partitions will be built. Thickening of the slab in a ground-supported floor or double joists in a suspended floor may be required for heavier partitions. The relationship with the external wall will depend on the role of the wall in taking load from the floor. If the wall is used to take the load from a suspended solution, the details will be similar to the upper floors. For the ground support solution the walls may contain the floor during the construction process but will have little interaction during the operational life of the building.

There will be interaction with the incoming and outgoing services: water, electricity, gas, communications and drainage. Each of these will come from different depths below ground and will have a minimum radius for the bend between horizontal in the ground to vertical in the building. They may also have to be maintained or replaced during the life of the building. These requirements are fairly easy to accommodate in the suspended floor but have to be detailed very carefully in the solid ground-supported slab. The provision of ducts cast into the slab for incoming services will need to be included and outgoing drainage will have to be incorporated in the slab before it is cast. While this is relatively easy, special care has to be taken in keeping the continuity of the DPM around these services penetrations. Particular care has to be taken if the membrane is specified to be gas proof, with the use of pre-formed top-hat units to seal around service pipes.

Site influences on choice of ground floors

Analyses of behaviour under load, environmental control and durability show that there are a number of solutions for a ground floor. The choice of the most economic is likely to be determined by the site. Many house builders have standard designs for the house but will have more than one option for the ground floor, which will be selected only after the site conditions are known.

The site conditions include not only the ground and the slope of the land but also the access, as this will affect the production operations. Much of the site investigation will be undertaken to determine the choice of foundations, and this will be introduced in Chapter 20. It is necessary to understand the production operations in order to undertake this aspect of the analysis before the choice is made.

Production operations for ground floors

Both the suspended and the ground-supported floor are constructed after the foundations have been completed, with the walls constructed to damp-proof course (DPC) level. This includes any internal walls with their own foundation, including sleeper walls for the suspended floor. The initial excavation will have removed the topsoil and reduced the level within the building to the underside of the construction for any ground-covering construction for the suspended floor or the hardcore fill layer for the ground-supported slab. All the operations to construct the floor therefore take place from the ground and within the external walls of the house. However, materials have to be prepared, stored and moved around the site.

The full construction of the floor will not take place in one continuous sequence. The basic structure will be completed before the external walls are built up to first-floor level. The finishing work will be undertaken after the building is weather-tight. These later operations are not considered in the broad analysis provided below.

Materials movement and storage varies with each solution. Timber is relatively light to handle but requires specific storage to ensure it is not overstressed and that its moisture content does not increase. Composite boards must also be kept dry and insulating materials, while they may not be directly affected by the wet, will be easily damaged. The concrete components of the beam and block floor are more robust but can still be overstressed if not stored the right way up and are heavier to transport and position. The fill material for the ground-support slab will need to be spread, levelled and compacted in layers and then blinded, for if the surface is not fine enough it may damage the DPM. This should allow the DPM to be laid easily, but care must be taken in sealing the DPM at joints and around the services and then at the edges to ensure continuity with the DPC. The concrete for the in situ slab has to be mixed, poured, spread and levelled and then cured.

Each of these operations requires different skills. The timber joist requires carpenters, while the beam and block requires perhaps a more limited set of skills associated with ground workers, although they may be installed by individuals with more traditional skills, such as bricklaying, to give continuity of work between walls and floors. Ground workers may construct the concrete slab or perhaps a dedicated concreting gang may be employed, although their skills are more associated with structural frame concrete operations and are unlikely to be used for more straightforward house construction.

All these operations will require temporary access for both machines and operatives. If possible, the two should be kept separate to minimise the risk of accident. The type of machinery will vary. Heavy transport such as ready-mix concrete lorries will need to get close to the building so they can discharge directly to the slab of a ground-supported floor. Most materials will be offloaded from the delivery transport and then moved on site by smaller transport, most likely based on the forklift. A suitable running surface will need to be provided for each of these transport and handling options.

As the work will take place at ground level scaffolds will not be required, so the main hazards will be associated with handling the materials. Only the wet concrete would be considered a hazardous material, requiring protective equipment for the operatives over and above standard personal protection. Collision with materials handling plant has been identified and is perhaps the greatest risk as the harm caused is likely to be great. Manual handling risks are perhaps greatest with the beam and block floor, where the beams weigh around 35 kg per metre run, making not only back injuries but also trapped fingers probable hazards.

The final choice

It is now possible to make the choice based on both performance and production criteria to match to the site conditions. However, one other consideration may influence choice. This is the scale of operations. For the single house the builder often had a particular mix of skills and may want to employ a more versatile machine to be used for many different operations. On a site with a large number of houses the opportunity for standardising and more economic scheduling of labour and machines will affect the best mix of resources.

Summary

1. Upper floors span over the rooms below, providing a clean, stable surface and contribute to privacy in reducing sound transmission between rooms above and below. They will be used to route services but will have few other functions within the single dwelling being considered in this case study.
2. The two general forms for upper floors, joisted and slab, can be considered, but currently the timber-joisted solution is likely to be the economic choice given the performance requirements within a dwelling.
3. The joist is traditionally a rectangular section in a structural grade timber with strutting to give stability to the floor. Boarding is most likely to be a sheet material such as flooring grade chipboard specified as moisture-resistant in potentially damp areas.
4. The boarding has to be joined together to make a continuous surface and then fixed securely to the joists. The connection of the joist to the supporting wall can be achieved by building the joist into the wall or being held in a hanger.
5. Openings in joisted upper floors that are greater than the distance between the joists have to be trimmed. Holes to pass services horizontally through the joists are limited in both size and position to maintain the structural integrity of the joist.
6. For longer spans, 'I' section joists constructed from laminated flanges and a board web can be used.
7. The joist and boarded floor provides a surface for the finish for the rooms above and provides support to fix a board material, normally plasterboard, to the underside to form a ceiling. The floor is now sufficiently stable to provide a diaphragm to support the external walls. Internal partitions below may require head fixings, and partitions above may require a double joist if running parallel with the joists.
8. Ground floors have different performance requirements, mainly from damp control and resisting the passage of heat, and are in a more aggressive environment with respect to deterioration of the material.
9. These ground floors can be suspended joisted solutions or ground-support slabs. If suspended solutions are adopted, concrete beam and block offers an alternative to the traditional timber form. Ventilation for the void below the floor is an important detail for suspended construction.
10. Insulating materials to resist the passage of heat can be specified in many forms that vary in the stiffness and compressibility of the layer, and this is important if the insulation has to take the imposed loads from the room above. The position and thickness also have to take into account condensation risk and cold bridging.

11. The ground-support slab constructed with in situ concrete, possibly reinforced, will be laid on a layer of compacted fill or hardcore. It will require a damp-proof membrane (DPM) normally placed under the slab. Insulation can be introduced in a number of positions with detailing at the edge important to eliminate the potential cold bridge. Gas-proofing may also need to be considered.
12. In addition to performance requirements, site conditions and production considerations will influence the choice of construction.

18 Roofs

This chapter focuses on the pitched roof, although flat roofs are introduced. The general forms of both waterproof coverings and roof structure are discussed, but the chapter provides a full analysis of concrete interlocking tiles on a trussed rafter structure as the most likely solution for the case study.

Pitch and span

The choice of construction for a roof is dominated by the pitch and span. The span (distance between supports) influences the structural solution, the rainwater collection and disposal required and the movement to be accommodated, particularly in the waterproof coverings. The pitch (angle of the roof surface) has a major influence on the specification of the waterproof covering, will dictate the form of the structure and have a significant effect on the appearance of the building.

Roofs are normally considered either flat or pitched. The use of the term flat roof is not a very good description as the angled surface of most pitched roofs is flat and even a flat roof has to have a fall (i.e. is not level) in order to ensure that rainwater is effectively shed from the roof surface. However, it is a useful distinction in so far as 'flat' indicates that the fall is so small that it is not perceived as pitched, so the structure can be treated as a floor and the waterproofing covering has to be continuous with all the joints sealed.

For houses designed for the UK the level of rainfall and the strength of the wind are likely to make the pitched roof the dominant form as this can employ economic waterproof coverings that are more reliable, making it a long-lived and generally more robust solution. It does, however, make the requirements of the structure more complex than the simple joisted (or slab) solution that can be adopted for a flat roof. While both the design of the house and the availability of an economic covering will dictate the actual pitch, the widespread adoption of the pitched roof is part of the characteristic appearance of the UK house.

Other functions

So far, two parts of the roof have been identified: the waterproof coverings (with rainwater disposal systems) and the structure. These two are linked by the need of the structure to support the chosen covering so that it maintains its waterproof function. In most designs there will be a need to provide a ceiling for the room below that will also be supported by the structure. The roof that is just waterproof and has a ceiling is very unlikely to provide sufficient resistance to the passage of heat, so consideration will have to be given to the thermal performance.

While the roof space has been used to house service tanks and has by some householders been used for storage, it is increasingly likely that designs will call for living space to be incorporated in at least part of the roof. This will not only influence the structure and the detailing of the ceiling and insulation but also introduce the need for light, through either windows that extend above the roof line or roof lights installed in the plane of the roof surface. These will call for holes to be formed in both structure and covering. Even if light is not a requirement, holes may be required for chimneys and other service pipes, but these are likely to be smaller than for windows and roof lights. These aspects will not be covered in detail in this text.

When sustainable development is considered, the roof, particularly when south facing, can provide a surface to collect renewable solar energy, in the provision of an area for both solar panels that warm water to heat domestic hot water and arrays of photovoltaic cells to generate electricity (see Chapter 21). These units have to be fixed to the roof above the coverings and will require services (pipes and cables) to pass through the roof.

In providing physical security and privacy the specification and detailing for the other functions will probably give sufficient performance unless local conditions prevail, for example sound transmission if the building is under a flight path. However, the risk associated with fire has to be considered in all terraced and semi-detached designs that require separating walls (see Chapter 19) between the dwellings. Fire can spread along the surface of the roof coverings over the separating wall. The separating wall can be extended above the roof line, but this creates the need for weathering details with initial expense and a possible maintenance requirement. If the covering passes over the separating wall, it must be chosen to resist the spread of flame and detailed to ensure the fire is stopped over the top of the wall. This will also apply to the detailing at the top of the wall to ensure that sound transmission loss is maintained across the separating wall.

Pitched roofs

Before any discussion of pitched roofs can take place, it is necessary to understand the vocabulary that is used to describe the configurations of pitched roofs that are required to cover a range of plan arrangements. The basic configurations and terminology used are given in Figure 18.1.

Eaves are edges at which water is collected for disposal, normally in gutters connected to downpipes to take the water to underground disposal systems. Verges represent edges where water is not expected to collect and are most usually associated with gable ends. Ridge, hip and valley are all formed where the pitched surfaces meet. While the ridge and hip shed water away down each of the roof surfaces, the valley effectively collects water. It has to be constructed to ensure the water on the roof is taken to the eaves. If the tiles themselves cannot be

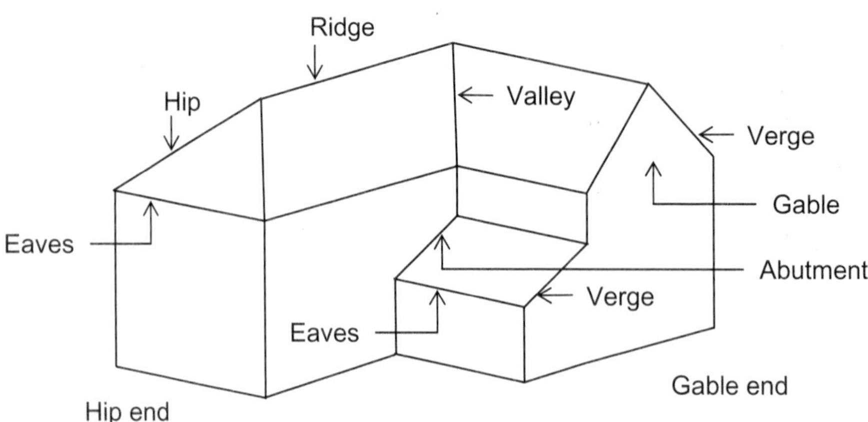

Figure 18.1 Pitched roof – configuration and terminology.

'swept' around the valley to maintain the weatherproofing of the covering, a gutter has to be formed to collect and direct the water to the gutters and downpipes at the eaves. Abutments are the meeting between roof surfaces and walls, so the joint between them has to be sealed so that the water runs to the eaves and not into the roof space.

These junctions may need detailing with special tiles and components in other materials, for example flashings and soakers in lead at abutments. Gutters formed at valleys have traditionally been in timber and lined in lead, although pre-formed sections in plastic materials can be used, fixed directly to the roof structure.

Coverings

There are three generic types of roof coverings, each of which has been developed into a number of common forms for which details for specific buildings can be developed. These are given in Table 18.1, which indicates the following forms:

- Continuous coverings
- Large sheet coverings
- Slates and tile coverings

The continuous coverings are all made from impermeable materials where all but asphalt can be formed into thin, flexible sheets to be laid on the roof, making them light in weight. Asphalt is an in situ, hot-laid process. All these coverings will require continuous support and fixing against wind uplift and, if used on pitch roofs, against sliding. Fixing can be achieved either by adhering the material to the support base or by mechanical fixing. The joints and edges have to be sealed to be weather-tight, either by adhesion as with bitumen for the built-up felts or by dry folding of seams on the metals. This action of folding the materials, forming welts and drips, relies on bringing the materials into close proximity on the folds to overcome both gravity and capillarity, which would draw the water into the unsealed joint. Asphalt is spread onto the continuous support in layers to a total thickness of around 18 mm, making this covering significantly heavier than the thin sheet alternatives. The joints become sealed in the cooling process, although joints are also staggered between each layer.

Table 18.1 Alternative roof coverings.

Generic forms	Common types
Continuous covering	Built-up felts Single-ply polymers Sheet metals Asphalt
Large sheet	Corrugated metals Corrugated plastics Corrugated cement compounds
Small tiles, slates or shingles	Clay or concrete tiles Natural and synthetic slates Timber shingles

This combination of impermeable materials and sealed joints means that these coverings will work on flat as well as pitched roofs, although asphalt will slump with time and is therefore rarely specified for pitched roofs. These coverings have commonly been applied to flat roofs for small domestic buildings (built-up felts) and to commercial buildings (asphalt, metals and single-ply polymers) as well as to pitched roofs on more prestigious buildings (copper and other metal sheets) including some housing.

Large sheet coverings are also made of impermeable materials and therefore in relatively thin sections. These sheets are, however, stiffer by the nature of the material, its thickness and the shape into which they are formed. The sheet is given strength to span between supports by shaping the material into corrugations. Although corrugations were originally created based on the curve, giving the characteristic shape of corrugated iron, many of the metals now have folding based on the rectangle. Whatever the shape of the corrugations, the sheets need to be fixed to the supports against wind uplift and sliding. These sheets are normally laid with both side and head joints lapped, so that they need not be sealed so long as the lap is large enough and the pitch sufficient to maintain adequate drainage run-off. The corrugations are used to assist weathering on the side lap as they are lapped on the crest of the corrugation. On lower pitches, head and even side laps will need to be sealed with strip mastic as the sheets are laid. Modern sheet materials can be colour-coated as a direct finish.

Tiles and slates are made from materials of variable permeability. Natural and artificial slates can be considered impermeable and can therefore be used in thin sections. The material is rigid but cannot be shaped, and therefore the strength of the material alone will limit the span between supports. Because the slate is flat, the side laps have to be butt joints, although lap joints can be used at the head. Supports also have to be close together as the slate will only be of limited size and each will require support. The thin material limits the weight of the individual slates, but because they are flat and have butt side joints they have to be laid as a double lap, as illustrated in Figure 18.2.

The other materials such as clay and concrete are permeable and therefore will have to rely on their thickness to 'store' water during rain and release it during dry periods. Tiles of concrete and clay are often shaped, not to provide any significant gain in strength but to provide weathering at the side (and in some cases head) laps so that the tiles can be used in a single lap. The single-lap patterns and typical forms of tile are shown in Figure 18.2.

The joints between tiles or slates are not sealed so they can only be used on pitched roofs. The design of the side and head laps will determine the minimum pitch to which each design will work. Although relatively heavy, the tiles and slates will have to have some fixing against wind uplift and must be held to prevent sliding down the sloping face of the roof.

In the UK, tile and slate roof coverings have become the dominant forms of use for small-scale structures such as housing. The generic form lends itself to the use of many locally found materials that can be economically produced for a local market. Indeed, in many areas the traditional houses are characterised by the dominance of the locally available roofing materials. With larger-scale production, particularly in concrete, the local feel is often retained in the finish, colour and possibly the shape of the tile.

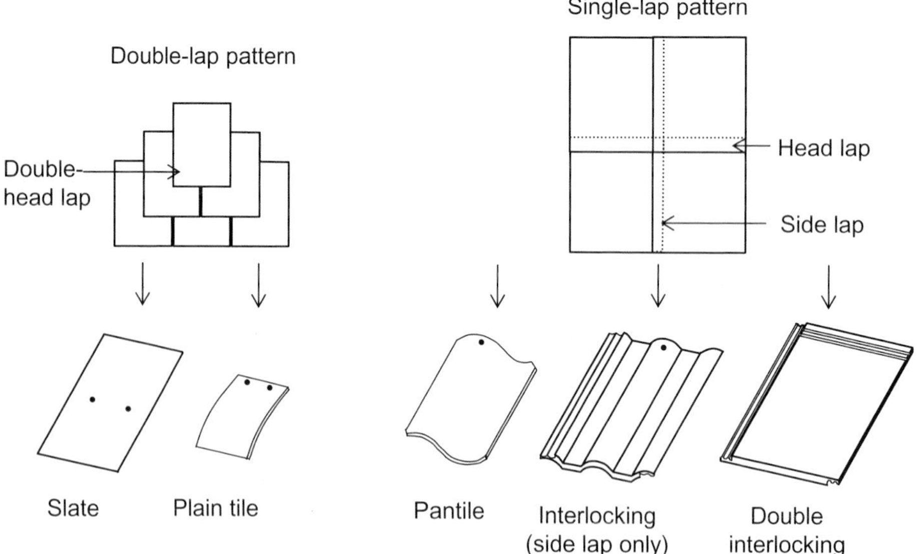

Figure 18.2 Slates and tiles – patterns and types.

More traditional materials such as stone, thatch and turf have not been discussed, but they fit into one of the three generic forms. Stone is most often used like a slate, only its higher permeability leads to thick and heavy coverings. Thatch and turf are variants on the continuous covering, but unlike the majority of currently used materials they are permeable and hence often have considerable thickness. They rely not only on 'storing' the rain till dry weather returns but also on drainage taking place within the material down the slope of the pitch with an underlay (traditionally bark) to form a surface over which any residual water can drain to the eaves of the roof.

There is a potential for reuse of old tiles and slates for refurbishment projects, but often old roofs are replaced because the tiles are weathering or deteriorating to the point where the reliability of the old roof is in question and therefore the old tiles are seen as of limited value on new work.

Interlocking concrete tiles

Other than in specific conservation areas or on more bespoke or prestigious designs, the concrete interlocking single-lap tile has emerged as a common choice of weatherproof covering for new housing in the UK. It provides an economic solution, with variations in finish, shape and colour available. In many cases traditional shapes of clay and even flat stone tiles and slates have been reproduced in concrete, which, while more regular than their traditional counterparts, can provide a local look to the roof covering.

The concrete tile has to be cut to form hips and valleys, and a range of special tiles are needed to create the ridge and hips, including flue terminal and ventilation tiles to work with eaves ventilation to limit condensation in unheated roof spaces. Some care is required in setting out the tiles to ensure that whole tiles are used at the verge as well as ridge and eaves, as part tiles are vulnerable to wind uplift, a particular risk at the verge.

The concrete single-lap interlocking tile is manufactured in a variety of styles or profiles that may influence the size. They are, however, normally around 420 × 330 mm and weigh around 50 kg/m² when laid on the roof. Tiles made to look like traditional shapes, such as the pantile, will maintain the traditional size, in this case around 380 × 230 mm, but the overall weight will be similar. Each tile has two nibs to hang them to the batten and most have the provision of a single nail hole to fix the tile down, as shown in Figure 18.3. Where nail holes are

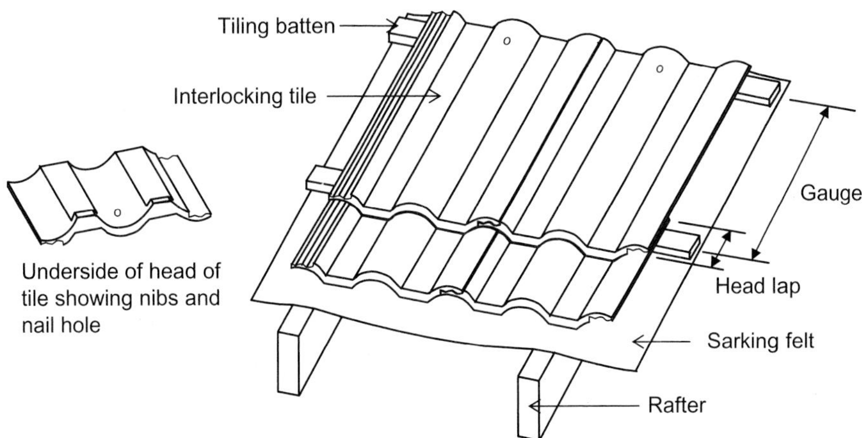

Figure 18.3 Interlocking tiles – fixings and felt.

not provided, the tiles will need to be clipped against wind uplift.

The battens are normally 38 × 19mm for support at up to 450mm centres and 38 × 25mm if support is only provided at 600mm centres. For the 420 × 330mm tile, battens at a gauge of 345mm will give a head lap of 75mm. While all edge tiles have to be nailed or clipped, the exact requirement for how many other tiles should be fixed will depend on the local exposure conditions, but some proportion of the tiles will probably need to be fixed down. Additional clipping to the front edge of these tiles, particularly at the eaves, and external verge tile clipping may be specified to overcome wind uplift forces.

Like all roof coverings, each tile type will have a minimum pitch at which the tiles will remain weatherproof, which can, to a limited extent, be varied by the size of the head lap. The style of the tile determines the minimum pitch, with some profiles being suitable down as low as 17.5° with the increase in head lap; others are only suitable for 30° pitch. Manufacturers' recommendations must be consulted for a particular specification, including any maximum pitch. The head lap will also determine the weight of the covering and the gauge of the battens.

With the specification so far there remains a risk of water ingress. In periods of prolonged or heavy rain and wind some water may get through the tile layer. Wind-blown snow can penetrate the laps in ways rain cannot. Any damage to tiles will also lead to a loss of the weatherproofing function, which could cause internal damage even in the short term during a storm, before any repair is effected. To provide a secondary line of defence against these circumstances the battens should be fixed over a layer of sarking felt. This is shown in Figure 18.3. This material is a thin impermeable sheet traditionally felt-based (hence the name) with a bitumen-impregnated hessian layer but now is more likely to be a plastic-based material. Based on polyester, reinforced polythene or spunbonded polypropylene, these underlay materials are more durable and have greater tear resistance. They can now be obtained as a breathable material that remains waterproof but allows the movement of water vapour in order to reduce the overall vapour resistance on the cold side of the insulation, necessary in unventilated roofs. Although lapped to ensure continuity of weather protection, the laps are not sealed as the level of exposure beneath the tiles is very sheltered and large volumes of water are not envisaged. The felt sags between the rafters to form a drainage channel under the battens leading to the eaves. At the eaves, care must be taken to detail the felt at the gutter to make sure that any water that does get past the tiles can be effectively passed into the gutter. This detail is discussed later in this chapter.

It is not the intention of this text to give details of all the options for roof coverings. If other options are to be investigated, it is necessary to be sure that the basic waterproofing mechanism is understood, and how the covering is supported, jointed and fixed to the structure, together with some analysis of the risk of damage in the event of failure of the primary waterproofing layer. All of this will determine the performance of the structure, the loading and the movement that will be allowable without loss of performance of the weatherproofing function of the roof. For instance, the use of plain tiles is a traditional double-lap system, as shown in Figure 18.2. The tiles are 268 × 165mm with two nibs and two nail holes for fixing to the battens at a gauge of 100mm. Although a small tile, its double-lap pattern means a loading of around 80kg/m^2. The tile is cambered to help break the capillary path between the tiles; however, its size means that it still only works on a minimum pitch of 35°, but it works well vertically (all tiles nailed) as a cladding material.

Structure

As the floor was based on the joist (Chapter 17) so the pitched roof is based on the rafter. Figure 18.4 shows the basic arrangement and variables of the rafters. As with the joist for the floor the typical size of the house generates limited spans, so the rectangular-section timber rafter is the initial potentially economic choice. However,

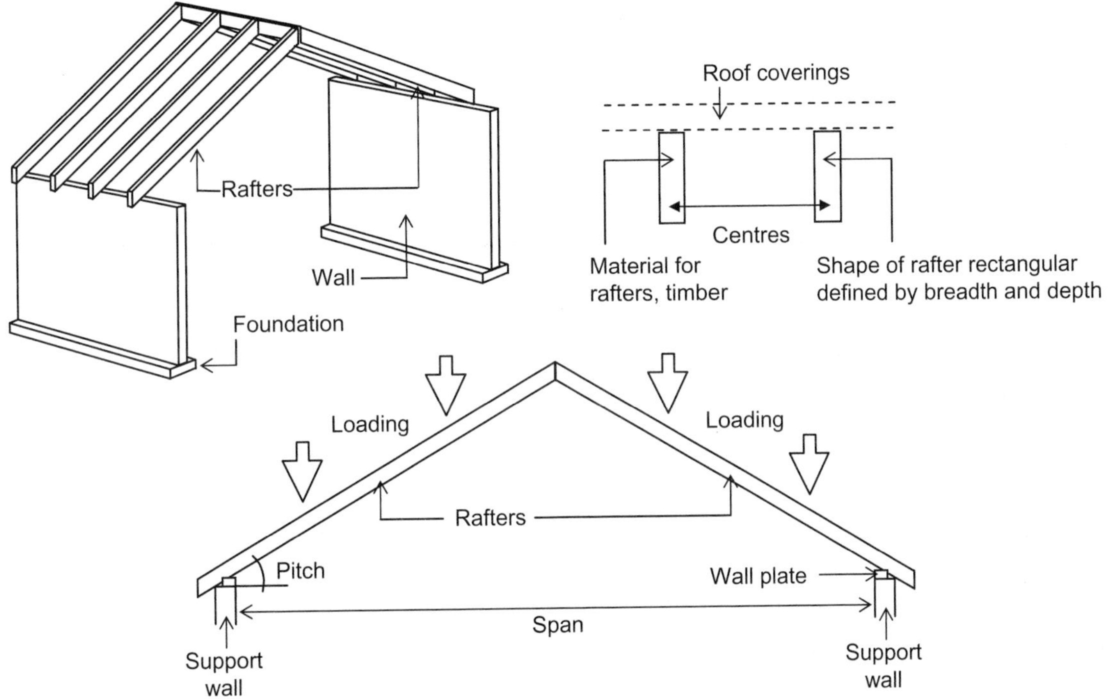

Figure 18.4 Timber rafters – arrangement and variables.

unlike the floor the rafter has to be supported at the required pitch. For the dual-pitched roof a pair of rafters provides mutual support at the ridge. With this loading and support pattern there are six possible modes of structural failure, illustrated in Figure 18.5:

- Overturning of rafters
- Thrust at the wall
- Deflection of the rafter
- Lateral torsion buckling in the individual rafters
- Failure of the rafter in bending or shear
- Wind uplift

The pairs of rafters are very vulnerable to overturning, so even on the smallest spans some longitudinal members will be required. While the tiling battens act in this longitudinal direction, specific timbers associated with the roof structure need to be provided. The simplest of these is the ridge board inserted between the two rafters running the full length of the ridge up to the gable walls. This arrangement of rafters and a ridge board is known as a couple roof and is shown in Figure 18.6. The ends of the rafters at the wall are fixed (birdsmouth cut and nailed) to a wall plate. The wall plate is a timber around 100 × 75 mm bedded flat in mortar on the top of the wall. It provides a bearing for the end of the rafter and level support to maintain a good line on the end of the rafters for the eaves.

With an increase in span of the roof and/or slenderness of the wall the thrust soon becomes a major instability, limiting the span of the couple roof to around 3 metres. While it is possible to solve the overturning of the wall in the design of the wall itself with, say, buttressing, it is more economic to tie the feet of the rafters together with a ceiling tie. This forms a structure known as a close couple roof, which is shown in Figure 18.6. The ceiling tie also acts as the ceiling joist as it provides support for the

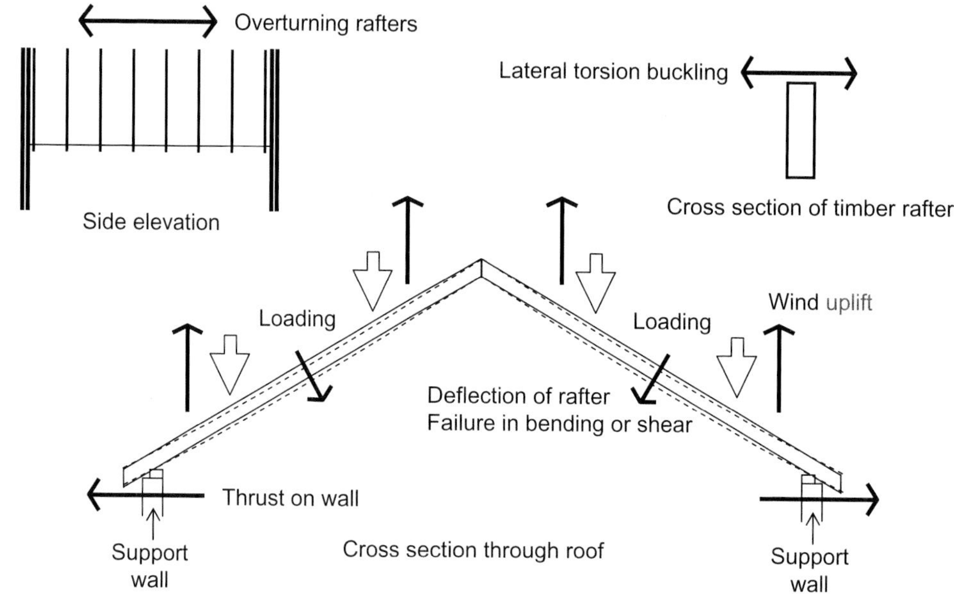

Figure 18.5 Timber rafters – modes of structural failure.

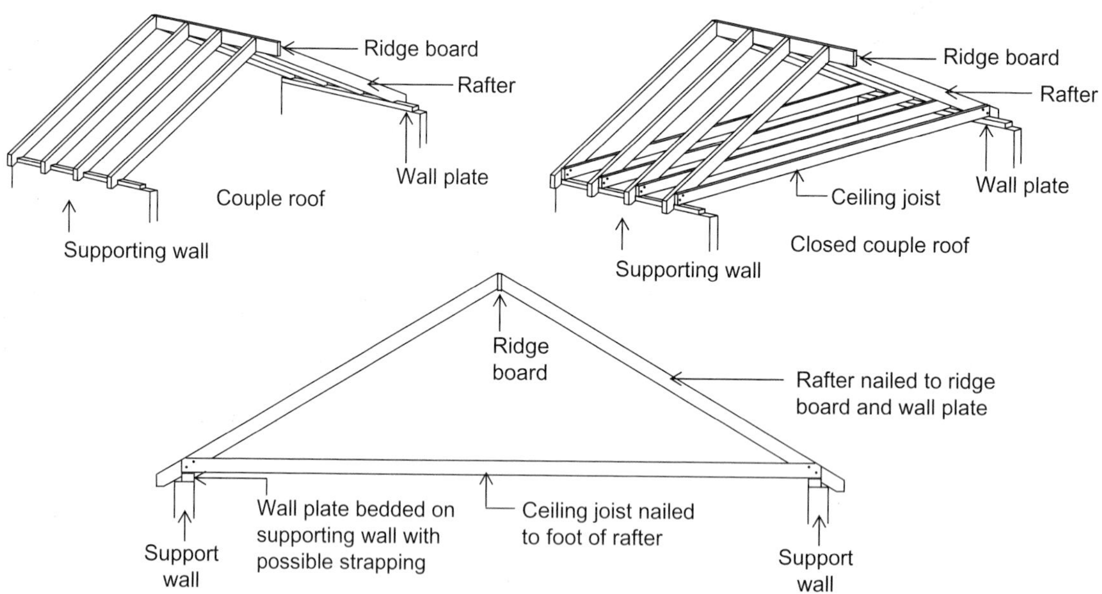

Figure 18.6 Roof structure – couple roofs.

ceiling and any storage or maintenance activity in the roof. This induces bending, but the primary structural behaviour for the roof as a whole is in tension. For the structural action in tension to develop it is necessary to fix the rafter and ceiling tie together, which can be achieved by nailing. The size of the ceiling joist is determined by the bending action and will therefore appear similar to floor joists. It is possible to reduce the size of the joist by breaking up the

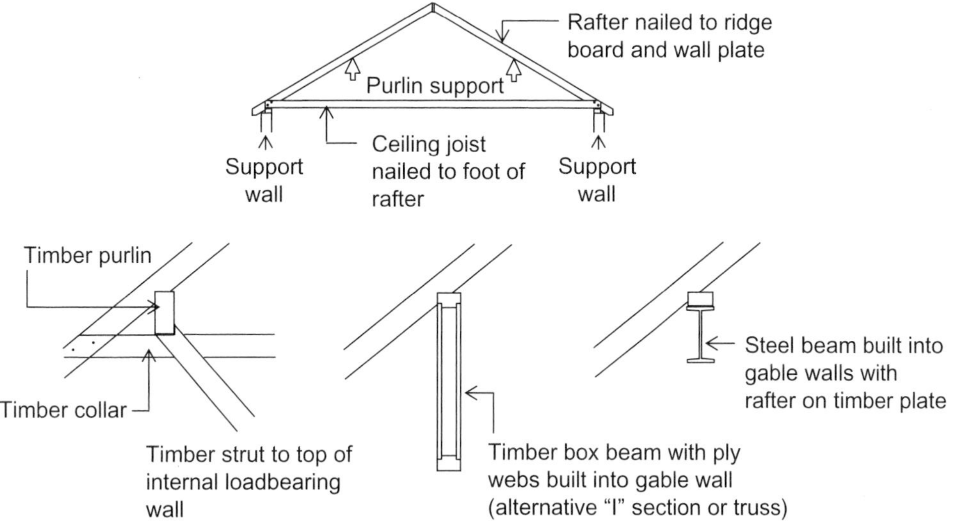

Figure 18.7 Roof structure – purlin support alternatives

span with binders if suitable support can be found for the binders.

As the span or the pitch increases the length of the rafter (and ceiling joist) increases and hence so do the bending forces. Deflections increase, as does the potential for lateral torsion buckling and the ultimate failure in bending or shear. Shear failure is unlikely, as the loading is relatively light and the span relatively long. This condition often makes deflection the design criterion even before a bending failure is likely. To overcome either deflection or bending failure additional depth of rafter is the most effective way to provide resistance to the bending forces. However, the width of the rafter is also significant in that it limits the tendency to lateral torsion instability. The need to increase both the depth and width of the rafter increases both self-weight and cost. The economic limits for this closed couple structure with its simple pair of rafters and a ceiling tie is around 5 to 6 metres.

It is possible to identify two alternative ways to provide a structure with a longer rafter and hence provide economic roof structures with over 6 metres in span:

- The purlin roof
- The trussed rafter

The purlin is a longitudinal member that provides intermediate support to each rafter, as shown in Figure 18.7. The purlin in turn has to be supported, the frequency of which will determine the span of the purlin and hence its cross-sectional size. The purlin is more heavily loaded than the rafter and can therefore only provide support over relatively short spans if the purlin is to be constructed with a rectangular timber section. While rafters are likely to be around 50 × 125 mm, the purlin is likely to be around 75 × 200 mm and will require support, at between 1800 and 2400 mm centres. This will require the use of an internal loadbearing wall to provide a support for the struts from the purlin, as shown in Figure 18.7, although the ends of the purlin can be built into a gable wall.

It is possible to use either a steel beam section or a deep timber purlin made from a composite timber member in a box, 'I' or truss form to span between gable walls and hence eliminate the need for the internal loadbearing wall. These options are also shown in Figure 18.7. While the purlin in steel would remain a relatively small section, the composite timber solution can become quite deep, making this an option that restricts movement in the roof space. Another option to eliminate the internal loadbearing

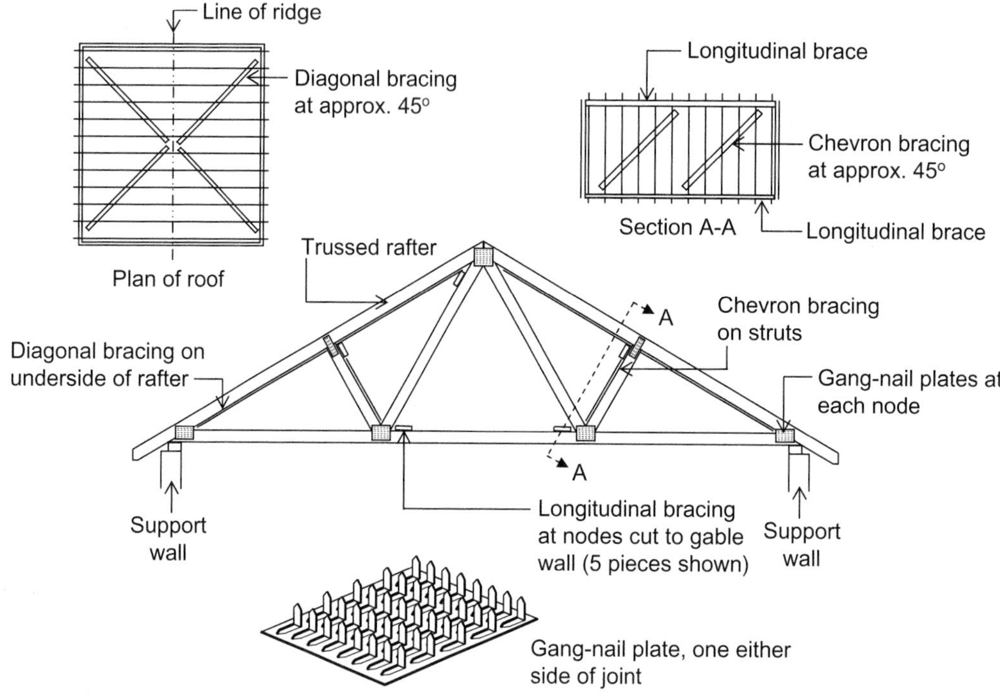

Figure 18.8 Roof structure – trussed rafter roofs.

wall is to use trusses at every fourth rafter to support the purlin with common rafters supported by the purlin in between. This adds expense and again limits the space available in the roof. The use of the truss and common rafters is a very traditional solution used for open barns and halls since medieval times but does not prove economic for houses.

However, in more recent times the idea of the truss has been revived in the introduction of the trussed rafter. This has been made possible by the development of the gang-nail plate to make the joint between members. This is illustrated in Figure 18.8. It provides a fixing that allows butt jointing of the timber members and has the ability to develop sufficient tensile strength in the joint to construct a lightweight pin-jointed frame (see Chapter 11) for each rafter. Although it is possible to fix gang-nail plates on site, they gain their full economic advantage in factory assembly processes even with the additional transport costs of delivering the trussed rafter as a whole.

The structural efficiency of this system leads to slender sections. This creates the need to not only handle the trusses with care during transport and assembly but also reassess the stability of the roof structure as a whole. It is not possible to use a conventional ridge board and the slender timber sections are more susceptible to lateral distortion, particularly those in compression. This leads to a requirement for a number of longitudinal and diagonal bracing members, as shown in Figure 18.8.

Even with the need for care in handling and the additional bracing, the trussed rafter currently provides the most economic solution. Its popularity has led to a wide range of components to allow hip and valley roofs to be constructed and alternative truss configurations to allow rooms to be built into the roof space. Standard designs for roof spans of 12 metres

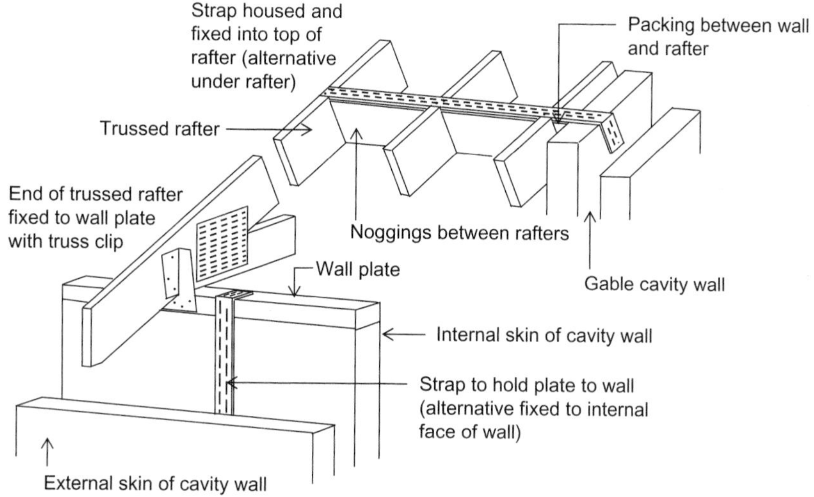

Figure 18.9 Roof structure – strapping to external walls.

and over are now available and most manufacturers offer to design and manufacture specialist design options.

Having specified a structure to support the finish to ensure its performance from the vertical loading, it is necessary to consider the proposed solution under wind-loading uplift conditions. If the uplift is sufficient to overcome the dead load, failure is possible. Given that the fixing of the finishes is sufficient to resist their separation from the structure, the effect will be initially to relieve the load and then, with increasing wind speed, the lifting of the roof as a whole. If this is a possible design condition, the roof structure must be strapped down to the walls. This is a common condition for the flat roof, where lightweight coverings are used and strapping is an essential detail. With a pitched roof, the weight of the finish makes uplift less likely in all but the most onerous UK wind conditions. Where strapping is required, the straps should be at 1500 mm centres and be at least 1000 mm long, with the bottom fixing to the wall in the last 150 mm of the strap. The feet of the rafters need to be located where nailing or galvanised steel brackets can be used to fix the rafter to the wall plate. If nailing is used, care has to be taken not to damage the gang-nail plate and reduce its structural effectiveness.

If the house has a gable end configuration, the gable wall is also subject to these wind forces along the edge of the roof, which exerts sideways forces likely to damage the top of the wall. The top of the wall is adjacent to a fully braced and stable roof structure, so it is possible to strap the top of the wall to the roof structure to stabilise it against these horizontal forces. These two sets of strapping – roof to wall at the eaves and wall to roof at the verge – are shown in Figure 18.9.

It is worth noting that if uplift is suspected, and the roof structure is successfully fixed to the walls, uplift will reverse the action on the structural members and joints. Joists on flat roofs and rafters on pitched roofs will reverse bending. The purlin roof, whose joints rely on compression, will be lifted apart. The truss will have upward forces, reversing the forces in many members. If this is suspected, members in tension under normal loading may become vulnerable to lateral instabilities, as they will not be braced for the compression forces that develop in high winds.

Internal finishes and other elements

The internal finish associated with the roof is a ceiling finish the same as that below a floor and the solution is likely to be the same. Analysis of the ceiling joists or the bottom chord of the truss shows a background the same as the floor, as discussed in Chapter 17 under the same heading of 'Finishes and other elements'. The solution of a plasterboard finish to receive paint is therefore likely to be the same choice.

The early part of this chapter introduced the idea of solar energy collection devices becoming part of the design. These will have to be connected to the roof structure and their loads need to be taken into account in the design. These fixings and the services connections will penetrate the tiling, requiring careful detailing if the weatherproofing is not to be compromised.

Thermal performance

The proposal so far provides waterproof and stable construction with a finish to the ceiling for rooms below. It gives very little resistance to the passage of heat and would, as so far considered, prove to be a large proportion of the overall heat loss from the building. This can be overcome most effectively with the addition of insulation.

This is best provided at ceiling level so as not to heat the roof space, so long as it is not being designed to be habitable. This will require any water services or tanks in the roof to be lagged in their own insulation materials to prevent freezing in cold weather. With the introduction of insulation at the ceiling level the roof space may not be receiving heat from the living areas below, but it will still be receiving water vapour. This cold roof increases the risk of condensation on the timbers and on the underside of the sarking felt. On the timbers this may be a durability risk, while on the sarking felt the condensation may drip back onto the insulation. While it may seem sensible to specify a vapour-control layer below the insulation, the detailing and installation difficulties make the provision of an effective layer unlikely. The alternative approach to condensation control, ventilation, is more likely to be reliable.

This ventilation must ensure a good flow of air throughout the roof. The wind will provide a cross flow if ventilators are introduced at eaves level. However, to ensure a full flow of air right up to the ridge level it may be necessary to introduce ventilation at the ridge or in individual tiles near the ridge.

Insulation is likely to be placed between the joists on the ceiling plasterboard and can therefore be in the form of bats or quilt or even loose materials. If insulation in excess of the depth of the joist were specified, access into the roof space could be difficult as the safe areas to tread on the joists might be obscured. If the roof space were to be occupied, the insulation would have to be at the rafter level where insulating sarking board systems might be used.

Lifecycle considerations

The construction of the roof considered so far includes concrete, timber and metal components. These components are subject to various exposure conditions, but none gains the full protection of the internal environment. Clearly the tile finish is the most exposed, where the concrete components are probably most vulnerable to frost attack and colour degradation from sunlight. The tile is fixed with nails, and these will be subject to corrosion and are therefore normally specified in a non-ferrous metal. Aluminium alloy or copper nails are suitable for normal use, but in aggressive environments silicon bronze or stainless steel may be specified.

The roof finish will experience a wide range of temperatures on an annual basis. Exposed directly to the sky, it will suffer from surface temperatures more extreme than the air temperature due to radiant exchange, reaching high temperatures under the midday sun in the summer and low temperatures under clear night skies in the winter. This will induce dimensional changes that have to be accommodated without loss of performance. While this is of concern with continuous finishes, the number and type of joints and fixings on tiling and slating allows individual tiles to move over each other, limiting the chance of stress developing and leading to failure of the components.

Perhaps the most exposed timber component is the tile batten and this should be specified to be treated softwood. Again, its fixing nails should be a non-ferrous metal. The sarking felt will be protected by the tiles from most sunlight but will tend to become more brittle with age. This makes it most vulnerable where it is bent, particularly at the eaves where it is dressed into the gutter. While this is not so much of a problem with most of the plastic-based underlay that will be UV-stabilised, eaves protection strips should be used with more traditional felts.

The main truss is now protected from the direct effects of the weather and, with adequate ventilation for the roof space, free from the risks of condensation. The roof space is, however, open to insect attack, the most severe in the UK being the house long horn beetle. This insect is limited to only a small area in the UK but has the potential for such damage that roof timbers in these areas have to be treated with preservative. The trussed rafter gang-nail plate is made of galvanised steel, but the 'nails' are punched out after galvanising. While the softwood should not interact with the exposed steel nail edges and ventilation will limit condensation that may promote corrosion, buildings in humid or exposed conditions, particularly coastal areas, may give concerns for the long-term corrosion of the plates. If this is the case, it is possible to specify stainless-steel plates or some other additional protection.

The ceiling construction will experience internal conditions and will not be exposed to a high risk of damage. Some forms of insulation above the ceiling may suffer from bedding down, reducing the insulating effect, but should not be subject to major deterioration mechanisms given the conditions within the ventilated roof space.

Wind-blown dirt will be stopped to a great extent by the felt under the tiles, but roof spaces do become dusty places. This will not inhibit infestations of birds and insects and even rodents that can climb using drainpipes and plants growing on or close to the walls. Some care has to be taken in detailing, particularly at eaves, to limit access if infestations are to be avoided.

Modern concrete tile roofs have a life expectancy of around fifty years. In this time the roof finishes and structure should need little regular maintenance, perhaps the repointing of ridge or hip tiles. The major maintenance is likely to be around the eaves, depending on the choice of materials. Following extreme weather events, particularly high winds, minor damage may occur, so access will have to be arranged to undertake repair. It is unlikely that there will be any provision for access to the roof in the original design, relying on the need for temporary works access and health and safety measures to be provided at the time of the repair. However, it is possible that during the repair and maintenance activity individual tiles will be stepped on and therefore they must have sufficient strength to span between the batten and the tile below on which it will be resting. This does need to be foreseen in the design of the tile and will have to be one of the standard tests on tiles.

The fifty-year life means that replacement of the finish if not the whole structure is likely during the life of the building. Once a safe environment is established for working at height and the old roof is removed, the replacement is the same procedure as the original installation, other than the house being in occupation during renewal. Tiles may be sorted, with some being saved for reuse but most will be weathered with surface deterioration so could be crushed to be recycled as clean fill. The condition for the battens is likely to be very variable, but a range of recycling options may be available rather than sending to landfill for disposal.

Edge and junction detailing

The range of configurations introduced at the beginning of this chapter indicates the need for

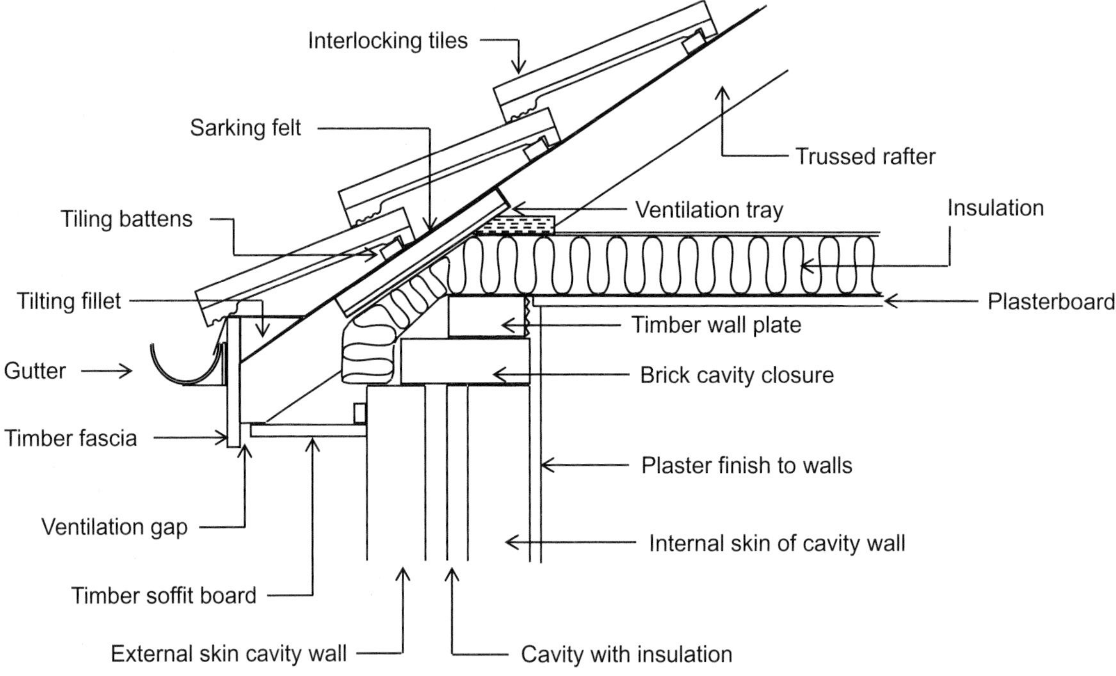

Figure 18.10 Edges and junctions – eaves detail.

a range of edge and junction details to be available, not all for one building, but the components and details need to be sufficiently developed to provide a reasonable choice for a wide range of buildings. Details and junctions are required at edges such as eaves and verges as well as where roof surfaces meet at ridge, hip, valleys and internal gutters and at abutments where a roof meets a wall. These are then needed in some combination for penetrating structures such as a chimney or dormer window.

It is not the intention of this text to detail all of these but to take a range that will illustrate the key features of these assemblies. The following will be discussed:

- Eaves (edge with rainwater collection)
- Verge (edge without rainwater collection)
- Ridge (meeting of surfaces at top of roof)
- Abutment (junction of roof with wall)

Figure 18.10 shows a typical eaves detail for a roof with concrete interlocking single-lap tiles on a structure based on the trusses rafter with insulation above a plasterboard ceiling. Perhaps the first decision concerns the size of the overhang. This most obviously affects the appearance but also the weathering to the top of the wall. Traditionally buildings in areas of high exposure will have wide overhangs, but even in less-exposed conditions eaves with little or no overhang are vulnerable to damp, particularly as the age of the roof increases.

The detail is shown with an overhang of around 300 mm with both a facia and a soffit board. The facia board, while providing a visual edge to the roof, gives a surface to attach the gutter to and a continuous straight and level support for the edge of the tiles. The top of the facia is fixed above the line of the rafters to lift the edge of the tile. All the tiles on the roof, except the one at the eaves, rest on the tile below, and the facia has to mimic this line of support to keep the pitch of the tiles the same all the way to the eaves. This creates a step behind the facia that would allow the sarking felt to sag, creating a potential for ponding behind the facia rather than a clean discharge of water into the gutter. The detail in Figure 18.10 shows a timber fillet to provide a solid surface for the felt that is taken over the top of the facia board and dressed into the gutter. The tile then overhangs the facia to discharge the rainwater into the gutter. Tiles with a bold or deep profile then present relatively large gaps above the facia board, giving potential access for birds that may damage the felt. Some filler piece, fixed to the top of the facia, may be necessary to eliminate this risk.

The facia has a primary role in weathering the eaves and providing the effective discharge of the rainwater into the gutter. The soffit board hides the ends of the rafters and seals the eaves. While providing protection from birds and larger insects, it also has the effect of limiting airflow into the roof space, which is required to provide the ventilation to minimise the condensation risk on the cold side of the insulation. The detail in Figure 18.10 shows a ventilator associated with the soffit board and a tray fixed to the top of the rafters over the insulation to ensure a path for the air into the roof space. Taking the insulation right into the eaves is important in order to limit cold bridging at the junction of the wall and the roof.

Both the facia and soffit board are shown constructed in timber that would have to have an applied finish, usually paint. Although the whole detail would be considered exposed, the facia and particularly the soffit are relatively well protected, but maintaining these boards still has to be considered. This has led to a number of alternatives, often plastic, being used for the boards, but the initial cost and ease of fixing makes the timber solution the common choice for new houses.

The transfer of load from the roof structure to the wall also takes place at the eaves and therefore has to be resolved in this detail. Each trussed rafter rests on the wall plate. This is a timber, bedded level on mortar, on the brick closure that seals the top of the cavity. This provides not only a level surface to ensure the line of the edge of the roof but also an even transfer of load to the wall. In Figure 18.10 the wall plate and closure are on the inner skin of the cavity wall, effectively transferring the roof load only to this skin of the wall. This will have to be taken into account in the choice of the detailing and materials of the wall considered in Chapter 19. For the sake of clarity, Figure 18.10 does not show the fixing of the trussed rafter to the wall plate or any strapping, which would be required to hold the roof against wind uplift provided at the eaves, as illustrated in Figure 18.9.

Figure 18.11 shows a typical detail of a verge over a gable end wall. The detail shows a bedded verge with a 50 mm overhang without barge or soffit board. The bedding is cement mortar, a wet process to seal the edge of the tiling from water ingress. Some tile systems are manufactured with special overhanging tiles or other covering components to provide a dry verge. These have been developed to overcome the long-term maintenance potential of the cracking and loss of the bedding material from between the tiles.

It is possible to provide an overhang detail similar to the eaves, but this requires some additional structure to provide support for the tiles over the wall. The problem arises because the last effective fixing to support the batten is the rafter on the inside of the wall positioned some 50 mm from the wall. If the tiles are to be finished with only a small 50 mm overhang on the external wall, the battens extended over the wall can provide sufficient support. If a larger overhang is required at the verge, a gable ladder has to be fixed to the last rafter and extended over the wall to provide support to both a bargeboard (the equivalent to the facia board) and a soffit board. A bedded or dry verge detail can then be chosen but this time with the 50 mm overhang formed over the bargeboard, not the brick wall.

The main detail in Figure 18.11 shows the last rafter with the sarking felt taken over the wall and the batten extending about halfway across the external skin to provide support for the last tile. In this detail the brickwork of the external skin has to be cut to a good line to provide a surface on which to lie the undercloaking. This is a thin rigid board, often a fibre-reinforced

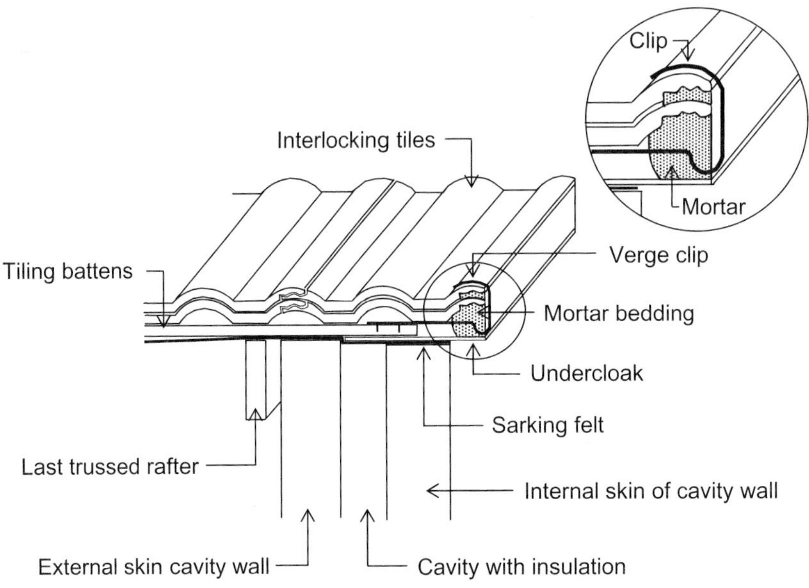

Figure 18.11 Edges and junctions – verge detail.

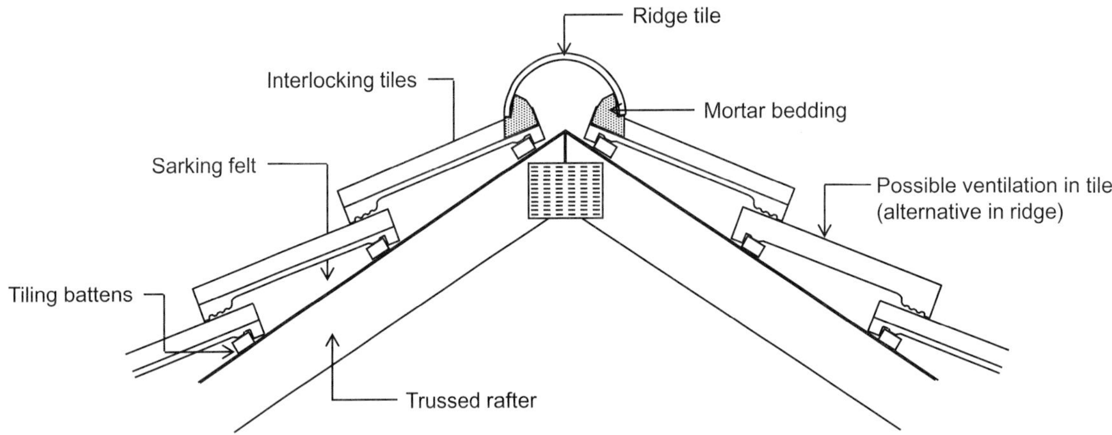

Figure 18.12 Edges and junctions – ridge detail.

cement material, positioned between the felt and the batten. This undercloaking provides the clean visual line to the top of the wall but also the support for the mortar material, so it must be a dimensionally stable material to maintain the integrity of the seal to the edge of the tiles if the bond with the mortar is lost.

To give extra protection to the tile against wind uplift, clips are fixed to the battens and then the tiles are bedded and nailed, with the bedding material struck and finished to give a clean edge to the tiling. To help with the setting out and to avoid cutting, half tiles are available, and the overhang can be varied between 38 and 50 mm. With some profiles, left-hand tiles are produced without the left-hand (upward-facing) interlocking grooves that would otherwise show at the verge edge.

Figure 18.12 shows a typical ridge detail. Like the verge, it shows a traditional bedded ridge, although dry ridge systems are available for some tile profiles. The figure shows a half round ridge tile (without nail hole) that is bedded on mortar on top of the tiles that will be nailed to the batten. Each ridge tile is 450 mm long with butt joints that have to be solidly bedded at the joint to maintain the waterproofing and provide a full fixing. If the profile of the tile has a deep role, the thickness of bedding may become excessive and liable to slumping and shrinkage. This is prevented with small pieces of tile known as dentil slips inserted into the mortar and normally showing as a feature on the tiles at the ridge line. If ridge-level ventilation is required, special ridge tiles can be used. The sarking felt, normally continuous over the ridge, will have to be cut and dressed up around the short internal duct that connects the ventilation terminal with the inside of the roof space.

The last detail, the abutment, is shown in Figure 18.13. The additional components in this detail are the flashings. Normally formed in lead, they seal the junction between tiles and wall. For tiles with a deep profile typical of the concrete interlocking single-lap tiles illustrated, it is usual to use a cover or overflashing where the lead is dressed on top to the tile, over the roll on the tile and onto the natural drainage path formed by the tile's profile. Where tiles do not have a deep profile, as with plain tiling, soakers or a secret gutter detail will have to be used (not illustrated).

However the flashing is formed in the plane of the tiles, the flashing has to be sealed against the brickwork. This is achieved by stepping the top of the flashing and turning the horizontal edges into the brick joints that have been raked out to a depth of 20 mm to receive the lead, which is held in with lead wedges and then pointed to match the rest of the brickwork. If the wall facing material does not provide natural joints to turn the lead into, a groove has to be provided.

Under the lead flashing the felt is turned up the wall and the tiles are butted up to the wall. Tiles may be cut to maintain setting out if necessary as long as the tile is fixed and the flashing covers the roll.

Production operations

Roofs have to be constructed at height and therefore there is a need for a safe working

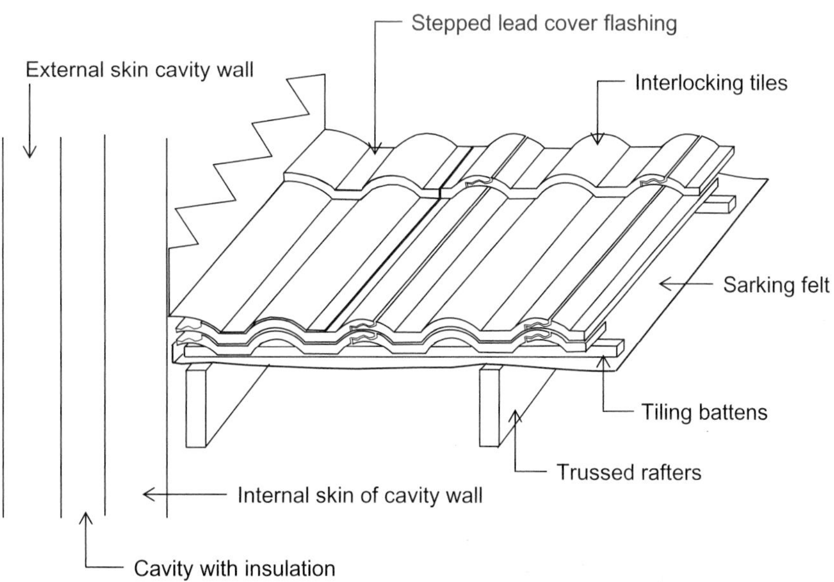

Figure 18.13 Edges and junctions – abutment detail.

platform for the operatives to prepare and fix the materials and components into the final assemblies. There then needs to be safe access for both operatives and the materials and components.

For the roof work to start, the walls will have to be complete up to the wall plate. Strictly only the inner skin needs to be complete. In a masonry wall it is not sensible to build the skins independently, although this is not the case in timber-framed construction where the whole structure is complete before the outer skin is built. The scaffold used to construct the wall should be in place, giving safe access and a working platform for operations completed at the edge of roof.

The basic sequence of work is to erect the structure, complete the timber construction at the eaves and verge and then undertake the tiling. The ceiling will be completed along with the other finishes. Insulation will have to be installed after the ceiling is fixed.

All the components will be delivered to site with some level of preparation. Trussed rafters will be factory-assembled, but the ends of the rafters will have to be cut to suit the eaves' overhang detail. Other timbers such as facia, barge- and soffit boards will be delivered in lengths to be cut to fit, as will the tiling battens that will be factory pressure-impregnated with preservative. The felt will be delivered in rolls as will any lead for flashing. Tile and guttering systems will be delivered with any special components and fixings, but the main components (the tiles and lengths of guttering) may still need to be cut on site to maintain fit and layout for some details.

All these components will need to be handled and stored on site. The trussed rafters are perhaps the most vulnerable. Excessive stress can be put on the joints during storage but particularly during handling. If the truss is manhandled, lifting should be at the eaves joint with the truss in the vertical plane. This ensures the stress in the truss, as it will be the same as when it is supported in the final position. This is also the case for storage in the vertical position, where supports should be level and at the eaves joints, although flat storage is possible if care is taken in lifting back to the vertical plane. Mechanical lifting often cannot be from the eaves joints but should always be arranged to be at the structural node points with a guide rope if a crane is used.

With the other components, damage and waste should be avoided with good storage conditions and careful handling. In particular the hazards of falling from the roof and designated areas for machines transporting and lifting the components should be considered in planning site procedures.

Flat roofs

Given the case study approach being used in these chapters, it is unlikely that flat roofs will be used on a house in the UK at the beginning of the twenty-first century, so they are not going to be dealt with in detail in this text. While the flat roof possibly offers a lower initial cost, this is only achieved with the less expensive continuous roof coverings, such as three-layer felting. Other continuous coverings are more expensive and make the overall flat roof more costly than a pitched equivalent. The three-layer felts have a lower life span of around 15 to 20 years and are more susceptible to damage due to movement and general exposure, making them less reliable and having a high maintenance cost for the owner. Any lower initial cost is, therefore, a false economy if funds are available at the time of construction.

However, a well-detailed flat roof with an appropriate specification can provide a workable solution. As indicated at the beginning of this chapter, the flat roof has a joisted structure similar to a floor but has to be provided with a fall to ensure effective removal of rainwater into the gutters. This is achieved by either sloping the joists (giving a sloping ceiling) or placing tapered battens known as firring pieces on top of the joists. The boarding, to receive the waterproof finish, should be specified to be damp-resistant. Some protection can then be given to the surface of the waterproof covering against direct weathering and surface temperature effects, with a white reflective topping layer often provided by light-coloured chippings.

Insulating flat roofs has to consider the condensation risk, as the waterproof layer is also a vapour-resistant layer and will effectively seal all vapour in the construction. If the insulation is laid between the joists, a 50 mm ventilated gap above the insulation is essential. The difficulties of fixing and edge detailing a vapour-control layer under the joists will make it relatively ineffective. The roof with insulation between the joists is known as a cold deck roof as the boarding is on the cold side of the insulation. An alternative, known as warm deck solutions, can be adopted where the insulation is above the boarding. Now a vapour-control layer laid on the boarding can be made more effective and the joist space can be fully ventilated. A solution with even less risk of condensation is known as an inverted roof, where the insulation is above the weatherproofing layer. The entire construction now remains warm and the condensation risk is eliminated. However, the insulation is now outside, so its water absorption and impact strength for maintenance loading will have to be considered, limiting the type of insulation material that can be used. These options are discussed in Chapter 29 in the section on roof construction.

Summary

1. The pitch and the span of the roof dominate the choice of both roof coverings and structure.
2. The roof will need insulating to resist the passage of heat and may have some fire resistance requirements for the coverings. If it is used as part of habitable space, lighting with windows and roof lights may be required.
3. The pitched roof is the most likely option for this case study as this represents the traditional look of the UK house, and recent problems with flat roofs have led to the pitched roof being seen as the more robust construction.
4. While continuous and large sheet coverings can be used, tiles (and slates) have dominated the traditional pitched roof in the UK, with the concrete interlocking single-lap tile being the most economic solution in the case study.
5. With the variety of colours and profiles, some based on traditional tile shapes, together with their suitability for a variety of pitches, the concrete interlocking tile can be used in most circumstances.
6. The structure is based on the timber rafter. There are a variety of ways to hold the rafter safely at the required pitch for a range of spans typical of the house in the case study. While solutions based on the purlin have been used in the past, the trussed rafter provides an economic solution as long as the required bracing, handling and fixing details are followed in the production process.
7. The fully braced roof structure provides a background for a plasterboard ceiling finish and support for the external walls at gable ends. This structure may still be vulnerable to wind uplift and may need strapping down to the walls at the eaves.
8. Insulation will most probably be provided between the ceiling joists, where the cold roof space can be ventilated to reduce the risk of condensation.

9. The roof finish is unlikely to last for the full life of the house and will need to be replaced. Detailing will have to ensure conditions that maintain durability and barriers to infestation and excessive dust and dirt that may enter during the life of the building.
10. The variety of roof configurations to suit all building plan shapes requires a number of edge and junction details. Details have to be devised for eaves, where water is discharged into gutters, and verges, the edges without rainwater collection. Details also have to be chosen for ridges, hips and valleys where two roof surfaces meet and at abutments, the junction between a roof surface and a wall.

19 Walls

This chapter considers external cavity walls, separating (party) walls and internal partitions. It introduces their performance requirements and provides the analysis of their construction in both masonry and timber frames, leading to typical details and specifications for each.

Wall types by function

Walls will play a major part in the construction of the house in the case study, providing a wide range of functions dominated by the structural and environmental roles they are asked to perform. Walls are given different names depending on the role they play in the overall design based on the broad set of functions they perform.

We talk about loadbearing and non-loadbearing walls normally based on the direct analysis of the load paths from dead and imposed loads. This is a limited definition of being part of the structure, as overall stability associated with wind loading and component instability often relies on walls that may not be carrying the direct loads from the construction and the occupation of the building. Many walls that are called non-loadbearing are part of the structural system and therefore are subject to stress under certain conditions.

Environmentally we talk about external walls, separating (party) walls and partitions (internal walls), each having an associated set of functions that will determine the choice of construction. Given the exposure conditions in this case study, external walls are dominated by the needs of weathering and insulation to resist the passage of heat. Openings for windows and doors are a major influence on the design in so far as they fulfil the functions of access, light, views, security and ventilation required of the wall. However, the creation of the opening is complicated by the need to maintain the function of the wall as a whole. The components associated with windows and doors will have limited performance in strength and insulation, functions that were all carefully incorporated into the basic wall construction.

Separating walls are between dwellings in terraced or semi-detached designs. They will have no openings and no weathering requirements but will need to have functions associated with security and privacy in the need for fire and sound resistance. The other internal walls within the dwelling are known as partitions. There is a need for openings, and the dividing function will again be dominated by privacy and security. Acting as a visual barrier will not be sufficient as the wall will need some fire and sound resistance and must be robust and capable of supporting finishes, cupboards and shelves and integrating services.

General forms

We also refer to the general forms of walls in a number of ways. In housing, one distinction commonly made associated with weatherproofing is reference to the external cavity or solid wall (not to be confused by the two skins sometimes used in a separating wall for sound resistance where the 'cavity' will be designed using different criteria). The external cavity wall has

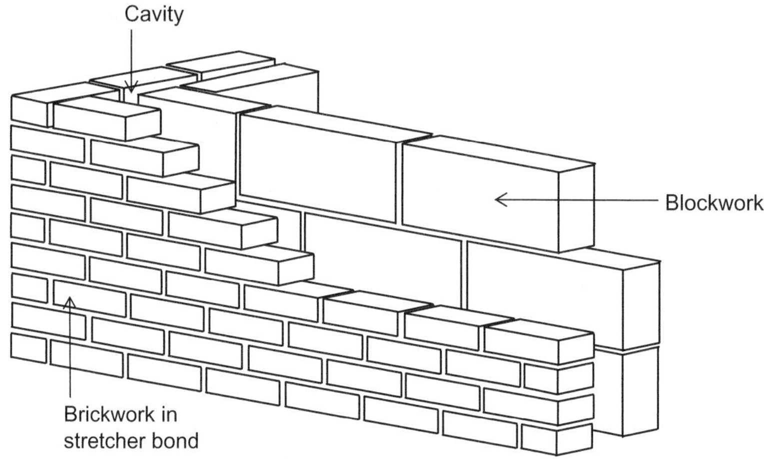

Figure 19.1 Masonry walls in brick and block.

gained almost universal acceptance, although the need for increased insulation and other energy-saving ideas and perhaps a wider acceptance of rendering as an external finish have revived interest in the solid wall, but it has yet to become widespread in practice.

For either cavity or solid walls, three generic forms can be employed in the construction of walls:

- Masonry
- Panel (framed or solid)
- Monolithic

The basic distinction is in the size of components and their spatial arrangements, which determine the types of materials, jointing and production process involved. All walls need to provide a continuous form. If they are loadbearing walls, they need to be capable of distributing loads evenly throughout the structure. They also need to provide continuity for environmental control plus the capacity to have openings formed within the construction.

Masonry is characterised by the heavy materials such as stone, fired clay or concrete that lead to components such as bricks and blocks that are normally relatively small compared to the dimensions of the wall and often regular in shape to form the so-called bonding patterns that give stability and continuity. The choice of materials and the size and regularity of the components vary greatly in their properties, appearance and cost. Given their weight, locally available materials have in the past been chosen on economic grounds based on the cost of transport, and this has led to geographically recognisable styles. With modern manufacturing methods and the relatively low cost of transport, fired-clay bricks and concrete blocks are the dominant choice in most areas of the UK. An external cavity wall constructed with a concrete block inner skin and a brick external skin is shown in Figure 19.1. The bricks and blocks have a modular size that includes the 10 mm mortar joint and that allows the bonding patterns (stretcher bond shown) and the use of bricks and blocks in the same wall. This basic relationship is shown in Figure 19.2. For the brick this modularity is extended to its third dimension. The modular width is 112.5 mm, half the modular length, which allows thicker walls to be built by laying the brick with the header to the face of the wall. This full modularity of the brick is shown in Figure 19.2. Blocks are not modular in the third dimension, being available from 75 to 200 mm thick.

Local availability may still influence choice in conservation areas and now local materials

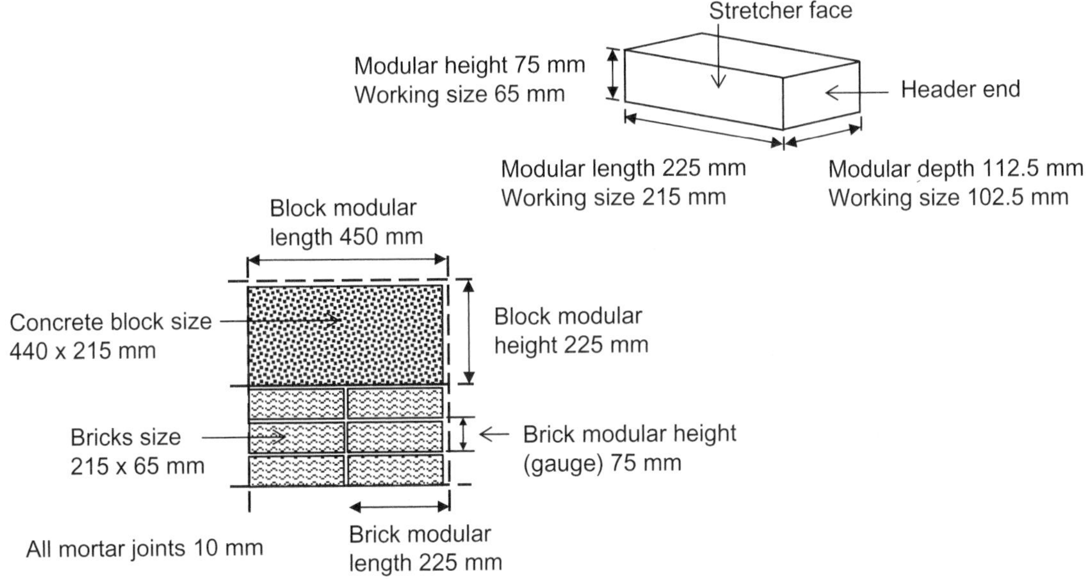

Figure 19.2 Masonry brick and block modularity.

with low transport needs can reduce energy use (rather than cost) and can therefore contribute to sustainable development.

Panel walls are storey-height constructions that can be constructed in position or prefabricated in panels in modular widths to be combined into walls of various lengths to suit the design and the site constraints. Framed panels in either timber or lightweight steel sections form the basis of many prefabricated systems that can be ordered from the factory and erected and jointed on site. The basic form of a framed panel wall in timber is shown in Figure 19.3. Solid panels can also be prefabricated, normally associated with concrete but more recently developed in sandwich construction, where high insulation values are achieved without cold bridges by forming rigid insulation materials between timber composite board materials. Particular attention has to be paid to jointing solid panels.

While panels form the basis for some of the most advanced factory-produced prefabricated systems for housing, they have in the past been fabricated on site and lifted into position, and this may still have some merits in low set-up and transport costs but is not now often adopted.

The component of the monolithic wall, as the name suggests, is the whole wall of the house. This has to be an in situ process where any joints that are necessary are dictated by the pace of the production process. These are known as day joints as they are determined by how much wall can be built in a day, but when cured the wall must act as a continuous material and this may dictate the position of these joints. Now made of concrete, with the use of temporary formwork and falsework, monolithic walls have in the past been associated with soil construction. Some soil walls used formwork (pisé), although this was not always necessary where higher clay contents were found in the local material (cob) where the wetter clay–straw mix held the material together during construction. Currently monolithic walls are rarely used in the UK, but again some of the environmental ideas, particularly using walls for thermal capacity and with the low embodied energy associated with soil construction, may suit this form of construction.

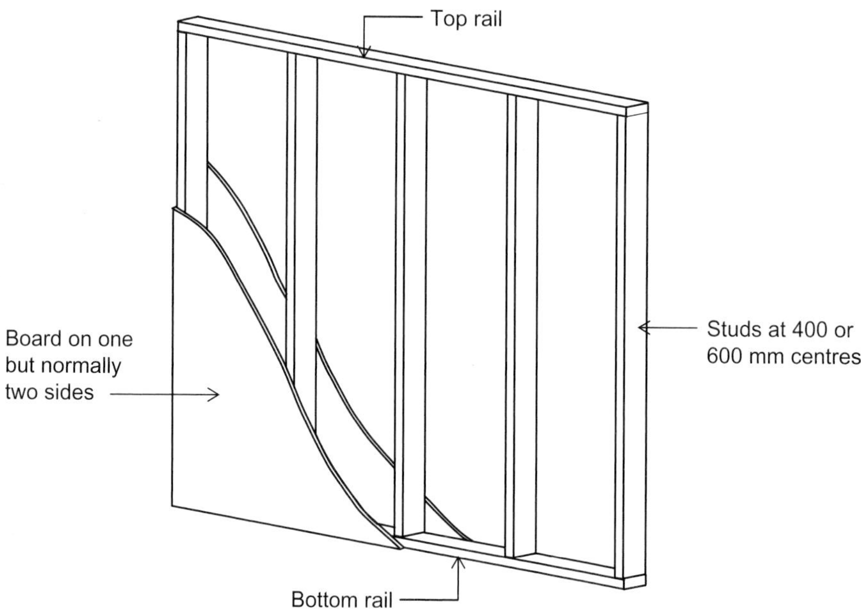

Figure 19.3 Panel wall in timber frame.

Walls in the case study

For the house in the case study the initial suggestions to be considered will be:

- External wall – cavity with masonry external and internal skins
- Party (separating) wall – in masonry
- Partitions – in masonry or lightweight studwork

After the analysis of this masonry solution the use of timber frame will be considered, as this is a major alternative in the UK at the beginning of the twenty-first century.

Identifying which of these walls are loadbearing or more generally acting as part of the structural system can only be achieved for a particular design. This case study is not that specific, but it is reasonable to assume that the external walls will be loadbearing and at least some of the internal walls will have to carry loads from upper floors, and this may well include the separating wall. These and possibly other walls will then have to take the racking forces from wind loading.

Masonry cavity external walls

The cavity wall was developed to limit the risk of damp penetration, which could be a problem with solid walls in exposed conditions, particularly via the joints, aggravated by poorly maintained pointing and with the introduction of strong cement mortars. The principle was to provide an outer skin that limited but did not have to eliminate moisture penetration under all conditions. The cavity, a clear air space, provided the guarantee of damp-proofing as long as the cavity was not breached. The inner skin then remained dry and could be used to support the finish without fear of damp within the building.

The simple principle of the cavity has proved very robust, so long as attention is paid to certain details in the wall and at openings (see discussion later). A number of materials can be used for the outer and inner skin, but for the case study the initial suggestion is to choose brick for the outer skin and concrete block for the inner skin. From experience the thickness of the internal and external skin can be expected to be around 100 mm. Given the scale of house construction, this would seem to be a reasonable estimate, but the actual size will have to be determined by the performance requirements and the materials chosen. If the thickness deviates far from this size, the additional material will have to be justified from additional performance or economy and details will have to be reassessed. The width of the cavity is the other key part of the specification of this form of wall. The cavity only works if the width of the cavity at all points can be maintained within tolerances and kept clean. These are both production problems so the production process dictates a minimum size for the cavity. These are, therefore, related to the materials specified for each skin; in this case both are masonry for which a 50 mm clear cavity would be the usual proposal. It is not possible to provide an analysis to check this width until the whole wall specification is chosen. The size will also be determined by the introduction of insulation, so this also has to be established before the size of the cavity can be determined. It will be necessary to return to confirm this initial suggestion of 50 mm when the wall specification is more advanced.

The external skin

The initial suggestion for the external skin will probably be based on its appearance. Clearly this not only will be determined by the style and status of the building but also will be a major factor in the building's response to its context. It may even be determined directly by the client who may have very strong views on what the house should be made of. Dominating the UK market is brick, although locally stone might have been historically used, so reconstructed stone blocks have been adapted for cavities. Renders might have been commonly used where local quality bricks were not available or in extreme exposures, such as coastal areas, and this again can be easily adapted for the cavity wall where concrete blocks can be used in the external skin as the backing for the render.

Exposure is a key factor in selecting the material for the external skin, less for damp penetration as the cavity has this function but for durability and potential maintenance. If masonry is to be used, the exposure to rain at times of low temperatures will make frost action a significant risk. Frost resistance will have to be considered not only in the brick but also in the choice of mortar. Other risks of change over time include discolouration. This change in colour may be desirable as weathered materials are often more mellow and attractive than when new. However, some discolouration of, say, pastel-coloured renders may not stand the test of time in the UK climate. Detailing to ensure limited pattern staining with good even washing from the rain needs to be considered.

Overall, the facing brick provides a good range of options, most of which can achieve frost resistance and have a wide range of colours. If local materials are preferred, the choice may be more limited. The modular size of the standard UK brick and block was illustrated in Figure 19.2. The brick itself is 102.5 mm thick so, if laid, stretcher bond can form an outer skin, around the 100 mm originally estimated. This 'half brick' skin will limit moisture penetration to the cavity but will not eliminate it, so detailing in the cavity will have to take into account the presence of some water on the inner face of the brickwork's outer skin.

The structure and the internal skin

It is not possible to proceed to accept this specification for the external skin that was based primarily on appearance, or start to think about the internal skin, without being clear as to how the loads are distributed across the wall. This will be determined by the detailing from the connections of the floors and roof. Figure 19.4 shows the cavity wall with these details (discussed in the two previous chapters) to identify the dead and imposed loads together with the wind loads. Inspection shows that the internal skin takes the dead and imposed loads. No strength requirements for these loads have to be considered for the external skin, so the facing brick

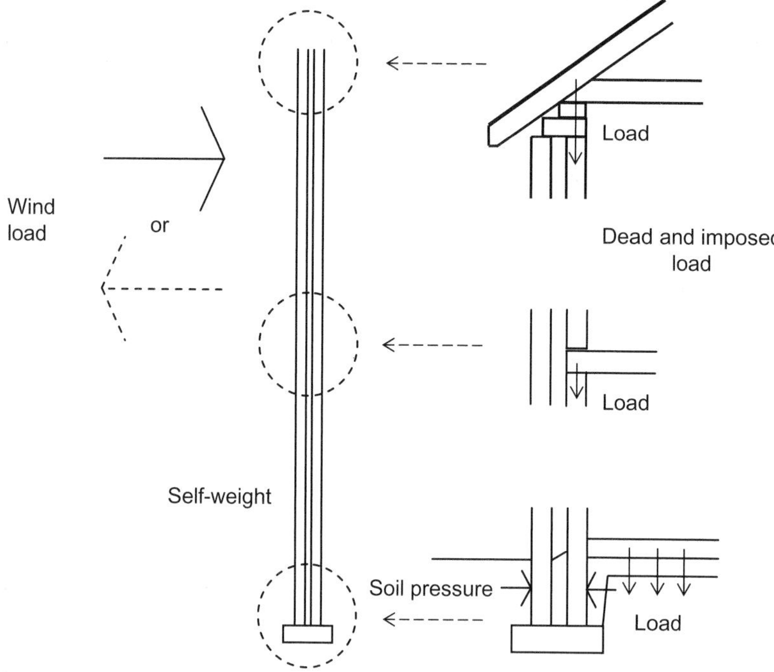

Figure 19.4 Loads on the external cavity wall.

should be sufficient. Durable bricks will have reasonable strength, so the external skin should be able to support its self-weight and deal with any local crushing from the lintels over openings. If large openings are included in the design, the strength of the brick may become an issue.

It is clear that one of the key aspects of the specification of the block for the internal skin will have to be strength. The most highly stressed areas of the internal skin will be between the ground-floor windowsills. Given a reasonable distribution of windows in both size and spacing apart and taking the skin to be 100 mm thick, it should be possible to use aerated concrete insulating blocks at 2.8 N/mm^2 (these may have a minimum thickness of 115 mm) for dwellings up to two stories high. These blocks are normally more expensive than blocks made with lightweight aggregates that are denser and therefore offer less insulation value but more thermal capacity. If high thermal capacity is required, dense concrete blocks may be a better choice as they are less expensive, and certainly strong enough, but will have poor thermal insulation properties. It is therefore best to leave the final specification of the block until the thermal behaviour of the wall is considered.

The structure – overall stability

Returning to Figure 19.4, another observation would be that the cavity wall is slender, making it fundamentally unstable. Given that there are adequate foundations, two other instabilities can be identified. The first is overturning by wind forces and the second is buckling of the internal skin under compressive forces from the dead and imposed loads.

The mortar jointing and the pattern of bonding in both the bricks and the blocks ensure that the wall acts as a panel. However, as presently discussed, the external and internal skins are independent. Wind loads on the external skin are not transferred to the internal skin and

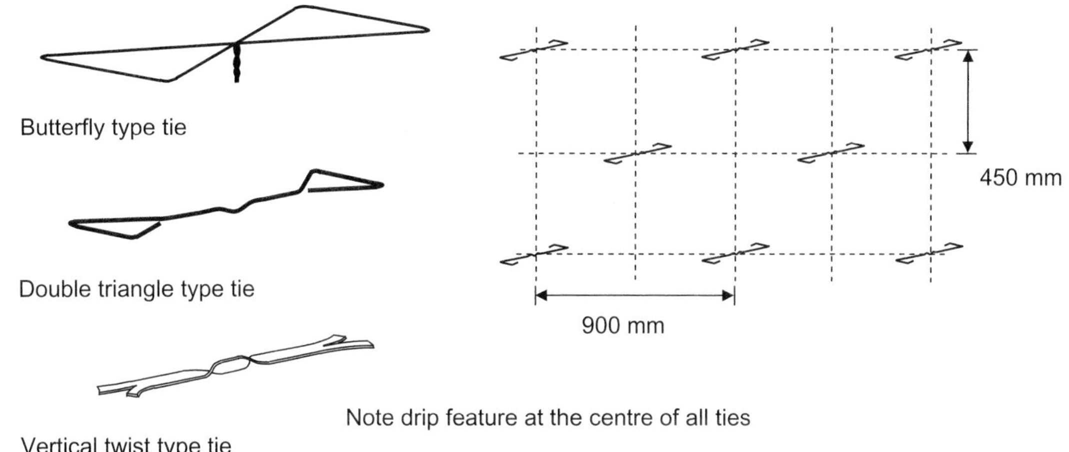

Figure 19.5 Cavity wall ties – types and spacing.

the external skin does not help the buckling of the internal skin. Mutual structural support is required. This is normally provided by wall ties (although under most conditions they will probably act as struts). The form of these ties and the spacing required are shown in Figure 19.5. The width of the cavity will determine the type of tie that can be used, with the vertical twist being the only form used in wider cavities over 100 mm. On all types of tie there is a drip, a device to stop water travelling horizontally, bridging the cavity across the wall tie and causing damp in the inner skin. Ties would now normally be specified in austenitic stainless steel, although galvanised steel and plastic forms have been used. With the introduction of wall ties the wall now acts as one panel for both overturning and buckling but is still unstable.

Any return walls will act as buttresses to resist overturning. If the walls of the house form a cellular pattern on plan, they will receive mutual support. If the walls are predominantly in one direction as in a form known as cross-wall construction, piers will have to be formed, normally at both ends of the cross wall.

This leaves one potential area of wall vulnerable to overturning. The wall forming any gable ends will not benefit from the return walls to give stability against the wind forces. A chimney may form a pier, but in the absence of any masonry support the wall should be strapped to the roof to gain additional stability. These details are given in Chapter 18. Straps should be provided at the top of the gable wall and then at not more than 2000 mm centres down the slope of the roof.

It is also necessary to use the support of the upper floors to reduce the effective dimensions of the wall against buckling. This is achieved by providing joist hangers with restraint or strapping, details of which are given in Chapter 17. Straps should be placed at not more than 2000 mm centres for walls over 3000 mm long.

Rising damp

The cavity provides resistance to penetrating damp from rain falling directly on the surface of the wall. The materials chosen are permeable, so moisture can enter the building by rising from the ground where capillary action within the materials will draw water into the bricks and blocks, creating a path for damp. The solution is to provide a damp-proof course (DPC) across each skin. In a cavity wall these DPCs are independent and can therefore be at different levels. For the external skin the DPC has to be at least 150 mm above ground level. This protects the wall from rain falling on the ground, which will splash back against the wall. For the

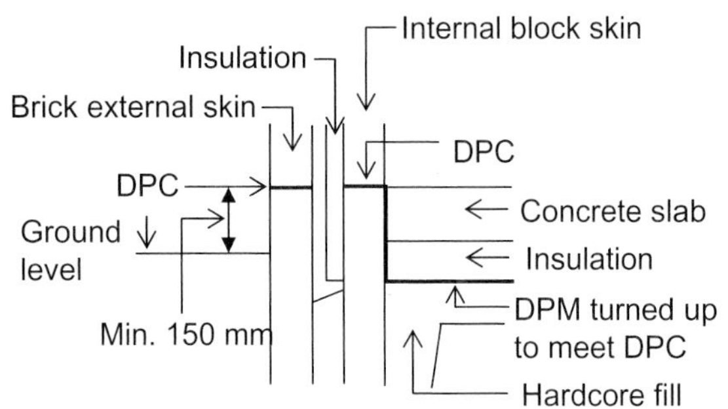

Figure 19.6 Masonry cavity wall – DPC arrangements.

internal skin the level depends on the internal floor level and the type of floor construction. For a suspended ground floor the DPC must be lower than any part of the floor construction that would be vulnerable to damp, while for a ground-supported slab the DPC must be continuous with the damp-proof membrane (DPM). This is shown in Figure 19.6, which is based on illustrations from Chapter 17.

For these DPCs a variety of materials can be chosen. In the past materials such as slate and even bricks with very low permeability known as engineering bricks have been incorporated in the wall, but thin, flexible sheet materials that can be laid in the bedding joint are now used. Metals such as lead and copper can be used but are expensive. The less expensive bitumen DPC, and in particular the polymeric (bitumen polymer and pitch polymer) sheets, are the most widely used. The polymeric sheets are also used for other DPC applications such as cavity trays, which will be discussed later in this chapter. The polymeric materials provide a tough material that will not extrude under load, give good adherence to the mortar and have good durability even where bent to form gutters. Although naturally black, it is possible to obtain the polymeric material in mortar-matched colours to reduce the visual impact of the black line of the DPC in the mortar joint.

Thermal performance

With concerns for energy saving regarding winter heating in houses, the initial focus for thermal performance has been on insulation properties to resist the passage of heat. The focus is on the elemental U value or thermal transmittance of the wall as an individual element. Still air provides good insulation, so the unventilated cavity not only gives the weatherproofing that is its main role but also contributes to the overall thermal resistance of the wall as long as it is at least 25 mm wide. This can be enhanced by providing a low-emissive surface facing the cavity to limit radiation transfer. It is also possible to increase the resistance of the wall by careful specification of the block. Concrete blocks vary in their density. It has already been established that higher densities give greater strength and thermal capacity but lower insulation values. Selecting a low-density block can significantly affect the insulation value of the wall but still provide sufficient strength for a two-storey house. However, even with the cavity and the low-density blocks, the wall will not have a thermal transmittance or U value sufficient to meet either the current regulations or the expectations of wider environmental concerns.

It is possible to increase the thickness of the most effective material (in this case the block, as increasing the width of the cavity beyond 25 mm will have little effect) or introduce a layer of material specifically for its insulating properties. The most common solution is to include an insulating material in the cavity. Either partial fill (still leaving a cavity) or total fill solutions can be achieved with bats or boards, and total fill solutions can be

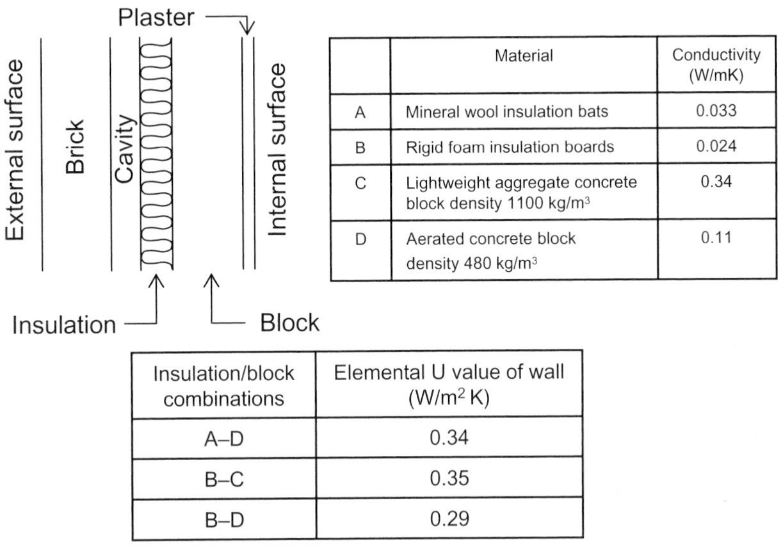

	Material	Conductivity (W/mK)
A	Mineral wool insulation bats	0.033
B	Rigid foam insulation boards	0.024
C	Lightweight aggregate concrete block density 1100 kg/m^3	0.34
D	Aerated concrete block density 480 kg/m^3	0.11

Insulation/block combinations	Elemental U value of wall (W/m^2 K)
A–D	0.34
B–C	0.35
B–D	0.29

Figure 19.7 Masonry cavity wall – partial fill insulation U values.

achieved with loose materials or in situ processes installed after the walls are completed. Many combinations of block and insulation within the cavity can achieve elemental U values of 0.35 W/m^{2}K and better with a 100 mm block and around 50 mm of insulation. Figure 19.7 shows some typical combinations of block and insulation bats and boards used as a partial cavity fill retaining the insulating contribution of the cavity.

It has been recognised that the wall itself is only one part of the construction through which heat is lost. Door, window, roof and floor elements all have different U values and it is possible to trade off high insulation value in one element for another as long as the overall heat loss from the house will not be affected. This leads to the specification of target U values that can modify the required performance of the wall.

Although the performance for the wall was initially set in terms of resistance to the passage of heat with its focus on limiting fabric losses, the wider purpose of insulation is the conservation of energy. Walls and their insulation are only one aspect of energy saving for the building as a whole. Elemental U values are still important but are increasingly only part of the scheme to limit the energy use in a house. This has been discussed as part of passive design in Chapter 15. Aspects of ventilation rates, water heating, internal gains and solar gains are considered along with fabric losses in energy-rating schemes that look at the energy consumption of the building as a whole. The external wall may now have performance requirements beyond insulation. The openings will influence ventilation rates, as will the detailing of joints and junctions where airtightness is now a consideration. The thermal capacity of the fabric can be used to harness the heat gains to reduce the overall heating requirements for the house. As thermal capacity becomes significant so the position of the insulation in the wall has to be considered.

To understand why this becomes significant it is necessary to consider the wall not in steady state but to see the temperatures on either side of the wall changing. The effect of heat gains raising the air temperature will be limited by a wall with a high thermal capacity, because the wall will take a longer time to heat up towards steady state. Further, when air temperature starts to fall, the heat stored in the wall will be given back to the air, delaying the need for additional heat inputs. A similar cycle can be used for summer cooling in larger commercial buildings.

The position of the insulation will also affect the condensation risk. The thermal behaviour of the wall for condensation risk, response times and thermal capacity issues are explained in Chapter 10.

The other major aspect of the wall resisting the passage of heat is the concern for thermal bridging. With masonry construction, this occurs at openings and junctions with the floor and roof. These details are discussed under these sections in this and other chapters.

Size of the cavity

It is now possible to return to confirm the size of the cavity that was initially set at 50 mm wide. It was suggested that in order to act as a barrier to penetrating damp the cavity has to be wide enough to be kept clean, and this will be influenced by both the inclusion of insulation and the production process. The cavity should be at least 25 mm if it is to achieve the thermal insulation potential of the still air providing good surface resistance on both sides of the cavity, so the 50 mm is satisfactory for the thermal function.

The influence of the production process can now also be identified. The process will involve bringing up the inner skin while protecting the cavity from mortar dropping as the work proceeds. If insulating boards or bats are to be installed in the cavity, these are then placed and restrained against the inner skin (quilts will not be stiff enough to give a stable face inside the cavity). It is then possible to lay the external bricks. Two aspects about laying the external brickwork now need to be questioned. How easy is it to clean the mortar that squeezes out from between the bricks as they are tapped to level and how accurately can the width of the cavity be maintained? While it was easy to clean the mortar from the blocks forming the inner skin, the bricks are forming the cavity and the bricklayer has to clean the mortar with the trowel from inside the cavity. If the cavity is too small, there is the danger that the mortar protrudes into the cavity and cannot be cleaned or falls down into the cavity where it will build up at the bottom of the wall, at window heads and even catch on wall ties, providing a damp path from external to internal skin. This will be aggravated by any narrowing of the cavity to keep the external wall in line, given the accuracy that was achievable in laying the internal block skin first. Experience has shown that it is only fair to the bricklayer to have a cavity around 50 mm in order to keep the cavity clean to achieve the insulation value and eliminate the risk of penetrating damp.

Many options for insulating the wall involve the cavity in different ways, some of which have been introduced in the section on thermal performance in this chapter. As an example, a wall could be specified with a cavity with partial fill insulation of 50 mm and a clear cavity of 50 mm, giving an overall distance between the inner and outer leaf of 100 mm. Given that the wall tie is often in compression acting as a strut, some thought has to be given to buckling as the distance between the skins increases. Wall ties were originally designed for clear 50 mm cavities with no insulation. They have to be embedded in the masonry joint around 60 to 65 mm and have to be stiff enough to resist buckling across the free length in the cavity. Ties for 100 mm wide cavities have therefore to be longer and thicker, so the correct type associated with overall cavity width has to be specified.

The wall so far

The variety of options for thermal insulation, its position and type of material, with its thickness in conjunction with the specification for the block and its influence on the cavity, have already been identified. Appearance and exposure conditions lead to a variety of performance needs that influence the choice of the external skin. Making a single choice for this case study is therefore not possible, but Figure 19.8 shows one option adopting the partial fill cavity insulation given as an example in the paragraph above. This wall has been chosen to achieve

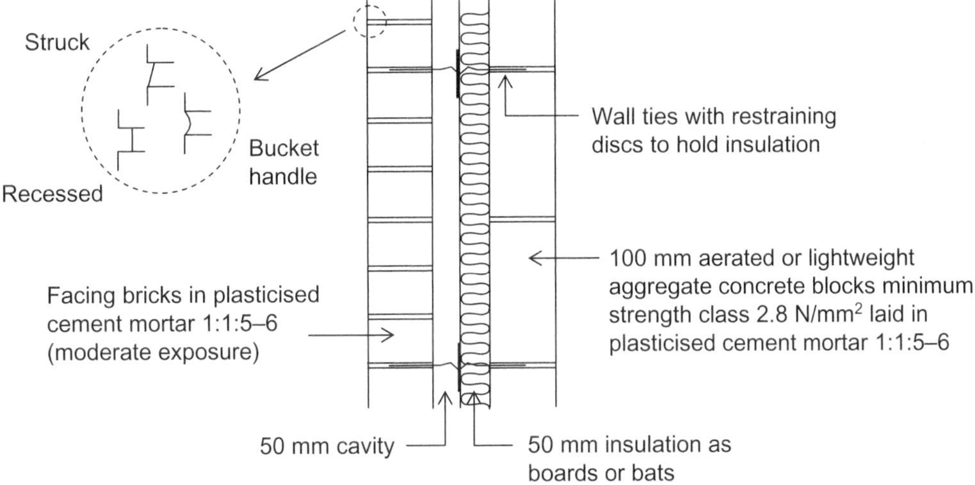

Figure 19.8 Masonry cavity wall – basic specification.

weathering, structural and thermal performance. As specified, it will provide a good level of sound insulation as it has both mass and a discontinuity with the cavity, and the material chosen will give good fire resistance. The basic structural and environmental functions will be fulfilled by this construction.

Internal and external finishes

The external appearance has been set by the choice of facing brick, although detailing and the use of more than one brick and the colour of the mortar or the use of renders can give design options. The shape of the front of the mortar joint is also sometimes used as a design feature, but it also has a significant effect on the weathering and durability of the wall. These alternative joint finishes are shown in Figure 19.8. The struck or weathered joint is a tooled joint (it has the surface smoothed after the mortar has started to cure), which promotes the run-off of water and protects the brick, while the recessed joint has the opposite effect, leaving the soaked edge of the brick vulnerable to frost attack. The struck or weathered joint takes time and skill, and therefore the more straightforward, tooled bucket handle joint is often specified. A fourth joint finish, the flush joint produced as the work proceeds without tooling, does not have good resistance to rain penetration and may not produce a good edge on uneven or open-textured bricks.

Internally, concrete blocks can be specified for a fair-face for direct painting, but this is unlikely to be satisfactory for a house. The options will be between a wet process, probably lightweight gypsum plaster, and a dry plasterboard system.

The blockwork provides the background or substrate for the finish. The characteristics of the background introduced in Chapter 17 for the floor finish are the same for the walls. It is necessary for the background for the finishes to provide:

- Sufficient stable support
- Suitable fixing opportunities
- Compatible flatness tolerances
- Matching for future movement
- Avoiding deterioration mechanisms

If plaster is used, the background characteristics of the blockwork will be satisfactory for the applications of two-coat work. The wall is clearly stable, as it is part of the structure. The density of the blocks is significant in that it determines the suction during plastering that makes the bond that provides the fixing between the block and the plaster and this will influence the specification of the plaster. The tolerances on the wall should be well within the capability of the first coat of plaster, normally about 12 mm thick to bring the surface to a flat plane. The final skim coat of plaster will then provide a tough, smooth surface for decoration. Plaster shrinkage should not be sufficient to break the bond with the wall, although shrinkage cracks may appear at the top and bottom of the wall. If blocks are inert, there should be no interaction causing deterioration of either block or plaster. However, if the blocks are laid and finished before they have completely cured and continue to shrink, cracks may appear in the plaster.

The plaster surface, while tough enough for most domestic situations, takes only limited-impact loads and is not suitable for external exposure. If either of these conditions is experienced, renders should be used. Based on sand and cement with a plasticiser, traditionally lime, these finishes can be used externally even in severe exposure conditions with the appropriate specification. If renders are used externally then, for economy, a concrete block could replace the brick external skin. As with plasters, the density of the block chosen would determine the strength of the background. The first coat of render would have to be specified to match the strength of the block, with the final coat(s) determined by the exposure conditions. Finishes on external renders can vary from smooth trowled finishes that can be painted to thrown aggregate finishes using decorative aggregates to provide a weathering surface.

Dry finishing systems for internal walls will most likely be based on the use of plasterboard. As a background the blocks are not sufficiently flat to apply the boards directly to the wall. The wall is, however, stable and ready to accept a range of fixings. Two options are possible: to provide timber battens levelled to receive the board to be fixed with nails or screws, or to use dabs of a plaster-like material to both fix and level the board to the wall. With both these systems the boards still have to be jointed. While square-edged plasterboard (fixed showing the grey face) with the joints reinforced with scrim can then be skimmed with plaster, the use of tapered edge boards allows the joint (and screw holes) to be finished without having to plaster the whole wall.

Interaction with other elements of construction

The interactions with the floor and the roof in both load transfer (roof/floor to wall) and the provision of stability (wall by roof/floor) have been identified. Services also interact with the wall and therefore may influence the specification and detailing. There may have to be some hiding of wires below the plaster in chases in the internal skin, but most pipes will be surface mounted and then boxed in if there is a wish to hide them. Radiators and other components will be fixed to the wall, demanding some consideration of the strength of the wall and the ability to identify suitable fixing devices to take the loads involved. This will also be true of the need of most occupiers to have shelves and cupboards fixed to the wall, particularly in kitchens. A range of fixings is available for a variety of block walls and even the low-density blocks have the capacity, with the correct fixings, to take these loads, so the specification so far suggested will meet these performance requirements.

Lifecycle considerations

The choice of the external material with concern for the exposure conditions to ensure durability has been discussed, as has the choice of the internal finish. The insulation and the internal skin should remain dry, so, in the absence of infestations in the cavity damaging the insulation, there should be little deterioration of these materials, although full cavity loose fills may compact with time. Maintenance requirements should be minimal if the external bricks are laid in cement mortar. Modification and adaptation will have to be undertaken with care, as the wall is a loadbearing structure. Any changes to

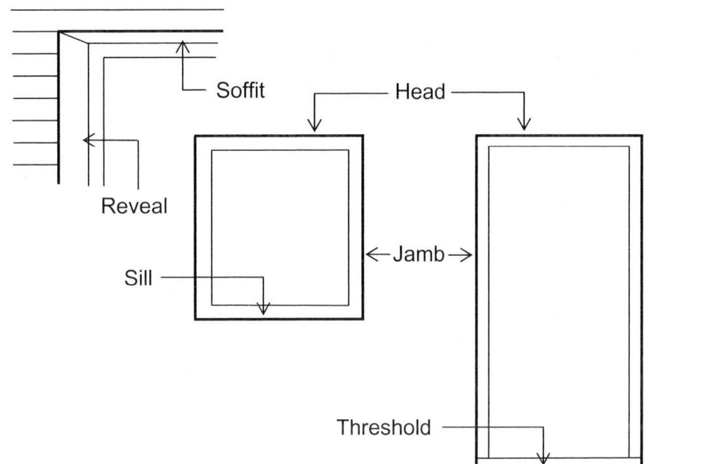

Figure 19.9 Openings in walls – terminology.

the external wall may also have difficulty in matching bricks at some time in the future, but internal finishes are easy to reinstate.

Consideration of the disposal of the masonry components at the end of their life indicates opportunities for recycling as fill, as the insulation should be easy to separate from the demolition waste. It is conceivable that the insulation may also be retrieved as it is only restrained in position, although it may have mortar from the blockwork stuck to the back that may be difficult to remove. The simple inclusion of a paper layer between the insulation and the block so that the insulation does not bond to mortar may be an inexpensive way to ensure the insulation could be reused. Metal wall ties could also be recycled and may be in sufficiently good condition to be reused.

Openings in the external wall

The wall considered so far does not provide access, light or ventilation, which will be required in an external wall. For these we need doors and windows. These, however, cannot take loads and will have less thermal, sound and fire resistance. They also create a safety and security risk and may limit privacy. This is normally resolved by creating an opening to maintain the integrity and performance of the wall while designing independent components for the doors and windows to complement the wall and ensure the function of the wall as a whole. To this must be added the need to provide fixing and jointing of the components into the opening. While the size of the opening will be dictated by the design requirements of the door and/or window, the basic detailing of the opening for the wall to the house is unlikely to vary.

The terminology used in openings is given in Figure 19.9. The detailing of the head, jamb and sill (threshold) will each need to be considered separately but are linked in the need to keep the depth of the reveal and soffit constant around the opening. The head will need to ensure that the loads are transferred to the jambs. The jambs will then need to be stabilised as they take the additional compressive forces. The sill has no structural requirements as it can be formed on the wall below, but the weathering requirements and the exposure conditions are more acute at the sill.

All this has to be achieved with an accuracy to ensure the doors and windows, now almost certainly components made off site, can be fitted and weathered into the opening. While some packing may be tolerated below the sill, the fit between the jambs has to be accommodated in the fixing and weathering details of component to reveal.

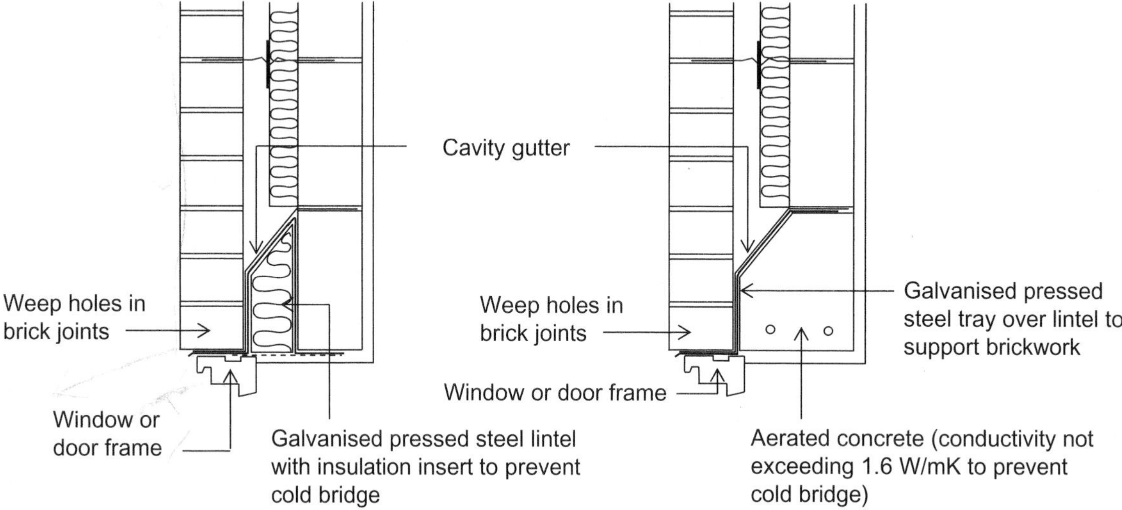

Figure 19.10 Openings in walls – alternative head details.

Detailing the head

Carrying the load across the opening is fundamental to the detailing of the head. In the past arches have been used because the basic material of the wall is weak in tension, so, with the basic structural action of the arch being in compression, the arch allowed the same wall material to be used to span the opening. With the greater range of materials now available the lintel (a small beam with a bending action) will almost certainly be used.

If the lintel is based on a rectangular cross section carrying both skins as a simple beam, the top of the lintel will collect water that can run back towards the inside of the wall, and the front of the lintel may need a finish to improve the external appearance and/or durability. The rectangular section has been used in the past for stone, timber and reinforced concrete lintels. All these materials will create a thermal bridge if the insulation is provided in the cavity, although this would not be the case if internal or external insulation were used.

While all these difficulties can be overcome, developments in pressed steel and lightweight concrete provide economic solutions that take all these concerns into account. Two alternative details of the head of an opening showing a steel lintel and a lightweight concrete lintel are given in Figure 19.10.

These solutions recognise that for the two-storey wall the loads are on the internal skin and that the need for strength to support the outer skin is limited. In the lightweight reinforced concrete solution the basic rectangular section remains to support the loads from the internal skin, although it is now shaped to form a cavity tray to direct water away from the inner skin. The gutter emerges at the steel shelf angle that supports the external skin, and weep holes formed in some of the perpendicular joints between the bricks help remove the water more efficiently. It is now possible to bring the brickwork across the opening to visually give continuity to the external finish and protect the lintel.

The pressed-steel solution gains its strength not from a solid material but from shaping sheet steel. The weathering details and appearance remain the same as the composite concrete and steel shelf angle solution. The major advantage is the low weight of the lintel, an advantage in the production process. The hollow section can be filled with insulation to reduce cold bridging and some means of providing a background for the internal finish has to be incorporated under

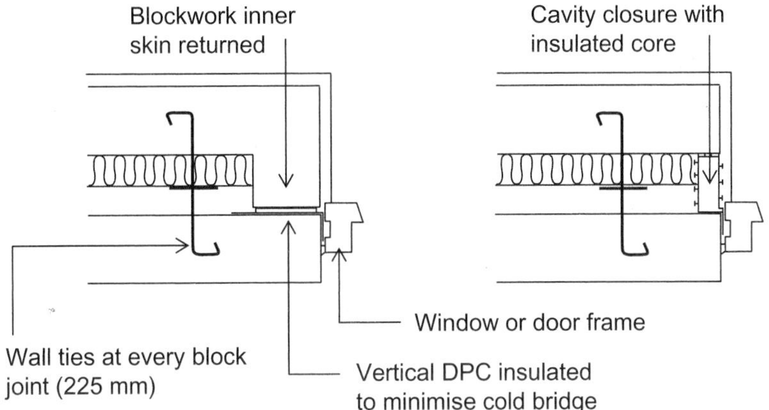

Figure 19.11 Openings in walls – alternative jamb details.

the lintel. The steel has to be protected from corrosion with a galvanising layer, but even this may be suspect over time in exposed conditions, particularly with the risk of damage during the production process. It is possible to specify polyester-coated lintels to provide extra protection.

One detail that cannot be seen from the cross sections in Figure 19.10 is that the cavity tray, forming the gutter from sheet DPC, should either extend beyond the lintel to minimise the risk of moisture running back off the end of the gutter and across the underside of the lintel or use stop ends to the gutter to prevent the water running off the ends and wetting the cavity insulation.

Detailing the jamb and the sill

The detailing at the jamb is relatively simple, and two options are shown in Figure 19.11. The first shows the return of one skin to close the cavity and then providing a vertical DPC to eliminate damp penetration across the jamb. Which skin is returned will determine how deep the reveal is, as the frame of the window or the door will be located to cover the DPC, ensuring no moisture reaches the internal finishes or the back of the frame if this is against the external skin. Returning the block (detail shown) will reduce the thermal bridging effect but will leave the frame fairly close to the external face. The reveal will be small. This means that the sill on the frame can be detailed to hang over the wall without the need for a sub-sill but brings the joint between the frame and the opening into a more exposed position. In order to give the joint better protection, the outer skin can be returned to gain a deeper reveal. A sub-sill will now be required, and the brick return creates more of a cold bridge, although insulated DPC can help reduce the effect. The second option is to use a cavity closure. This hollow plastic box section can be filled with insulation to reduce the cold bridge, act as a DPC and in addition can provide a fixing for the frame.

Structurally the edge of the wall has to take the load from the lintel but has lost the stabilising effect of the bonding, and this makes the jamb subject to buckling. The return skin or the cavity closure provides support if the skins buckle towards each other, but provides no resistance to them moving apart. To provide this resistance, wall ties are placed in every block joint adjacent to the returned skin. The cavity closure can incorporate ties that attach to the box section.

The detailing of the sill, with and without a sub-sill, is shown in Figure 19.12. The major concern is to shed the water that runs off the impermeable window or door surface away from the wall as effectively as possible. The top surface of the sill should be sloping, the lip should overhang the wall and the underside of

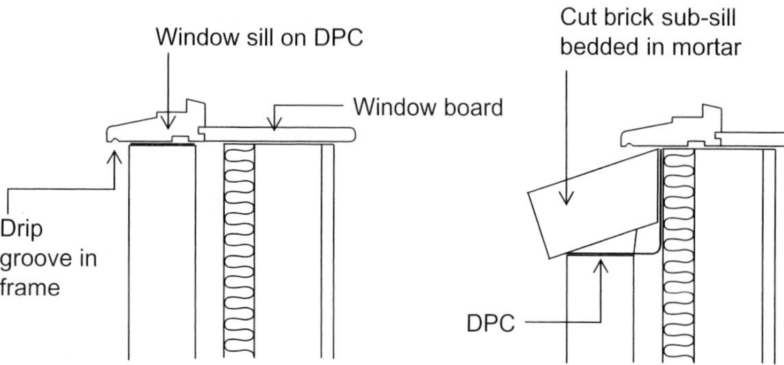

Figure 19.12 Openings in walls – alternative sill details.

the lip must be provided with a drip. This amount of water around the sill raises the concerns of durability so the material of the sill needs to be chosen with care. Moisture penetration is an additional concern, particularly at the ends of the sill at the reveal and under the frame, where wind-blown rain will be held and may even be under pressure rather than running off down the slope provided on the sill. If the cavity is closed in the sill detail, a DPC will be required to protect the internal finishes, although bedding the sill on a DPC is common practice to protect a timber frame sitting on the external skin. Care has to be taken to ensure that the edge of the DPC is not laid forward of the line of the front of the wall, or the drip may be breached and water drawn back under the sill. Cavity closure sections can also be used in the sill detail.

Doors and windows – the components in the openings

It is not the intention to consider the details of the components themselves. This has become a specialist area of expertise, where the manufacturing of sections and the factory assembly of the components have to be understood as well as the component design aspects such as strength, weathering, safety and security. All these are beyond the scope of this text. However, the design of doors and particularly windows with the joint and fixings into the openings needs to be thought through, particularly as concerns for sustainable development are leading to a re-evaluation of energy saving by considering the behaviour of the house as a whole. This includes a consideration of the orientation of the facade, the area or percentage of glazing, the airtightness of the joints and the shape of the windows, as well as the specification of the materials involved. While this has a significant impact on the look of the house and the views experienced by the occupants, it also has an impact on the energy efficiency of the house.

Given the predominant energy used in houses during winter heating conditions, any contribution that can be made to reducing heat loss or increasing gains needs to be considered. Windows have the potential to do both. Solar gain from south-facing facades can be usefully used for many days even in winter, where direct radiation from the sun will pass through the glass and heat objects in the room, adding to the heat gains. However, the glass is poor at retaining the heat from the air in the room, increasing heat loss. Heat loss can be considerably reduced not only with double (or triple) glazing but also by the choice of glass with coatings or gas-filled cavities which can reduce U values of double glazing almost to that of triple without the expense of the more complex frames and opening casements.

Using glazing for heat gains is only effective on facades facing in a southerly direction. In these facades larger areas of glazing can provide a net gain, but this is not true for other facades; here it is better to limit the area as heat gains are not possible and the wall construction can provide much higher U values to reduce heat loss. However, double glazing with coated glass can still make a significant contribution to reducing heat loss on these facades.

These choices to reduce winter energy use have consequences for both lighting in the rooms and heat gains during the summer. Light from the south can be very bright and cause glare, so larger windows may create discomfort or even disability during the summer, requiring the use of blinds or curtains. Similarly heat gains will be greater, possibly requiring shading. Conversely, while north light is good quality and more even, the smaller windows may make rooms dim during the winter, requiring artificial lighting and adding to energy use. Heat gain is unlikely to be a major problem from the north.

Currently the number of days when summer heat gain or poor lighting are likely to be of concern in the UK is limited as for many households daytime occupation is limited (unlike commercial property where this is a major problem, particularly with high heat gains from equipment and lighting). For the days when conditions are uncomfortable, ventilation by opening windows is the main mechanism offered to householders to control the internal temperature. If either climate changes or occupation patterns vary, this may have to be reconsidered, and designs including light-coloured walls, screening and even verandas that are used for houses in hot climates may be considered on some designs in the UK. Perhaps high thermal capacity construction with night-time purging being used in commercial property in temperate climates will be adopted for house construction.

Having considered shape and size along with any additional components such as blinds or shading, the choice of materials has to be considered. Glass has been discussed but not the choice of material for the frame. For domestic buildings, timber (soft and hardwood) or plastic (uPVC) are the main choices. Aluminium is also specified, but it has a higher capital cost than timber or plastic. Steel has in the past proved to have a high maintenance cost to control corrosion. However, galvanised, polyester coatings and stainless steel overcome these problems of the past but are not often considered for housing, given the viable options of timber and plastic. Aluminium can give slender sections and be produced with coloured anodising if this is required for appearance. The choice between these frame materials will include an analysis of the need for maintenance, potential life, ease of replacement and potential recycling and disposal options. Timber (particularly hardwoods) can have an impact on habitat and global climate, while plastics have pollution and energy considerations associated with their manufacture.

Having chosen the size, shape and material for the windows and doors and had them manufactured to give strength, be weather-tight and secure, it is necessary to consider how they are to be fixed and weathered in the opening.

Windows and doors can be built in as the work proceeds, being fixed back to the brickwork with frame cramps or secured to the cavity closure. They can also be installed after the brickwork is complete by screwing through the jambs. If they are fixed after the brickwork is complete, the bricklayer has lost any direct control on tolerances to maintain the gap between frame and brickwork where the appearance and weathering functions have to be achieved. This will have to be replaced either by careful measurement of opening size, square and plumb or with the use of temporary frame forms. Packing will then have to be used between frame and brickwork to ensure the jambs of the frame are not distorted as the window is secured.

However the frame is fixed, weatherproofing is normally achieved with gunned mastic, although consideration should be given to placing the window towards the back of the reveal to give greater protection, particularly in exposed conditions.

The panel frame option

The construction analysed so far has been based on an initial suggestion of masonry skins to the wall. These have proved capable, providing the required performance and being able to be detailed around openings and at the connections to other elements of construction. Given the context of the case study described in Chapter 16, a viable alternative to the masonry internal skin that may be considered is a wall based on the framed panel.

The panels can be formed on site but are more commonly formed in a factory and delivered to site as flat panels. Some developments have built the panels into rooms in the factory that are delivered to site as volumetric units. The basic form of the loadbearing panel in timber was shown in Figure 19.3. Similar construction can be achieved with light steel sections. The following text only considers timber, but the process of analysis could be applied to the light steel solution.

Timber frame construction

Suggesting that the inner skin is formed from timber frame panels rather than a concrete block affects nearly all the aspects of specification and detailing of the wall. It has such a fundamental effect that it is referred to as timber frame construction even though it is only the inner skin that has been changed in principle.

The origins of these fundamental effects can, for the most part, be attributed to the change already identified:

- The basic form (solid masonry to open panel)
- The material (concrete to timber)
- The production process (site operations to factory prefabrication)

The basic form

The component layers of a typical timber frame inner skin are shown in Figure 19.13. The basic form of open studwork with a sheathing board on one side allows the insulation to be introduced within the thickness of the skin. The temperature profile will now show the condensation risk on the cold side of the insulation, leading to the need for a vapour-control layer behind the internal finish. This vapour-control layer is likely to be a polythene sheet where, even with care on lapping and sealing edges, it can only be considered a control and not a vapour barrier. To ensure the control is effective it is necessary to ensure what is known as the five times rule. The vapour permeability on the warm side of the insulation (including the vapour-control layer) should be at least five times that of the rest of the construction on the cold side. It is

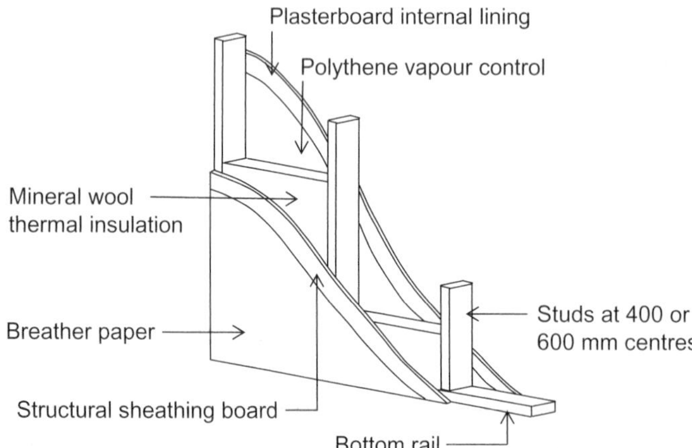

Figure 19.13 Inner skin of timber frame cavity wall.

therefore necessary to choose the materials on the cold side with care to ensure they have high vapour permeability if condensation is to be controlled. It is usual to specify insulation materials that can breathe, such as open fibre mineral wool bats or quilts.

For structural purposes in a typical panel for a two-storey house the studs need only be around 89 × 38 mm at 600 mm centres, although this limits the thickness of insulation if it is only to be provided between the studs. For increased strength and/or insulation 140 × 38 mm studs are common. For increased insulation 180 × 47 mm could be considered but using 'I' section studs, fabricated with timber flanges and oriented strand board (OSB) webs similar to the floor joists shown in Figure 17.6 may prove more economic.

The open nature of the studwork also affects the possibilities of providing fixings. Components transferring load, such as wall ties, must be fixed to studs and therefore have to be at centres to match the studs. Heavy items inside the building need to be identified because if they cannot be fixed to studs they may need to have noggins incorporated.

The open framework also dictates the finish. The studs do not provide a continuous background but do form a plane with good flatness tolerance, the timber gives easy fixings and a plasterboard finish is most appropriate. Screwing the plasterboard to the studs and tapered edge boards to save a skim coat is the most likely specification.

Influences of the material

Perhaps as fundamental as the open form of the panel is the change of materials. The change of materials brings with it changes in properties. Those most significant to the performance of the internal skin of the wall are movement, fire and some aspects of durability.

The external clay brick skin will tend to expand, while both concrete and timber will shrink. This differential movement in the masonry wall is not significant and wall ties and lintels that span both skins have sufficient flexibility to accommodate this movement over two storeys. This is not the case with the timber/brickwork combination. Wall ties have to be flexible or designed in some way to accommodate vertical but not permit horizontal movement. A clear 12 mm gap at the top of the brick wall to the roof structure resting on the inner skin is needed for a two-storey house (assuming the inner skin is resting on a concrete ground floor). Window and door frames should be fixed to the timber internal skin and then gaps have to be left between the brickwork and the underside of the sill and the head of windows. These gaps will have to have a compressible filler and flexible sealant to weather the joint.

The fire resistance of the timber will be less than the block skin. While the brick maintains the fire resistance of the wall as a whole, the fire may reach the cavity, and this provides a clear path particularly to spread the fire to other rooms and the roof. This requires the provision of cavity battens, the regulations and recommendations for which vary depending on location in the UK. They are required horizontally at the top of the wall to protect the roof and may be required at each floor level. They are normally provided around all openings and are required vertically at party walls. The cavity batten has to be in contact with both the cladding and the sheathing in order to act as a barrier to the spread of fire in the cavity, but this will bridge the cavity, providing a potential path for damp penetration. These cavity battens may be ridged (38 mm preserved timber battens on DPC) or flexible (50 mm wire-reinforced or polythene-sleeved mineral wool) with polythene DPC tray over all horizontal battens.

Consideration of the deterioration of the timber affects the detailing. Timber has to be maintained at moisture contents below 20% to inhibit decay. During production, finishes should not be completed if they might seal in timbers with values above this limit. Detailing should be devised to keep moisture below this level during the life of the building. Where there is a risk of this being exceeded, durable species or preservative should be used. As the timber likely to be chosen is unlikely to be a recognised durable species, preservative may be specified. With careful detailing it should not be necessary to preserve all timbers in the house. However, some timbers are difficult to protect with detailing alone. Preservative treatment is essential for sole plates, timber against DPCs and cavity battens, although for insurance reasons the preservation of all the external wall frame timbers may be recommended.

While for production reasons the cavity will still have to be 50 mm, if the cladding is brickwork it will have to be drained and ventilated. The build-up of moisture in the cavity is a threat to the timber, so a flow of air helps keep levels of humidity low. In addition, a breather paper protects the panel during construction until the brickwork is completed. The breather paper is waterproof to minimise the risk of moisture being absorbed by the timber but has good vapour permeability to ensure the five times rule is maintained.

The production process

The production process is determined by decisions about how large the panels can be to transport to site and handle into position. The basic floor and roof construction of timber joists and trussed rafters is still applicable, but it is now sensible to consider some level of prefabrication or at least preparation to coordinate all these timber elements with the walls. In the UK, storey-height panels are fixed above DPC level where floor cassettes are fixed at first-floor level bearing on the top of the panels. From this new platform the first-floor wall panels are erected where the roof trusses and bracing are fixed to complete the structure. It is also possible to prefabricate the internal walls and erect these at the same time. This storey-by-storey height process with floors resting on the top of the panels is known as a platform frame; a multi-storey-height wall with floors hung off the wall, known as a balloon frame, has been developed but is not often used in the UK.

It is possible to conceive of a wide range of levels of prefabrication for the platform frame system for all the elements of the house. In the UK most wall panels are delivered as open panels. The basic stud and sheathing board panel often with breather paper is factory-produced. The panels can be easily nailed together and the breather paper laps arranged on site. It is possible to think through the prefabrication of closed panels, where insulation, services, vapour control and wall linings are fixed in the factory. This makes site process connecting panels more complex and the need for site protection more acute.

Whatever the level of prefabrication, it will require some coordinating dimensional discipline to take full advantage of the factory process and this will have to be applied to the design of the house and the brickwork outer skin. It will be necessary to establish a grid on plan and decide on some vertical heights. This idea of a dimensional reference framework was discussed in Chapter 4.

All of these structural elements (walls, floors and roof) are erected in one set of operations, with the outer skin or cladding being added later. This is a shorter site operation compared with completing the structure with masonry walls. Scaffolding has to be erected to eaves level before the erection procedure is started and a crane will be required throughout the procedure.

This production sequence means that the brick skin has to be complete with the whole of the internal skin in place. The brickwork is now highly dependent on the accuracy with which the panels have been erected. Normal brickwork tolerances are ±10 mm in 5 metres' length and ±20 mm in 6 metres' height. The timber panels should be erected within these limits, in which case the bricklayer can maintain the width of the cavity to ensure that wall ties, cavity battens and any lintels attached to the panel should all be installed within their tolerances. Erecting the timber frame panels within tolerances to the dimensional discipline grid is also important to allow the brick bonding to be set out without excessive cutting of bricks or variations in the thickness of the joints.

Alternative cladding options

Render is possible with a timber frame, but it is not likely to be applied to a block as it was in the masonry wall but to metal lathing fixed to preserved battens fixed directly to the studs. If the lathing is unbacked (render can squeeze through to behind the lathing), the cavity has to be 50 mm to keep it clean, but with backed lathing systems the cavity can be reduced to 25 mm wide.

Other claddings such as tile hanging and board cladding can be adopted, again fixed to preservative-treated battens. With tile hanging the battens are horizontal so there is no vertical cavity, but with lap jointing each tile can drain independently. With boarding the joints cannot drain and a 10 mm wide cavity should be formed. In this case the insulation value of these cavities is not very great, but they will fulfil the waterproofing function.

Care has to be taken in specifying these alternatives. The bricks provide the major part of the fire and sound resistance to the whole wall. Changing the cladding may leave the building unprotected in these areas. Some cladding may be combustible and have poor surface spread of flame. Sound insulation is dependent on other elements such as windows and doors, but the contribution of the wall has to be considered.

Party (separating) walls

Unlike external walls, these walls have no weathering function and will have no openings, making their construction generally more straightforward. However, they may be loadbearing (including racking forces from wind loading) and will need greater performance in both fire and sound resistance, and this will determine most of the construction specification and detailing. The basic construction options are shown in Figure 19.14.

In masonry construction both fire and sound insulation can be better achieved with dense concrete rather than the lower-density blocks chosen for their thermal properties for the external wall. The cavity is not necessary for weatherproofing but does provide sound reduction, so both solid and cavity construction can be used. For fire resistance the wall has to extend into the roof space and be fire stopped up to the underside of the tiles.

For a solid party wall the total mass should be at least 415 kg/m^2, and this can be achieved

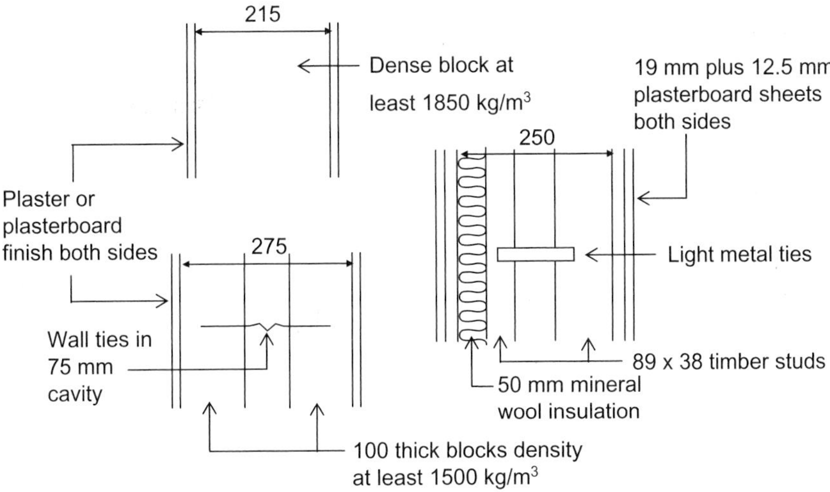

Figure 19.14 Alternative party wall options.

with a 215 mm wide dense block or the wall can be formed by laying thinner blocks flat to form a 215 mm thick wall. Both of these can then be finished on both sides with lightweight plaster or a single sheet of 12.5 mm plasterboard as discussed for the internal finishes to external walls.

If a cavity wall is used, two 100 mm skins with a 75 mm cavity (with wall ties) can be formed in lighter blocks with densities around 1500 kg/m^3, again finished with lightweight plaster or plasterboard both sides. In both these specifications it is important to have the mortar joints fully filled to ensure the full performance is achieved.

Party walls can be constructed in timber frame construction as cavity walls. In this case it is difficult to achieve mass and fire resistance directly from the structure of the timber frame panel. The wall has to be constructed of two stud frames with a cavity between. The finish will have to be two sheets of plasterboard, one 19 mm and one 12.5 mm (staggered joints) on either side of the wall, and with at least 50 mm (80 mm recommended) of mineral wool quilt with a density not less than 10 kg/m^3 in the panel(s) facing the cavity. At floor level a fire cavity barrier should be provided, and a row of lightweight metal cavity ties is required to stabilise the two skins for each storey. Fire barriers are also required vertically where the party wall meets the external walls while maintaining the sound discontinuity to avoid flanking paths. This construction relies on discontinuity and absorption in the cavity for its sound insulation, with the plasterboard providing some mass and the fire resistance between the dwellings.

This construction of the party wall has no sheathing board on the timber frame, which was a major component of the structural action of the panel, particularly against racking. The plasterboard is now fulfilling this function and may not be sufficient. If additional stiffness were required, this could be gained with diagonal bracing inside the cavity, or a sheathing board might have to be reintroduced into the construction.

Partition walls

Like external and party walls, partitions can be constructed in either masonry or framed panels. The basic construction is the same as the walls shown in Figures 19.1 (similar to block internal skin) and 19.3 (timber or steel stud), as are the options for finishes with plaster or plasterboard, as appropriate. Proprietary partitions where panels are delivered to site finished both sides ready for decoration are available but are generally more expensive.

Partitions are likely to have lower performance demands than either external or party walls. Walls immediately below the roof will not have to carry loads if the roof is constructed with trussed rafters. Walls below floors, however, may have to carry loads from the joists. There may be a need for internal walls to act as buttresses against overturning from wind loads, in which case they will need to be bonded or otherwise securely tied to the external wall over their entire height. Otherwise, the partition can just be built up to the inner skin. When partitions are built against timber-framed party walls, both sheets of plasterboard on the party wall have to be fixed first to ensure no weakness in either the sound or fire performance of the party wall.

Given no structural role, the choice of partitions within a dwelling will be dominated by the functions associated with appearance and privacy. They do have a nominal half-hour fire resistance requirement, but this should be satisfied with masonry with plaster or plasterboard, or a framed stud partition with 12.5 mm plasterboard on either side for the finish.

Appearance demands are normally limited to plain surfaces that can be achieved with plaster or plasterboards, and the basic construction will provide a complete visual and light barrier. Sound may become the limiting performance for the choice of partitions. There is a requirement for the construction of partitions between bedrooms and between rooms with a water closet and any other room to achieve an R_w value of 40 dB. This cannot be achieved in a stud wall without including a sound-absorbing quilt at least 25 mm thick between 75 × 38 mm studs. Block walls to achieve this sound rating will have to have a mass of 120 kg/m^2 or more.

The thickness of partitions in a house is likely to be between 100 and 125 mm. This should ensure stability for normal domestic floor-to-ceiling heights with head restraint for lightweight framed panels, as discussed in Chapter 17. If it is required to fix items to the block partition, this should not be a problem, but for the frame studwork the need for noggins (marked on the outside of the boards) for fixings for heavy items will be necessary. The interaction of partitions with other elements, particularly services, is the same as discussed for the external wall in the section above. Indeed, the finishes may well be chosen to be similar for all internal surfaces of the rooms whether they are to external, separating or partition walls.

Summary

1. The three wall types discussed – external, separating and partition – each have a different set of functions to perform.
2. Each of these wall types can be built either as masonry, panel or monolithic, which in practice are normally constructed in brick and block, timber or steel frame, or as in situ concrete, respectively, although other options are available.
3. The cavity wall, chosen for its weatherproofing action, is the most likely option for the external walls for the house in the case study, specified as either masonry or timber frame construction.
4. The cavity that provides the weatherproofing is formed between an outer skin that determines the appearance and an inner skin that normally takes the loads. The cavity and the choice of lightweight blocks for the inner skin provide some thermal insulation, but this has to be augmented with a layer of insulation to achieve what would now be considered a reasonable U value.
5. To provide all the functions of an external wall, windows and doors are required. To accommodate these components openings are formed in the wall, which have to maintain the performance of the basic wall. The components are then fixed and weathered into the openings.
6. The components are now normally designed and manufactured by specialists and require only detailing to fix and weather the units into the openings. However, the size of the opening, the performance and operation of the components and the materials from which they are made must be chosen as part of the overall environmental design for the house.
7. Timber frame construction is based on creating the internal skin of the cavity wall in a timber-framed panel rather than in blockwork. This change to panel construction in a different material, together with the changes this brings about in the production process, has a profound effect on the rest of the construction details, requiring a re-evaluation of most aspects of the structure and envelope of the house.
8. Separating or party walls between dwellings have both fire and sound performance requirements that dictate most of the choice of materials and detailing.
9. The options for internal partitions are also based on either masonry or studwork. They may be loadbearing downstairs and may have sound transmission requirements, particularly between bedrooms. The requirements for the integration of services and the finishes for partitions are similar to those for external and separating walls.

20 Foundations

This chapter considers the construction up to damp-proof course (DPC) but not including the ground floor. It introduces the influence of the soil and the site on both design and production for a range of foundations that may be considered for the house in the case study.

Basic function

It is the function of the foundations to maintain the integrity of the structure and fabric of the construction above ground. This text considers the foundations to comprise all works up to the DPC for the wall, but not including any construction associated with the ground floor that is taking loads from the activities on that floor (this is covered in Chapter 17). The foundations, therefore, have no environmental functions. They are structural and have some small appearance considerations where the wall shows above ground up to the DPC. Analysis of the lifecycle is necessary as maintenance targets should be zero and the foundation will affect the possibilities in redevelopment of the site when the building is modified or demolished.

Structural integrity

Given that the primary structural role is to maintain the integrity of the building, two things need to be known about the structure and fabric above ground:

1. The distribution, magnitude and direction/point of application of the loads
2. The stiffness of the construction above ground

For the house, all the loads come from walls, arriving at the foundations as uniformly distributed strip loads. Each wall will have a different load per metre run depending on which walls take the roof and floors. Wall loading for housing also depends on the self-weight of the wall, which can be a significant part of the load particularly if constructed in masonry and with few or small openings. While the loads from external and party walls vary, the difference is unlikely to be great and therefore it is usual to specify all foundations to these walls based on the greatest loading. This will theoretically lead to different settlements under each wall, but in practice, on reasonably good bearing soil, these variations in loading do not lead to significantly different settlements that would cause distress to the superstructure. However, the loads on internal partitions will be considerably less and the wall will be thinner, and so smaller foundations would be chosen.

House construction will produce vertical loads at the foundations and, although theoretically most of the imposed load will be from the internal skin, the load will be taken to be acting through the centreline of the whole wall.

The stiffness of the building, its structure and fabric, is a measure of how much differential settlement it can take before it suffers any excessive distortion or cracking that would lead to a failure of performance. Initially, distortion or cracking may only affect appearance, but as they become more pronounced, weathering (serviceability) and structural stability may be threatened. All buildings distort and crack to some extent and much detailing is designed to accommodate these movements. It is this detailing and the choice of materials that determine how much movement can be accommodated without affecting performance. In house construction a major consideration is the stiffness of the block/brickwork. As the stiffness of the structures increases, critical strains (the movement that develops ultimate stress and cracking) will be reached with smaller differential settlements. The use of cement mortars for strength makes the wall relatively stiff and less able to take differential settlement than walls in weaker, less rigid lime mortars. The nature of the connections of the floors and roof to the wall being pinned allows some rotation but they are susceptible to lateral movement, which may involve a reduction of the bearing.

Soil as part of the structure

The soil under the foundation will be stressed and is therefore part of the structural system of the building. The foundation of the building must act to transfer the load in such a way that neither the foundation nor the soil is overstressed. The analysis of the behaviour of the soil under load is given in Chapter 11. It identifies that the compressibility of the soil below the foundation to a depth of around 1.5 times the width of the foundation will determine the settlement, while the soil to the side of the foundation will be involved in a collapse failure dependent on the soil's shear strength. All this is dependent on the absence of instabilities in the mass of soil outside this area of stressed soil caused by volume change from moisture movements or soil collapse from slope instabilities or voids caused by natural or mining activity.

This analysis identifies what needs to be known about the soil and the extent of investigation that is required to establish good information on which to base the design of the foundation.

The basic soil properties of the shear strength to determine the ultimate bearing capacity and the compressibility (including time-dependent behaviour) have to be known. In addition to these properties that are related to the capacity of the soil to resist the loads it is important to know about the soil's volume change with changes in moisture content.

It is characteristic of clays (and the more cohesive silts) that changes in moisture content are accompanied by changes in volume. Removing water from these soils below foundations leads to shrinkage that would cause settlement additional to that expected from the application of loads. Increased water content in these soils causes swelling. If additional water becomes available, the suction forces in the very fine structure of cohesive soils are great enough to overcome the weight of the building and the ground will heave, lifting the building up. Once it is established that the soil on the site is subject to these volume changes, it is necessary to ask to what depth these moisture changes can be expected.

There are two major causes of moisture change: seasonal changes and the vegetation (mainly trees) around the building. In the yearly cycle between summer and winter the soil near the surface will dry out (desiccate) from the sun and take up moisture again with the increased incidence of rain. In the UK these seasonal changes are variable, but over the life of the building some dry summers and wet winters will be experienced. Past records have shown that these are unlikely to affect the soil below about 900 mm so any foundation below 1000 mm should not experience this seasonal volume change in the life of the building. Vegetation, however, can extend its influence below this level. The effect is very dependent on the species of the tree, its thirst for water and its mature height, and how far it is from the building. Mature trees will draw water from deeper and wider in dry summers, leading to shrinkage and settlements. If mature trees are removed, they will cease to take water and the soil will return to a higher water content, causing swelling and heave. The depth of soil affected has to be assessed for each site and incidence of trees but can be as deep as 3500 mm or more, much greater than seasonal influences.

There is one other volume change associated with the presence of moisture, but this is not from change in moisture content but in the freezing of any moisture in the soil. Soils susceptible to movement under these conditions include silts, chalk, fine silty sands and some lean clay. This can cause heave if freezing takes place below the foundation. For housing this is a seasonal change associated with very cold winters with prolonged sub-zero temperatures. Again, the winters in the UK are very variable in this respect, but past records have shown that in areas adjacent to occupied buildings depths of up to 450 mm have been affected, so foundations below 500 mm should not be affected. For specialist installations such as cold stores, greater depths may have to be considered.

Site investigation

It is now possible to identify what needs to be known about the soil to a depth that includes the zone of stress and moisture changes if the soil is expected to be clay. It is also necessary to be clear how these can vary under the building, including the possible shear zone around the building. If the position of the building is not known, the whole site must be considered in the investigation. Beyond this, other information is necessary to consider the possibilities of subsidence or other ground collapse, which may be a potential failure to be allowed for in the design.

In establishing a site investigation it is also necessary to gather information important to the production process. Work in the ground carries with it risks directly related to the ground conditions. These are not only technical risks but also health and safety and financial risks, which tend to increase the cost of works below ground. Selecting foundation solutions that minimise these risks is part of the process of choice. The soil properties identified for the design will give information on the risk of collapse of exposed excavations and the need for temporary support, but additional information, such as ground (standing) water and access possibilities, needs to be established.

Information for any environmental analysis must also be collected at this time. This could be concerned with habitat and the local ecosystems on a green-field site, but needs also to be concerned with previous land use and the possibility of contamination on brown-field sites. Care must be taken in making this distinction between green- and brown-field sites. Some areas of the UK now considered wild and unspoilt might have had an industrial past from Roman and even medieval times. Previous development may indicate the existence of old foundations or other construction below ground or contamination. It may also reveal antiquities or other archaeological remains.

Not all of this information can be gathered in the same way. This leads to most investigations being based on some or all of three processes:

- Desk study
- Walk-over study
- Ground investigation

The desk study uses existing published information. This can include climatic data, but it is most likely to be based on geological and topographical maps and aerial photographs. This study will yield information to assess the wider possibilities of subsidence from both geological and man-made landforms that could make the site unstable. This would therefore include records of maps or old mine works that would show previous land use. The names of features or settlements are often clues to previous use. The desk study would also include surveys of the utilities – water, electricity, sewers, etc. – which can be obtained from the appropriate authorities.

The desk study would indicate what might be found on the site and help to inform the other two phases of on-site investigations. Along with any local knowledge, it will give some idea of what land features and vegetation would be visible from above ground and what soil type, previous development and contamination might be below.

What is considered contamination could come from two concerns: the adverse effect it may have on the health of the occupiers (and constructors) of the building and the possible aggressive effects it may have on the materials to be chosen for the construction. This will determine what could be considered sufficient levels of concentrations in the soil or ground water to be considered contaminated. For these reasons, contamination surveys on brown-field sites are often undertaken independently of the engineering soil survey.

The walk-over survey gathers information from above ground. This would include the topographical and vegetational features and the conditions of adjacent development. It may be possible to just mark these on an existing site plan plus a photographic record, but it may show the need for a more accurate land survey if the positions of features or the exact levels of the land were important to the proposed building, its form and position.

The scope of the ground investigation will depend on what the desk and walk-over study identified and the size and possible construction of the building. For house foundations, simple trial pits will probably be sufficient, with the soils being identified with the use of soil classifications from visual inspection. If the site provides good bearing conditions, simple strip foundations can be designed to satisfy the Building Regulations from a table in the Approved Documents, where only soil type is needed to select a foundation width. Simple visual classification will also be sufficient to determine the presence of the shrinkable clays important to specifying sufficient depth to ensure volume stability. Trial pits will give an indication of standing water and will be sufficient to take water and disturbed soil samples for chemical analyses, such as sulphates, if these are suspected. Where visual identification of soil type is not sufficient, in situ tests and the removal of undisturbed samples for laboratory tests, possibly from boreholes, may be required.

It is important to consider a fourth phase that takes place during construction where the conditions anticipated by the site investigation and used for the design need to be continually compared to the information revealed by excavations for the building. This will also lead to inspections prior to significant events, such as pouring concrete for the foundations themselves.

Foundation design

Whether the foundations are chosen from tables or are the subject of a structural design, four conditions must be satisfied:

1. The load has to be distributed over a sufficient area to limit settlement and avoid collapse.
2. The centre of gravity of the load has to be over the centre of the area of the foundation.
3. The foundation has to be stiff enough to press evenly on the ground over the whole area of the foundation.
4. The foundation has to be founded at a depth below ground to be resting on a sufficiently stable soil mass.

For the house under consideration in the case study the loads are relatively low, and many sites in the UK have soil with a good bearing capacity and low compressibility, although it is more than likely to be shrinkable clay in England, particularly in the south and east. If there are no major previous developments or geological or typographical difficulties, the simple strip footing will be the economic choice.

The strip foundation

The basic form of the strip foundations is shown in Figure 20.1. The external cavity wall rests on a DPC to stop rising damp. The DPC has to be at least 150 mm above the ground to avoid soaking of the wall from splashing back from the ground. The first part of the wall below the DPC has to remain as a cavity wall to ensure that the cavity remains clear and can drain below the DPC. The external skin has to remain brickwork for at least two or three courses below ground to maintain the appearance, but this assumes that the brick will remain durable in this zone where saturation and freezing are

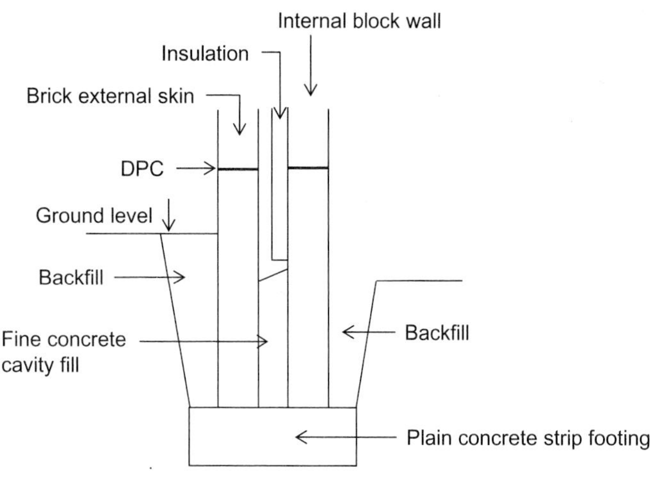

Figure 20.1 Plain concrete strip foundation.

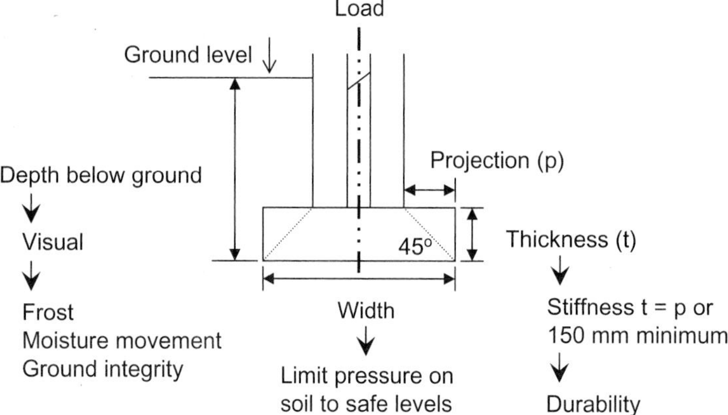

Figure 20.2 Design considerations for plain concrete strip foundation.

more likely. Below this, the external skin can be constructed in blockwork. Nearly all blocks are suitable for construction below ground, so a change from the insulating block used in the inner skin above the DPC can be made as long as this does not create a cold bridge at the edge of the slab.

From 150 mm below the lowest DPC the cavity has to be filled, as the wall will be subject to pressure from the soil tending to collapse the two skins together. It is possible to use a single trench block along the total width of the wall to avoid having to build the two skins and then fill the cavity.

The bottom of the wall then needs to be stiffened and probably made wider to transfer its load evenly to the soil. If the width does not need to be too much wider than the wall, a strip of plain (unreinforced) concrete can be used. This is the case for many sites in the UK, where foundation widths between 300 to 600 mm will be sufficient. This concrete will not have to have great strength but will need to be durable and therefore the specification needs to include the minimum cement content (and type if sulphate resisting is required), as this will be governed by durability considerations.

Figure 20.2 shows how the design principles given above are observed. First the load and type of soil will determine a width to provide sufficient area. The wall must then be detailed centrally on the strip. To provide a stiff base to

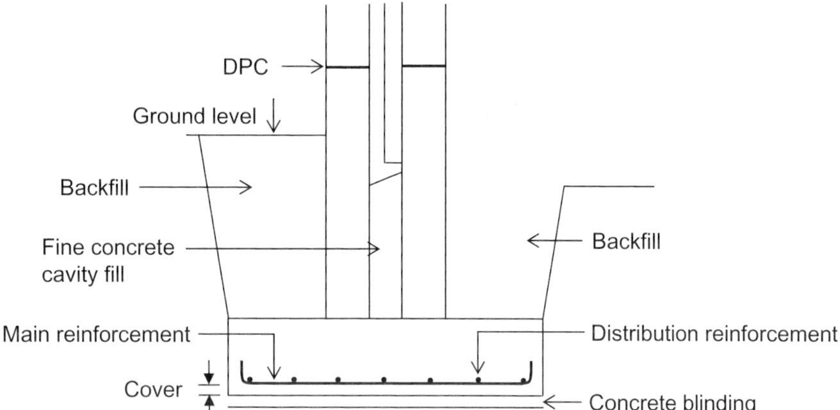

Figure 20.3 Reinforced concrete 'wide' strip foundation.

the wall in its length the footing must be at least 150 mm thick. However, to ensure the load is transferred across the whole width of the strip the thickness must be equal to the projection. This is related to the choice of the material as plain concrete. The strip will transfer the load to an area under the footing as long as the material remains in compression inside the 45° punching shear line.

This requirement for the thickness to equal the projection will give thicker foundations as the width increases. As the thickness increases it may be worth considering the possible change to the material whose structural behaviour is not dictated by the 45° punching shear failure in plain concrete. This can be achieved by introducing steel reinforcement into the concrete to provide a 'wide' strip foundation, as illustrated in Figure 20.3. The foundation can now sustain a bending action, ensuring the load is spread across the full width of the strip. The introduction of reinforcement requires the specification of cover. This is the thickness of concrete required to gain a bond between concrete and steel to develop a composite action in bending and to protect the steel from corrosion (above ground this may also be determined by fire requirements). For foundations this is determined by corrosion protection and will be 40 to 50 mm, depending on conditions. To ensure this cover is achieved in the production process a concrete blinding is required and the reinforcement will be supported on spacers standing on the blinding. In this way the reinforcement will not be trodden into the ground during the concreting operations.

Returning to Figure 20.2, the last parameter is the founding depth below ground to ensure the foundation is resting on a sufficiently stable soil mass. Given that the top of the strip has to be at least two or three brick courses below ground, it is unlikely that the depth of the foundation (to the underside of the strip) will be less than 300 mm, so it should be below the level of the topsoil (normally around 100 to 150 mm in the UK). However, this is not below the frost line, requiring depths of 500 mm, or the influence of seasonal moisture variations on clay at 1000 mm, so these may determine the depth on soils susceptible to movement under these conditions. Deeper foundation may have to be considered if there is vegetation on shrinkable clays or on any site where geological, old industrial or topological features have to be taken into account.

The simplest topological feature is a sloping site. Even if it is not sufficient to make the site unstable, it will affect the foundation depth. The specification of depth below ground will have to be for the lowest part of the site, so, if the

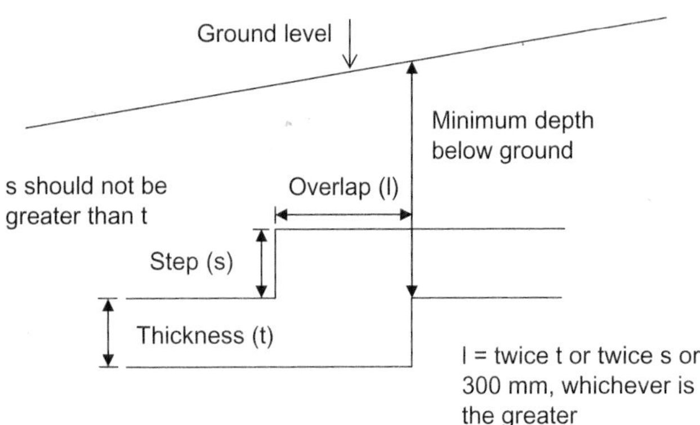

Figure 20.4 Steps in plain concrete strip foundation.

whole strip is laid to the same level, the depth of the foundation will increase at the rate that the level of the ground rises.

As the depth increases the production risks increase. There is a greater need for working space and temporary earthwork support. This will need larger machines and require the removal of more soil. All of these increase the health, safety and environmental risks, hence both cost and time will increase. These concerns need to be taken into account beyond around 1200 mm below ground level.

To limit the depth on a sloping site the foundations can be stepped, as shown in Figure 20.4, where the dimensional requirements are for a plain concrete strip.

Foundations for poor surface conditions

Apart from sloping sites, other conditions lead to the need to consider deeper founding depths; these include:

- Potential volume change in shrinkable clays
- Variability of the soil across the site
- Surface soils with poor bearing capacity

The production costs and risks will have a great influence on devising solutions to provide deep foundations. Most systems for deep foundations have been developed to be installed from ground level to limit the health and safety risks, construction time and cost. However, the answer may not be to provide deep foundations at all but to support the building at the surface in such a way as to limit the pressure on the ground and stiffen the foundations to limit the effect of differential settlement on the structure and fabric of the building.

Solutions available fall into three categories:

1. Foundations to depths around 4500 mm
2. Foundations to greater depths
3. Surface 'floated' rafts

Deep foundations

Solutions for depths to 4500 mm have been mainly developed for sites with problems associated with shrinkable clay, but they can be considered for any sites with good bearing ground within this distance from the surface.

In clay soils, or more generally soils that can be excavated without the need for earthwork support within the first few hours of excavation, trench fill foundations can be considered. These are shown in Figure 20.5, where, however deep the founding level, the concrete is filled up to a level that allows the safe and economic laying to the wall up to the DPC. In this type of foundation the production conditions are the key. The trench has to be able to be excavated accurately (line, level plumb and width) without the risk of collapse. This has to be achieved in a reasonable time to excavate sufficient length of foundation, confirm the founding depth and pour the concrete to return the ground to a safe condition. Generally this cycle of operations will have to be completed within one day. These conditions can be fulfilled in most clay soils, but the time factor may limit the depth to which this form of foundation can be undertaken, as the longer the trench is left open, the greater the risk of collapse as the moisture content of the clay changes. As long as the trench can be cut cleanly and accurately the quantity of concrete can be controlled, limiting the financial as well as the health and safety risks.

If there is then a risk of heave (expansion) in the mass of soil contained under the building between the foundations, a compressible material should be introduced to the face of the foundation, and the ground floor possibly constructed with a void as a suspended floor.

An alternative to trench fill is to provide piles to take the loads down to the safe bearing soil with a beam across the top of the piles to support the wall. Up to depths of around 4500 mm this can be achieved with what are known as short bored piles, as illustrated in Figure 20.6. Here short and bored are determined by the production requirements. Piling is achieved either by removing the soil and then replacing it with concrete (replacement piling) or by driving the pile to displace the soil (displacement piling), and these can be achieved in timber, steel or concrete. In replacement piling the method of removing the soil is by boring or drilling a shaft into the ground. This process will produce a circular shaft, the diameter and length of which will determine the size of plant (piling rigs) that will be required. By limiting the length to around 4500 mm and the diameter to around 350 mm, piles can be achieved with

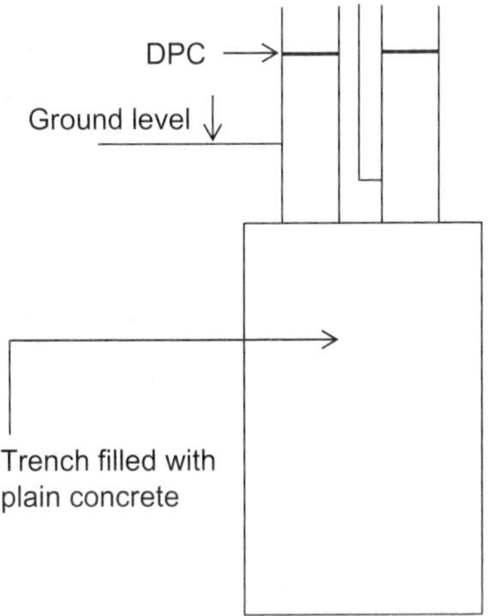

Figure 20.5 Plain concrete 'trench fill' strip foundation.

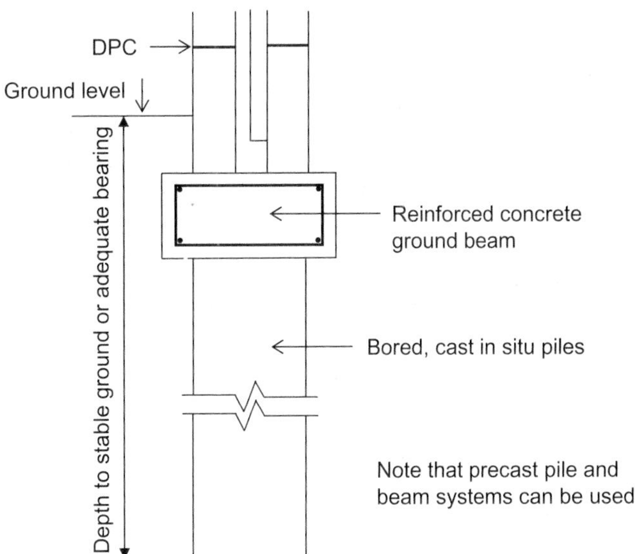

Figure 20.6 Pile and beam foundation.

relatively small rigs. If greater depths and/or larger diameters are required, full piling rigs are required.

Even with a circular shaft with a limited diameter, some soils may collapse into the hole before there is a chance to fill the shaft with concrete. In this case the shaft must be created with a tubular lining normally recovered as the pile is concreted. Displacement pile systems with precast piles and beams have also been developed where this risk of collapse is considerably reduced so long as founding depth is well established and verticality during driving can be achieved. Both of these conditions can usually be found on clay sites.

The beam has to span between piles to support the wall. Piles are placed at relatively close centres of around 3 or 4 metres, limiting the need for a deep beam. In addition, piles are placed at corners of the building and at the ends of door and wide window openings. Production operations to create the beam can be achieved in a trench near to ground level. Here the beam has to be wide enough to carry the wall so the characteristic shape of the beam is wider than it is deep, not the usual proportions of a beam. If these foundations are in shrinkable clay with the beam near the surface, shrinkage and heave could take place under the beam and around the top of the pile. The beam must be cast on a compressible layer, and in some cases there may need to be a slip membrane around the top section of the pile.

Where founding depth is greater than 4500 mm, the need for longer piles necessitates the use of full-size piling rigs, but the principles are the same. With the relatively light loads from houses, the need for deep piling is unusual. There are a number of options of both replacement and displacement systems to create the piles. These are discussed in more detail in Chapter 28.

The idea of the short bored pile can also be achieved with a pier and beam foundation. This involves the excavation of pits for pad foundations where masonry piers are constructed to take the beams that support the wall. This requires men to work at the founding level with the necessity to provide working space and health and safety measures.

Raft foundations

The alternative to providing deep foundations is to investigate the use of surface 'floating' rafts. These combine the ground floor and the foundations into one continuous structural unit. All the loads combine and are spread out under the raft so the entire footprint of the house is the foundation bearing area. This will create very low pressures on the soil under the raft, although it will stress the soil to a greater depth, albeit that they will be low stress values. This gives a safe solution on soils with limited bearing capacity and has the potential to accommodate differential settlement without transferring the strains into the superstructure. If differential settlement occurs, the raft can be designed to tilt as a whole with no strains induced in the superstructure (this can be successfully used in areas subject to mining subsidence). This can only be achieved if the pattern of loading is relatively even to give the centre of gravity of the loads at the centre of the area of the raft and if the construction can produce a raft stiff enough to act as one foundation.

The stiffness or rigidity of the raft is the key on sites where there is variability in the soil strength and/or compressibility under the building, giving the possibility of differential settlement. As the foundation to a house is a limited area with a relatively high perimeter/floor area ratio it should be possible to achieve this rigidity by stiffening the edge where the main loads will be taken from the external walls. Thickening may be necessary under internal loadbearing walls. As the ratio of perimeter to floor area decreases it may become economic to stiffen the internal area of the slab with beams rather than just increase the thickness of the slab.

For housing, the slab is likely to be between 150 and 300 mm thick with an edge stiffening detail dependent on the soil conditions. For soil

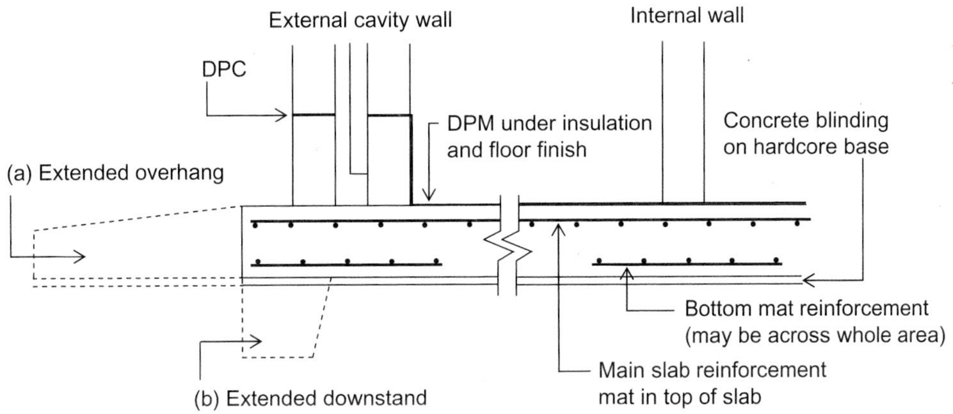

Figure 20.7 Plane raft foundation.

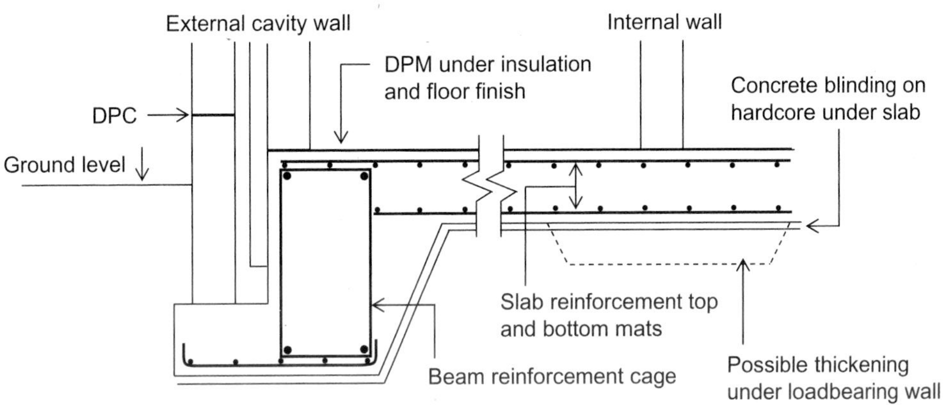

Figure 20.8 Edge beam raft foundation.

of medium to low compressibility that is consistent under the whole raft the edge detail may only need to be a reinforced extension beyond the external face of the wall. These details are shown in Figure 20.7. This extension is particularly significant when the raft is constructed close to the surface. Relatively small extensions can give some protection from frost heave but may have to be greater if the raft is on clay, with the risk of shrinkage and loss of support under the edge of the raft. On granular soils a small downstand provides the best protection, as erosion may progressively undermine the edge of the raft over time.

On soils of high compressibility but particularly where the soil is variable over the site effective support may be reduced over some sections of the foundations, and there is a need to provide additional bending capacity as well as protection. Here the downstand beam detail shown in Figure 20.8 will be required.

All rafts are reinforced with mats for the slab that is usually provided with pre-formed mesh reinforcement. For light edge downstand details pre-bent mesh cages may be considered, but for edge beam raft design the beam cage will need to be specified with bars and links that may be fabricated on site. As with the 'wide' strip foundation the reinforcement will need to have sufficient cover, and to ensure this is maintained in the production process there is a need for 50 mm of concrete blinding and the use of spacers and chairs to maintain the position of the top mat in the slab. The edge of the raft will have to be formed using formwork with some support falsework to hold the form in position during concreting. This will determine the accuracy of the line and level of the concrete that will determine the ability of the bricklayer to achieve the tolerance of the external brickwork.

Economics of foundations

The evaluation of costs for foundations with the elements of risk and variety of site conditions can be a major determinant of the choice of foundation method. Depth is not the only factor. If the production process for deep foundations could be carried out from the surface with little risk, they might be less expensive than a raft with formwork, reinforcement and concrete making up the major cost. For the strip foundation options when to use wide strip or trench fill rather than the simple strip will again be dependent on site conditions and the risks involved.

To this can be added the possibility of using ground improvement techniques to increase the bearing capacity of the soil particularly in non-cohesive soils. This may make simpler, less expensive foundation viable, offsetting the cost of the ground improvement process. Techniques such as dynamic consolidation, the installation of stone columns and the introduction of cementatious materials into the ground can improve ultimate bearing capacity to economic depths for strip foundations to be used for housing. These techniques are discussed in more detail in Chapter 28.

Basements

Basements have not found favour with UK households, where underground rooms are more associated with cellars and coal holes than habitable rooms. The construction of the basement has to deal with horizontal ground forces as well as the loads from the construction above. It also has to have high performance levels of waterproofing, which have been poorly achieved in the past. Basements are discussed in more detail in Chapter 28 but only in the context of commercial buildings. While the principles are the same for houses, the smaller scale of construction can achieve similar performance with alternative specification and detailing. There are now established construction methods for housing that can achieve high levels of performance, but they are beyond the scope of this case study.

Summary

1. The basic function of the foundations is to maintain the integrity of the structure and fabric of the construction above ground.
2. The foundation has to take account of the pattern and size of the loads from the superstructure and the stiffness of construction above ground as this determines its ability to respond to differential settlement.
3. The soil needs to be investigated to determine not only the structural properties of the soil but also its moisture stability and any contaminations over an area and to a depth governed by the scale of construction and any suspected natural and man-made features that may affect the stability of the foundations.

4. Foundations will have to have sufficient area, have loads placed centrally, be stiff enough to transfer loads evenly to the ground and be at sufficient depth to limit settlement and avoid collapse.
5. The simple strip of plain unreinforced concrete will be sufficient for most houses considered in the case study.
6. In site conditions where the construction of the simple strip carries higher risks and costs due to the required site operations, trench fill, piles or a raft 'floated' on the surface could be considered.

21 Services

This chapter looks at the active technologies, the building services that provide environmental and operational support to the household. It provides the analysis for a number of domestic systems, with particular reference to the range of emerging systems to support the growing environmental concerns of sustainable development.

Introduction

Chapter 10 discussed the requirements for the choice of construction to create and maintain environments in buildings. It introduced a group of technologies that are active in this role which are part of a range of systems known generally as building services. These technologies have grown in importance in the contribution they have made to achieve environmental performance levels over the last 100 years. They are now subject to further changes from the growing need to consider sustainable development as a factor in making choices. They significantly affect the use of energy and other resources such as water, as well as producing waste and potential pollution during the occupation of the building. These are typical factors being used to measure the environmental rating of a building.

This evaluation against sustainability criteria is leading to changes at all levels of analysis of services. It has involved changes in the efficiency of part of the system, as with central heating boilers, as well as the introduction of completely new systems, as with solar heating and grey water systems, which will be discussed later in this chapter. Perhaps the most radical has been the change in design of the whole building, integrating the active services with both the passive and active fabric, as with the design of natural ventilation systems explained in Chapter 15. With the use of motorised windows and blinds in office developments the distinction between active services and fabric is becoming blurred.

For the house being considered in the case study the most likely services that would be included are identified in Table 21.1. All these systems contribute either directly or indirectly to the quality of the internal environment to maintain the well-being of the occupants as well as the objects and equipment in the house, as discussed in Chapter 10. However, in this text only four systems will be examined in detail:

- Hot and cold water
- Heating
- Drainage (above and below ground)
- Lighting and power

Table 21.1 Typical domestic services systems.

Supply (utility) services	Input services systems	Disposal services systems
Water Electricity Gas Telecommunication	Heating Hot and cold water Lighting Power Security	Drainage Air extract

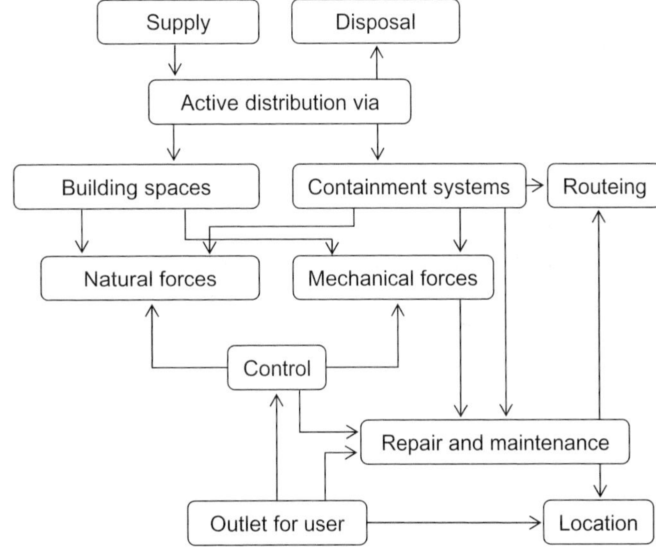

Figure 21.1 Aspects of active services systems.

Analysis of services systems

It was suggested in Chapter 10 that services have to be thought through in a different way to the passive fabric. The aspects identified are shown in Figure 21.1, a repeat of the figure given in Chapter 10. The essence of a building service system is active movement. In the environmental services this involves either supply or disposal in order to maintain a safe, healthy and comfortable environment for the user. These are shown at the top of the diagram. The system must provide active distribution via a containment system (pipes, wires or ducts) or the spaces in the building itself using either natural or mechanical forces. Once a basic system is established, control has to be chosen and decisions taken on the provision to effect repairs and maintenance. It is then possible to think about location of the parts and establish the routeing of any containment systems within the building.

As indicated at the beginning of this chapter beyond the aspects shown in Figure 21.1 cost and sustainability are significant at all levels of choice, influencing both details and the introduction of completely new systems.

Hot and cold water

From the assessment of the source, through the whole process of choice associated with water supply and distribution systems, two factors dominate the details and specification:

- Quality of water
- Quantity (rate) available

The choices are made within the context of resources, cost and environmental demands, but the impact of these has to be assessed against the criteria of quality and quantity to ensure that basic health and hygiene standards are maintained. Without water of the right quality in sufficient quantity, sustainable development cannot be achieved. This may involve education and a reassessment by users of their expectations and their behaviour that sets performance level, but the supply of clean water remains fundamental to maintaining healthy communities.

For each household it is possible to identify at least four different quality categories, only some of which have to be provided as hot as well as cold water:

- Drinking and food preparation
- Cleaning and personal hygiene
- Disposal of waste
- Use in services systems

Drinking and food preparation (potable water) is likely to be the highest standard and therefore the source must be of this quality. For most households in this case study the source will be from a mains supply through a pipe laid in the street outside the property, but there are still some wells and springs used in rural areas in the UK. All sources are subject to regulation to maintain their quality. Typical measures of quality will include standards related to:

- Biological content
- Chemical content
- Particulates
- Appearance
- Taste
- Odour

Consideration of this list will show that, while all of these need to be of a high standard for drinking water, lesser standards may be possible for the other uses of cleaning, disposal and services systems. Some care is needed in setting these 'lower' standards, as exemplified by the incidents of legionnaires' disease. Spread by droplets of water in the air, two sources have been identified. From air conditioning cooling systems it can be controlled by disinfectant, as the water does not have to be drunk or otherwise come into contact with the skin. However, the same pathogen spread in the droplets of water from showerheads has to be controlled by heating the water, as the quality is associated with personal hygiene rather than a service system.

Even the regulation associated with the quality of the source may not provide the quality required by the user. Many households will filter the water for drinking (or buy bottled water) and will have to reduce the calcium carbonate in hard water areas for any water that will pass through an instant hot-water heater in order to limit the build-up of scale on the heat exchanger. Both of these are primarily to change the chemical content of the water.

Water for drinking and food preparation for a house has to be supplied to the kitchen. This must be taken directly from the supply via a distribution system that will not contaminate the water or allow cross contamination from other parts of the system carrying water of a lesser quality. Many of the regulations governing the materials and specification of these systems are concerned with contamination and cross contamination. There are many requirements and it is not the intention of this text to give them all, but an example is the requirement of the use of leadless solders for any jointing of the pipework that is to carry water of this quality. Lead is one of the chemicals controlled at source, as it is harmful to health; any use of lead (which is soluble, particularly in soft waters) in the system may compromise this standard.

It is common practice (although by no means universal) to store water in a tank (cistern), normally in the roof space. This water cannot now be considered to be of drinking quality but may be used for all the other water needs of the household. Still some control is necessary, such as covering the tank and maintaining a minimum distance from the inlet to the water level to stop cross contamination.

Storing water in a tank has another effect, not on the quality but on the potential quantity available at the tap or other outlets. Important to the user is the rate of delivery, or how long it takes for the quantity required to be delivered at the outlet. The quantity passing through the pipe is measured in litres per second. This will be dependent on the size of the pipe and the pressure that is providing the force to move the water through the pipework distribution system. From the mains the water authority has to provide the water at a pressure within limits to ensure an adequate rate of delivery. Once this water is stored in a tank, the pressure comes from the level of the tank relative to the outlet. This is why the tank is in the roof space, to give a reasonable pressure to outlets on the upper floor. Generally the pressure from the tank will be lower than that from the mains. Mains water is said to be the high-pressure system and water from the tank on the low-pressure system. This is why washing machines and dishwashers may have to be taken off the rising main. They do not need the quality of water but need the pressure to operate successfully.

Basic system

The system identified so far is shown in Figure 21.2. It has two pressure systems and only two qualities of water. Hot water is provided from a tank, although the heating source is not yet identified. Hot water is only required for cleaning and personal hygiene and so can be supplied from the tank on the low-pressure system.

The figure shows the basic parts of the system:

- The supply identified as an incoming main
- Range of outlets required: sinks, baths, toilets, etc.
- The pipework and tanks of the distribution network
- Operational control arrangements: taps and float valves

Analysis of the rate of delivery at the outlets shows that the greatest rate is required to the WC pan, where the flushing effect requires a relatively high volume in a short space of time. Even on the high-pressure system this would require a large pipe to be able to achieve the cleansing and disposal effect required using a wet WC system. The alternative that is shown in Figure 21.2 is to deliver the water to a cistern at a relatively low rate to be stored and released in a short space of time to flush the pan.

The outlet requiring the next highest rate is the bath. Owing to the large quantity required

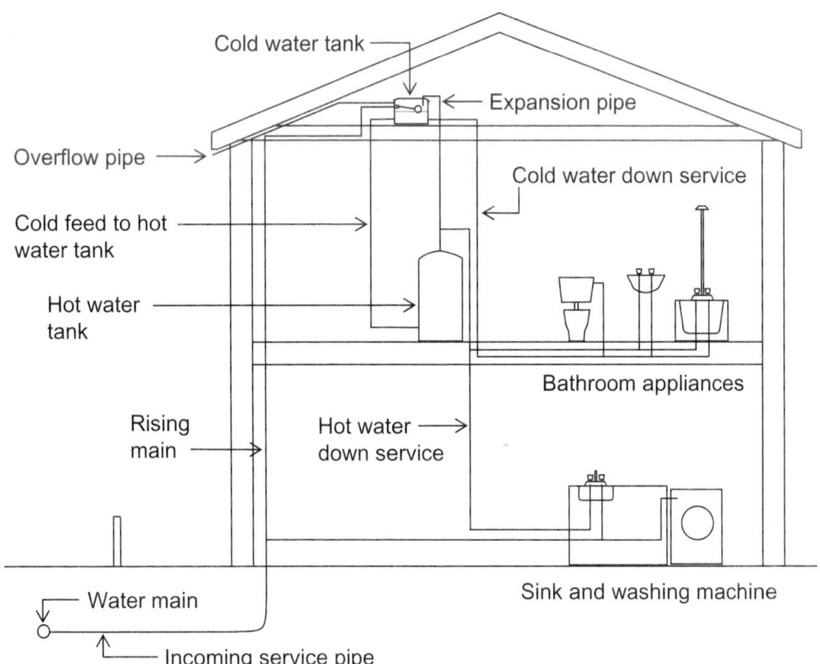

Figure 21.2 Hot and cold water – basic system and pipe work.

to fill a bath, a reasonable rate of delivery is required. This can be achieved, even on the low-pressure side, with a 22mm pipe. All other outlets can be supplied using 15mm pipes, including the WC cistern. Pipework of this size can be created using copper pipes that provide a safe and economic solution, having a wide range of fittings to configure the network.

Within a house this system should be able to maintain a rate of delivery, even if more than one outlet is being used, but performance will drop as more outlets are in use at the same time. All systems have to be designed for some simultaneous demand, but it is unnecessary to design for all outlets to be used at once, as the probability of this happening is very low. However, different systems will have different capacities to provide for simultaneous demand and this must be one of the criteria against which performance will be judged.

The tap provides basic manual control for hot- and cold-water systems, although cisterns need to have a float (ball) valve to cut off the water when the required level of water in the cistern is reached. Electrical appliances such as dishwashers and washing machines will have electromechanical devices to admit water at appropriate points in the cleaning cycle.

Provision has to be made for any taps and valves that remain on and overfill the basins or tanks. This accidental discharge can be the result of wear of the washers in float valves in cisterns, or taps being left on at the same time as the plug is in the sink or basin. Sinks and basins have an overflow provision built in but cisterns will have to have overflow pipes normally directed to outside to discharge the water safely and act as an alarm. Some WC cisterns now have an internal device to overflow into the pan. While this provides a safe discharge, it may lead to water wastage if action is not taken to repair the valve.

Maintenance, location and routeing

Once the basic requirements of the system – its supply, outlets, active force, distribution system and operational control – have been identified, it is necessary to consider the following:

- How repairs and maintenance will be accommodated
- The location of the outlets and the routeing of the distribution

The system considered so far, even with its control for everyday operations, will be difficult to maintain and repair once the system is full of water. Essential to any work that has to be undertaken on the system once it is in operation is the need to isolate sections of the system by turning off the water in order to carry out the works. It is then an advantage to drain off the water in a controlled manner so that the pipes are empty before the work starts. How small a section of the system needs this individual isolation is the next decision. When systems were first introduced into houses, it was common practice to provide only two points of isolation and often no facility for draining down the pipes in a controlled manner. These would have been a stop valve in the pavement outside the premises, belonging to the water supply company, and a stop valve on the incoming (rising) main as it entered the house, belonging to the owner of the house. If a storage tank were provided, a gate valve would normally be installed to isolate the low-pressure side. These two stop valves (and the gate valve from the tank) are still the primary points of isolation for a hot- and cold-water system. The water authority's valve is also now used for any metering and the valve on the rising main will have a drain tap for the controlled emptying of the system.

It is now good practice to provide more isolation points. Deciding on how many isolation points and where they should be comes from an analysis of where there will be a need for maintenance (and how often) and the need for fast isolation in case of failure, to minimise damage before a repair can be arranged. There is then a consideration of keeping as much as possible of the system operational while the maintenance or repair work is carried out on defective parts.

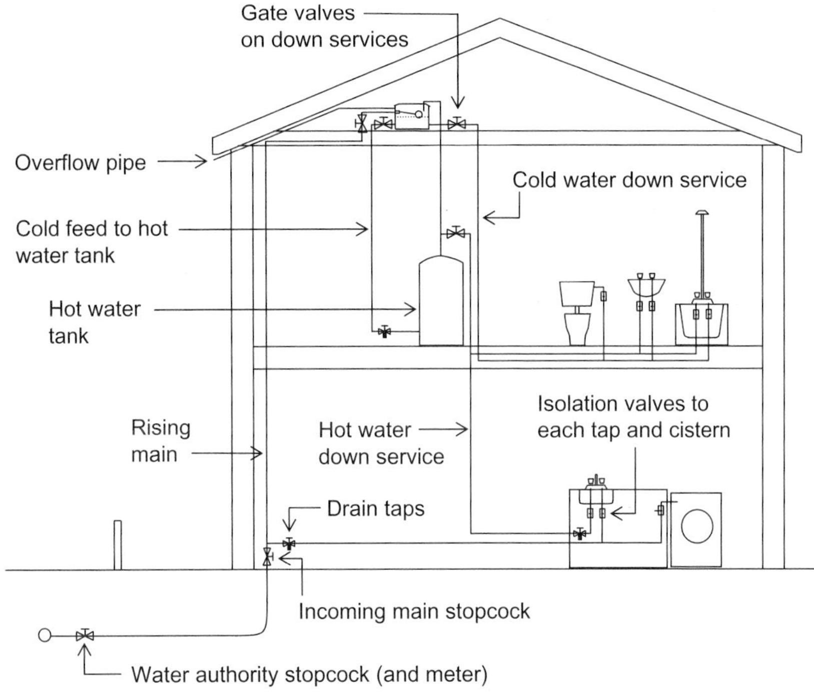

Figure 21.3 Hot and cold water – isolation and drain down.

A typical arrangement of isolation and drain-off is shown in Figure 21.3 superimposed on the basic system. The most common maintenance requirement is to change the tap (or cistern float valve) washer, and therefore an isolation valve is installed just before the tap or cistern. This allows the work to be achieved quickly without affecting any other part of the system or having to drain down any water. Having isolated each tap and cistern, more major section isolation could be considered. If a tank is provided in the roof space, it will include the gate valves from the tank to the low-pressure systems for hot and cold water. It may then be worth considering isolation and drain down for each room (kitchen and bathroom) for more major works in these areas, such as future refitting.

These points of isolation must be given good access, as must the items that need maintenance and/or replacement. This is particularly true of any isolation that has to be reached quickly in the case of failure where water damage needs to be minimised. This has to be taken into account in the location and routeing decisions for the system.

In contrast to this need for access is the requirement for concealment, which comes from a concern for appearance. It is now common for householders to wish not to see the services, so concealment becomes another key aspect of the location and routeing decisions.

Concealment cannot be separated from space competition, as the most obvious space for concealment is the construction zones, where structure and fabric for other performance requirements are already taking up space. Any distribution that cannot be accommodated in existing construction zones will have to take up usable space in the building, altering the room size and shape.

The two large items in the system are the hot- and cold-water tanks. The location of the cold-water storage cistern has already been identified as being in the roof space in order to provide a sufficient head to create the pressure to maintain a required rate of delivery to the outlets.

The gate valves for the low-pressure hot- and cold-water systems from the tank will therefore also be in the roof space. The location of the hot-water tank, however, can be at any level, as it is the location of the cold-water storage cistern that will dictate the pressure (less pipe losses, but these will not be significant in a small system for a single dwelling). The hot-water tank needs to be fairly central, often in a cupboard somewhere in the house, to ensure no outlets are at the end of long pipe runs, thus avoiding excessive delays in the hot water reaching the tap. If a roof space is not available for the cold-water cistern and space is limited to provide a central cupboard for the hot-water tank (common in flats and increasingly the case in houses), systems can be specified that eliminate the need for one or both. One such system is the use of a combination boiler, which will be introduced later in this chapter with the heating system.

The pipework system, while taking less space than the tanks, needs to be more widely distributed. It is now necessary to take account of design and operational efficiency. The grouping of outlets to minimise pipe lengths and bends reduces losses in the system and the cost of pipes and fittings. This is more of a consideration in a large commercial installation. In housing, the drainage system will have a larger influence on the need for grouping or stacking wet serviced areas (kitchens and bathrooms), but this will tend to also concentrate the hot and cold-water supply systems.

Even with this concentration of wet areas, some vertical and horizontal distribution will be necessary. Options for horizontal distribution have already been considered in Chapter 17, where the use of the floor void was discussed. There are limits to where joists can be cut for pipes to pass, and the introduction of sound insulation has implications for the free space available in detailing to ensure that sound performance is maintained. Vertical distribution may have to be boxed into ducts in the corner of rooms. They may have to be buried or hidden behind finishes, but this will make access more disruptive and accidental damage more likely.

Lifecycle and sustainability

Because of the active nature of services, lifecycle issues like cost and sustainability are particularly important considerations. There is a need to consider the lifecycle in the following stages:

- Design
- Installation (production)
- Operation (running)
- Replacement (maintenance, repair and refitting)
- Disposal

For cold water, the use of a natural pressure head means that running costs in fuel and energy terms are low, but there is a need for saving water, particularly high-quality drinking water, as an expensive and precious resource. Saving water is worthy of consideration even if it increases installation requirements. For hot water, the heating options do have fuel and energy implications.

There is a concern for the increasing demand for water and the impact of its disposal after use. Both are incurring increasing infrastructure costs that have to be passed on to the consumer and are having a high impact on the environment. The choice of system for each individual house can make a considerable difference to future costs and impacts.

Designing systems that use less to do the same job will contribute to sustainability, and this can be exemplified by the provision of showers (to limit the need for baths) and in the use of low-volume, dual-flushing cisterns for WC toilets. While this reduces the quantity of water used, they may still use water of a high quality even though this is not strictly necessary.

It has been recognised that the volume of water used for cleaning and personal hygiene by the average household equates with the volume of water used for flushing the WC(s). Grey water systems intercept the cleaning water before it enters the drains. The used water is filtered (for particles and other suspended matter that affect clarity), disinfected and then pumped to a separate tank in the roof. This is then connected to the toilet cisterns to be used for flushing. This has to be a parallel system with the ordinary water supply, as, for health reasons, the household cannot be left without flushing water if the grey supply cannot be maintained. Similar green water systems can be installed to capture (harvest) rainwater, which needs less cleaning, but the system has to accommodate much larger storage provision in order to smooth out supply (rain) and demand (flushing WCs). These grey and green water systems clearly increase installation and maintenance costs, but, with metered water supply, will show a saving in running costs and will have a low impact on the environmental needs for both water sources and disposal.

For the running of hot water, solar heating, even at the latitude of the UK, can make a significant contribution to the annual requirements for heating water. This would have to be a dual system with conventional heating sources, but again it can save on both running costs and environmental impact.

These examples are of system choice that affects the operational use of water and energy and can therefore make a major contribution to sustainable development. Using less, recycling and using renewable energy sources achieve this. All these have the effect of reducing pollution and helping to maintain habitat by reducing overall demand on the infrastructure. Some of these could only be achieved by introducing new systems that have to be accommodated in the building, paid for at the time of construction, maintained and disposed of at the end of the life of the system. The initial cost might be recouped over the life of the system if water is metered, but the environmental impact, significant to the community and future supply (and hence cost), is less easy to calculate for each individual household. The choice comes for a commitment to all aspects of sustainability – economic, social and environmental – and not just to the notion of an economic lifecycle cost.

There are sustainability issues in the choice of materials and components for the actual systems, their manufacture, transport and disposal demands. The sanitary fittings will be stoneware, and the plumbing systems are dominated by metal components, mainly copper, although plastic can be substituted for some of the components. Analysis of these is important, but currently systems that save water and fuel and have lower demands on the sewer system are seen as making the largest contribution to sustainable development.

Heating systems

There are a number of ways of providing heating for a house. This can be based on individual heaters in some or all of the rooms, perhaps with a variety of heat sources and/or emitter types to match the occupation pattern. However, in the case study being considered in this text, a full central heating system is likely to be the expectation of the user, with high levels of insulation to reduce component sizes and running costs. The most usual expectation of a central heating system is radiators as heat emitters in each room, supplied with hot water pumped through pipes heated in a boiler most probably fired by gas, if this is available.

In a domestic central heating system the main control will be based on a time clock (to set heating periods) and a single room thermostat to turn off the boiler (and pump) when the house is up to temperature during the required heating periods. Some of the individual radiators can then be fitted with thermostatic valves to give control over temperature in some rooms during heating periods in order to match occupation pattern. Even where thermostatic valves are not fitted, there will be a valve on the flow to each radiator for the occupier to turn the radiator off completely by shutting off the flow of water. There will also always be a valve on the return side of the radiator, but in this case it is not a control for the day-to-day operation of the system. Known as a lock shield valve, its purpose is to balance the system to ensure a fair distribution of water to all the radiators and is set at the time of commissioning when the system is first brought into operation. The need for control, not just for operational purposes but also for commissioning, is a feature of many services systems.

This expectation for a central heating system of the form outlined above is the case for the mass housing market, although alternatives, often based on more sustainable solutions, are emerging for individually designed houses.

The description of the typical central heating system indicates five major parts to the system:

1. A heat source – the gas-fired boiler
2. Heat emitters – the radiators
3. A distribution network of pipes carrying water
4. A pump to move the water (heat) from boiler to radiators
5. Control from time clocks, thermostat and radiator valves

Figure 21.4 illustrates a basic system fed with water from a low-pressure feed and expansion (of the water when heated) tank in the roof space. This also shows that the same boiler will heat the hot water in a hot-water storage tank. However, because the pattern of demand for hot water is different from heating, it will require separate demand (motorised valve) and temperature controls (tank thermostat) and may even have an alternative heat source such as an electric immersion heater. This also complicates the control of the boiler, which is normally resolved in the programmer. The pipes are shown as a two-pipe system with a flow (to the radiators from the boiler when the water is hot) and a return (back to the boiler after the water has cooled, having been through the radiator). The pump will maintain an adequate flow around the system. On a domestic system such as that shown in Figure 21.4 the pipes will be 15 mm diameter up to each radiator, with 22 mm pipes for the flow and return, carrying the water from several radiators to and from the boiler.

Maintenance, location and routeing

In the system shown in Figure 21.4 the most likely parts to require maintenance and replace-

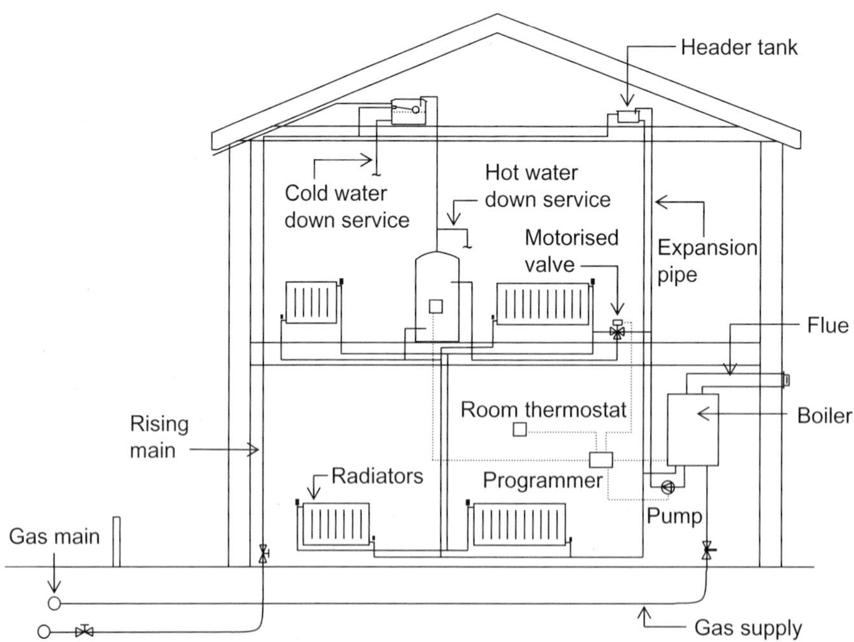

Figure 21.4 Central heating – conventional system.

ment are the boiler and the pump. Not only does this mean that access must be available, but, in the case of the pump, isolation valves on either side allow it to be removed without draining down the system. The radiators may need to be removed, not for wear in the system but for changes and redecoration of the room. The valves that are provided at either side of the radiator for control can also be used to isolate the radiator so it may be removed without draining down the whole system.

The location of the boiler has to take into account a number of factors. There is a need to connect the boiler to both a fuel source and a flue to safely take away the waste products of combustion. Some types of flue can be restrictive. Boilers with balanced flues have to be on an outside wall, but fan-assisted flues can take the gases horizontal for a limited distance, allowing a more internal location for the boiler. The concentration of pipes and connections around the boiler requires space and may be best concealed in some rooms but may be left exposed in others. Boilers do create noise directly from the unit and the noise may be transmitted via the pipe system. This should also be taken into account in locating the boiler.

The other parts of the system requiring careful consideration for position are the radiators. Their position may have some effect on the comfort of the individuals using the room. Radiators act mainly as convectors but do have a radiant component of heat exchange. The convector action promotes a circulation of air in the room to warm the space. The radiant heat acts over a limited distance, as the surface temperature of the radiator is limited for safety reasons. However, its influence, both radiant and the warm convective current rising above the radiator, is opposite to the cool feel of the windows with falling cold air when the temperature is low outside. Although this is less important with double and triple glazing, the positioning of the radiator below windows tends to balance the effect of the cool glazing, so this has been the traditional position for radiators. It is also true that there is a convenient wall space below the window that is less useful than other walls in the room. With higher levels of insulation and smaller radiators this position is less critical so long as a convective current can be achieved and no sedentary activity is too close to the radiator, where the radiant component can be uncomfortable.

With radiators in each room, particularly if located under each window, the pipes will be widely distributed. Vertical distribution is likely to be above the boiler and horizontal distribution in the floor, but on a pumped system this is not critical. There will probably be a preference to conceal pipes, but some discreet exposed pipework may be acceptable. All controls and valves must be easily accessible.

Lifecycle and sustainability

The five major phases in the lifecycle (design, installation, operation, replacement and disposal) were identified in the section on hot and cold water above. In contrast to the water systems, the heating system is a major consumer of energy during its operational phase. This not only affects the economics of the choice of system over the life cycle but also has major implications environmentally.

Consideration of lifecycle issues has to take into account the specification for the passive construction as well as the specification of the active heating system. Both insulation and the heating system itself have a capital cost and affect running costs and energy consumption. Concern is now not just for the cost of the fuel but for the type of fuel and its environmental impact, particularly CO_2 emissions as a major greenhouse gas. The use of biofuels and heat pumps for heating systems discussed in Chapter 15 (under renewable energy sources) have been applied to more bespoke houses but are not yet a widespread choice for mass housing. With limited options on energy sources that are not based on burning fossil fuels, the emphasis is on minimising the loss of energy in the conversion process from burning the fuel to delivered heat.

Insulation is recognised as the best way to save energy. Well-insulated designs, that also take into account heat inputs from solar gains, the occupants and equipment in the building, provide homes that need heating on fewer days per year and/or less hours in the day. Heating systems' efficiencies can also help in the move towards lower energy consumption when the heating system is operating. The use of thermostatic valves in some rooms that are less used or do not need such high air temperatures is an example of a system control measure that saves energy. Boiler efficiency is another way of minimising energy consumption. Boilers are now rated for energy efficiency, where the best gas boilers are currently based on condensation mechanisms where heat is recovered from flue gases to increase the effective heating obtainable from each cubic metre of gas. Running systems at higher flow temperatures is more efficient to increase transfer of heat into the room by both convective and radiant processes, but this makes the temperature of the surface of the radiator high, introducing safety and comfort considerations for anybody who gets close to the radiator.

Alternative central heating installations options

Installation costs (and possible running costs associated with water heating) can be reduced with the use of a combination boiler. The central heating is now on the high-pressure side of the water supply and the boiler becomes an instant hot water heater. An installation using a combination boiler is shown in Figure 21.5. This system does not need a low-pressure feed and expansion or a hot-water storage tank. This also simplifies the control (programmer) as the hot water is now on demand from the taps and only the heating has to be controlled from the clock and room thermostat, independent of the hot water.

Combination boilers, that include the pump, were developed for flats where a feed and expansion tank was difficult to position, and distances to hot-water taps were short.

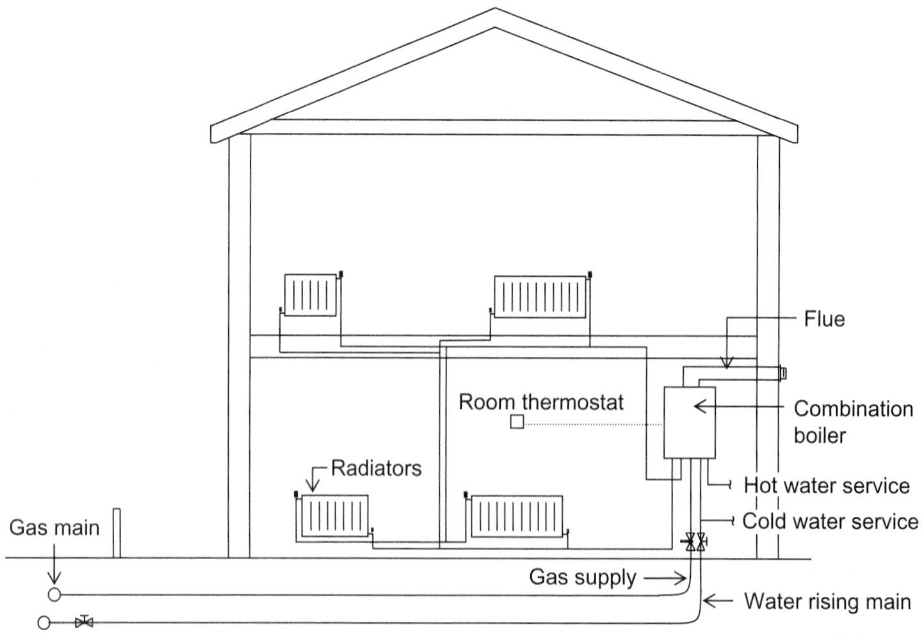

Figure 21.5 Central heating – combination boiler

While hot water is always available (no time delay in reheating water in a tank once it has been used), it is not so good at simultaneous demand, as the boiler can only provide a limited quantity (litres per second) at the required temperature. If the boiler is not central, it may take some time for hot water to reach the tap, wasting both water and energy to heat the water left in the pipe to cool.

To overcome these problems with the hot water in larger households high-pressure boiler systems (no feed and expansion tank) can be used to heat a tank of hot water. These are known as system boilers and need to be provided with the control and programmer facility of the original low-pressure installations. The hot water is often still a low-pressure system being fed from a cold-water storage tank in the roof space. However, it is possible to use unvented cylinders connected to the rising main, making the whole installation high pressure and taking away the need for a cold-water storage tank in the roof space for the hot-water supply.

Drainage

As a disposal system, drainage is concerned specifically with removing unwanted water from the house. Drainage deals with unwanted water from two sources:

- Foul water: comes from cleaning and personal hygiene that would pose a direct health hazard and a threat to the fabric of the building if it were not removed
- Surface or rainwater: collects on hard surfaces such as roofs and paths, which if not removed increases the possibility of damp conditions that would threaten the fabric and create potentially unhealthy environments

Both of these have systems above and below ground that, while based on similar design principles, will require different materials, jointing and support arrangements for the pipes with different maintenance and routeing considerations.

This text will introduce the principles and consider foul water drainage systems. Surface water or rainwater systems are almost identical, although the risks of escaping odours and blockage are less.

Basic system

As the drainage system takes water from collection points above ground to pipework below ground it is possible to use gravity to move the water away from the building. In nearly all cases the final discharge point is low enough compared with the levels around the house to arrange for the flow of water from collection to discharge without any resort to pumps or ejectors. To achieve this flow, pipes have to be laid to a fall or gradient to keep the water moving through the pipe.

Pipes do not run full but act as channels carrying the water and any solid material that is present in the foul water. The air that remains in the pipe will become unpleasant, with a smell, generated by the waste, which will accumulate as a gas in the pipe. The system has to ensure that the solids are carried away and that the gas is not released where the smell would be offensive.

To carry the solids the water must maintain a self-cleansing velocity. If this velocity is not maintained, the solids are not washed away by the water and remain in the pipe, building up and causing a blockage. Typically, for a domestic underground drainage system from a single dwelling, flow rates are very low, so the minimum pipe size for foul water of 100 mm would be specified and a fall of around 1:40 (1 metre dropped for every 40 metres travelled) would be required. Shallower falls would be possible but rely on greater accuracies in production to ensure falls are maintained throughout the pipeline. As well as pipes with these relatively shallow falls a vertical pipe is also self-cleansing. Drainage systems are, therefore, made up of pipes with relatively shallow falls

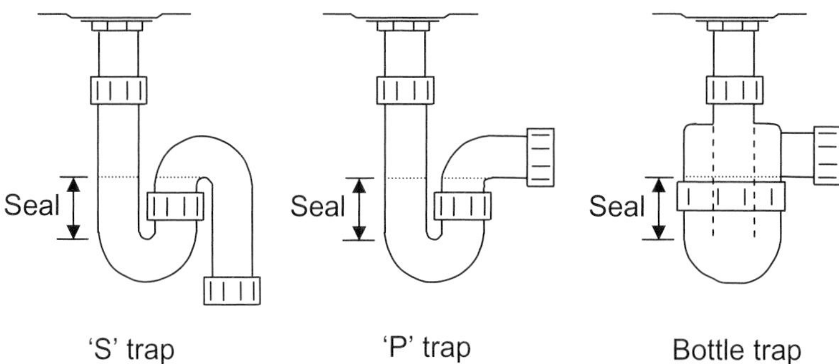

Figure 21.6 Trap fittings for plastic waste systems.

and vertical drops to connect sinks, toilets and other appliances to the disposal system.

Gas (smell) will be released back into the room if the connections that are made at the appliances are made straight into the drain. This escape of gas back into the house can be stopped with the simple device of the water seal. The seal is moulded into appliances such as toilet pans but will have to be provided by fittings, known as traps, under sinks and other waste outlets such as washing machines and dishwashers. Trap fittings for plastic waste systems are shown in Figure 21.6. The bottom section of these traps can be removed for cleaning, as traps are prone to blockage.

The act of sealing the system at the point of collection creates the conditions for differential pressures to be generated in the system when water is flowing away. These conditions tend to remove the seal when the system is in operation. This can happen to both the seal at the appliance being used (self-siphonage) and then to other seals as the water passes through the system (induced-siphonage and back pressure).

It is necessary to have the system open or vented to normal atmospheric pressure so that these operational pressure changes are minimised and quickly restore an equal pressure on both sides of the water seal. The opening to the air also lets the gas out of the pipework and has therefore to be located where the smell cannot be offensive to anybody either using or passing by the house.

Above-ground system

A typical above-ground domestic installation known as a single-stack system with pipes, seals and venting is shown in Figure 21.7. In this system each appliance is individually connected to the main vertical stack. There are limits on distances and falls related to the diameter of the pipe and depth of seal in traps, which minimise the risk of siphonage, examples of which are given in the table in Figure 21.7. Making the main stack a soil vent pipe is sufficient to prevent siphonage, with pressures being restored quickly enough, so long as the basins and baths have relatively flat bottoms for the last of the water to drain slowly to refill any loss of water in the seal due to siphonage. In these domestic systems venting of individual appliances is not necessary.

The pipes will most probably be uPVC, although cast iron may be used for the vertical stack if there is any danger of damage, perhaps if the pipe is located in a garage. Plastic is easily cut and will have a range of fittings such as traps, bends and connectors to the stack, with joints that can be formed either with cold solvent welds or, more usually, with rubber ring seals with push fits or hand-tightened collar nuts. The traps in Figure 21.6 are illustrated with hand-tightened collar nuts.

Below-ground systems layouts

While above-ground single-stack systems will be similar in nearly all houses, underground

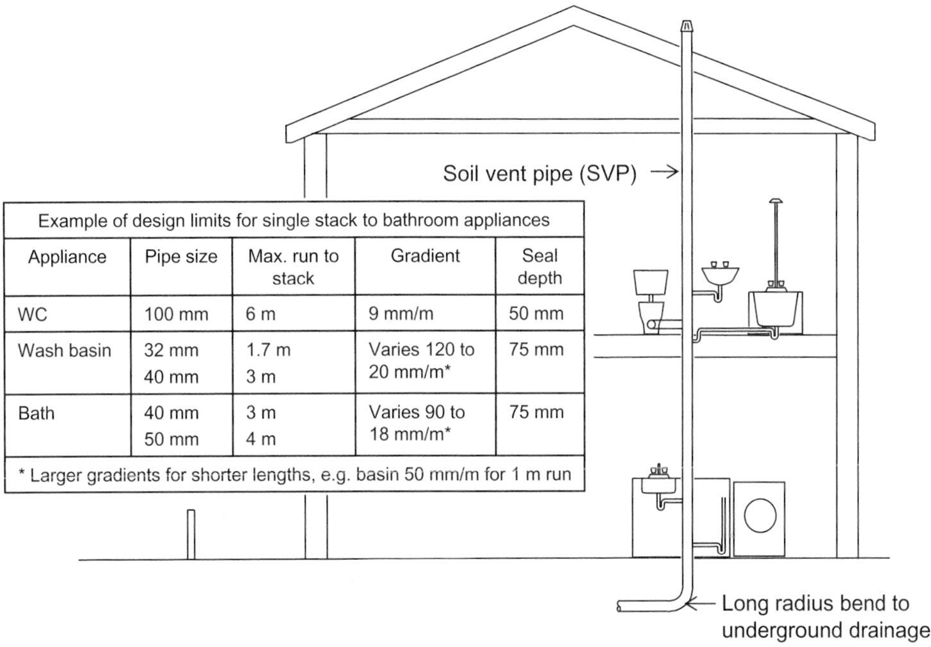

Example of design limits for single stack to bathroom appliances				
Appliance	Pipe size	Max. run to stack	Gradient	Seal depth
WC	100 mm	6 m	9 mm/m	50 mm
Wash basin	32 mm 40 mm	1.7 m 3 m	Varies 120 to 20 mm/m*	75 mm
Bath	40 mm 50 mm	3 m 4 m	Varies 90 to 18 mm/m*	75 mm
* Larger gradients for shorter lengths, e.g. basin 50 mm/m for 1 m run				

Figure 21.7 Above ground drainage – domestic single stack system.

drainage is governed by the site and the availability of a discharge point. The pipes still have to have limited falls or be vertical to be self-cleansing, but it also becomes important to think about ensuring a streamlined flow at bends and connections so as not to stop or slow down an existing flow at bends or when a second discharge joins another pipe. With limits to falls vertically and consideration of bends and junctions on plan, this makes the pipe system a three-dimensional layout below ground. It is important to relate this to ground level to ensure an economic system, as depth below ground has a considerable influence on the initial cost and on the ease and cost of any subsequent maintenance and repair.

It is necessary to identify how and where the waste can be safely discharged into an existing system. For most development this will be a public sewer located in the road outside the property. If there is no sewer available an economic distance from the building, cesspools (holding tanks below ground on the property that need to be emptied when full) or in more rural areas septic tanks (where some treatment of the liquid takes place) can be considered. Small-scale treatment plants can also be considered for isolated single or small group developments where road access to clear cesspools and septic tanks may be difficult.

Even if public sewers are available, the systems may vary. The most common is a separate system where foul and surface water have to be discharged into separate sewers, creating the necessity for two separate underground drainage systems to be installed on the property. Older sewer systems may be combined, taking both foul and surface water, or even partially separate, where independent foul and surface water systems have to be created on the property but a limited amount of surface water can be put into the foul drain to simplify layouts.

An example of a below ground foul system for a single house is illustrated in Figure 21.8. The plan shows two above-ground stacks collecting from bathroom and kitchen appliances and a utility room with sink and washing machine connections. The illustration assumes

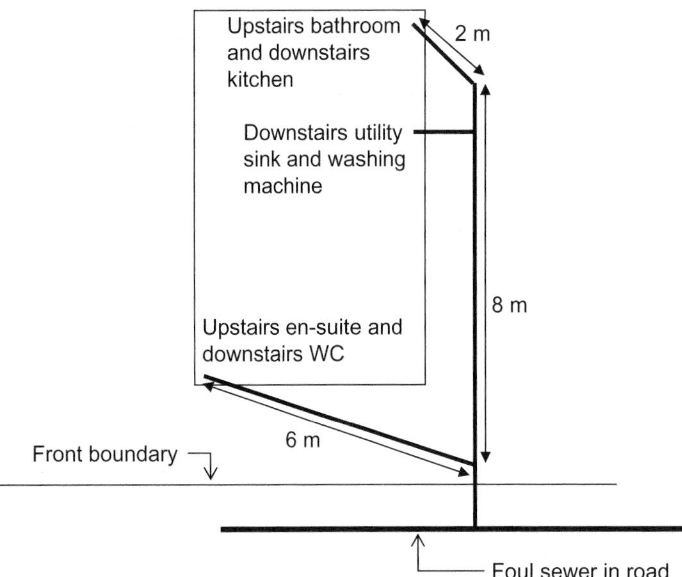

Figure 21.8 Below ground drainage – plan of foul water layout.

a public sewer on a separate system, although the surface water drain is not shown.

The need to maintain a streamlined flow at bends and connections to minimise the risk of blockage at these points is achieved by taking all the pipes and connections towards the sewer. No connections or bends go against the flow. Indeed, all the connections other than the short connection from the utility room have pipe bends and connections at angles greater than 90°, the angle being influenced by the fittings available.

The frequency and position of access is part of the decision-making for the pipe layout. While the streamlining of flow is done to minimise the risk of blockage, there is still a risk often associated with misuse (flushing items into the drains for which the system was not designed). While even this may be a rare occurrence, the consequences are unpleasant and place a high risk on health. Provision has to be made for the speedy clearance of a blockage with minimum disturbance to the property.

To understand where access is required and what form it may take, it is necessary to know the processes available to clear a blockage in a drain. Traditionally the process involved rodding the drain, but there are now other more flexible clearance methods, such as water jetting, commonly available in the UK. These new methods require less space within the access, particularly where the drain is relatively close to the surface. The four basic types of access are shown in Table 21.2. There are also limits on the distance between access points to ensure no long lengths of drain are installed where a blockage may occur beyond the reach of the drain clearance process, but these are unlikely to be exceeded on a domestic layout.

The system illustrated in Figure 21.8 is shown in Figure 21.9 in three dimensions, giving the type of access related to depths below ground. Assuming a relatively level site, the building outline is shown at ground-floor level. The highest point of the drain is at the connection from the kitchen/bathroom stack and this must therefore be the vent pipe being open to atmospheric pressure, normally the soil vent pipe taken out through the roof. It has been assumed that the depth of the sewer has given the final point of access on the property an invert level (level of bottom of the channel formed by the pipe) of 98.950 metres and that this will fix the levels of the drains back through the system.

Table 21.2 Below ground drainage – access types

Recommendation on dimensions and depths of access types					
Type	Depth to invert	Internal size		Cover size	
		Rectangular	Circular	Rectangular	Circular
Rodding eye	2 m max	–	100 mm min	–	100 mm min
Access small[a]	600 mm max	150 × 100 mm	150 mm	150 × 100 mm	150 mm
Access large[a]	600 mm max	225 × 100 mm	150 mm	225 × 100 mm	150 mm
Inspection chamber	600 mm max	225 × 100 mm	190 mm[b]	225 × 100 mm	190 mm[b]
– shallow	1 m max	450 × 450 mm	140 mm	450 × 450 mm	140 mm
Inspection chamber	1.5 m max	1200 × 750 mm	1050 mm	600 × 600 mm	600 mm
– deep (manholes)	Over 1.5 m	1200 × 750 mm	1200 mm	600 × 600 mm	600 mm
	Over 2.7 m[c]	1200 × 840 mm	1200 mm	600 × 600 mm	600 mm

[a] Large access gives greater maximum distance to next access point
[b] For drains up to 150 mm diameter
[c] If used with shaft 900 × 840 mm rectangular or 900 mm circular

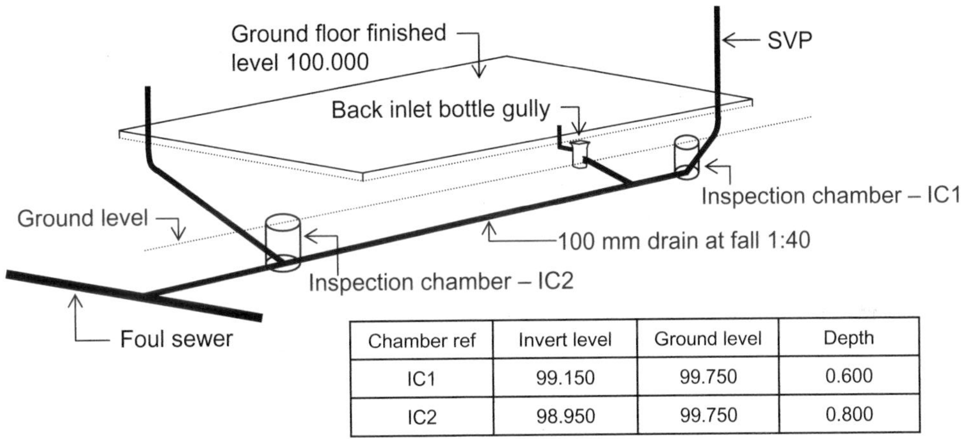

Chamber ref	Invert level	Ground level	Depth
IC1	99.150	99.750	0.600
IC2	98.950	99.750	0.800

Figure 21.9 Below ground drainage – foul layout with access.

Laying the 100 mm pipes at 1:40 gives inverts and depths below ground, as shown in Figure 21.9 on the drawing and in the associated access schedule.

These depths below ground have determined the choice of the type of access required. The utility room connection can be an access fitting with a direct connection to the drain. The soil vent connection has been chosen as inspection chamber 1 (IC1), which, being only 600 mm below ground, could be smaller than inspection chamber 2 (IC2), which is 800 mm below ground, as indicated in Figure 21.9. If the foundations were at 1000 mm below ground, all these connections can come out through the wall above the foundations. Since all the connections will be 100 mm pipes, the hole in the wall will have to be formed with lintels with a gap above the pipe to the lintel, so that foundation settlement does not put pressure on the pipes.

This layout with these depths below ground will provide an economic solution with limited excavation and relatively inexpensive access provisions in that deep inspection chambers (manholes) are not necessary. The system is, however, sufficiently far below ground not to be damaged or subjected to excessive pressure on the pipe from above-ground loadings as long as the pipe and bedding are chosen carefully, as discussed in the next section.

The level of the sewer in the road established the invert levels of the drains in the example. If this sewer had been deeper, it would lower the levels throughout the system, making the depth below ground greater, so the access would have to be reconsidered and excavations would be deeper. To keep the system on the property economic it may be better to make the last access (IC2 in the example) a backdrop manhole. This keeps the levels and access arrangements the same up to the backdrop, where a vertical pipe takes the sewage to a lower level to make the final drain connection to the sewer at a reasonable fall.

Below ground pipes and pipelines

The two main materials used for drains in domestic systems are clay and plastic. Clay gives a rigid pipe that will break with a brittle failure, relying on a crushing strength to avoid damage. The pipes are made in short lengths up to 1.5 metres. Plastic (uPVC) makes a flexible pipe that deforms under load or movement without fracture, so the design needs to ensure that this is limited for the joint to remain watertight. These can be manufactured in long lengths (and are light enough to handle) of up to 6 metres.

Both rigid and flexible pipes make flexible pipelines when using jointing systems based on 'O' rings, which allow some angular movement without leaking. This will allow the pipeline to move with small settlement and other movements in the ground. The traditional sand and cement joint used on clay pipes will make the whole pipeline rigid, with the most common failure being at the joint. While these sand and cement joints were commonly used in the past, they are not now usually specified. Both clay and plastic pipes can be obtained with spigot and socket ends or as plain-ended pipes with a sleeve to make the joint.

This distinction between rigid and flexible pipes is important in deciding on the bedding and the depth of pipes in order to maintain support and protection. The pipe needs to be laid on a stable and accurate surface to lay the pipe to a regular fall and provide sufficient support. The pipe may be vulnerable to uneven loading unless a selected material evenly surrounds the pipe and fully fills the trench in which the pipe is laid. This has given rise to a range of bedding classes offering different levels of support for different soil types and depths as well as depending on the type of pipe. Some of these options for bedding are shown in Figure 21.10.

Rigid pipes with a nominal 100 mm diameter can be laid to within 400 mm of the surface to the top of the pipe using any of the bedding options, although care has to be taken with construction traffic and compaction of trench backfill this close to the surface. If pipes have to be laid very deep (greater than around 8 metres), the strength of the pipe and the bedding specification become structurally important where bedding with granular material has to be specified.

The 100 mm flexible pipe needs to be at least 600 mm below ground to the top of the pipe for fields and gardens (900 mm under roads). Its performance under load is also determined by the width of the trench. The trench needs to be as narrow as possible, with a minimum of the diameter of the pipe plus 300 mm to ensure adequate sidefill. Deep drains are often laid in a trench with working space cut to within around 100 mm of the top of the pipe, with a narrow trench cut to contain the pipe and its bedding. If flexible pipes are laid within 600 mm of the surface (900 mm under roads), a concrete slab capping has to be provided at least 75 mm above the top of the pipe, as shown in Figure 21.10.

All pipes can be laid close to the surface (or close to building foundations) where the loading and the chance of damage are high by bedding

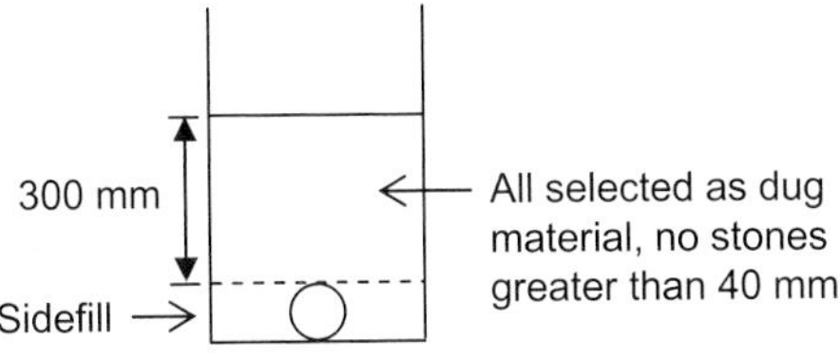

Trench bottom capable of being trimmed and suitable as dug material available (Class D)

300 mm
Selected as dug
as above
Graded all in
granular material
Bedding 50 mm for
sleeve coupling, 100 mm
for socket pipes

As dug material available but poor trimming to trench bottom (Class N)

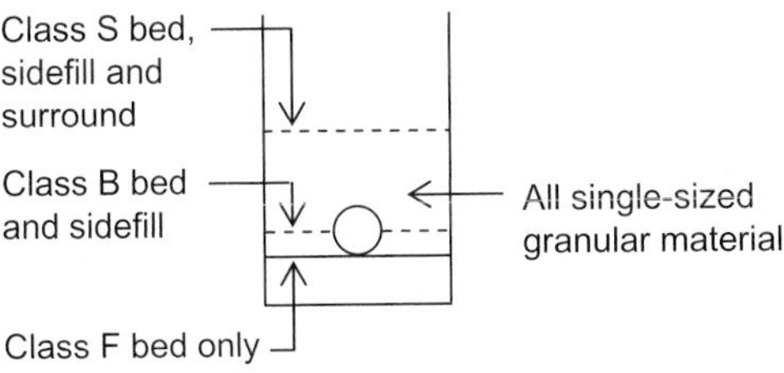

For sites where ground conditions and depths require higher 'bedding factors'

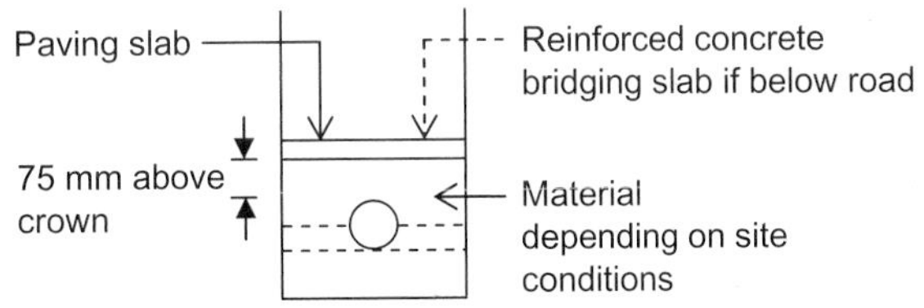

Flexible pipes laid close to ground

Figure 21.10 Below ground drainage – trench bedding.

and surrounding the pipe in concrete. This, however, makes the whole pipeline rigid, independent of the pipe material or jointing system.

Lifecycle and sustainability

Most of the lifecycle issues have already been introduced. The major maintenance task is clearing blockages, so some investment in access can save damage and expense during the operational life. Making the pipeline flexible with the choice of joints between pipes minimises the risk of damage to the joints due to minor movements in the ground, although below ground there is little deterioration in the material as they are designed to carry water.

Drains laid to falls use no energy to move the water, but they do carry the potential for pollution and flooding risks, placing demands on infrastructure works and threatening habitat. Measures to limit the use of water in the house have the greatest influence on the discharge from foul drainage systems, although treatment using septic tanks and/or reed beds would limit the discharge into the public sewer system even further.

Discharge from surface water systems places particular strains on the infrastructure and environment in that they have to be designed to cope with intermittent high flow rates (storms), which increase the risk of flooding from sudden high discharges. Permeable, slow-release systems from hard surfaces, known as sustainable urban drainage systems (SUDS), used on commercial building, particularly on car parks, have application on large-scale housing developments.

Electrical systems

Electricity provides the user with power to run household equipment (e.g. entertainment systems and kitchen appliances), thereby satisfying operational needs. Directly involved in lighting and in supplying power to heaters and the central heating boiler and pumps, it also provides an environmental service.

These various uses have different power demands and patterns of use, so it becomes sensible to provide a number of circuits in the house to serve these different purposes:

- Lighting has relatively low power demands, but the majority of lights may be on at the same time (normally taken as 66% of total demand).
- Power for general equipment may include some appliances with high power ratings such as fan heaters and low power ratings such as home entertainment systems. The exact pattern of simultaneous use of these different appliances will be highly variable, but there will be a limit to the number of pieces of equipment being used at any one time for any given floor area in a house.
- Power for single highly rated equipment such as cookers, hot-water immersion heaters or instant hot water heaters such as shower units. These require their own circuits.
- Power for units where the pattern of demand is for electricity at night, when demand is nationally low, and it is possible to take advantage of off-peak lower-cost tariffs. This can be used for storage heaters, where separate metering arrangements may require separate circuits.

Electricity is dangerous to life. This is not only from direct contact but also from the potential for fire caused by defective systems. It can also damage the appliance or equipment itself. To minimise these risks, circuits have to be provided with the following:

- Earthing to provide an alternative route for the current in the case of individuals or other conducting objects coming into contact with a live circuit. This will lower the risk of serious injury and fire from such contacts.
- Fusing to provide a weak point in the circuit that will fail if a circuit carries too much current by either too high a demand or a fault. This will lower the risk of damage to equipment and the subsequent chance of fire.
- Insulation resistance around wires and at connections to ensure that any contact with the system's components will not result in electric shock or short-circuit.
- Correct polarity to ensure that any switching is provided on the live wire and not the neutral so equipment does not remain live when switched off.
- Sufficient cross-sectional area of the conducting wires to carry the current without overheating causing premature breakdown of insulation or fire.

Supply to the house

The electricity supply for a house will be a single phase providing nominally 230 volts across a pair of wires, the live and the neutral, with a third wire, the earth, to provide continuity back to earth through the supply system. The arrangement for supply and the separation into the required fused circuits is shown in Figure 21.11. The supply company protects its distribution system from excessive demand from a number of houses by installing in each house a fuse that will break the supply if the household takes too much power from the supply at any one time. This is housed in a sealed unit known as the service head, normally provided with an 80 or 100 amp fuse. After the service head, the supply company installs its meter(s) connected to the live and the neutral, and then two wires known as the meter tails are taken from the meter to the consumer unit. The earth is taken to a main earthing terminal.

The earth wiring is kept separate from the live and neutral at this stage to allow earth connections from all potential sources of electric shock. It is possible to get an electric shock not just from direct contact with the live conductor wires but also from indirect conducting systems that can become live from contact with the live wires. The casing around electrical appliances can become live but so can services systems such as the water, heating and gas supply copper piping. The earth wire in the cable will be connected to the casings, giving protection to appliances, but bonding will be required to

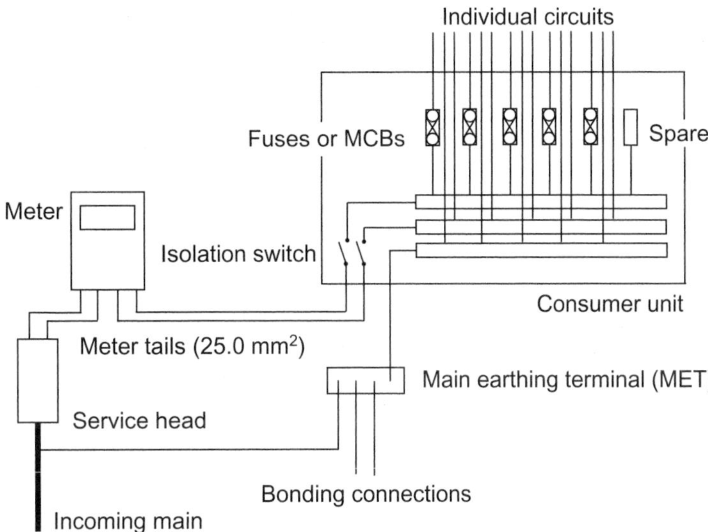

Figure 21.11 Electrical systems – supply and circuit distribution.

the copper pipe systems (with appropriate cross bonding between the pipes to sanitary appliances in wet areas). All these connections are made at the main earthing terminal. The direct earth wire is taken to the consumer unit to provide a connection for all the earth wiring in the circuits. Bonding wires are then taken to all potential conducting service systems.

The consumer unit provides a double pole (both live and neutral) isolation switch, which is usually a residual current device (RCD) or earth leakage circuit breaker (ELCB), which are sensitive to any earth current and will quickly switch the whole system off automatically if even a small earth current is detected. After the main isolation switch, provision is made to create separate circuits, each with its own overcurrent protective device (fuse) on the live wire. Originally a rewirable cartridge, the fuse will now almost certainly be a miniature circuit breaker (MCB). The size of the fuse is determined by the maximum current the circuit is designed to carry.

The power drawn from a circuit by equipment and appliances is rated in watts. Given that the supply is provided nominally at 230 volts, a 1000 watt (1 kilowatt) appliance will create a flow of current of around 4.35 amps (watts = volts × amps). Not all appliances will be used at once, so circuits can be designed on the basis of diversity factors that give domestic circuits fuse sizes as follows:

- Lighting circuits (one for each storey of the house): 5 amps
- Power for general equipment, provided via a ring main serving no more than 100 m² of floor area: 30 amps (radials are possible, see later in text)
- Power for individual appliances (normal or off-peak tariff) depending on rating of appliance: e.g. cooker 30 or 45 amps, hot-water immersion heater 15 amps

A cartridge fuse in the plug should then protect each item of general equipment connected to ring main circuits, which in the UK are rated at up to 13 amps, although 5 and 3 amps should be used for lower-rated equipment. Arrangements for both fusing and switching of the larger rated individual appliances (cookers, immersion heaters, etc.) will vary depending on their actual rating, position and the risks associated with wet areas.

Lighting circuits

Figure 21.12 shows a simple lighting circuit with no two-way switching and with each light

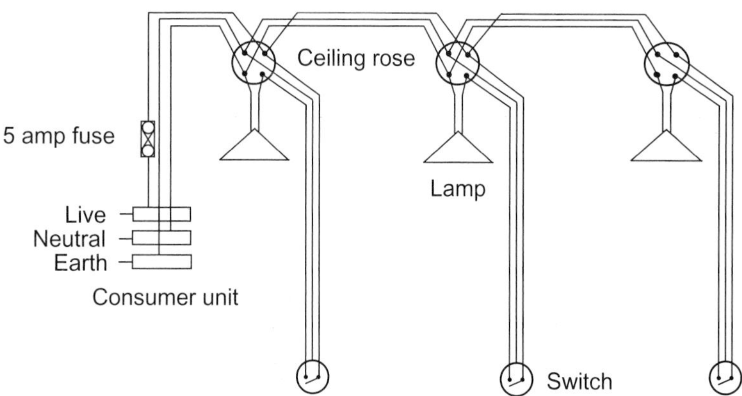

Figure 21.12 Radial lighting circuit.

having an independent switch. The circuit is a radial circuit with the wires taken to connections at each light from the consumer unit out to the last light on the circuit. The live is not taken directly to the lamp at each light connection but via a switch to control each light independently. The earth is also taken to the switch and connected to the switch box to ensure earth continuity if the switch box should become live. These connections at each light are made to a connection strip on the ceiling rose that is mounted to the ceiling at the light fitting. There are enough connections in the ceiling rose to accommodate two-way switching and multiple lamps on a single switch. An alternative to making the connections in the ceiling rose that is potentially more economic on cable but makes access to the connections more difficult is to use junction boxes (not illustrated). Lighting circuits are on 5 amp fuses from the consumer unit with wires of 1 mm^2 for PVC-insulated cable.

Ring mains and radial circuits

Figure 21.13 shows a ring main with, as the name implies, wiring returning to the consumer unit from the last socket outlet. This means that, unlike the lighting circuit, the first wire does not have to carry the full current of all the appliances and equipment on the circuit. Able to flow in both directions, the power will balance itself out in the circuit, depending on what equipment is on each socket at any one time. The sizing of the fuse (30 amps) and the wires (2.5 mm^2 PVC-insulated cable) is not dependent on the number of sockets but on the probability of the current from equipment associated with a maximum of 100 m^2 of floor area. Any number of outlets may be installed to ensure the convenience of an outlet for each appliance (or group of equipment as occurs with computer systems) and limit trailing wires. It is possible to gain some economy in the wiring by providing a spur from any of the sockets (or a junction box), but there should only be one spur for each socket outlet on the ring main. Socket outlets may be single or double, will normally be fitted with a switch and are designed to receive a fused plug from the appliance (maximum fuse rating 13 amps). Some fixed equipment may need to be wired in directly without a plug but still must be fused and switched. These are known as fused spur box outlets.

The ring main is economical where sockets are widely distributed as the wiring can be arranged with the first and last socket near the consumer unit. However, radial circuits, where there are no return wires, similar to the lighting circuit, can be used if economies can be made in the wiring. Radial power circuits can be installed with an unlimited number of sockets using 2.5 mm^2 PVC-insulated cable, but can only serve 20 m^2 of floor area and need only be protected with a 20 amp fuse. An alternative to serve a larger floor area up to 50 m^2 with a radial circuit is possible but needs to be protected with

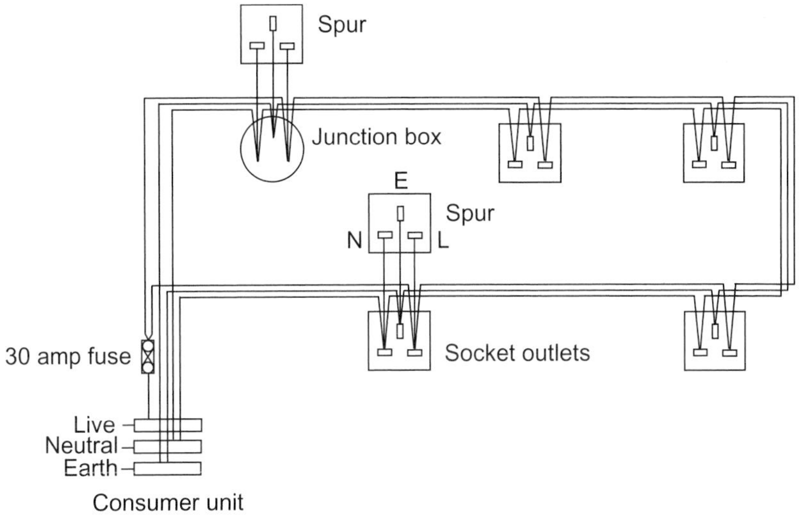

Figure 21.13 Power ring circuit.

a 30 amp fuse and should be wired with 4 mm^2 PVC-insulated cable.

Cables and routeing

Decisions on cables and routeing are mainly influenced by safety considerations, although maintenance and cost are also factors in making the choice. The cables are insulated to protect users from shock and short circuit, colour coded to ensure correct polarity and with wires with sufficient cross-sectional area to carry the current in the circuit without overheating when protected by the correct fuse. For domestic installations (lighting and power) cables would normally have PVC insulation around solid copper conductors. The live and the neutral will have individual colour-coded insulation. The colours have been subject to change and it is vital not only to understand the current colour conventions when wiring a new property but also to know past colour coding when working on existing installations.

Maintenance on electrical systems is normally associated with the consumer unit and outlets as this is where all the fusing and connections are made. Although power outlets are usually easily accessible, the consumer unit is normally located high up, out of reach, for safety reasons. Cables are then probably concealed running horizontally in floor and roof spaces. Vertical cabling is then normally buried in a chase in masonry and covered with plaster or placed in the cavity in plasterboarded partitions. Any surface-mounted cables are usually placed in ducts to maintain a tidy appearance. Once cables are hidden, some mechanical protection against accidental damage has to be considered, but, other than the protection of the ducting for surface-mounted cables, this is not normally provided in domestic installations.

Another consideration when concealing cables in floor and roof voids is their positioning with respect to insulation material. If cables are buried in or held up against solid construction by insulation, their temperature may rise when carrying current. It may be necessary to increase the size of the conductor wires in the cable in order to limit any potential temperature rise.

Lifecycle and sustainability

Considerations of safety and concealment tend to place the components in dry and dark environments, keeping deterioration to a minimum.

The design features that limit the risk of fire will minimise the chance of overheating and damaging components with electrical faults. Rewiring may be necessary if insulation efficacy reduces. This is most likely in the flex associated with the drop to the light fitting as this is exposed to daylight, embrittling the PVC.

Extending the system with additional outlets is not a problem so long as the floor area served by the circuits is not exceeded. The ability to provide new circuits will be limited by the capacity of the consumer unit. Most installations are provided with a consumer unit with spare fused circuit positions.

Although a major system for the distribution of energy, the circuits discussed so far are not consumers of energy, other than the losses in the cables, which can be minimised by limiting cable lengths. This will also have an impact on cost and materials used (PVC and copper).

The major contributions to sustainability are in the possibilities for the generation of the electricity from renewable sources and the efficiency of the appliances and equipment specified. This efficiency is not only in the design of the equipment but also in the control over the pattern of use. Installing low-energy bulbs in lighting is one step to sustainability, but switching them off when not required may have more impact on overall energy demands. In commercial installations, automatic detection of the need for lighting with movement and light-level sensors that switch lights on and off makes a considerable impact on energy use often because the lighting is required during the day as well as the night. The energy consumption from lighting in domestic systems is less, because daytime lighting is generally not required. Energy saving relies on the occupiers switching off lights when they leave a room. Here education, and the provision of information about energy use, has been found to be a major stimulus to changing behaviour. Information can now be provided via smart meters that give not only immediate consumption figures but also information on patterns of consumption associated with time of day and type of appliance being used. Education of the users has to extend to buying appliances with low energy (and other environmental) ratings. If the appliances are specified with the original building, low-energy use should be one of the factors associated with choice.

Having minimised energy and other environmental impacts in the choice of equipment and controls, it is possible to consider generation for an individual property.

Electricity can be generated using photovoltaic arrays, preferably south-facing. For an urban property this is most likely to be on the roof, as other positions on the site may be limited. These units generate direct current and therefore an inverter is necessary to provide the AC supply required for the household appliances and equipment.

Photovoltaic units will generate in cloudy conditions, albeit at considerably reduced output and in cold weather (where their efficiency is slightly increased), but they cannot generate at night. Their pattern of supply will not match the household demand and therefore a mains connection is still required. However, there will be times during the day when they are generating more electricity than is needed by the household and this over-generation can be sold back to the electricity company. This is achieved through the use of a net meter, where the meter runs backwards when electricity flows back into the supply grid. It is possible to have battery backup to cover loss of supply as, if a net meter is used, the voltaic system has to be shut down, with loss of supply to protect from power surges. To increase the sustainability credentials of photovoltaic units it is claimed that they are made predominantly from silica in a form that is non-polluting and that in good generating conditions (probably not possible in the UK) they can generate the equivalent of their embodied energy in 4 or 5 years and will last for at least 30 years.

Electricity can be generated from wind energy as another renewable source if the site is suitable. It is also possible to look at waste energy to generate electricity in the use of combined heat and power (CHP) systems. These have been mainly developed for larger community-based energy supply but are available as domestic units.

Summary

1. Services are active technologies involving movement through distributions systems that can be controlled. The distribution has to be routed through the building, often concealed, but requiring access for maintenance, repair and the renewal of defective parts.
2. The hot and cold-water systems have to deliver water to a variety of appliances at the required quality and at a sufficient rate of flow (quantity). This will determine most of the basic design of the system. The need for isolation for maintenance and repair has to be considered. New systems to support sustainable development include low-usage appliances and grey water systems.
3. The usual expectation of the householder in the case study is for central heating. In the current fuel supply economy this is likely to be gas-fired, pumped hot water to heat radiators in each room. Insulation provides the most effective energy-saving investment for a house, although appliance efficiency and controls can make a significant contribution.
4. The disposal of both foul and surface water is important to health and maintaining a dry environment. Both above and below ground drainage systems are based on moving the water by gravity, by providing falls on the pipes. This restricts the routeing and positioning of pipes in order to maintain self-cleansing flow and requires particular attention to be paid to access, to clear any blockage that may occur.
5. Electrical installations provide power to enable other equipment and appliances to operate. As one of the most dangerous of the service systems, due to its potential to electrocute and start fires, safety determines much of the detail and specification. While equipment and controls can reduce energy consumption, it is possible to consider local generation from renewable resources such as photovoltaic arrays.

Part 3
Choice – Commercial Construction

22 Applying the Framework to Commercial Buildings

This chapter provides the link between the first part of the book, which introduced the framework for analysis, and the nature of making choices for commercial building. This is seen as different from the approach taken to housing, mainly because of the greater diversity of scale and use of commercial buildings that leads to a much greater range of solutions. These solutions vary not only in the mix of technologies used but also in the variety of detailing and specification. Housing tends to a common form within which a limited range of details arise, as introduced in Part 2 of the book. The greater variety in commercial buildings created the need to see the technical choice in two stages. First, broad options need to be identified in the early stages of the design. The mix of technologies for structure, enclosure and services has to be chosen. These choices have to be made with the confidence that specific details and specifications can be developed in the second stage of detailed design, that a building can be fully detailed to perform and be constructed at reasonable cost and to fulfil the design vision.

The story so far

Part 1 of this text suggests a general approach or framework to making choices for the construction of buildings. It identifies that the process of choice starts with a suggested specification or detail and that this is evaluated by applying a series of analyses based on the two key questions 'Will it fail?' and 'Can it be built?' The areas for analysis involve a series of physical behaviours, the implications for appearance and then production issues, including economic and social concerns to ensure a safe solution in the environmental and social context pertaining to that building. Here context is most important as it will, often in subtle ways, determine the specific solutions that are adopted for each building. This general approach is outlined in Chapter 1, and each aspect explored in some detail in the subsequent chapters of Part 1 of the text. This approach will be maintained in this part of the book and should therefore be familiar to the reader.

The subsequent chapters in Part 1 suggest that a solution can be seen in a number of ways. Initially the building can be represented as a series of flows and transfers through the building and through the fabric and services systems, as outlined in Chapter 7. It is then necessary to see the fabric and services as fulfilling a series of functions with the function of the parts for the construction derived from the function of the building as a whole. Each function will then need to have a performance assigned to form the basis of choice. While the function of the building as a whole is determined by the purpose of the building, the function of the parts will be determined by the design. This will lead to a mix of technologies, the most significant of which has been identified as the mix between the role of the fabric of the building and the contribution of the building services.

In Chapter 3 it was suggested that the technologies can be seen as broad options based on either generic forms or common forms of solutions. These are appropriate ways to think about construction at the concept design stage and in the early stages of planning the building production process, as broad production options are associated with each broad technical solution.

Broad options have to be chosen with some confidence that the specific solutions can be derived from materials and detailing that will not fail and can be built, that any detailing can be achieved with little risk of failure and in an economic and timely manner.

At different times in the process of realising a building it may be appropriate to think about the construction as:

- Experiencing flows and transfers (the dynamic conditions it will experience)
- Fulfilling a number of functions (what it has to do) and its performance (how well it has to do it)
- General (generic) forms based on the actions or mechanisms being used to achieve the function (how it is going to do it)
- Common forms/broad options that indicate the materials and the sizes required (what it could look like) to fulfil the performance
- Specific solutions with materials specified and details prepared with all dimensions, jointing and fixings specified to achieve the required performance (what it is going to be)
- A production process initially associated with the broad option but eventually to be followed through in the detailed design (how it can be realised)

All these ways of thinking about construction will be required in the design of commercial buildings.

Although commercial buildings are created in the same physical and social environment that was outlined in Chapter 16 for housing, the demand for commercial building as represented by the clients, those who commission and pay for buildings, is diverse compared to housing. Broad social changes from manufacturing to office-based employment and the growth of retail and leisure, including tourism, as economic drivers has created the need for new buildings with new functions. Regeneration led by investment in infrastructure and iconic buildings has given a wider range of contexts (city centre, business park, edge of city) and the opportunity for design innovation. This has led to a change in the resource base with the use of components and factory-based production with site processes of assembly relying less on traditional craft skills and capitalising on global trade.

Another difference between housing and commercial buildings that influences the range of solutions adopted is the way buildings are procured. Houses are seen as a good, long-term investment, and there is a culture in the UK of individuals purchasing houses where they are primarily concerned with the quality of life they offer. Even the provision of social housing in the UK is so regulated that standards and quality of life are key drivers that, even in the absence of a commercial interest, lead to a similar choice for the construction. The pressure to reduce cost, or more specifically increase the differential between cost and sales price, comes from the house builder (developer). This method of procurement – developers building for individual sale – leaves the choice of technology in the hands of the developer and, given a relatively homogeneous market, there is a convergence of solutions. Any pressures for change come from advantages in production costs as the market sets the land costs and sales price for the home. This gives the house builder an interest in prefabricated systems but only if the set-up costs can be regained in sufficient volume of houses of the value range for which the systems are developed. The developer is such a strong part of the market that the resources availability follows the demand and therefore even more bespoke housing is likely to make similar technological choices, as the resources are available at reasonable cost.

The way commercial buildings are procured is more variable, and this influences the way choices are made not only for the broad technological options but also for the final specifications and detailing. It is not the role of this text to explore in any detail the influence of this variety as it involves an analysis of contract and risk as well as the type of organisations employed at each stage in the technical decision-making. Suffice it to say that each procurement method determines who has the main responsibility for technical decision-making and the risks they take in those decisions.

The focus of this text is the analysis of solutions, whoever suggests them, to ensure buildings will not fail and can be built against performance criteria. Use and size of buildings are perhaps the greatest determinants of broad options that could be chosen, as is the vision of the design along with the context in which it has to be built. All these factors lead to a greater range of common forms being available and specific solutions that have to be designed for each commercial building.

The need for the integrated approach outlined in Chapter 16 at the beginning of Part 2 of this book is perhaps more important for the choice of commercial buildings than for housing. It may be worth re-reading that section before continuing with this chapter.

The way forward

With the greater diversity of use and form, studying commercial buildings as a series of different types of buildings with their different functions could miss the point that they employ many common forms of construction between them. More importantly, the process for choice for each can be seen as a common process and it is therefore best to retain the broad idea of commercial buildings.

Although these building still need floors, roofs, walls, foundations and services, this is perhaps not the best way to think about making choices for this range of size and types of buildings. There is far more of a need for specific technical solutions for each building. There is far less convergence of details and specification than there is for housing. The analysis of one or two details and/or specifications for each element that provided the key solutions to mass housing in the UK at the beginning of the twenty-first century is not repeatable for commercial buildings. This would require the analysis of hundreds of details, few of which could be used as specific details in the future were the pace of change to continue.

With little variation in the overall operation of the house or its size and appearance there has emerged a common form for the whole building that needs little or no analysis before it is adopted for the next housing project. The focus of technical choice becomes the details and the specification of materials and their implications for the production process. This was therefore the focus of the analysis in Part 2 of this text.

For commercial building the diversity means that the process of choice has to begin with a questioning of which of the common forms of construction would be appropriate to make the building work as a whole, before exploring too deeply the issues of detailed design. There is a need to explore some broad options at the concept design stage that can be developed into details and specifications at the time of the detailed design.

For most commercial buildings there is probably no immediately obvious common form for the whole building. For any one client brief and site there will be a limited set of options, but even these may then be achieved in a number of different materials, making the number of possible combinations quite large. There will need to be an additional stage in the process of choice, very early in the design process, associated with the development of the design concept. The nature of these broad options will have implications for the production process as sequence and methods are best initially analysed from a broad understanding of the overall technological approach to the construction. There is a need to make broad choices of a mix of technological solutions that would work for each individual building.

Ultimately the success of the building will rest on the choice of detailing and specification, and this task will have to be undertaken with the same care and using the same process of analysis as for housing. However, the need to identify broad options for construction for the whole building before getting too involved in detailed design demands an understanding of general and/or common forms and their potential and economic limits.

To make the link between the function of the building as a whole and the final choice for the technical solution expressed as specification and details is a big a step unless solutions have had time to become established, as they have in house construction. In much commercial development either the design of the building or the emergence of new production approaches may make existing detailing, however well thought through for a previous building, potentially inappropriate for the next. Some intermediate steps in the process of choice are required to develop specific solutions as new forms evolve.

Starting at the concept design

For commercial buildings the client's brief will establish the range of accommodation to be provided and the scale of the building required. If the site is known, the context will have been established and the building can start to have form and size. This will be driven by the design concept (Chapter 7) or the vision of the building, and this will give an early indication of the aesthetics that will determine the appearance of the building. Once the brief and the context have established the use, size and vision, it is possible to start to make suggestions for what the technical solutions could be.

At this stage it is only possible to suggest broad options in line with the emerging design concept. However, there has to be some confidence that the suggestions will be able to be detailed and specified at the detailed design stage to realise the design concept. The dialogue between design concept and technical choice has to start at this very early stage in the design process.

Broad options at the concept stage

The evaluation of these broad options starts during the development of the design concept and gets refined through the detailed design process. During the concept design stage, options are still fluid, changes in design are still emerging, economies can still be achieved, detailing can still potentially be simplified. There needs to be sufficient understanding of these options to establish a level of confidence so that it will be possible to take the option forward to full detailed design without incurring excessive risk or cost.

This need to consider broad options involves a knowledge and understanding of a range of common forms not for the whole building but for parts of the building, the sorts of materials and sizes that are technically possible, their economics and, possibly most important, how they work. It is not possible to start with an overall description of a common form for the whole building as it was with housing. The diversity of commercial buildings requires a range of common forms at the elemental level to be considered to achieve the best mix for each building depending on use, size and vision as established by the context in which the building is to operate.

Choosing components and detailing to ensure that the building works as a whole requires analysis at a broader level than individual components but not at the level of the whole building. This view leads to the notion that what all buildings (including houses if more fundamental changes in design or production methods are considered) have in common are three types of elements, each with their own set of broad options, namely:

- Structure
- Enclosure
- Services

Each of these has a body of knowledge and its own design approach, and there is a danger that, if these are pursued separately, the building will fail to operate efficiently and economically and may not work well as a whole. It should be a constant concern when making choices for the parts of a building that they contribute to the design concepts as a whole. Perhaps as important as the study of these three elements is the analysis of how they interact with each other: the design and technical solutions at the interfaces.

- Enclosure–Structure
- Enclosure–Services
- Structure–Services

Each of these has a different characteristic. These will be explored further in Chapter 24, but broadly the enclosure–structure interface is predominantly a question of detailing to ensure a continuity of performance while the enclosure–services interface is defined in the environmental design strategy and is therefore less physical and more conceptual. This is a very clear example of how the function of each part has to be defined by the overall design concept. Each of the elements can be detailed physically separately, but they are bound together by the overall environmental strategy driven not only by concerns for the quality of the environment to be achieved in the building but also by cost in use and limiting environmental impact, which are three key objectives in providing a sustainable building.

The structure–services interface is often characterised by competition for space, particularly in the horizontal distribution of services in floor zones. When considering the vertical distribution of services (and the location of plant rooms) and the need for the vertical transfer of structural loads, the space competition is normally with the usable and circulation space demands of the user. Although predominantly about physical space, aspects of the low environmental impact design can usefully use structure as thermal mass, illustrating that these interfaces can be as important a focus in choice as the function of the individual elements themselves.

If the building is to function as a whole, there is a need to start to identify the technological mix that may be employed for each of these three elements at the concept design stage of the building.

The technological mix

The technological mix has to be established at the design concept stage. As the design concept establishes how the building will function as a whole it is necessary to start to see how the elements of the construction may contribute to the overall performance of the building: to start to identify the function of the parts. Ultimately each part of the construction will have a function to which can be ascribed a performance and this will allow the choice of materials and detailing to be confirmed at the detailed design stage.

Identifying the technological mix needs knowledge of potential broad options, how they work as a general form and the common forms of construction that these might take. Viable alternatives will be driven by the use and size of the building and from the vision for the building.

Perhaps the broadest technological options are between the building fabric and the building services. While some of the fabric can be used actively (opening windows), most is passive in its response to the changing conditions during the operation of the building. It allows neither intervention (control) nor any energy inputs to perform. In contrast the building services (and the active elements of the fabric) have both energy inputs and controls as they operate only when required to bring conditions back into comfort or operational levels. Active elements can be turned on and off and up and down in response to changing conditions. Their energy inputs mean that achieving many of the sustainability and low environmental impact objectives will be established by the initial decision on the technological mix.

Once the mix of active and passive has been established, it is possible to start to make broad decisions on the passive construction of the structure and enclosure and to think about the interfaces to ensure the building performs as a whole.

Identifying broad options

Broad options can be identified as being a generic form that is defined by the way the construction works to fulfil functions. This generic form will imply the properties the potential materials must have to ensure economic sizes to achieve performance. An example of a generic form of external wall to fulfil weatherproofing would be a semi-permeable wall. Broad options can also be usefully thought of as common forms of construction. Each common form will have been developed for a range of sizes and types of building for which the resources to both design and construct are available within the society in which the building is to be built. An example of a semi-permeable common form would be the cavity wall. It is either these general forms or the common forms of construction that are the basis of the broad options that need to be identified at the concept design stage.

Broad options will have potential associated materials and a range of economic sizes. For the cavity wall the usual materials are masonry, but the internal skin can be timber-framed. In its domestic form the skins and the cavity are likely to be around 100 mm each. The type of materials and approximate sizes are sufficient at the design concept stage, so long as they can form the basis of the final technical solution as all the materials specification and details can be finalised in the detailed design.

For the structural elements this is predominantly based on their behaviour under load. However, the enclosure elements have a greater number of functions, each of which has a number of generic forms. Therefore the range and combinations of broad options are more varied and more complex to analyse. The complexity of the analysis is compounded if some parts of the construction can fulfil a number of functions in order to achieve certain desired economies without compromising the brief. These economies are only possible at the time of considering the broad options. Cost reductions later in the design stage often involve reducing performance.

As an example an external wall will have some structural functions. This is clear if the structural solution uses loadbearing walls, but even as the enclosure of a framed building the wall has to transmit wind loads and may need some sub-framing to support components that cannot span directly to the elements of the frame. The external wall as an enclosure element will also have functions associated with maintaining environments. Each of the functions will have general forms. These have been introduced in Chapter 11. For waterproofing, an impermeable external layer can be established or the semi-permeable identified above or a rain-screen approach can be adopted. Each of these carries particular issues for the detailed design. For the impermeable surface it is jointing to achieve similar levels of water resistance to the impermeable materials, and for the cavity it is eliminating bridging and offering drainage and, on some rain-screening systems, maintaining pressure differences. Understanding how each general form works suggests the properties of the materials that could be used and indicates the important aspects of detailing. These general forms have, over the years, been developed into a number of common forms of construction. The cavity wall has been adapted to work with framed structures. Cladding, curtain walling and applied facings have been developed in a variety of materials. Each can now be called a common form for the construction as the components and resources to construct these forms are now available within the industry.

Testing the broad option for the specific solution

The extent to which these broad options can just be accepted without any further thought at the concept stage will depend on how well established the technological form is for the scale and use of building being considered.

Each common form will have a range of solutions. These solutions may vary in a number of ways:

- The materials that could be used
- The size and shape of the components
- The way the components are connected together
- The interaction with options being considered for other parts of the building

- The production process options
- The resources required for the scale and timing of production
- The level of risk in adopting that solution

It is necessary to be aware of all these factors in exploring the suitability of a broad option at the concept stage. Indeed, it may be one of these factors that determines the common form that may be most applicable. If for aesthetic reasons an external facade needs a particular material in a particular component form, this may dictate the general form for the whole external wall. If the client is looking to use the building by a particular date, the production process and the availability of resources may influence the broad option.

It is important to realise that if poor choices are made about the broad options at the design stage there is little chance of resolving the design at the detailed design stage. This is no different from the design process for the building as a whole and it is therefore not surprising that some thought about the technological mix for the building has to be introduced at the concept stage.

Summary

1. The two key questions of 'Will it fail?' and 'Can it be built?' apply to commercial buildings.
2. The diversity of the scale and use of commercial buildings has led to a range of technical solutions that introduce the need to make broad decisions on the technological mix of general forms for any one specific building.
3. Broad choice has to be made early in the design stage but with the confidence that the detailing and specification can be resolved in the detailed design stage.
4. An understanding of the general forms of construction that could be used is required along with their typical detailing to judge the potential for the building under consideration.
5. General forms need to be chosen for structure, enclosure and services elements of the building.
6. Interfaces between each option need to be analysed to ensure the building will work as a whole.

23 Common Forms and Emerging Technologies

This chapter explores the need to be clear about change and the provisional nature of knowledge and know-how about some technical solutions. This implies that any knowledge from a precedent will also be indirect as the actual detailing and specification will not have been in service for any length of time. Many of the technologies used in commercial buildings are well established and so details and specifications can be proposed with confidence of performance and production success. It is important to recognise when this is not the case and emerging technologies are being suggested. This is explored using a brief history of technological development in the twentieth century, exemplified by the introduction of concrete, and then the main drivers for change at the beginning of the twenty-first century are identified.

Introduction

It has been established in the previous chapter that the technology and its detailing and specification for commercial buildings is currently relatively diverse. Another measure of technology is how fast it is changing. What can be considered common forms and what should be identified as emerging technology? As the name suggests common forms can be identified as being widely used where practice over a number of buildings has established experience in both detailing and production. This has generated knowledge and know-how at both the design and production stages that have led to resources that lower both the technical and financial risks in adopting the technology. The buildings might not have been in operation long enough to reveal any premature failures, so care needs to be taken in using these solutions until there is experience of a significant proportion of the lifecycle. Emerging technologies lack knowledge gathered from experience, demanding more analysis and evaluation, which will still leave risks to be identified and accepted. Perhaps the most important ability during the process of choice is being able to recognise the difference between a common form and an innovative technology when either is proposed.

At certain times there are periods where common forms are widely used and convergent solutions become the norm, but there are also times when emerging technologies are adopted. The drivers for change in the form of buildings and their technological solutions come from three main sources:

- Client requirements, design trends and contract organisation
- Material processing, manufacturing and site procedures
- Social pressures, most recently concerns for the environment

The expectations of the client (either directly or a proxy for the user) are driven by social and economic change. Economic change affects the activity of the client and therefore the type of building they require but also the resources that are available for the manufacture and assembly process. The economic system and the organisation of commercial activity associated with the design and production of buildings is also significant in responding to and promoting change. This is expressed in procurement and contractual arrangement for the realisation of the building. The most significant aspect of this for the choice of technology is that of who makes the choice and what risk they carry in making that choice. This will lead to individuals being motivated to promote change either in design or production methods, both of which have implications for the technical solution.

To these pressures for change has been added the social concern. In the past this has been expressed as public health and then health and safety. It has also been for social justice and then social inclusion, which has made demands on greater performance levels for all buildings. Many of these social concerns become the subject of legislation and regulation. The most recent concerns are for sustainable development, the new concerns being specifically about the aspect of sustainability requiring low environmental impact. At the beginning of the twenty-first century our knowledge about incorporating this requirement into buildings is provisional and we only have emergent technologies to build into our designs.

UK commercial buildings through the twentieth century

Structure and facade

The rise of the corporation, particularly in America, created a need for large corporate buildings. The development of the framed structures at the beginning of the twentieth century facilitated this need to achieve both greater heights and longer spans. These possibilities led to developments in both design and technological solutions. The frame provided a freedom to change the external enclosure, particularly the walls, which no longer had the function of carrying the dead and imposed loads from the building and its occupation. It was now possible to use a much greater range of materials and their component forms to express the facade of the building. Both the use of frames for structure and the range of facade treatments are now commonplace in our city and urban landscapes.

When these forms were emerging in the early 1900s, design often called for traditional materials such as brick or stone to be used with these frames. While established detailing might be used for the wall itself, the connection to the frame required new solutions. Also with walls of greater area and taken to greater heights more thought was required with respect to the choice of materials, movement, weathering and maintenance access costs. However, designs also called for new facade materials. Among the new materials to be used in developing new aesthetics perhaps the best examples are concrete and glass.

Concrete as the facade material has been used both as precast panels or exposed in situ reinforced concrete. The process of precasting gave a regular, panelled appearance where individuality and response to context led to decisions on different panel sizes and surface treatments. The precasting industry had to respond and develop new knowledge and skills but so did the design in the detailing of the enclosure–structure interface. A new range of fixings and tolerance demands had to be resolved at the connection to the frame, and the joint between the panels became the focus for fit as well as weatherproofing. This led to the development of both closed and drained joints with the need for new materials such as mastics. Developing materials and detailing solutions for connection to the frame and jointing between the panels provided the opportunity to make panels in materials other than reinforced concrete such as glass reinforced cement (GRC) and glass reinforced plastic (GRP) but these had their own technical difficulties which, although largely overcome, proved more expensive than traditional reinforced concrete.

In situ reinforced concrete construction for the structure and then the facade required technological developments in temporary works as well as development in the specification of the concrete mix itself. The temporary works represent a major proportion of the cost of concrete, making it a requirement for the designer to understand the production process and 'design in' many of the features that might have been seen as the decision of the engineer and constructor. This was particularly true when exposed concrete became part of the design. Decisions on day joints, heights of pours, the spacing of the formwork ties and sizes of formwork panels all become exposed along with the finish. These now had to be specified by the design. Designs incorporating exposed in situ concrete also used coloured concretes and textured finishes, such as timber boarding, which put additional demands on the production process to provide a consistent quality required by the design.

Concrete as a facade material has shown that choice associated with an evaluation of its behaviour over time is important. As a facade material its weathering characteristics had to be considered. This put even more significance on the design detailing and choice of finish that would ensure the quality of the appearance with age without a commitment to maintenance and cleaning.

With the introduction of framed structures the limitation of window size in loadbearing walls was lost. Walls could be constructed without the appearance of openings, and facades could be made of materials in sheets or panels to provide flat facades. These materials had to be impermeable and the detailing for weatherproofing developed to seal the joints to provide the same barriers to water ingress as the facade material itself (although developments in rain-screen walling have changed this demand on the joints). Although metal panels could be used, it is perhaps glass that allowed the greatest design expression of the technical form. The use of many facade materials requires a sub-framing. For glass this was initially provided by storey-height frames or curtain walling to support and seal the glass (and other components). It is now possible to construct walls of glass, known as structural glass.

Many of these frame and enclosure solutions can now be seen as common forms, and their specification and detailing are discussed in later chapters. However, they were developed essentially for rectangular buildings. As buildings have changed so too have aspects of common forms. This has necessitated new technical solutions, details and materials specification and hence more evaluation to ensure that new risks are not introduced.

Building shape and form

Designs that call on components and materials to achieve different shapes for the whole building may have a profound effect on the ability to use common forms of established technology without modification and even innovation in materials, component forms and the detailing of assemblies as well as in production methods.

Buildings have traditionally been rectangular in both plan and elevation and, unless the site dictated otherwise, based on right angles. These shapes can be generated in technical solutions where structural loading is regular and vertical and components and connections control fit based on horizontal and perpendicular control grids. Many designs have refined this basic rectilinear format to develop aesthetic forms of great quality, and for these designs the technical solutions are often relatively well established.

At the turn of the twenty-first century curves both in plan and on elevation were required by a growing number of designs. Roof structures like domes and shells had already been developed to cover large spaces and buildings based on the circle on plan had been constructed, normally based on the loadbearing wall, and were not curved in elevation. Designs for curves on buildings that were to use the now common forms of frames and facade treatments initially were based on facets. The components remained flat and rectangular but the connections and jointing were at angles other than right angles. While this can produce smooth curves with small components as it had with bricks in the loadbearing structures, the larger flat components did not give a smooth curve particularly when viewed from the proximity of the urban landscape. The connections required detailing that could allow for adjustment not only in line level and plumb but also in the junction angle. The components and the joint detailing could remain repetitive, keeping manufacturing and assembly economy.

One established technology that lent itself to curving on plan was in situ concrete for the frame. The formwork to edges of slabs could easily be curved and slabs could be designed to either overhang column lines or have columns arranged along a similar curved profile. A series of manufacturing developments driven by the curve as a design element led to the components for both steel structures and cladding panels, including glass, to be curved to give designers even more ability to push the limits of non-rectilinear buildings.

Other forms are the angular or even shattered shapes, which demand fresh technical solutions. Gaining designs that maintain the cost advantage of repetition and regularity becomes a major challenge in the design. It is possible to devise and design structures and cladding systems to create these complex shapes and provide performance and production possibilities. However, detailing junctions to achieve performance such as weatherproofing has its challenges and cost implications. The cost is not necessarily in the material but in the design and production effort, which may require innovative thinking in achieving both safe design and quality in production.

Manufacture and assembly

Where design requires modifying, or new, technical solutions there are normally production and cost implications. However, even with the same demands from similar designs, competition within the supply industry will drive changes in technical solutions to gain cost and speed advantages in the production process. An example of this in the last half of the twentieth century was the change in the form and common detailing of framed structures.

The demand associated with the size of buildings was for an increasing number of storeys and for larger clear spaces covered by a single roof. The main technological development allowing the design of multi-storey buildings was the elevator (or lift), and this had been introduced at the beginning of the twentieth century, putting some pressure on the need for structural developments. This development needed a base of knowledge of material (concrete and steel) and manufacturing methods to allow safe and economic designs. The structural response had been largely developed by the middle of the twentieth century with the skeletal frame for tall buildings and the truss or girder for long-span structures. New economic pressures in the second half of the twentieth century from increasing commercial competition between steel and concrete as structural materials for skeletal frames led to further developments mainly focused on new technologies for structural slabs.

During this period, many forms of beam/slab combinations were tried. There was an interesting set of developments driven by the time and cost arguments between steel and concrete. While the whole story is beyond the scope of this text, the outcome has been the emergence of new common forms for skeletal frames for both steel frames (composite deck) and in situ reinforced concrete frames (flat slab) that would not have been recognised in the middle of the twentieth century. These, now common, forms are outlined at the end of Chapter 11 and explained in some detail in Chapter 25.

These now common solutions have the same performance levels as their earlier forms but have been shaped and developed by the twin pressures of cost and time through attention to the analysis of the production process. There is already evidence that these structural forms may evolve further because of the pressures to provide low environmental impact solutions for the building as a whole, particularly in the application of thermal mass. While these are not production-led changes, they will have an impact on cost and time that may well lead to changes driven by the competition between concrete and steel as structural materials.

The story of the development of the long-span roof frame has been different. As the dominant material for these structures has been steel, with some success for aluminium and even laminated timber for some forms, concrete is rarely used, so the competition to drive down cost and time has not pertained to these frame forms. However, manufacturing advances have led to the exploitation of new forms for long-span structures.

The common solution that had emerged to accommodate larger clear spaces covered by a single roof by the beginning of the twentieth century was the plane frame, predominantly pin-jointed truss or girder on columns, mainly in steel based on hot-rolled angles and bolted gusset plate connections. This simple form was straightforward to design and economic to fabricate. This form has few aesthetic qualities but that was of little consequence for the industrial manufacturing and warehousing that was required up to the middle of the twentieth century. Another form of the plane frame, the moment or portal frame, was understood in theory but apparently required large sections and was therefore used only over relatively small spans. The need for large sections in steel was the outcome of the predictions of a long-standing structural design approach known as elastic design. This was the prevailing design method that was used to prove safe designs. The development of a new structural design approach, plastic design, showed that much smaller sections would not fail, and this proved to be the case so the steel portal became economic over larger spans and rivalled the pin-jointed truss and girder solutions.

Changes in manufacturing methods allowed economic production of the truss and girder with welded sections made from tube or box sections that considerably improved the appearance. These advances in manufacturing also allowed the development of two-way spanning roof structures. By constructing a roof with a structural action across both horizontal directions, the structure was inherently stiffer, reducing depth of the structure without excessive deflections, and gave more stability against wind loading. These factors allowed a greater variety of forms in both shapes and materials to be developed. While the new manufacturing process made these new forms possible, it often led to more costly solutions, but the resulting structures often had enhanced aesthetic qualities. These forms will be discussed in more detail in Chapter 26, but the general developments were in grid structures such as the flat-space frames and the curved domes. The idea of curving (or folding) the slab form of a spanning structure led to the development of shell roofs that, while competitive on price and time, did not find a place in building designs in the last half of the twentieth century, possibly because of their aesthetic.

The structural design, manufacture and production techniques and resources for both skeletal frames and long-span structures identified above are now well established and can be incorporated into buildings with little technical or financial risk for a great range of building sizes.

Prefabrication

Another production movement that recurred over the last century with the increased manufacturing capability and transport infrastructure was the movement towards component manufacture and the assembly of components off site in prefabrication. Manufactured components as parts of the building that arrive on site with no expectation to change their size and shape before incorporation into the building are now widely specified and the detailing to ensure fit is well established in most cases. The history of prefabrication, where a number of components are assembled off site to make sections of the building to be connected together on site, over the course of the twentieth century was characterised by a series of emergent technologies, few of which have gained the status of a common form. There is a potential for high-quality finish under factory conditions that can be exploited at the high-value end of construction. However, the high set-up costs and the need for a guaranteed market has made prefabrication expensive, as changes in user requirements and design movements have frustrated the potential for long-run economic production. Towards the end of the twentieth century manufacturing methods in other industries were developed based on computer control that allowed a direct link from design information to production control, and this more flexible manufacturing arrangement could be exploited for buildings. At the beginning of the twenty-first century this more flexible manufacturing capability has been recognised in the MMC (modern manufacturing construction) initiative to provide a large number of low-cost houses.

Another emerging form based on prefabrication at the beginning of the twenty-first century is the pod, or volumetric solutions, for repetitive rooms such as bathrooms and hotels bedrooms. These are not always based on flexible manufacturing techniques but may be organised with traditional trades completing these small parts of the building in the controlled environment of the factory.

Technical solutions that incorporate components are now common and the detailing for fixing and the achievement of fit is broadly understood. This allows a wide range of standard and purpose-designed components to be specified and detailed with little technical or financial risk. Prefabrication, however, still requires a great deal of thought with design and manufacturing integration to resolve cost and performance issues as well as the accommodation of deviations in the detailing to ensure fit.

Low environmental impact

The requirement of low environmental impact is part of the sustainable development agenda that emerged at, and has gained considerable impetus since, the end of the twentieth century. As sustainability includes economic and social development the scale of building is unlikely to change and therefore the demands for structural integrity and the maintaining of internal conditions is unlikely to change. This indicates that, at least for the immediate future, low environmental impact building is likely to be achieved by the more imaginative design concepts based on the exploitation of current building forms. While there are some radical solutions such as earth-sheltered houses that should be used where they can, these are unlikely to be able to be developed into widespread common forms if the scale of current building activity to support social and economic development for an increasing population is to continue. There are, however, some emerging passive technologies as well as active technologies such as energy micro-generation systems that can be incorporated into buildings. They do, however, have to be considered at the design concept stage for new buildings and communities to gain the full impact of the emerging forms.

The concerns and some approaches to low environmental impact designs have been explored in Chapter 15 based on the key concerns for pollution and resource depletion. The aim is one planet living, where each individual lives within their share of the planet's ability to support them. We then have to generate a quality of life within these limitations. This has become focused on energy, as the immediate threat is seen as global warming where the use of fossil fuels pollutes (greenhouse gas, CO_2) and is depleting stocks for future generations. The need is for convergent solutions that use less energy as well as to look for energy from alternative sources. It appears that saving energy is the greatest contribution building can make in these early days of concerns for the environment. Added to the concerns over energy consumption are the more general issues of buildings as consumers of material and creators of waste that contribute more generally to energy use but directly to resource depletion and pollution increasing our environmental footprint above the aim of one planet living.

Perhaps the challenge as new technical solutions emerge is the provisional nature of the knowledge we have to predict the behaviour of the fabric of the building to achieve the performance demanded. Many solutions to energy saving involve using the passive fabric in what has been termed natural processes as opposed to the mechanical and environmental services. This involves an understanding of the dynamic interaction between the internal environment and the fabric of the building, where quite small effects are being exploited but are vital to the success of the whole system. As this knowledge becomes more certain the final forms of specification and detailing will emerge. This has become evident in the many competing arguments for each of the main materials used in buildings to claim green or environmental-friendly credentials.

Even if the knowledge we have for specification and detailing is still emerging, there are some established principles that are being employed and are putting demands for change on both design and production. Perhaps the most widely emergent form is the atrium or street design outlined in Chapter 10. To take full advantage of this design it is not just used for natural ventilation and lighting but becomes the mechanism for summer cooling, explained in Chapter 15. For this to work much of the detailing has to change. An example is the passive role of the structural slab for thermal mass associated with the automatic control of opening windows for summer night purging.

The implication for control to make the system work automatically and gain efficiency in the operation of the building is another emerging technology. Poorly operated systems may not take advantage of the full potential of the design, significantly increasing energy use of the building. Part of the technical solution must include a control strategy including levels of automation and its integration into an overall building management system (BMS). These are emergent technologies that have had no previous equivalent technical form in building. Many are made possible by IT technologies that now have to be built into the design. They become a new (or at least a variant of) the communication services in the building. Control systems are currently based on wires, but wireless technologies are now possible and this may make repair, maintenance and extension easier.

It is now becoming clear that environmental impact and the sustainability agenda can be best served by considering not individual buildings but communities to provide local services. These might in the past have been either provided nationally, as in power, or self-contained within individual buildings, as perhaps with heat. The full implications for this approach on technical choices is not yet clear but already small-scale local heating and power plant installations can show significant resource and pollution reduction. These and other technical solutions are the new generation of emerging technologies.

Low environmental impact and existing buildings

Many of the emerging technologies have applications to the existing building stock. This is a significant area of application if we are to reduce our environmental footprint in a relatively short period. Most of us live and work in old buildings, constructed before the widespread concerns for low environmental impact. This text focuses its examples on new buildings, but the underlying framework for analysis and the process of choice is just as applicable to refurbishment and retrofitting schemes. In particular, if progress is to be made on reducing CO_2 then attention will have to be paid to upgrading existing buildings, and this will include developing new technical solutions and installation techniques to suit a range of previous building forms.

Understanding the behaviour of the building both before and after retrofitting will be important to ensure not only that the new performance levels are reached on energy use but also that the building continues to perform as it was originally intended. The early attempts to introduce insulation into existing cavity walls that caused damp problems showed the importance of understanding the materials and site control procedures and serves as an object lesson for failing to carry out a full evaluation of emerging technologies.

Emergent technologies and risk

At any one time (and in any one place) there will be a set of common forms and some emergent technologies. Perhaps one of the key aspects in the choice of a technical solution is to identify how well established it is and if it is being applied into new design formats or near the edge of its performance range. If it is emergent and cannot yet claim to be a common and well-understood solution with a significant track record of being tried and tested, or it is being applied to new circumstances, there will be a greater need for analysis and evaluation to reduce the element of risk.

This technical risk is one of loss of performance. If either the specification or detailing is not sufficiently thought through, it may fail. Alternatively if the resources, particularly the knowledge and skill, are not available, there may be an increased risk of failure due to inappropriate design or production practices.

As for all risk assessment the potential failure has to be identified against the consequences of that failure and the probability it will occur. Here the notion of safe technical solutions is the same as for health and safety with the identification of the hazard (potential failure) and the harm it will cause (consequences) and a chance it will occur. Risk can then be established on a scale of high to low. Performance failure that has serious consequences should be treated as a high risk and steps be taken to reduce the risk. If the consequences are less but the probability of the failure is high then, again, changes in the specification, detailing or resource availability should be sought.

While the concept of harm in health and safety (or even environmental risk assessment, which takes the same approach) is normally clear, the consequences of technical failure will have different impacts on different individuals associated with the building. At one level the harm is the same: collapse or even some serviceability failures will endanger life and health. Some failures, however, affect the efficiency of the operation of the building. These can have cumulative effects on the users, the running costs or the energy consumption anticipated in the design.

There is also a link to the financial risk carried by the client and/or those involved in the design, production or operation of the building. If the technology is emergent, there is not just a financial risk of failure to the client but the uncertainty to the designer or constructor as to the cost (or time) to produce the safe solution. Who carries this risk is set out in the contract. All parties to the contract need to understand the status of the technical solutions associated with the understanding and experience that can be drawn upon in the design and realisation of specific details and specification. Innovation may be actively sought by agreement between the parties and the resources made available to reduce the risk, but this will come at a price.

There is perhaps one other risk and that is aesthetic. The technical solution and its detailing will take a particular form that will determine the appearance of the building. This is not only true for the overall appearance of the building when viewed from a distance but when seen in detail by those using the building passing close by. In this respect the choice of technology is particularly significant to achieving the vision for the design.

Summary

1. When suggesting a potential technology, it is important to recognise whether this is well established or new and/or emerging.
2. Developments over the last century have established many of the technical solutions we now use that can achieve the health, comfort and safety levels required.
3. There are still drivers for change that come from design in new building shapes and form and from production, particularly from prefabrication.
4. Perhaps the greatest driver for change is the environmental agenda, which raises technical challenges that have to be considered at the very beginning of the design process as well as in the specification and detailing of materials and components in the detailed design.
5. Risk has to be evaluated to include not only technical failure (both physical performance and appearance) but also health, safety and environment as well as the implication for the cost and time aspects of the contract.

24 Interface Design

This chapter considers the nature of each of the interfaces between the major elements of the building that have been identified as structure, enclosure and services. Before considering the major technologies available for the elements, it is necessary to have a view on how they will interact with other parts of the building. This is a significant aspect of deciding on the technological mix at the early stages of design in order to have the confidence that the detailing can be resolved for the building to work as a whole.

Introduction

In practice it is necessary to choose technical solutions for elements of the building with defined performance requirements, and these have been broadly characterised in this text as enclosure, structure and services. For most buildings these will be selected from common forms that offer the best opportunities to achieve performance and then need to be detailed for the specific building. One aspect of this selection process is to ensure that the emerging mix of technologies being considered for the building are compatible and will work with each other to ensure the performance of the building as a whole. This compatibility is not only in performance but can also be in achieving detailing and specifications at the detailed design stage.

Chapter 22 introduced this idea of the need to focus on the design of the interfaces as well as the elements to ensure the building works as a whole. The design issues around these interfaces vary and have to be considered at different stages in the overall design of the building.

The interface between enclosure and services is predominantly conceptual and needs to be considered at the very early stages in the design where the strategies for passive environmental design have to be established while the shape, orientation and configuration of the building itself are emerging. Taking advantage of these fundamental elements of design to provide passive environmental design needs a very early identification of the possible technical solutions to the enclosure (and structure) as these will have an impact on the type and extent of services that will be needed to ensure the provision of the required internal environment. There may be some physical and detailing interactions between services and enclosure where services pass through or are fixed to enclosure elements, and these will have to be resolved at the detailed design stage.

The interface between enclosure and structure is predominantly one of physical connection, which has to be resolved at the detailed design stage. The most complex of these connections is between the external walls and a framed structure. Some care is needed in resolving connections to the horizontal enclosing elements such as floors and roofs.

The structure–services interface is often about space and the need for vertical and horizontal distribution in shafts and voids but, as with enclosure, services have to pass through and be fixed to the structure where some services components are large and may include additional loading. However, the consideration of the interaction of structure–services design is increasingly involved with passive design, where the thermal mass of the structure can be used to store and release heat at appropriate times in the diurnal–nocturnal (24-hour) cycle.

The enclosure–services interface

There has always been a need to consider the contribution of the enclosure and the environmental services to creating and maintaining the internal environment of a building. At the beginning of the twenty-first century concerns for the global environment have caused a reassessment of this balance between passive and active technologies. It is no longer acceptable to just make the enclosure watertight and then solve internal comfort conditions with active, energy-thirsty, services. The move to 'zero' CO_2 and low environmental impact building has a considerable influence on the development of technical solutions for both external enclosure (roofs and walls) and environmental services.

The inclusion of environmental impact as a criterion of design decisions has moved the focus to an approach characterised as passive design. The idea is to use the passive fabric to its maximum, as, by definition, it needs no energy to moderate the environment during operational conditions. This involves using the passive fabric of the building to moderate the flows and transfers between inside and outside to maintain a quality of the internal environment throughout the year. While this approach seeks to limit the use of energy during the operation of the building, the construction to achieve this is not without environmental impact, and this has to be taken into account. All this has an influence not only on the choice and detailing of the fabric but also on the provision of environmental services within the building.

The enclosure and passive design

Environmental design using passive flows and transfers is considered in Chapter 15 as part of the more general need to provide buildings with a low environmental impact. Many of the ideas introduced in that chapter are relevant to the selection of enclosure and its relationship to services. To understand the performance and selection of the technical solutions for the enclosure there is a need to return to consider what make a quality environment. Particularly important are the elements of the thermal, light and air quality, as these are aspects of the services solutions that may have high-energy inputs with the potential to contribute to CO_2 emissions if the energy source has to include the conversion of fossil fuels. These services solutions include systems of heating, lighting, ventilation and air conditioning. Systems vary in their dependence on energy but any reduction or elimination of the need for these systems will have a considerable impact on the need for energy and hence global CO_2 emissions. The technical mix that may be appropriate for an individual building must be considered at the concept design stage, where the available fabric and services technologies need to be considered and matched to achieve economic, effective and environmentally acceptable solutions.

Passive designs for thermal environments vary depending on the daytime heat gains in the building. For dwellings these are small and there is little contribution to winter heating and a limited risk of overheating in the summer. In these circumstances the enclosure needs to restrict the loss of heat in winter, normally with insulation and controlled ventilation, while maximising the opportunities for heat gains during sunny winter days. If the enclosure is effective in capturing radiant heat gains (solar gain) in winter, it may overheat in summer when heat gains are potentially high. This has to be considered in the choice of enclosure, particularly windows and roof lights facing south (in the northern hemisphere) where heat gains are greatest. Windows are a key element in providing light, so the requirements of light, heat loss and utilising heat gains from the sun makes them an important aspect of the design. Some form of shading may need to be incorporated in the design of the window, particularly on the south elevation. For both summer and winter conditions it may be useful to consider storing heat to smooth out the effect of the day–night cycle where external temperatures vary. Capturing heat during periods of heat gain (normally daytime) and then releasing it during the night (into the building during winter and out from the building in summer) will require the fabric to have high thermal capacity. This requires some consideration of air movement and ventilation.

For buildings with high daytime internal heat gains (e.g. offices, retail and leisure) the same basic mechanisms are exploited but the focus is on limiting heat rise in summer to eliminate any need for air conditioning (easier to achieve where the natural climate does not include high humidity). This requires limiting heat gains through the enclosure, perhaps with thermal capacity and with shading to the south-facing windows, and then storing the daytime heat to be released to the outside during night-time, when air temperatures are lower. Ventilation is now important. During the day this will limit heat rise and during the night it is required to draw out the heat from where it is stored in the fabric. However, these two ventilation patterns may not be the same.

Typically the daytime ventilation needs to be around body and head height while at night the air flow needs to be across the fabric storing the heat, normally the soffit of the structural slab at ceiling level. In addition it may be possible to allow the occupants to control the ventilation in winter (opening windows), but the summer night purging will have to be automatic with windows opened automatically on a time and/or temperature basis. Ventilation can be provided mechanically as one of the active building services. However, this requires energy, and therefore a passive ventilation system would be preferable. The mechanisms are discussed in Chapter 15 where both cross-low and stack-effect work on relatively low pressure differences, which will be significantly affected by the width and height of the building.

During the winter, daytime heat gains may not be sufficient to heat the building, although this will be dependent on the levels of insulation provided by the fabric. Heating can be supplemented with heat gains from the sun. This will be achieved through the windows, so the summer shading, designed to inhibit heat gain, must be ineffective during the winter. This can be achieved by taking the variation of the elevation of the sun above the horizon between summer and winter. Lower in winter, the sun' rays need to be able to enter the building below or between the shading provided for the summer, when the sun is higher in the sky.

This consideration of the thermal environment involves the use of the windows. These are, of course, the major part of the passive fabric involved in lighting the building. Lighting is also a major user of energy, particularly in commercial buildings if light has to be supplied by the building services. The other aspect of lighting is the depth of the building, the distance from the external wall to the back of the room to be lit. Depth has to be useful for the activity in the building, and so large windows, particularly with high head heights and views of the sky, are required if the need for artificial lighting during the day is to be limited. Quality of light varies between the north and the south. North light is more even, while high levels of light in the south make glare a significant problem.

Services systems

The services systems that are required for the technological mix within a passive design must also achieve the design criterion for CO_2 emissions and other aspects of low environmental impact. This can be achieved with alternative energy sources with no-carbon, carbon-neutral and low-carbon options, which were discussed in more detail in Chapter 15. Efficiency in the services equipment can help, as can automatic sensors and management systems. Automatic control of heating systems with thermostats and light sensors for lights are examples of sensors but even these can be more effective with zoning. Heating zones on north- and south-facing parts of the building and lighting in banks back from the windows are examples of effective zoning to match the need for the system to become operational based on environmental need.

The enclosure–structure interface

The nature of this interface is significantly different, dependent on the basic form of the structure. If the structure is loadbearing, the structural elements become part of the enclosure and have to be solved together. For structural frame solutions the interface is a physical connection or junction between the structure and enclosure and is concerned almost entirely with joints and fixings and the shaping of components to accommodate them. Each element is likely to be created from components in different materials that only act as a whole if care is taken in thinking through how each is fixed and how joints are formed to allow the process of assembly, movement and give to the continuity of performance.

One of the characteristics of buildings with framed structures is the separation in design and performance terms of the structure and the enclosure elements, particularly the external walls of the building. This separation allows a great range of facade treatments to be considered but all will be dependent not only on their ability to modify environments but also on their connection to the frame. This makes the focus of the enclosure–structure interface one of connection, a combination of fixings and joints to allow assembly on site and ensure the performance of the building as a whole.

With the separation of enclosure and structure the external walls are often termed non-loadbearing, but this is perhaps too simplistic. It is true that the external walls do not take the dead and imposed loads from the building. However, they are subject to wind loading, and this has to be safely transferred back to the structural frame. These external walls do, of course, have the more obvious performance requirements of appearance implied in the word 'facade' often used for these walls and the environmental performance where the term 'skin' is used to indicate the external protective and environment-modifying nature of these walls. All of these, particularly load transfer, have to be taken into account in the connection.

The main concerns in providing this connection to the frame for most of the common forms of external wall will be structural and environmental performance. To this has to be added not only the major considerations of the joints and fixings to facilitate assembly at the production stage but also behaviour over time, particularly movement. For some forms of external wall such as structural glass the appearance of this connection can become a significant aspect of the design.

Of almost universal concern in providing these connections are:

- Load transfer (support and restraint)
- Environmental performance continuity
- Accommodation of movement (inherent deviations)
- Manufacture and assembly tolerances (induced deviation)
- Assembly resource and sequence, including health and safety risks

Each of these concerns is discussed in general terms below, and its application to specific systems will be covered in Chapter 29, where the options for external enclosure walls will be discussed.

Load transfer (support and restraint)

Support has to be provided to ensure that the weight (dead load) of the wall is taken safely to the foundation as the first condition of stability. Restraint is then concerned with horizontal forces, mainly the wind loads. However, there is accidental loading, such as explosions.

It is possible to visualise the external wall supported on its own foundations with restraint provided by the frame at appropriate intervals all over the height of the wall. While this option is often used with any heavy wall construction to a single-storey industrial building, multi-storey walls would normally transfer their dead load to the frame at each floor level.

If the support and restraint is to be provided at each floor level, the connection of the enclosure elements to the structure has to be designed to take these forces. The connection must ensure the stability of the enclosure wall and transfer the loads to the structure. The nature of this connection depends on the structural behaviour of the wall. If a masonry or stud wall is specified, the wall has the strength to act as a panel spanning between the beams and columns, although additional column supports are sometimes necessary, where they are known as wind posts, acting to transfer wind forces to the structure but not carrying the vertical loads from the building as true columns. Figure 24.1 shows a schematic arrangement for the connection of this type of infill wall to the structure. It shows the simplest of arrangements for the structural element of the wall as an infill standing on the edge beam of the building structure. It is possible to conceive of this infill wall being the facade material, but it leaves the beam exposed to view. It is possible to add support components to the front of the beam to take the external leaf of the wall to cover the beam and make the facade material continuous. It is also possible to use the infill wall to provide a backing to some form of facing material. In all these cases there is a need for restraint at the top of the wall normally from the underside of the structure, where it is usual to provide a movement joint, for the reasons discussed below. Restraint will also be required at intervals in the length of the wall from the columns and any wind posts that have to be provided.

Figure 24.1 also shows the wall constructed from what are normally known as cladding panels. Normally pre-formed, the panel either hangs or stands on the beam. Spandrel panels (not full height) are restrained at the columns or short wind posts, while storey-height panels are restrained at the beams and to the adjacent panel. The third option shown in Figure 24.1 is to create a framing to support elements of the wall that cannot in themselves span between the building's structural elements and therefore need some intermediate support. With this type of wall the interface is between the structure and the framing not requiring the continuous support of the infill and cladding panels. This

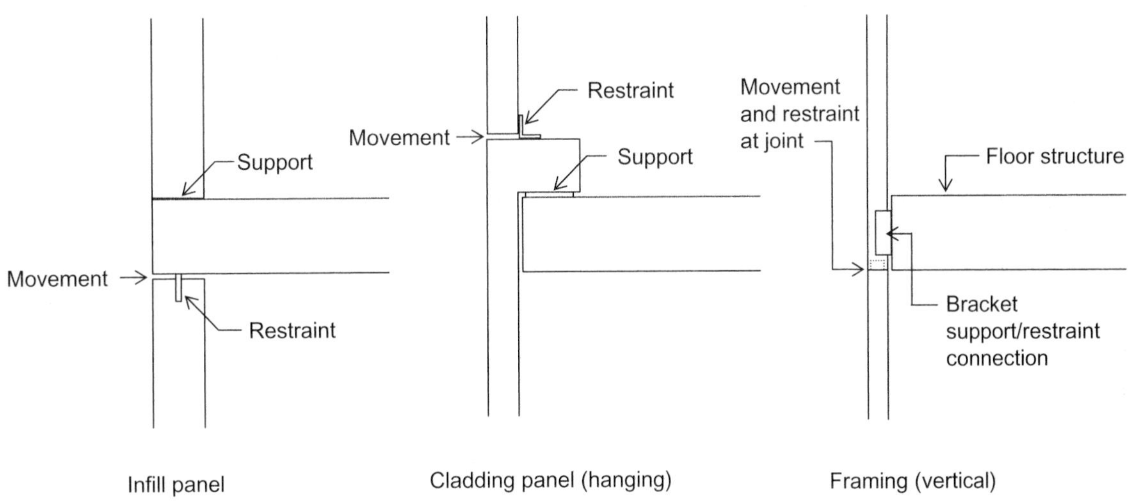

Figure 24.1 Load transfer options from external enclosure to structure.

point connection is most likely to take the form of a bracket capable of attaching the framing and transferring the loads to the structure.

Environmental performance continuity

The external wall of the building will have many environmental functions that may be compromised at the connection to the structure. These will include the fundamental weatherproofing requirements, particularly around damp-proofing if the basic wall construction is attached to the structure as with the normal detailing of the cavity wall. In this detail the fundamental damp-proofing is associated with the cavity and if this is interrupted at the structure, which it often is, cavity trays and weep holes have to be introduced. If the damp-proofing is associated with walling systems such as rain-screening designs, attention to detailing will be associated with the connections to the backing wall, and this will become part of the detailing of the wall system more generally.

When analysing the interface connection, all environmental performances need to be considered. For a wall with high insulation values, if the structural connection brings the wall and structure physically close, there is a potential for cold bridging reducing the overall thermal performance of the wall. Detailing to hold the wall away from the structure to allow insulation to be fitted across the structure may have to be considered. Another environmental performance that could be compromised by the rigidity necessary for structural connections or the space between the wall and structure is noise flanking paths. These are most acute where a physical gap is generated in systems such as structural glass walls.

Accommodation of movement (inherent deviations)

The requirements of the structural connection will inevitably require some points of fixity where relative movement cannot take place. Any change in dimensions or position in the structure will take the wall with it and any change in dimension of the wall will be restrained, putting stresses into the wall, the structure but perhaps most significantly the connection itself.

The changes or movements that take place in the structure and the wall come from different origins and can take place in different directions often exaggerating relative movements. The structural movements and distortions come primarily from the loadings. Beams bend and columns shorten. While most of this movement takes place as the loads are applied and only becomes a problem if loads are applied after connections are made, some movements are time-dependent. Many foundation settlements take place over time and concrete will creep over time even with constant load. Cracking or loss of connections may not take place immediately but may be failures at some time during the operational life of the building.

The wall will experience changes in dimensions due mainly to environmental factors, although some materials will have time-dependent equalisation changes such as bricks and blocks that take some time expanding (clay) or shrinking (concrete) and may continue after they have been installed if used too soon after manufacture. There is a particular risk if a concrete frame is clad with clay brickwork. The frame shortens and shrinks while the brickwork expands. This requires positive movement joints immediately below the support of the brickwork at each floor level.

The environmental changes include both moisture and temperature changes, particularly for the cladding materials. These will depend on exposure, where perhaps the greatest risk is on the south face where the radiant heating from the sun will create surface temperatures considerably higher than the air temperature. Different materials have different properties associated with their inherent dimensional deviations. The thermal expansion of metals, particularly in thin sections where they heat up quickly, may need attention in the detailing on large areas of facade. These movements can be significant in the length of the wall as well as between the floor interface connections.

Manufacture and assembly tolerances (induced deviation)

The wall structure connection is a major point of assembly in the construction process. The structure is often produced by a different process from that of the wall components as extreme as in situ frame to large prefabricated panel walls. The influence of these processes on the achievable deviations is great and the differences have to be accommodated in the detailing of the connections. The wall has to be attached to the frame in a way that maintains the line, level and plumb of the wall and ensures the components of the wall can fit together to maintain the performance of the wall.

The approach to identifying tolerance and the significance of joints in this process and the importance of detailing the fixings to allow assembly and adjustment are discussed in Chapter 4.

Assembly resource and sequence including health and safety risks

As a major point of assembly the connection between wall and structure is the focus of many of the site operations. It also takes place at the edge of the building and often at height. It can involve large, heavy and possible awkward and even fragile components. This can pose significant health and safety concerns leading to major temporary works and lifting plant. The detailing of the whole interface will determine the sequence of operations and where an individual will have to stand to make the connections, and the direction from which the components will have to be brought up to the structure to be fixed. All of these have a great influence on safe working in the most cost-effective way, both in temporary works and the time to complete the works to a safe and permanent stage of construction, minimising the need to return to complete operations at a later date.

Poor detailing associated with this need for resources cannot be rectified by good management at the production stage. The options for sequence and resources are set by the detailing. Analysis of the assembly process at the design stage and modification of the detailing of this interface can make a significant contribution to safety and cost of the assembly process.

The structure–services interface

The structure–services interface has always been about space and support. The most convenient place for the horizontal distribution of services systems is in the floor zone often below the floor and above a suspended ceiling. This interacts with the floor structure, specifically with the beams. If all services run below the beams, the floor zone can become deep, particularly if ducting is involved. Flat slabs, where there are no downstands below the soffit of the slab, give a clear area for services. Where steel beams hang into the void above the ceiling, it is possible to create holes in the beams using specially fabricated cellular, or castellated, beams although these are a deeper option than standard universal beam sections. If standard sections are used, holes can be detailed, although some care has to be taken in the size and position of the hole and particularly in areas of high shear where web stiffeners may be required.

The vertical distribution of services takes space away from the usable space in the building. However, the vertical distribution also requires holes or openings in the floor structure, and these are often close to columns where the usable space is already compromised. This can produce the need for openings in areas of high shear and will give some detailing difficulties, for example with some forms of precast flooring systems.

Other concerns around the interface between structure and services are the need to provide fixings and the more general issue of loadings. Most services systems add little to loading, but some of the plant components can be large and heavy, although from the point of view of the services design they are best placed on the roof.

Reference to the section above on passive design will identify that the structure can be effectively used for thermal capacity to hold heat during the day for release at night. This is yet another example of the interaction between the major elements that need to be considered at the early stages of design when the broad technological mix is being decided.

Summary

1. The enclosure–services interface is essentially conceptual and needs to be considered very early in the design process.
2. This interface is concerned with gains and losses flowing across the enclosure that need to have balanced inputs from the services to maintain a comfortable and healthy internal environment.
3. While the combination of passive fabric and active services has always needed to be balanced to maintain the internal environment, the rise of passive design has made the analysis of interaction more significant.
4. The enclosure–structure interface is one of physical connection that has to be resolved at the detailed design stage.
5. Not only is the enclosure–structure interface a focus for performance with load transfer, environmental behaviour and the accommodation of movement but also is a major connection at which production issues are resolved in terms of resources, health and safety, and tolerances.
6. The structure–services interface revolves mainly around space, particularly in floor zones but also when considering floor slabs and the position of columns that can affect vertical distribution.
7. Passive design is now considering thermal mass that can be provided by structural components.

25 Structural Skeletal Frames

This chapter introduces the performance, detailing and production aspects of the skeletal frame. The frame is based on columns and beams with slabs to support the occupants and wind-stability members to stabilise the whole structure. This simple pattern has been developed using the two major high-strength, quality-controllable materials of steel and concrete, which are used in all major frame types. The performance aspects of reinforced concrete are considered along with the production of both in situ and precast frames. This approach of performance and then production aspects is also adopted to explore the nature and form of the structural steel frame. The chapter concludes with an exploration of the choice of structural form based on the overall design and integration into other elements of the building.

Materials development and structural options

In the modern era the success of the skeletal frame structure has been dependent on the development of two high-strength, quality-controllable materials: concrete and steel. Indeed, much of the debate around the choice of structural frames is not whether a frame is the right structural form for the building but what material should it be built with. Should it be concrete or steel?

In truth the use of the labels 'steel' and 'concrete' for frames is a little misleading as nearly all frames use both concrete and steel in their construction. These labels are often shorthand for structural steel and in situ reinforced concrete frames. The labels are, however, useful in so far as they give a clear indication of the form and production process involved. It may be better to see frames as being characterised by the production process. This gives a good indication not only of the connection between structural members, and therefore the structural behaviour of the frame, but also of the cost breakdown. On this basis there is a strong link between structural steel and precast concrete. In both the production and structural behaviour of the joints between beams and columns, precast concrete is much more like steel than in situ concrete in material terms (the exception being fireproofing, where both concrete options have similar design criteria).

Frame components – beams and slabs

The concept of the skeletal frame is associated with the use of beams and slabs to provide a spanning structure to form the floor and then columns (and sometimes walls) to take the vertical loads to the foundations. There will then need to be some form of wind-stability element to complete the overall frame. Traditionally beams spanning between columns form the frames and these can be repeated through the building to create support for floor slabs and then vertically to create the required number of storeys in the building. These rectangular frames (if pin-jointed) are unstable against racking from wind but also against rotational collapse of the overall frame, so some individual frames have to be stabilised throughout the height of the building and the horizontal wind forces have to be transmitted through the floors to these stabilised zones. This basic arrangement is shown in Figure 25.1.

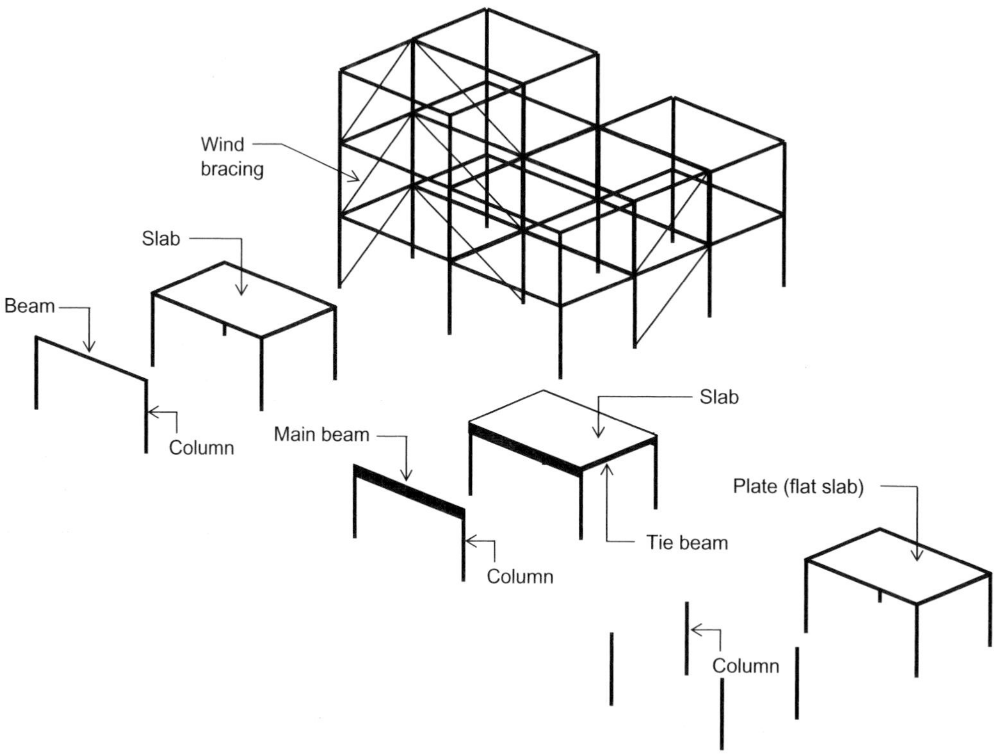

Figure 25.1 Basic components and form of the skeletal frame structure.

The description of the beams and column making an individual frame as the fundamental unit of the overall structure is still a good description of skeletal frames made from preformed structural elements in structural steel and precast concrete, but the in situ form of reinforced concrete is now likely to use a plate for its spanning structure. As the name implies there are no beams but a single slab reinforced to act as a plate held up by the columns. The relatively small cross-sectional area of the column creates relatively high punching shears around the column head but specialised, often pre-formed, reinforcement assemblies allow detailing to provide a truly flat plate. Although this form is called a plate floor in many parts of the world, in the UK it is normally known as flat slab construction. This option is also illustrated in Figure 25.1

While the plate floor may be seen as a completely different way of providing a spanning element for a framed building, it can also be seen as a logical development of the possible combinations of structural connections between beams and slabs, as illustrated in Figure 25.2.

In the first case separate slabs are simply supported on the beam. Most likely to be precast concrete, each element acts independently. This arrangement is perhaps the most straightforward for structural analysis and design and, if the structure is to be constructed of preformed elements, gives the simplest detailing. It does, however, give relatively large structural members. This does not necessarily mean the most expensive solution, as the simplicity of detailing will lead to economies that may outweigh any increase in material costs for the structural members.

The second condition is to provide structural continuity in the slab. This reduces the forces at the centre of the span of the slab (sagging) but induces forces over the beam support (hogging)

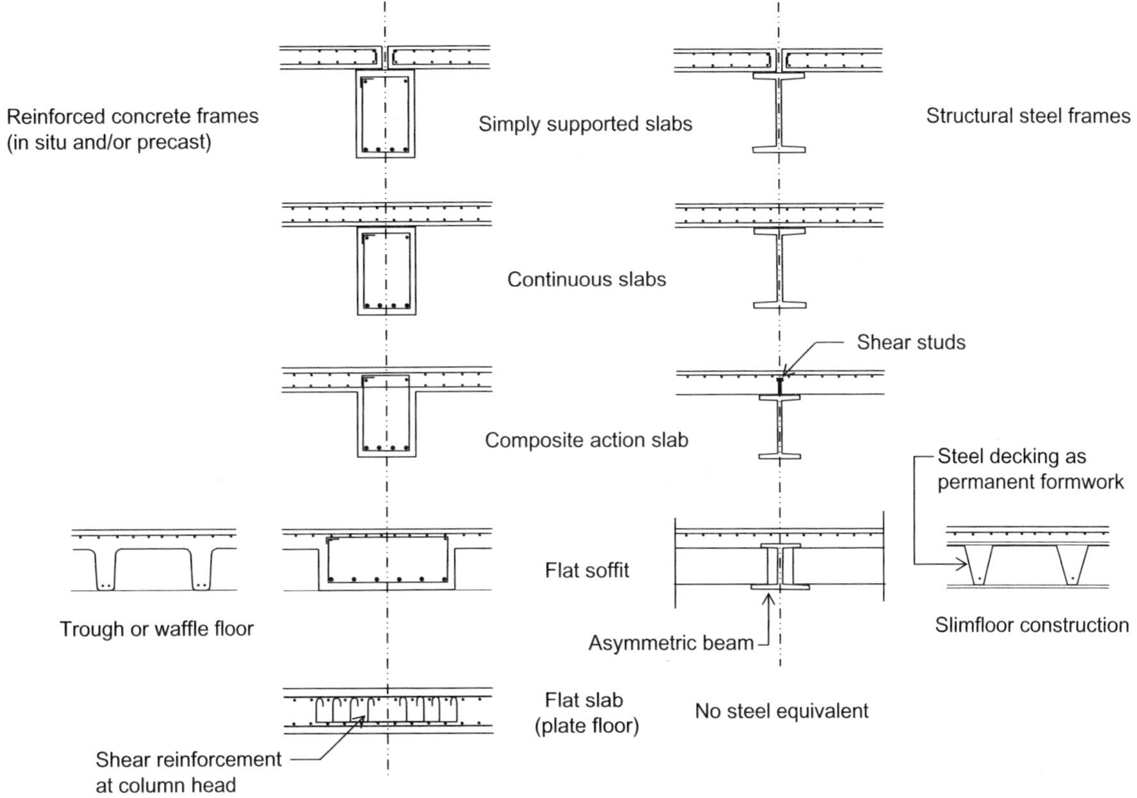

Figure 25.2 Structural connection between beam and slab.

where the tensile bending forces are in the top of the slab and compressive forces in the bottom. This will reduce the thickness of the slab but requires detailing across the beam to ensure the action of continuity and resistance to the hogging forces. In some solutions this continuity is easier to achieve than in others. With in situ concrete it is difficult to avoid continuity, and specific detailing has to be considered if simply supported slabs are required. With precast floor slabs continuity can be achieved with stitching reinforcement and structural toppings.

In the third case the continuous slab has now been connected to the beam to give a composite action. The spans of the beam and slab are at right angles. As the slab acts in hogging over the beam the beam is sagging under the slab. If the slab is connected to the beam, the slab becomes part of the compressive zone of the beam reducing the size of the beam below the slab. For this composite action to work any potential shear between the downstand part of the beam and the slab has to be resisted. While this is achieved naturally, in in situ concrete construction it has to be carefully detailed in structural steel solutions most commonly using shear studs.

The fourth option in developing beam and slab combinations that maintains both continuity and composite action was developed initially for in situ reinforced concrete. Developing a design where the depth of the beam is the same as the depth of the slab gives an opportunity to use flat soffit formwork, giving significant economies in the formwork and falsework as well as faster construction times. The absence of downstand beams is also an advantage in the distribution of services and the reduction of the depth of the overall floor zone, saving on building height. For these reasons a flat soffit construction has also been developed for structural steel frames, known as slimfloor. For both the in situ reinforced concrete and the steel slimfloor solution there needs to be a compromise. The slab thickness has to increase, increasing dead weight, and the beam depth has to reduce, reducing the effective bending resistance.

The thicker slab will need to have dead weight removed by taking out concrete from areas where the concrete will experience little or no stress. This is essentially through the centre of the slab (the neutral axis) and at the bottom where the tensile forces are taken by the reinforcing steel. Some care has to be taken to provide shear resistance towards the support of the slab and in the crown thickness to provide sufficient fire resistance. A common way of reducing the dead weight, as shown in Figure 25.2, is by creating ribs on the underside of the slab. This is achieved by placing standard trough moulds on the flat formwork for in situ reinforced concrete and by using pre-formed metal decking for the steel slimfloor.

The reduction of the depth of the beam has to be accommodated by increasing the width. The width of a beam is less effective at resisting the bending forces so small reductions in depth require comparatively large increases in width to provide a beam of similar strength and deflection characteristics. Beams in reinforced concrete, in these flat soffit combinations, are therefore typically wider than they are deep, not the characteristic shape of an independent beam but the flat formwork brings overall cost savings for in situ concrete. In the steel slimfloor it is only the bottom flange taking the tension that needs to be wider (the concrete in the slab takes a proportion of the compressive forces) so the characteristic asymmetric beam is used. This is achieved by either using universal sections with plates welded on a bottom flange or the use of purpose-rolled asymmetric beams. Both these options increase the cost of the steel frame.

This form of beam slab combination can be developed for in situ reinforced concrete to make a two-way spanning solution where the ribs run in both directions and the areas where the concrete is absent create a so-called waffle effect on the underside of the slab.

The final illustration in Figure 25.2 is the plate floor. There is now no beam in either the design or the detailing of the slab. This can only be constructed with in situ concrete or with a composite, precast and in situ, concrete frame. In the structural design two-way spanning column strips and middle strips acknowledge the beam-like bending behaviour over the columns, but both are designed and detailed as slabs. The plate experiences high punching shears at the column and this led in the past to detailing of conical column heads or thickening of the slab towards the column connection. This increased the cost of the formwork. Development of reinforcement techniques has allowed the shears to be resisted in the steel within the depth of the slab, making the formwork truly flat. This has emerged as an economic design for in situ reinforced frames and will be discussed in more detail later in the chapter.

Frame layouts – columns and walls

The spans of beam and slab combinations and plate floors have economic limits that create a need for internal columns (or walls). The production process and the detailing of the connections between structural members have led to the adoption of a small number of common approaches to the layout of slabs, beams and columns. These are shown in Figure 25.3. In this figure the span ranges shown are indicative of, and based on, economy for a normal range of commercial floor loading. Both larger and smaller spans are structurally possible, depending on the specific circumstances.

The simplest form is a one-way spanning slab supported on main beams that are connected to the columns. The beams that are provided parallel to the span of the floors take no floor loads but tie the frames together and may have to transmit horizontal wind loads back to the parts of the frames providing the wind-stability

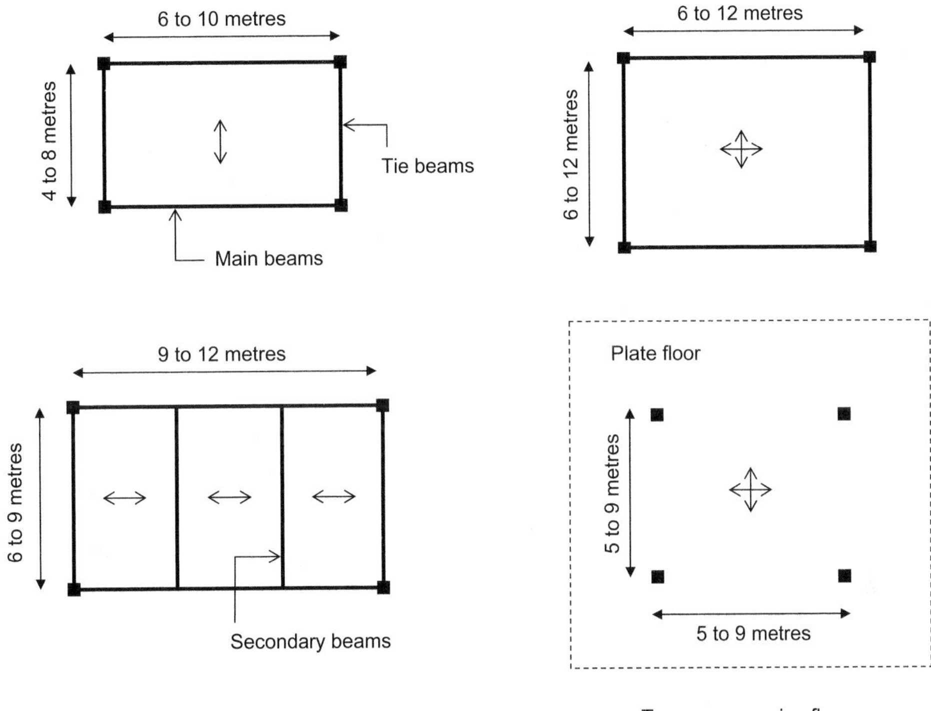

Figure 25.3 Approximate economic column grids.

elements (along with the slabs these can act as a diaphragm).

The economic dimensions of this layout are dictated by the span of the slab and the beam to which it passes its load. This simple layout was the basis of the early frames. Slabs would almost certainly have been (and still are) in reinforced concrete. In situ or precast solutions have been used for these floors. With current precast concrete floor systems it is possible to have floor spans well in excess of 10 metres, but this will place a large load on the beam, which will in turn have to be large. The need to limit beam sizes for both economic and space criteria means that the capability of the larger spans of these precast floors cannot be fully exploited in frames. Economic solutions for both slab and beam combinations are achieved at floor slab spans below 8 metres with beam spans up to around 10 metres, although larger spans are possible. This often leads to a rectangular layout for the columns with the main beams forming the longer sides of the rectangle. This layout would also be common for precast concrete frames.

The second layout shown in Figure 25.3 introduces secondary beams. This layout still uses one-way spanning floors but takes the slab loads to secondary beams that span onto the main beams that then take the loads to the columns. This allows shorter floor spans and longer main beam spans giving wider column spacing due to reduced dead loads from the floor slab and the point loading condition on the main beams. This layout is best exploited with in situ floors that develop continuity over the beams, particularly if composite action with the beam can also be achieved. This form has become a common form for structural steel frames with metal decking as permanent formwork for the floor slabs and shear studs to achieve composite action. This will be discussed in detail later in this chapter.

The third layout introduced the two-way spanning slab. By taking the loads out to all four sides where all four beams become main beams there is a potential for smaller beams where a squarer column layout can be achieved. Achieving two-way spanning slabs is relatively easy using in situ concrete with main reinforcement in both directions. The use of ribs in the design of waffle floors giving spans of 12 metres will be discussed later in this chapter.

The final layout shows the flat plate floor, which is only achievable with in situ concrete. With the plate spanning in both directions spans of around 8 metres will give slab thickness of around 300 mm for office loadings with a fire rating of 1 hour and will require internal columns of around 350 mm square.

While each of these layouts in Figure 25.3 is shown as rectangular, irregular grid layouts and even curves are possible. Irregular grid dimensions may be used even though the basic layout remains rectangular. Here the loss of economy in repetition will be limited and the risk of errors may rise. Non-rectangular layouts require more structural analysis and non-standard detailing that will increase fabrication costs but may be necessary on restricted and constrained sites. Curves have always been possible with in situ concrete, where curved edge forms are relatively easy to construct, but recently the process of curving standard structural steel sections has made smooth curves in steel frames possible.

Another variation on the frame layout is to incorporate a cantilever at the edge of the building. This is best achieved with continuity in the beam/slabs between the internal spans and the cantilever to avoid a moment connection with the columns. The economy of the cantilever is dependent on the ratio of the internal span to the length of the cantilever. Cantilevers at both ends of an internal beam around a third of the internal span actually give a structural advantage, reducing the bending moment at the centre of the span but producing high bending moments over the supports. The cantilever can be used to project parts of the floor out over the ground floor or perhaps to make an irregular floor edge off a rectangular internal grid. A variety of detailing options is available for both in situ reinforced concrete and structural steel. In the in situ solution the continuity developed naturally in the detailing will be accommodated in the reinforcement. However, the detailing in structural steel often requires additional fabrication, as standard connections are at best only semi-rigid joints. This inevitably raises the cost of the steel frame.

Frame components – stability members

There is a fundamental instability in a frame made of beams and columns (or plates and columns). In the previous section consideration has been given to the plan arrangement of floor spanning to columns. Looking on elevation at the four-sided shape bounded by the columns and spanning structures reveals a susceptibility to racking, as indicated in Figure 25.4. The forces creating this racking are from the wind and from a lateral instability of rotation of the whole frame from the overall vertical loading pattern. There are a number of ways to prevent this racking, the most commonly applied to building skeletal frames are shown in Figure 25.4. They fall into three basic approaches.

The first is to make at least two of the connections rigid, making what is known as a moment frame. The second is to break the four-sided shape into triangles that are stable, and this is known as bracing. This only needs one diagonal but this will be in tension if the frame racks one way, and in compression if it racks the other. As a potentially slender member this has to be designed in compression as a strut to allow for lateral buckling. If two diagonals are provided, they can be designed in tension. With no requirement to design the bracing for buckling the structural member is much smaller, meaning that even though there are two members the overall solution is less expensive. This use of two diagonals is known as tension–tension bracing. Another form of triangulation bracing is known as K bracing because of the shape it makes if the centre connection is made to the column (as illustrated), although the centre

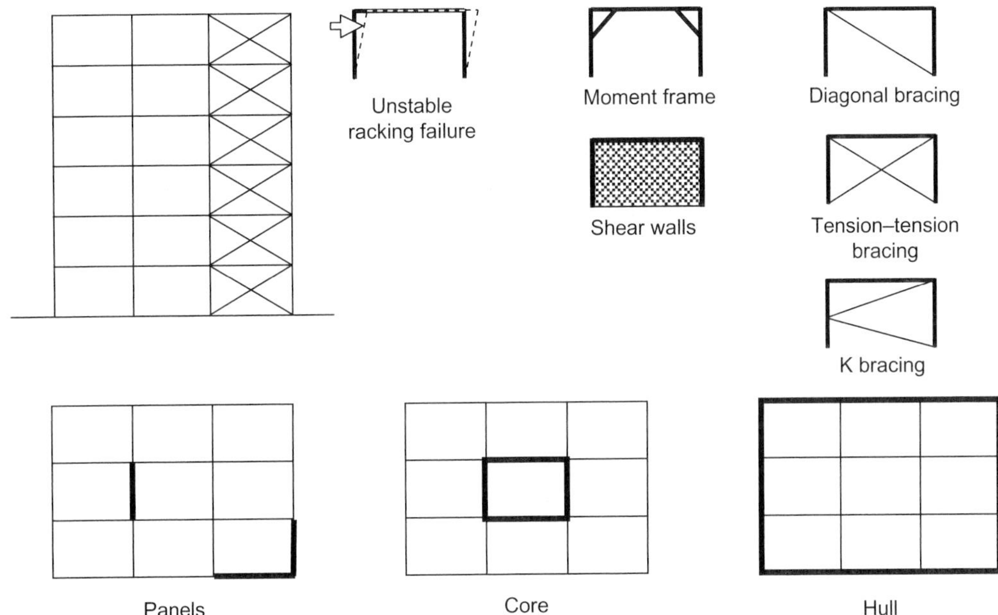

Figure 25.4 Wind-bracing arrangements.

connection can be made to the beam. In this bracing one member will be in compression and one in tension with the forces resolved at the centre connection without inducing bending in frame members. In this bracing both bracing members have to be designed as compression members but are shorter than the single full diagonal brace.

The third approach is to fill the four-sided frame with a wall where, if good contact can be ensured with the sides of the wall and frame members, the natural resistance of the wall to racking through its shear action will hold the frame stable. In many cases the frame members are replaced with a wall to take the racking forces. These solutions are known as shear walls.

Associated with the choice of the basic approach to stabilising individual frames is the need to consider an overall strategy to stabilise the whole building frame. Just stabilising one individual frame will not stabilise the whole building but neither is it necessary to stabilise all the individual frames. It is necessary to choose vertical panels of the frame to stabilise against racking for the full height of the building. This panel will support the adjacent individual frames but only in the same plane. There has to be another vertical panel in another plane, at right angles, to make the building stable to take wind forces from all directions. There is then a need for at least one more panel to resist a lateral rotational action on the whole frame from the vertical dead weight of the building and any wind forces acting at a distance from the stabilised panel that will also have a rotational effect. This third panel must not be in the same line as either of the first two or their action will meet in a point around which the rotation can take place. In many buildings three panels will not be enough as the effect of each panel reduces with distance from the stabilised area. In an elongated building stability will be required towards each end to ensure overall stability.

This pattern of panels is shown in Figure 25.4, which also shows how this can be developed by bringing the panels together either to the core of the building, often stair or lift shafts, or to the face (or hull) of the building, making the external walls the stabilising element. In the UK the option of the core is a common solution.

The choice of the basic approach to resisting frame racking (moment frame, bracing or shear walls) will depend on the frame material and the approach to the overall stability of the building (panel, core or hull). Common options include panel bracing (often tension–tension) for steel frames, and concrete shear walls for in situ reinforced concrete frames. Concrete core walls or steel-braced cores can be used for all frame construction. Hull solutions are not common in the UK but could include moment frames or an exposed diagonal-braced facade.

Fire and building structure

Another aspect that has to be considered in the choice of a frame, its specification and detailing is maintaining the performance of the frame in a fire. In this respect concrete and steel are very different. While both have reduced strength at elevated temperatures, steel heats up faster and can, within the duration of a building fire, lose sufficient strength to yield under normal loading conditions. In reinforced concrete the approach is different. Most structural members are sufficiently large (normally thick) for the bulk of the material not to heat up during the fire. The failure is most likely to be the loss of integrity of the surface of the concrete, known as spalling, which exposes the steel, causing bond to be lost and allowing the steel to heat up, both of which could cause the section to fail. Spalling will occur due to differential thermal changes in the matrix of the concrete that is a function of the design of the concrete, mainly the choice of the aggregate. In addition the steel reinforcement has to have a sufficient thickness of concrete covering the steel (specified as cover) to limit the maximum temperature of the steel during the fire even if the concrete surface does not spall.

Ensuring fire performance in structural steel sections has always been more problematic. Exposed steel is vulnerable in a fire and the approach can be to encase the steel in heat-absorbing or -insulating materials to protect it. More recently thin coatings known as intumescences that expand in heat to provide an insulating layer have been developed. These can be applied directly to the steel and even sprayed on the steel before delivery to site. Other approaches to modern designs include a structural design approach known as fire engineering and/or the installation of sprinklers (a building services provision) that are now being used to provide safe solutions to fire performance. These aspects will be dealt with in more detail later in the chapter when steel frames are discussed.

The development of frames

The history of the evolution of skeletal frames in both structural steel and reinforced concrete since the beginning of the twentieth century is fascinating in that it illustrates the dynamics of change in technical development. This is not only in the development of technical knowledge to be able to design, specify and construct with confidence to provide safe structure but also in the influence of the commercial, economic and even political environments in which technical development takes place. This is, however, beyond the scope of this book, in which the process of choice for technical solutions in the UK at the beginning of the twenty-first century is being considered.

Clearly the basic structural principles that were employed to design the first skeletal frames in concrete and steel remain the same, although design methods have changed. However, it is our understanding that the materials and changes in manufacturing and assembly know-how have changed the technical detailing and specification over the years and still continue to do so.

In this text material properties related to structural behaviour and production methods will be used to show the development of some basic forms of the components, their cross-sectional shapes and size range for some common forms of frames as they would be specified and detailed at the beginning of the twenty-first century in the UK.

Reinforced concrete – performance

Concrete has no structurally significant tensile strength and this will need to be provided by a reinforcing material if it is to be used as the elements of a frame. The reinforcing material has to provide the tensile resistance in structural elements that develop bending or shear forces, each of which has a significant tensile component. Reinforced concrete is, therefore, a composite of two materials that need to have complementary properties which, when acting together, provide the required performance characteristics of the single structural component. To act together there has to be a bond between the two materials so that any strain induced by the application of external loads, developing stress in the concrete, induces a strain in the embedded reinforcing material developing the stress associated with resisting the loading conditions. The most common material used for reinforcement in concrete is steel. Originally all the reinforcement would have been mild steel formed into plain round section bars but the development of high-yield steels allowed more economic sections, so all main structural bars would now be specified as high-yield ribbed bars. The ribbing gives greater bond strength to match the additional forces that may be applied to obtain the full potential of the higher-yield stresses. Mild steel is still used for reinforcement for shear and distribution bars.

As indicated above, concrete and steel only have the potential as a composite to provide a working section so long as there is a bond between the concrete and the steel. Here the property of the concrete that is significant is the shrinkage on curing that, when acting locally around the steel, provides the potential for the bond strength. This means that the steel cannot be too close to the surface of the concrete, it has to have cover, and this becomes part of the specification. Bond failure would be the reinforcement pulling out of the concrete before the working tensile stress in the steel is reached, reducing the ultimate failure load of the whole structural element. This will happen at the end of the bar and therefore requires detailing to ensure the full anchorage bond length

Now the type of bar and the strength of the concrete determine the resistance offered to the pull-out force that represents a bond failure, and this leads to certain detailing requirements. There is a need to consider any situation where the steel bar ends. At edges and bearings the steel has to be taken right to the end of the member possibly with a turned end, known as a bob. If continuity of members is required, bars have to be lapped to ensure full bond strength throughout the structural member. If a lap is not desirable, a physical coupling has to be used to maintain a full bond stress across the connection. It also has implications if bars are curtailed at the point where reducing stresses in a member needs less steel in the section.

Another approach to overcoming the limitations of the tensile strength of concrete is to pre-stress the concrete with a compressive force in the potential tensile zones so that when the loads are applied they take out this pre-stress, leaving only compressive forces in the concrete. This is normally achieved with a tensile force applied to steel wires within the concrete. A bond between the steel and concrete is now only required if the steel is pre-tensioned (tension applied to steel and concrete cast around wire which is released when the concrete is cured). Pre-tensioned systems are normally used for factory-produced precast concrete components. In situ pre-stressing will usually use post-tensioning, where the tension is not applied to the steel until after the concrete has cured, using the hardened concrete to anchor the wires and provide the jacking resistance to apply the pre-stress. In this case there must not be a bond between the steel and the concrete, although in some systems the wires are grouted in after the pre-stress has been applied to enhance durability. In the common forms of the building frame pre-stressing can be usefully applied to slabs but will not be covered further in this text. The analysis that follows refers only to reinforced concrete detailing.

Reinforcing areas of tensile stress

Reinforcement is required to resist the tensile forces within the structural member. These tensile forces are identified by the structural analysis taking into account the loading, support and jointing arrangement for each structural member (see Chapter 11). Steel also has significant compressive strength and so will contribute to resisting compressive forces if introduced into compressive areas. It is, however, the analysis of the tensile forces that gives the characteristic reinforcement patterns of cages for columns and beams, and mats for slabs and walls, to be discussed in more detail below.

In addition to the tensile forces developed in structural sections due to loading, support and jointing arrangements, the curing process of concrete is accompanied by a shrinkage, useful in developing the bond strength, but will also induce tensile forces across the section that can cause cracking. In most reinforced sections associated with skeletal frames this cracking is most likely at the surface, as the structural reinforcement will resist the shrinkage forces across the whole section. There are, however, design requirements for minimum areas of reinforcement to take into account shrinkage-induced tensile forces. The most vulnerable elements are the slab and wall sections. In large single pours the shrinkage forces can be sufficient to crack right through the section. This is certainly the case and the major design condition for large area ground-supported industrial floors where bending is not the major structural action. These floors are discussed in Chapter 28.

One other condition where tensile forces develop is during fire. Differential temperatures created by the fire as the surface is heated but the body of the section remains relatively cool can induce differential expansion in the surface and near-surface concrete which induces shears that could disrupt the concrete causing the spalling that exposes the steel to greater heating potential from the fire.

Balanced sections

Reinforced concrete elements are designed to be balanced sections. As loads are applied the section will deform where the strains in the concrete and the steel are linked as they are bonded together. In a balanced section both materials will be at their maximum safe working stress at the design condition. This gives an economic section, with both materials being used to their full potential. In practice it is not desirable to achieve the completely balanced section. It is important to ensure that, should the section fail, the imbalance will result in a failure in the steel. The reason for this lies in the different failure mode of each material. Reference needs to be made to the stress–strain curves for each material (see Chapter 11). Concrete is a brittle failure, while steel yields, creating a failure that becomes visually evident before actual collapse.

Material compatibility over time

Having ensured that the basic structural failure modes due to loading have been checked, there are other ways in which the materials in a composite have to be compatible. These are associated with the analysis of the behaviour of the structure over time. There must not be any chemical reactions between the two materials that would degrade one or the other of the materials or destroy the bond. This analysis is necessary as part of the evaluation of the durability of the structure.

The steel is potentially at risk from corrosion that, as a surface phenomenon, is a potential threat to the bond. However, when embedded in concrete steel gains protection from the alkaline environment of the free lime created in the hydration reaction of the cement. Corrosion is inhibited at the high pH values created in concretes made with Ordinary Portland Cement (OPC). While pH values are high in freshly cured concrete, they do reduce with time when exposed to the air through the process of carbonation. Initially this happens at the surface of the concrete but air will permeate into the concrete, increasing the depth of the carbonated layer. If this reaches the steel, the corrosion protection will be lost and the potential for bond loss is increased. Protection from carbonation is controlled by the specification of the depth from surface to reinforcement, known as cover, and controlling the air permeability of the concrete. Cover for durability is normally greater than that required to achieve bond strength and therefore this will determine the specification, but cover is also significant in fire resistance (as discussed above), and this may determine the specification of cover in some circumstances.

There is another chemical interaction associated with the high pH that is a threat to reinforced concrete but is not a reaction between the concrete and steel but is within the concrete matrix. It is a reaction between the cement and the aggregate. For the concrete to remain a homogenous material there has to be another bond; the one between the cement and the aggregate. Some aggregates have a surface reaction with the cement's high pH that destroys that bond and therefore disrupts the concrete, reducing its strength. This is an alkali silica reaction but is sometimes known colloquially as concrete cancer. The high pH that can protect the steel and its bond with the concrete may destroy the bond between the cement and some aggregates.

Another potential incompatibility that will cause a premature failure over time is differential thermal movement between the two materials, sufficient to break the bond and destroy the composite action. The coefficient of thermal expansion for concrete and steel is sufficiently close for the materials to be used as a safe composite.

Cost and sustainability

The discussion of the success of the composite has so far concentrated on the technical aspects of the two materials. However, success is also determined by the cost implication. In this respect concrete and steel make a good combination. Steel, while more expensive than concrete, is significantly stronger. The relative stress–strain characteristics give much smaller areas of steel than concrete, limiting the materials cost of even relatively bulky structural sections. The materials are not the only cost of a complete frame. There is a significant component of cost associated with the production process, and this is discussed in the next section. In commercial terms, however, the cost of the materials and production together keeps the overall component price for reinforced concrete competitive.

There may also be concerns with the sustainability of these materials. Both steel and cement require high-energy inputs in their production and removal of earth-based resources from quarries and the use of water during manufacture, all of which will have an environmental impact. They then require transport from quarry to processing plant and then to site.

Both materials are recyclable. Steel, after reprocessing, can provide a source of equivalent grade steel, while concrete can be crushed to provide a fill or aggregate for further use. This is, however, dependent on the ability to separate them at demolition, and again techniques have been developed to do this. Concrete can save energy during the life of the building if used in a passive environmental scheme as thermal mass. Concrete can also be produced with cement substitutes such as ground-granulated blast furnace slag (GGBS) or pulverised fuel ash (PFA), which are themselves waste products, although they do need some processing to be used in the concrete. These considerations are leading to new specifications for the materials and detailing of the components.

In situ reinforced concrete frames – production

The above sections have introduced the major properties that have to be considered in the selection of the materials for reinforced concrete frames for them to perform in operational conditions. They included strength, durability and fire characteristics, all of which show that combinations of concrete and steel have considerable potential to be used in the construction of structural frame elements. Concrete and steel, however, have very different production characteristics.

Having a good understanding of the production process is as important as understanding the materials and structural behaviour if an economic form is to be developed. Much of the potential cost and construction time will be established in the specification and detailing of the frame members in so far as they determine the production opportunities particularly associated with temporary works. The major aspect of temporary works associated with casting the concrete is the formwork and falsework, but the specification of the concrete for handling, workability and early strength can also influence the choice of plant and machinery.

Concrete is produced at ambient temperatures by mixing aggregates with cement and water. The water allows the whole mixture to flow, to be poured and moulded, known as its workability. It also starts the hardening reaction that limits the time available for moulding but will ultimately provide the strength required under operational loading conditions. In both specification and production the water/cement ratio is important, as water is needed for workability but needs to be limited for ultimate strength development and permeability that affects durability. Water/cement ratios that give useful structural strength produce a relatively stiff mix that has to be compacted with vibration to move the concrete to fill the mould and expel the air, to ensure the final density that is required to give the full potential of the design strength of the concrete. It is possible to increase the workability of the concrete without changing the water/cement ratio (or reduce the water to increase strength without loss of workability) with the use of additives known as plasticisers and superplasticisers. These add to the cost but can be helpful if compacting is difficult due to the shape of the mould or congestion of the reinforcement.

This basic production operation for the concrete of mixing and moulding at ambient temperatures makes it possible to use concrete as a site-based, in situ material. It also implies a mould normally referred to as formwork. For in situ work this formwork has to be held in the final position by falsework, which taken together are the temporary works that account for a significant proportion of the cost of producing a reinforced concrete frame.

It is also possible to precast the structural elements and then lift the pre-formed elements into position. This will change the requirements for formwork and falsework. While precasting saves falsework, it incurs transport costs if casting is undertaken in a factory rather than on site, plus arrangements for lifting into position on site. Precast concrete frames will be introduced later in this chapter.

The production processes involved with in situ and precast not only change the temporary works and transport arrangements but also fundamentally affect the nature of the structural connections, as indicated at the beginning of this chapter. The joints necessary for in situ reinforced concrete are known as day joints. While there are some structural reasons associated with the position of day joints (principally avoiding areas of high shear), they are fundamentally required for production reasons. While there are some concreting processes that are continuous (e.g. slip forming for core walls), the majority of concreting operations have a cycle where there is a need to complete the concreting of a section by the end of the day (hence the name, day joints). This requires decisions to be taken on where it is both economic in production terms as well as structurally desirable to form these day joints. The economics will be based on balancing the resources while maintaining the design assumptions that the structural members are continuous. If precast members are used, it is usual to cast each component in one pour, eliminating the need for day joints. The joints between the precast components are the physical connections between individual members; slab to beam, beam to column, etc., formed when the components are assembled on site. They have to transmit the loads as well as provide the basis for the assembly process. The simple, pragmatic jointing process is unlikely to provide continuity unless some in situ work is carried out around the joints to provide this action.

Another consequence of a process using moulds is that it dictates the economic cross-sectional shapes of the structural members. Generally simple rectangular shapes are the easiest and most economical to form. Columns will normally be square, although circular column formwork is available at reasonable cost. If beams are cast, they will normally be rectangular in cross section. Slabs can be shaped on plan but will normally be a constant thickness across the whole floor. This allows the formwork and falsework systems to be simplified, increasing the opportunities for reuse without much modification, to give increased economy and speed in the construction. The more the form of the frame is simplified, the greater the opportunities to use the same temporary works on many sites, encouraging manufacturers to provide systems that, while initially expensive, are fast and easy to erect and strike with multiple reuse, reducing the cost of the temporary works for each pour. It is possible to make formwork for special shapes particularly in precasting but, unless the formwork can be used many times, there will be a cost penalty over and above the cost of the concrete and reinforcing steel required for the component or element of construction.

Sequence, activities and operations

There is a basic set of operations associated with three activities – formwork, reinforcement and concreting – that need to be carried out in a particular sequence for the production of all concrete components and elements. This is shown in Figure 25.5.

Formwork

Formwork (and falsework) will be made from components, and like all component systems, temporary or permanent, the more repetitive in dimensions and construction detailing, the

Activity	Operation	Sequence						
Formwork	Make	■						
	Erect		■					
	Strike						■	
Reinforcement	Cut and bend	■						
	Fabricate		■					
	Fix			■				
Concreting	Mix, transport, place and compact				■			
	Cure					■	■	■

Figure 25.5 Operations in producing of reinforced concrete.

more economical they become. In the case of temporary works systems, the simplicity of connection between components for both assembly and disassembly is also significant. It should be quick and achieved with simple tools. In addition, the components will be subject to handling many times and the striking process needs to be considered as it takes place after the concrete has been cast and has hardened, often requiring some force to take it apart. This means that components must be robust or they may become damaged. Traditionally these components have been made from timber. Components now often use plywood or other facing materials for the form, and metal components not only for connections and fixings but also for the main structural components, often using aluminium for lightness in handling. Falsework and formwork systems are now incorporating temporary working platforms (traditionally provided by scaffolding) and other health and safety features such as attachment points for harnesses.

It is necessary to use the formwork and falsework several times to achieve an economic solution, so striking (dismantling) has to be possible with minimum damage to the components. This is another reason why the design of simple cross-sectional shapes for the structural members is important. It also requires careful consideration of the sequence of erection to ensure it can be reversed when the concrete has hardened. However, even if the sequence of striking is possible there is still a tendency of the concrete to adhere to the formwork surface. It is necessary to treat the surface with a release agent before every pour. These are sometimes referred to as mould oils, although many are not now oils but emulsions or chemical release agents. For some shaped plastic moulds, such as those used in trough and waffle floors, provision is made to break the adhesion with compressed air.

Two other key decisions associated with the economics and reuse of formwork are striking times and keeping as much of the assembly in one piece for striking and re-erection. Striking times refer to how long after casting before the falsework can be removed and how long before the formwork can be removed. These are not necessarily the same time. This will be discussed later when considering the curing and early strength control of the concrete. Maintaining the size of the assembly reduces not only cycle times but also the incidence of damage and subsequent repair that can be necessary, particularly as the number of reuses increases.

Formwork and falsework have traditionally been prepared in timber, now most often with plywood face and metal adjustable props for support. More recently, proprietary systems have become available with standard components, making robust, easily erected and dismantled assemblies with high reuse potential.

The sections that follow introduce the detailing and the production process bringing together the understanding of both the performance and production issues for each basic structural element for the superstructure.

Walls with traditional panel formwork

The wall will commonly start with the forming of a kicker. The kicker is formed in concrete of the same specification as the main wall. Its purpose is to locate the formwork in the correct line and hold the formwork in position during casting. It has a major part to play in dimensional and tolerance control.

It is usual to fix the reinforcement before the formwork, although in some cases where access to one side of the wall is limited, for example basement or core walls, one side of the formwork will be erected first. Reinforcement rods will have been cast protruding from the foundation or slab from which the wall is to be built. These starter bars are required to develop bond strength to provide continuity between frame members. The main reinforcement, normally the vertical bars, will be tied to the starter bars and then the horizontal distribution bars will be tied to the vertical bars to keep them in position, resist shrinkage and take the shear forces that develop in the wall. This gives the characteristic mat of steel on both faces of the wall.

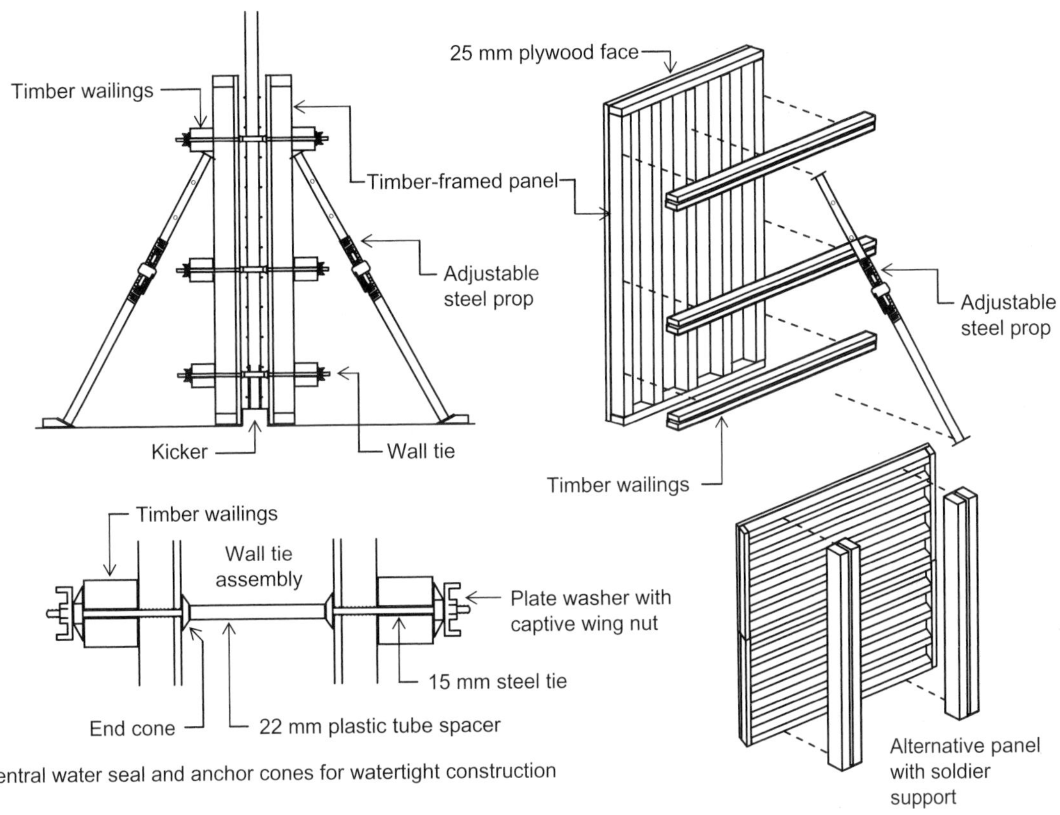

Figure 25.6 Traditional timber panel wall formwork.

The formwork will be based on panels. The traditional approach to making these panels on site from timber and ply is shown in Figure 25.6. The timbers framing the formwork must be close enough together to ensure the ply does not bow under the hydraulic pressure of the wet concrete, typically at 400mm centres for a 25mm thick plywood. The sizes of the timbers framing up the panel are normally 100 × 50mm, and these determine how often the horizontal wailings have to be positioned, as these timbers have to span between the wailings. The wailings are made from a pair of timbers normally 100 × 75mm at 900mm centres. The wailings span between ties at around 900mm centres that, on a two-sided form, receive their support from the wailing on the other side. The concrete pressure will put the tie into tension, but there must be some device to hold the formwork panels apart. One solution to the tie is shown in Figure 25.6, which also shows the use of vertical soldiers as an alternative to horizontal wailings. The timber framing of the panels now has to be arranged horizontally. Soldiers are used on tall pours as they stiffen the horizontal joint between panels. The timber sizes and centres given are for a rate of pour less than around 1metre an hour. This gives the concrete time to stiffen from the initial set that relieves the full hydraulic force from the formwork. Faster rates of pour will require larger sections and/or closer centres of wailings, particularly at the bottom of the formwork.

It is important to realise that the positioning of the formwork sets the dimensional accuracy of the frame. The props must not only be adjustable to set the position within the appropriate tolerances but must also hold the formwork in position during concreting. Production operational loads are dynamic and the wet concrete exerts hydraulic pressures on vertical formwork surfaces. In addition to the structural and dimensional control aspects there are also demands for health and safety requirements. Fixing reinforcement and placing concrete will require working at a height requiring a working platform.

With walls the formwork will promote curing and can help protect from frost, where in extreme cases it can be insulated. However, for economic reasons it is required to remove the formwork for reuse and this makes consideration of striking times important. With all striking times the concrete has to have achieved sufficient strength to support itself and limit the risk of damage to the surface of the concrete when the formwork is removed. For walls the main loading risk is from wind loading, although care must also be taken not to move the wall when removing formwork.

The climatic conditions are one of the main factors in determining striking times with the mean air temperature having a considerable influence. Air temperature not only identifies potential frost conditions but also has a significant effect on the chemical reaction that determines the rate of strength gain in the concrete. Added to air temperature, wind not only affects curing conditions but also adds load on walls that are vulnerable in the early days after casting, so formwork may have to be left up as indicated above. These climatic conditions make striking times very variable. Added to this variability is the specified 28-day strength of the concrete, where stronger concretes with higher cement contents can gain sufficient strength for striking earlier. To illustrate this variability, striking vertical formwork to OPC structural concrete such as walls, columns and beam sides with the mean air temperature around 15 °C can take place in around 18 hours, whereas, if the mean air temperature is only around 5 °C, this may take up to around 36 hours but could be even more if high winds are expected within 24 hours after striking, unless repropping is considered. Longer striking times may also be necessary for high-quality finishes.

Slabs with proprietary formwork systems

For a frame using any flat soffit slab such as plate floors (flat slab) the formwork and falsework is more likely to be a proprietary system. While there may be some parts of a frame where the dimensions will require making up pieces of formwork in a more traditional way, the basic components of the formwork will be manufactured with the formwork and falsework integrated in one system and assembled on site. These systems are designed to minimise the time taken to erect and strike by limiting the amount of dismantling required between strike and re-erect for the next pour. Once assembled for the first pour, either the entire, or major parts of the, assembly will be lifted by crane to the new position. An example of this proprietary approach to formwork is shown in Figure 25.7. It can be used for both flat slab (plate floors) and for trough or waffle floors where the soffit of the beam and the ribs are the same. Unlike walls, the reinforcement for these slabs will be fixed after the falsework and basic soffit form has been erected. The form becomes the working platform from which steel fixing and concreting will be carried out.

In a plate floor the reinforcement will be mats in both the top and bottom of the slab with greater concentration of steel in strips that run in line with the columns in both directions. At the head of the columns additional reinforcement is required to take the punching shears that develop in the slab at the top of the column. Because of the concentration of reinforcement from the bending steel, the shear reinforcement will often be prefabricated as heads, ladders or strips of studs. In this plate construction, where the proprietary formwork can be erected and repositioned relatively quickly, the steel fixing becomes a major proportion of the time taken in the concreting cycle. This has led to looking at prefabricating the reinforcement mats rather than fixing individual bars on the formwork, which is the usual practice.

For trough or waffle floors the main reinforcement is associated with the beams and the ribs. There will be a mat of reinforcement in the

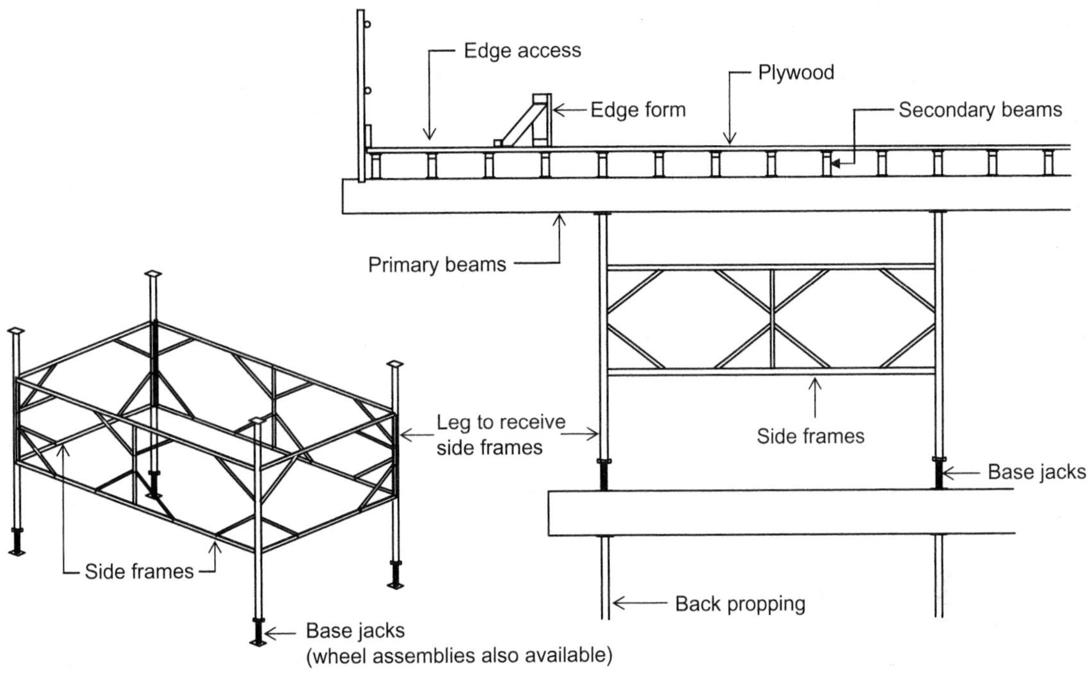

Figure 25.7 Proprietary slab falsework and formwork.

top of the slab but this will be relatively light as it is there to control shrinkage cracking and fireproofing. The main steel will be in the beam cages and steel bars in the bottom of the ribs. There will be a need for a form to create the troughs. This could be a void-filler such as lightweight concrete blocks or a mould that can be removed and reused on the next pour.

The falsework has to have adjustment to level the formwork (actually set to a camber so the loaded slab deflects so it is level) and has to be stable during steel fixing and concreting operations. It has to be easy to strike with minimum damage. It is also an advantage to keep sections assembled to be lifted to the next floor. This is the basis of the so-called flying forms, where the props and side frames, beams and formwork remain assembled and the base wheel allows the unit to be moved to the edge where the crane, fitted with a special lifting frame, can remove the falsework unit from under the newly formed slab and place it on top of the slab ready for the next floor cycle.

The time from concreting to striking the formwork is very significant in the cycle to get the maximum reuse from the temporary works to limit the amount of formwork and falsework components on site. Striking is dependent on the strength of the concrete as it cures, which is affected by a number of factors such as grade of concrete and the mean air temperature as described in the section on walls above. When the formwork is struck from the soffit of a slab, it will have to support itself and the construction loads from subsequent operations, including the support of the formwork for the next floor. To help release the formwork and most of the falsework components as soon as possible it is usual practice to leave props under the slab while the work above continues. Props may have to be left in for the construction of the floor not only immediately above but also for up to three slabs above. This is known as back propping. For the first few floors of a multi-storey building the load is taken through the back propping to the ground (that does not deflect significantly). However, on higher floors back propping shares loads on the new slab with up to three slabs below that have had longer to cure and can therefore safely take more load.

Striking times for the soffits to slabs (and beams) with props left under for slabs made with normal OPC structural concrete are around four days where the mean air temperature is around 15 °C, going up to around eight days where the mean air temperature is around 5 °C. For some projects it can be economic to take steps to reduce these striking times, and these can include insulating the slab in cold weather, increasing cement content or using rapid-hardening cements. It may also be worthwhile to monitor the actual strength development so that formwork removal can be determined by the actual strength, giving the earliest possible safe striking time.

The proprietary temporary works systems are likely to include access platforms and health and safety provision such as edge protection and even enclosure elements to protect from the worst of the weather. For concreting this is not just cold and wet but also hot and dry, particularly a drying wind where protection is afforded to both the operatives and the curing concrete.

Columns and beams

Walls have been used to introduce traditional formwork and falsework systems and slabs to introduce proprietary temporary works systems. There are, however, proprietary systems for walls, and slabs can be formed with traditional ply forms supported on timbers held in position with adjustable metal props with bracing from tubular scaffolding. Beams and columns can also be formed in traditional timber or be part of proprietary systems. Columns have many similarities with walls and beams, which require a similar approach to slabs.

Columns experience vertical compression, which induces shears that require the presence of reinforcement across the column section. This cage is often prefabricated on site (horizontal on stands at waist height) and lifted into position, where it is rested on the kicker and tied to the starter bars the same as a wall. By making the cage a storey high the top of one cage makes the starter bars for the next column. The formwork panels can then be held in position with cramps that form a ring to hold the panels against the concrete pressure, as shown in Figure 25.8. Props are then used to adjust the formwork. The kicker holds the form on line; the adjustable props are used for plumb. Level is controlled by filling the form to the correct level during the concreting operations, so forms are often a little taller than the column and then filled to a predetermined level. Column forms (like walls) have to be designed to a limiting rate of pouring and have similar striking times.

Beams are, by definition, in bending so there is a need for main reinforcement to take the tensile bending forces and reinforcement to take the tensile component of the shear forces. This gives the characteristic beam cage with the main reinforcement being at the bottom at the centre of the span and in the top where the beam is continuous over the support, normally the column. These cages are often prefabricated, sometimes on temporary supports on the soffit formwork as a working platform immediately above its final position but at waist height so that the steel fixer can stand and work with a good posture. When the cage is complete, the weight is held by a crane while the supports are removed, at which point it can be lowered into position. Beam cages can be prefabricated on the site and then craned into position with the continuity bars loose to be fixed once the beams are in position.

The beams soffit formwork has to be supported much the same as a slab. If the beam downstands from the slab, the formwork for this soffit has to be lower and, being narrow, the stability of the falsework will require additional attention. There is then a need to form the side of the beam either to the underside of the slab for internal beams or to the top of the slab for beams at the edge of the building. This is illustrated in the detail of an edge beam shown in Figure 25.9.

For upstand beams (or short walls) the soffit formwork can be flat, as described for flat slab

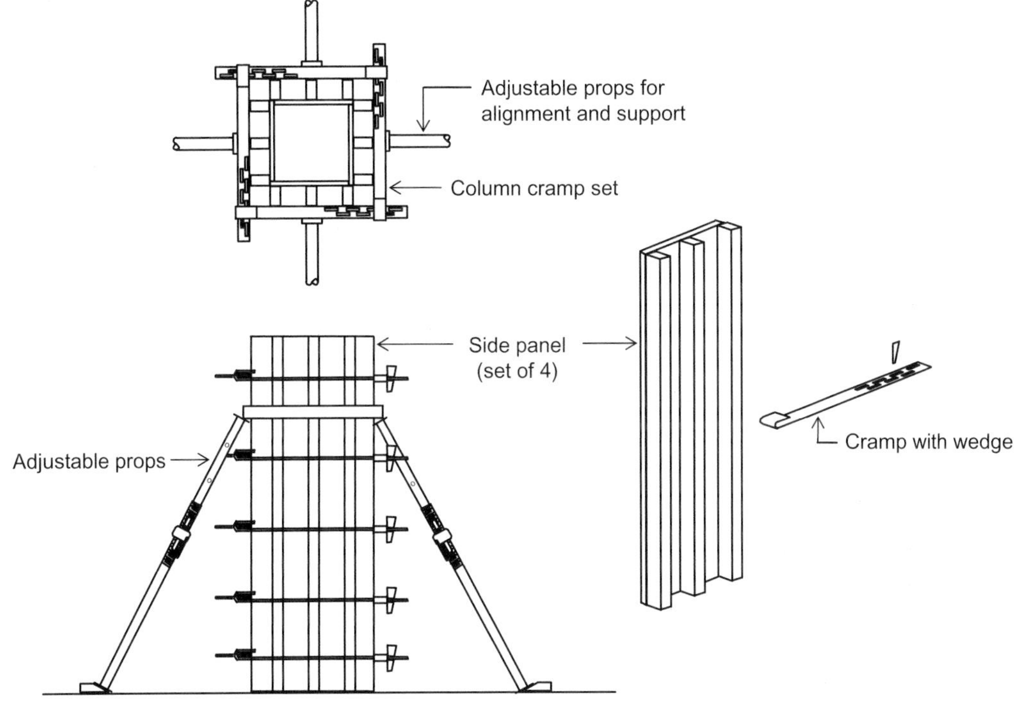

Figure 25.8 Traditional timber panel column formwork.

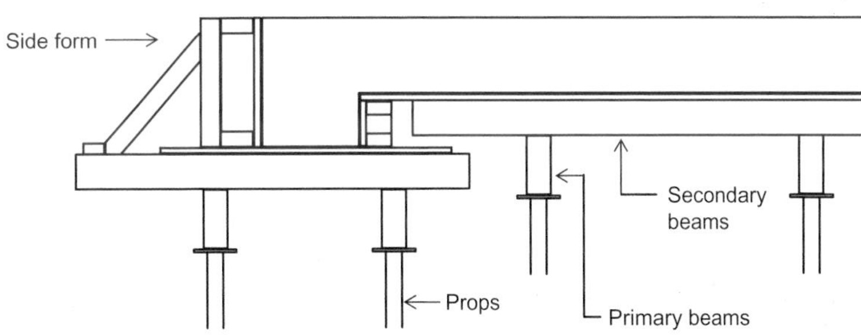

Figure 25.9 Edge beam formwork.

floors, and the part of the beam (or wall) above the slab can be cast separately, much the same as walls, subject to the same structural conditions. Although upstands are more economic as a production solution, there is a structural-efficiency penalty, as the slab cannot be used as part of the compression resistance at mid-span. Further, it is only really an option for edge beams, as a raised final floor finish would have to be constructed over the upstand were this solution to be used within the building. Striking times for beams sides and soffits are similar to walls and slabs given above.

Reinforcement

The most usual way to provide the reinforcement introduced above is in the form of bars of steel cut and bent to shapes ready to be fixed together as mats for floors and walls, and cages for beams and columns. Detailing of reinforcement has to take this process into account. Bars

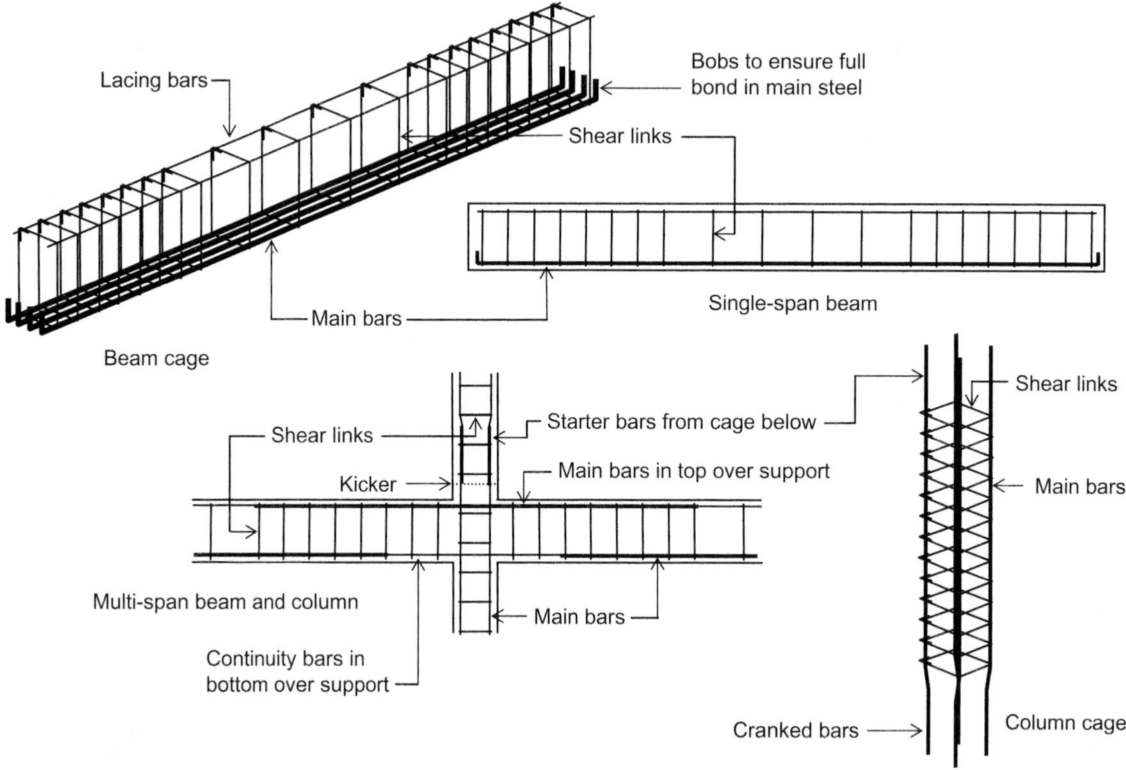

Figure 25.10 Reinforcement cages for beams and columns.

are normally cut and bent in a factory to standard bar code shapes that have been agreed with designers and are published in codes. This not only allows the designers and reinforcement suppliers (and the steel fixers) to communicate in a standard way but also allows the suppliers to organise the increasingly mechanised aspects of the process in an efficient and economic manner. It is, however, possible to cut and bend reinforcement by hand with very simple bench-mounted cropping and forming tools. This allows site production of a few bars where modification or additions are required, to maintain site production without having to order and wait for deliveries.

Individual bars are then tied together to form the mats and cages. With mats for walls and slabs this is normally done in position. However, it is possible in some circumstances to have these mats prefabricated, although this may increase the cost but will save time. This is particularly helpful in flat slab designs where reinforcement fixing can become a major part of the cycle time as formwork and concreting times reduce but reinforcement becomes more complex. Cages are often assembled on the site but adjacent to the final position and then craned into position, where a limited number of bars, normally those providing the continuity, have to be fixed. Figure 25.10 shows the nature of bars and cages for beams and columns, while Figure 25.11 shows the nature of mats for walls, which would be similar to the appearance of slab reinforcement. In mats the main steel will be in the directions of bending. For instance, in two-way spanning slabs, used in flat slab constructions, there will be main steel in both directions. When bending is in only one direction, the steel in the other direction is known as distribution steel and will be lighter than the main steel.

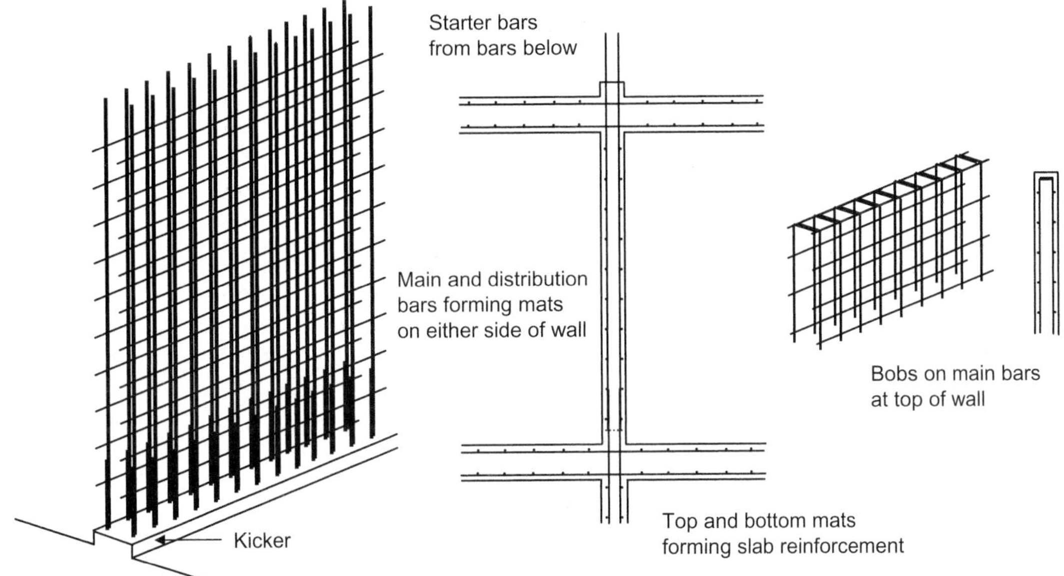

Figure 25.11 Reinforcement mats for walls and slabs.

Cover, the distance that the reinforcement has to be from the face of the concrete, has been identified as a significant aspect of the specification. It is important to achieving bond, limiting the risk of corrosion on the steel and achieving fire resistance, as discussed above. Controlling this dimension during the concreting process is vital. This is achieved using spacers, often preformed plastic components that clip onto the bar and rest on the face of the formwork. Examples of spacers are given in Figure 25.12. It is important that they be sufficiently closely spaced to limit deflection of the bars and have the strength to support the production loading (particularly in slabs where the reinforcement will be walked on during concreting operations). It may also be important that they do not show on the surface and do not form a pathway for air and water to reach the reinforcement, which can cause premature corrosion. It is important to select the correct spacer for each circumstance.

In slabs the specification of cover is also to the top surface and this will not have formwork, so the spacer is not appropriate. In this case it is usual to maintain the dimension between the top and bottom mat using 'chairs' made from light reinforcing rods or fabricated light steel spacers. Examples of these are also given in Figure 25.12. These again have to support the individuals carrying out the concreting operations without collapse, excessive deflection or damage to the top mat of reinforcement.

While some surface pitting from corrosion on the reinforcement is not detrimental to the formation of bond, loose rust scale is. Care also needs to be taken to ensure that oil or grease is not present on the reinforcement before concreting, if the structural bond is to be achieved. When both formwork and reinforcement are complete, the formwork should be cleared of production debris, often by blowing out with compressed air.

Concreting

The process of concreting involves a number of operations: mixing, transporting, placing and compacting, all of which have to be completed in the period where the fresh concrete remains plastic. This time is dependent on many things, but for normal structural concrete mixes with

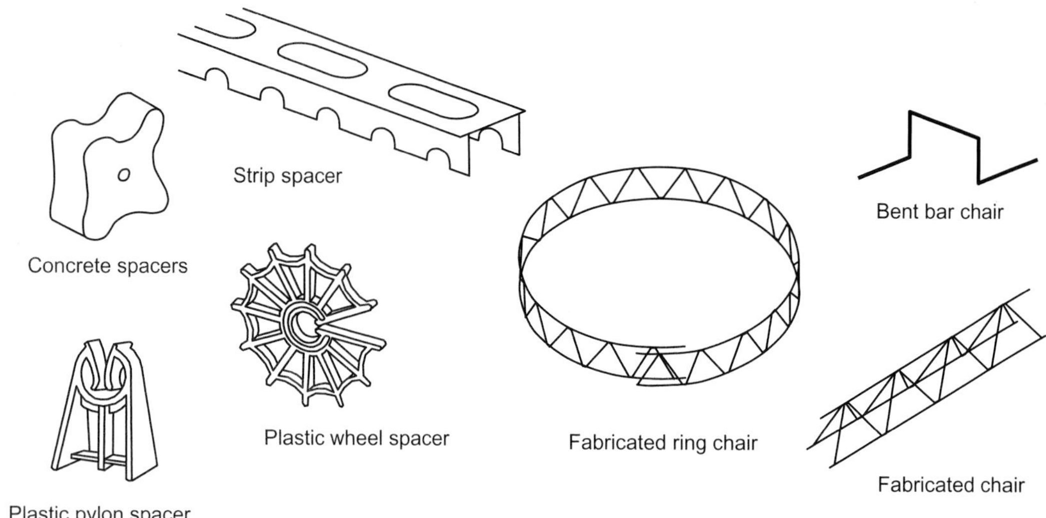

Figure 25.12 Examples of spacers and chairs.

OPC at normally ambient temperatures (around 20 °C) the recommendations are 1 hour without agitation after the mixing and 2 hours if the mix continues to be agitated until transferred into the formwork. This is the case with ready-mixed concrete truck-delivered to site where the drum revolves while on the road and on site until discharge commences. These relatively short times demand a significant level of organisation and contingency for breakdown in plant and machinery.

Transporting and placing will vary greatly depending on the site configuration. The amount of horizontal and vertical movement and the quantity and rate of delivery are perhaps the greatest determinants of the choice of plant, as introduced in Chapter 13. Compacting is a significant operation in concreting as it removes air in the mix, increasing the final density of the concrete, which is directly related to the final strength. It also ensures that the moulds are fully filled to give good sharp corners and surface finish. Surface finish not only is important for appearance but also determines the permeability of the surface to both water and air, which is significant to durability and the potential premature corrosion of the reinforcement. The compacting process will also increase the risk of a loss of grout (cement, water and in some cases very fine aggregate) through joints in the formwork. This loss will leave the surface of the concrete locally depleted of this fraction of the mix, leading to surface blemishes know as honeycombing. These areas are clearly a threat to the premature corrosion of the reinforcement. Interestingly the removal of water without the cement and fine fractions of aggregate will significantly reduce permeability and increase durability, particularly in aggressive environments. This is achieved with a technique known as controlled permeability formwork. All this indicates the importance of designing the formwork with tight joints to allow full compaction. While the consequences of under-compaction are well established, over-compaction is not normally a danger in a well-designed mix.

Once compaction and finishing are complete, the concrete is left to cure. Concrete takes time to develop its working strength, and this period is known as the curing period. The concrete is particularly vulnerable in the first few days following casting. The condition for curing in this initial period affects both the time taken to achieve its final strength and even the ability to reach this potential strength at all. Perhaps the most destructive condition is frost, where if, despite the heat of hydration, ice crystals form in the newly cast concrete they will reduce its strength and even cause complete disruption, particularly towards the surface. Even if frost action does not take place, low temperatures slow the chemical reactions, delaying the achievement of working strength and extending the curing period.

A more general threat to the achievement of working strength is premature drying, which can occur at any ambient temperature and is aggravated by wind, where evaporation at the surface is enhanced. Premature drying is the removal of water from the mix during the hydration period leaving insufficient water to complete the chemical reaction that will develop the full working strength. This is a particular problem on large open surfaces of freshly poured concrete such as slabs and is a risk on walls when the formwork is removed. Traditionally the concrete would have been covered with wet sacking or covered with polythene to create a still, high-humidity environment at the surface of the concrete. More usual now is the spraying of a curing agent to seal the surface to limit the rate of evaporation.

While there is an initial set in the first couple of hours that affects the workability of the concrete for compacting, the development of final working strength takes days. The strength of concrete is specified by its 28-day strength, but this does not mean that the formwork has to be left on the concrete for 28 days. This does, however, indicate that knowing the rate of development of the strength of the concrete is important for striking times for the formwork and the complete removal of falsework. There are many ways of speeding up the development of early strength. This will require quality-control measures to give estimates of this early strength to allow the safe removal of formwork at the earliest possible opportunity. This is significant as it reduces the whole concreting cycle, releasing formwork for the next pour to minimise the amount of formwork and falsework on site and maximise the reuse factors. Striking times for each of the structural components have been discussed above.

There is a danger that the removal of formwork may be seen as the end of the cycle of concreting operations. However, this cycle of casting and removal of formwork even in the case of floors can be achieved in days, long before curing to the 28-day specified strength. Consideration has, therefore, to be given to concrete sections right up to the 28 days. For quality-control purposes the concrete cannot be fully accepted until the 28-day strength is confirmed. It is usual to make a series of concrete cubes at the same time as casting to be tested for compressive strength at 7 and 28 days. The 7-day test gives confidence that the concrete mix has the potential to reach the specified 28-day strength. This in turn gives the confidence to continue building in the expectation that, given good site curing practice, the 28-day strength will be achieved. Receiving certification of the 28-day strength marks the end of the whole process for the structural members cast with that day's mix.

Precast concrete frames – production

Precast frames differ from in situ in both the production process and the structural behaviour of the joints. The components of columns, beams, slabs and wind bracing (often bracing is made from steel) are all made at ground level, often in a factory with all the implications this has for good dimensional control and potential quality of finish. It also opens the opportunity to cast the external components integral with the cladding, including the finish. Reinforcement will be mats and cages similar to in situ detailing but contained within the components, possibly with short exposed ends to form joints. The site process focuses on lifting

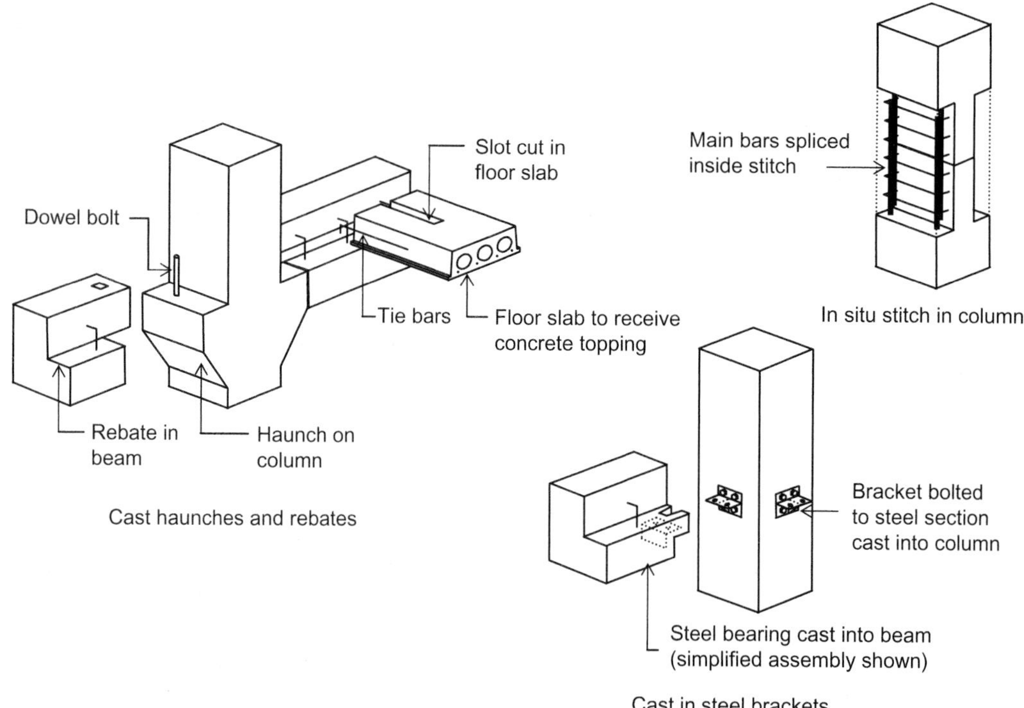

Figure 25.13 Precast concrete frames.

into position and connecting components. Now the dimensional control of the final frame is associated with the adjustment and tolerances in the connection design. This also determines the structural behaviour of the whole frame.

Connections can be achieved in a number of ways, as illustrated in Figure 25.13. The casting process can be used to produce haunches and rebates to provide bearing for subsequent components. The connection then normally includes locations pins that are grouted or concreted in when the frame is finally accepted for line, level and plumb. With the bearing surfaces being concrete to concrete, these connections often need protrusion or additions to the beams and columns, making the sections perhaps larger than required for structural purposes and adding complexity to the casting moulds. This can be simplified by designing the connection to have bearing surfaces in steel by incorporating brackets either cast into the sections or bolted on to the completed component. This detailing will provide simply supported connections. Some continuity can be achieved by either casting the components with exposed reinforcement that can be connected on site and then stitched together with concrete or with steel connectors that are bolted together and then cast into the elements, again on site, to provide the full connection.

Structural steel – performance

Slabs for structural steel frames will almost certainly be in reinforced concrete and require a similar understanding to that explained above but will be detailed differently. For an in situ solution there is no need for falsework, as the steel beams can be used to support the formwork, normally permanent metal decking, and (for structural efficiency) shear connections may be introduced to provide composite action between beam and slab. This detail, along with an option for a precast reinforced concrete slab, will be given later in this chapter.

The major components that give this frame option its name are the beams and columns. Unlike reinforced concrete, structural steel components can be made from the one material as steel has significant strength in compression and in tension. It is indeed significantly stronger than concrete, having compression strength around ten times greater, which means that less material is required for similar bending and shear forces. It is also heavier (denser) and more expensive than concrete. The overall size of the structural sections is dictated by the structural behaviour of the components as well as the material used. For beams the bending resistance is largely dependent on the depth of the beam, as this determines the separation (lever arm) of the material resisting the compression forces and the material resisting the tensile forces that provides the moment of resistance. The depth of a beam is also a major factor limiting deflection. For columns, the ratio of the width to the height (slenderness) will determine the buckling behaviour.

This use of a strong, heavy and expensive material makes a solid rectangular cross section of these structural members inappropriate. In a beam with a rectangular cross section the material towards the neutral axis experiences less stress from the bending forces but experiences increasing shear forces. For a typical load and span combination for a commercial building the shear forces are less than the bending forces. Two cross-sectional shapes can be considered to reduce the material content by concentrating the material where the greater forces are generated by the bending action. These are the 'box' and the 'I' section. Which of these is appropriate will depend to some extent on the shaping and fabrication process available for the material.

Steel can be formed into shapes when it is hot and can be cut and connected by various means when cold. When hot, the steel can be formed into wires, bars, sheets, tubes, angles, channels and 'I' sections. When cold, any of these forms can be cut, bent, drilled, machined, screwed, bolted or welded to form working components and assemblies. The hot processes have limits to the size of sections produced, while the cold manufacture of components and the fabrication of assemblies are limited by working, handling and transport considerations.

An analysis of typical building span–load combinations shows that the bending and shear force in a beam can be economically accommodated in an 'I' section, and these are now the common solution for structural steel frames. Part of this economy is achieved by limiting the range of sizes as these have to be hot-rolled. It is not possible to specify any size of beam, but a choice has to be made from those that are manufactured. Initially the rolling process was limited to small sections that were not sufficient to make whole building frames but were useful to construct joisted floors in commercial buildings that had heavy loading. These were known as rolled steel joists (RSJs) but now rolling can achieve larger sections known as universal beams (UBs), and it is these that are used in structural frames. The shape and range of sizes for UBs is given in Figure 25.14. The steel is concentrated in the flanges at the top and bottom of the beam where the bending stresses are greatest and where the full strength of the material can be realised. There is considerably less of the heavy, expensive material where the bending stresses are less towards the neutral axis. However, the shear forces are greatest at the natural axis and therefore there has to be sufficient material in the web to take the shear forces. The shear action will put both compressive and tensile forces in the web. The tensile forces will tend to tear the steel, but a more likely failure is the compressive forces buckling the slender web.

While the rolling process making a range of standard sections is part of the economy, it is still worth making savings in the amount of steel in each section. Known as serial sizing, the rolling process is controlled to vary the thickness of the web and flanges of the beam section. This changes the strength and the cost but also the weight of the beam and this fact is used in the specification. The full specification of a beam section is the width by the depth and a

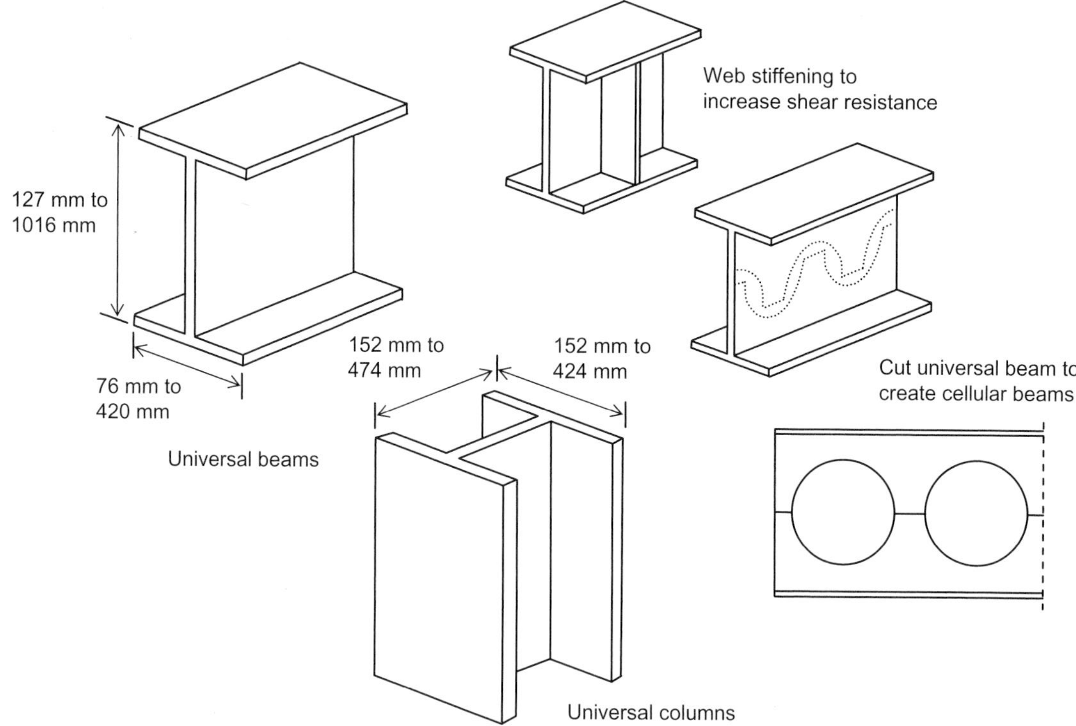

Figure 25.14 Structural steel sections.

third figure, the weight per metre in kilograms. Each size has a number of serial weights.

These serial sizes have been developed for the normal span and loading conditions of a typical commercial building to suit the potential failures in bending, shear and deflection. Short, heavily loaded beams experience comparatively higher shears and long, lightly loaded beams comparatively higher deflection. Under these conditions UB sections may have to be modified with web stiffening for shear and by forming cellular (or perhaps castellated) beams to give greater depth with the same weight of section for deflection. These are also shown in Figure 25.14.

Similar analysis is required to look for economic column sections, but the structural behaviour is different. In columns the induced bending from buckling rather than the direct bending in the beam becomes the driver for the overall section dimensions. Relatively small areas of this strong material can resist typical loads on columns. If these were provided by a square cross-sectional shape similar to reinforced concrete, the overall width of the column would be very small compared to the height for a normal storey-height building. Buckling in such a slender column would limit the load, thus not gaining advantage from the full strength of the steel. Again, the high cost of the material makes a hot-rolled, serial-sized section worth producing, but with the possibility of buckling in either axis a squarer shape is most efficient. This gives the characteristic 'H' section, where the material is more evenly distributed across the web and the flanges. Known as universal columns (UCs), the shape and range of sizes is given in Figure 25.14.

Wind-stability members may be provided as walls or bracing. Moment connections can be used but these change the beam–column connection detailing and are expensive compared to walls or bracing in most situations. Walls can be reinforced concrete or in some cases blockwork. These walls will normally be associated with the circulation cores but steel-braced cores are also a possible solution. While the reinforced concrete walls will have their own continuity, use of blockwork will rely on infilling between the steel columns and beams and the continuity at this connection is vital, requiring careful detailing and production control. For these reasons blockwork is not now often used.

Bracing with steel members keeps the jointing and fabrication process common and can be detailed to gain good structural continuity. Each of the three bracing forms (diagonal, cross and K) will need different sections, depending mainly on the need to allow for compression forces. Diagonal and K bracing both have to be designed to resist compressive forces so are susceptible to buckling. While loads are not great, the bracing is long, so potentially slender, and will therefore need a greater cross-sectional overall size, as previously explained. Here, while in the past a variant of the 'I' section, the angle, was used, now it is a variant of the 'box' section in the form of a tube that is most likely to be specified. Cross or tension–tension bracing can be formed with rods, or more likely flat sections where simple gusset connections can be used.

The structural members identified provide the strength and stability required of a building frame. Performance with time raises the two issues, durability and behaviour in the event of a fire, both of which have been discussed earlier in the chapter. These will need additional consideration to provide protection or some additional engineering that will be dealt with in the detailing later in this chapter.

Structural steel – fabrication and construction

So far the performance of steel has been considered and the basic structural sections from which the component of the frame can be made have been identified. It has been established that the basic structural sections have to be formed while hot in a rolling mill, an industrial process associated with the steel industry. It has also been implied that this is a costly process in both financial and energy terms, and so much thought has been put into the form and nature of the sections. The next stage is to work on the sections while cold: cutting to length, plating and drilling to form the connections. This is again not a site operation and equipment and handling facilities required make this part of the factory-based process normally known as steel fabrication. This opens up possibilities for a high level of automation, including the introduction of CNC (computer numeric control) machines with the possibility of this being directly linked to CAD/CAM (computer-aided drafting/computer-assisted manufacture) compatible systems to link production to the design drawings.

The steel will require protection from both corrosion and fire. Some specifications will involve applying a protective coating prior to delivery to site. For exposed conditions this may involve galvanising for corrosion protection. While many fire-protection systems involve site operations, intumescent coatings can be factory-applied, although this requires greater care in handling and carries with it the risk of damage during assembly.

Assembly is an almost entirely site-based procedure. It is not sensible for components of the main frame to be transported connected together, although it may be an option for the braced cores. Site operations involve lifting and making connections, normally by bolting, although site welding can be specified but is not usual on straightforward commercial skeletal frames.

This need to separate factory fabrication from site erection with its need for transport, site handling and safe working practices has led to common connection detaining. This normally involves plating the end of the sections to provide a flat surface for bolting to the next component. This arrangement can be seen in Figure 25.15 for the secondary beam–main beam, the main beam–column and the tie/secondary beam–column connections. Structurally these connections act as semi-rigid joints. They

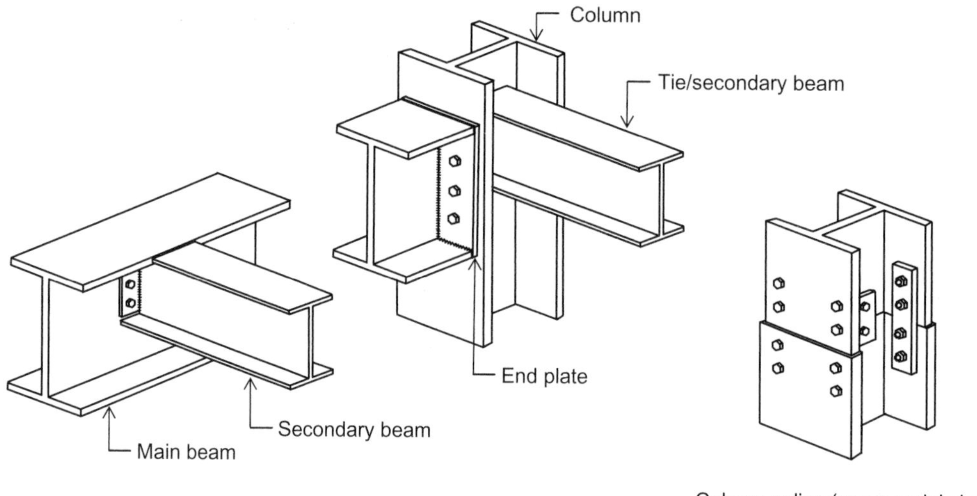

Figure 25.15 Structural steel bolted connections.

have some degree of fixity but cannot be taken to be rigid joints, so wind bracing is still necessary. Figure 25.15 also shows a column–column connection. The illustration shows the splicing arrangement for two columns in the same serial size. It shows the splice inside the flange as this is the constant dimension within a serial size. Columns of the same section size are often plated on the outside of the flange and columns of two different serial sizes will need to have end plates and the smaller packed to allow the splice plate again to be fixed outside the flange.

Factory fabrication determines most of the potential dimensional control of the frame. However, the initial position of the columns on line level and plumb are determined by the connection to the foundation. The foundations are most likely to be formed in concrete either as shallow foundations such as pads or pile caps if deep foundations are chosen for the site. Whatever the foundation, the dimensional deviations that can be achieved in the casting of the concrete foundation will be greater than those required in the erecting of the frame. The concrete–steel connection is shown in Figure 25.16. To provide adjustment to ensure dimensional control of the whole frame the holding down bolts must be able to move to give the adjustment for line, and the concrete cast low to allow packing on shims to adjust for level and temporary wedging to set plumb. When finally positioned, the connection is grouted to hold the whole connection in place.

It is normal to erect all the steelwork components from foundations to the roof in one operation. Stabilised sections with the wind bracing have to be constructed in advance or temporary bracing will be necessary. The use of braced cores for the circulation areas of the building, including the staircases, provides not only this safe erecting sequence but also access to the higher levels as the floor construction proceeds.

There have emerged two common forms for providing floors for structural steel frames, one using in situ concrete and one using precast concrete planks. These are shown in Figure 25.17. The in situ option uses a metal decking that will take the tensile forces in the floor and act as permanent formwork. If spans are limited to around 3 metres, this decking will be strong enough to take the construction loads and the slabs will be relatively thin, at around 130 mm for normal office loadings and fire requirements. These limited spans will require secondary beam layouts but can be used with shear studs to gain composite action between beam

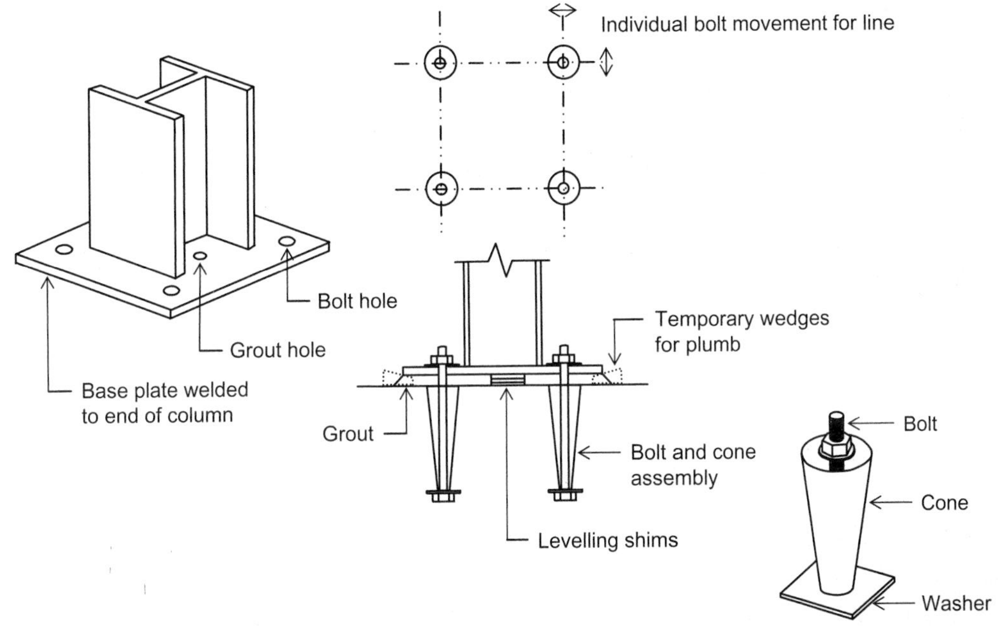

Figure 25.16 Column base connections.

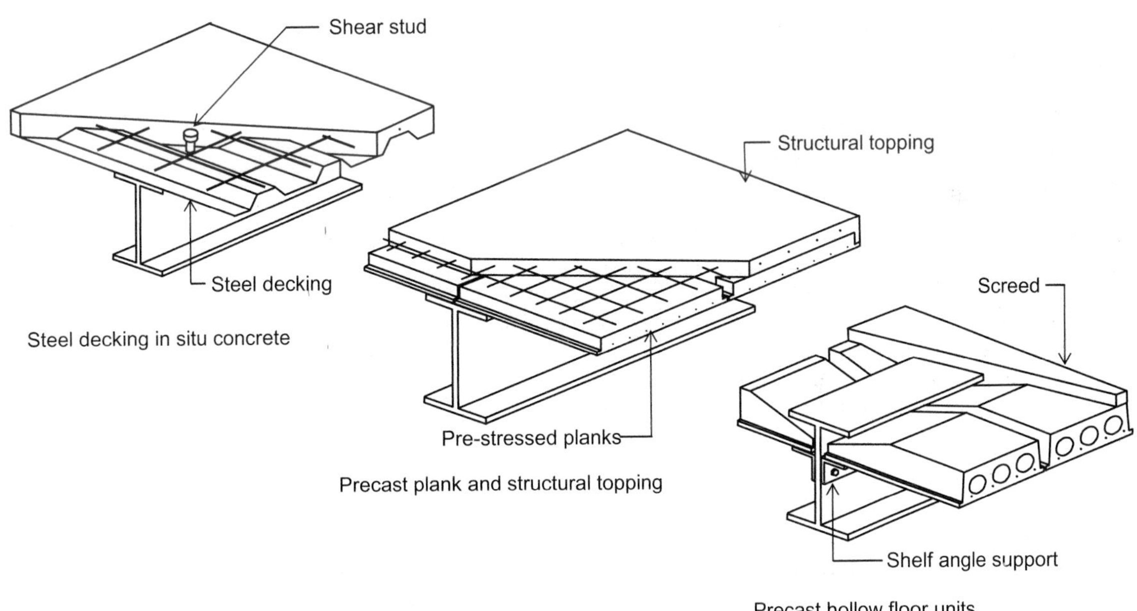

Figure 25.17 Floor slabs for steel frames.

and slab providing economy in reducing the size of the beams. Larger spans can be achieved with this construction, reducing the number of beams, but this may require temporary propping to the metal decking and will require thicker slabs, but this may provide a more economic solution for some buildings.

The economic combination of precast planks and beams are with spans for the planks of around 6 to 8 metres. While this gives larger main beams, there is no requirement for secondary beams and these spans give column spacing that can be accommodated in most commercial buildings. The simplest connection is to rest the planks on the top flange of the beam, but this can create deep overall floor zones. These can be reduced by providing shelf angles bolted to the web of the beam to provide the bearing for the plank so that the top of the plank coincides with the top of the beam. Detailing with continuity reinforcement and structural toppings can give continuity in the slab and some structural action between slabs and beam can be developed.

The frame is now fully stable and has floors, but unless it has been fire-engineered or intumescent coatings have been factory-applied, the frame is not yet fire-proofed and will need protection. Originally the common solution would have been to encase the steel in concrete but this is now only normally used for situations where damage is likely to occur to the less robust materials normally used for fire protection. Brick- and blockwork can also be used, and this may be economical for columns where the steel can be built into walls. More common forms of protection are to box in the steel or spray the steel directly with heat-absorbing or -insulating materials. The form of each of these along with alternative materials is shown in Figure 25.18. Boxing in creates a rectangular profile. Mineral fibre bats are not very robust and are susceptible to damage. They can be plastered but are often used above suspended ceilings, where damage and appearance are less problematic. Vermiculite boards and plasterboard are more robust but again could be plastered for a more substantial finish. The joints between boards need to be formed with rebates or internal cover strips to ensure there is not a straight joint between the boards. The pre-formed casings are made for a range of steel section sizes in either a board material or an insulated pressed sheet

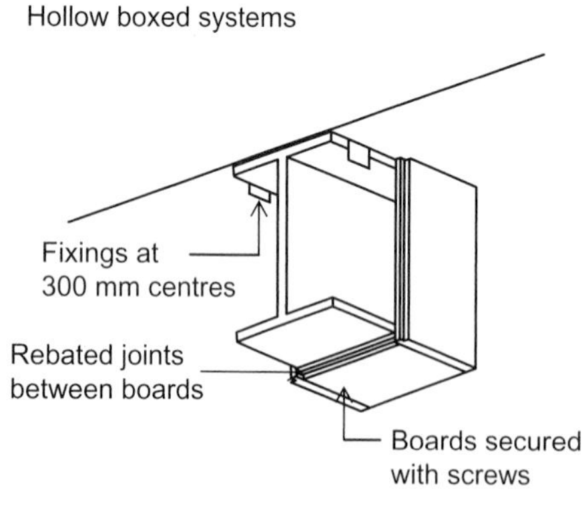

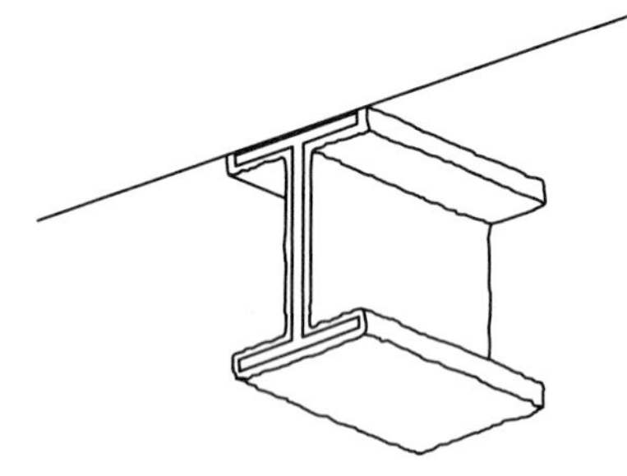

Figure 25.18 Fire-proofing for steel frames.

metal finish with interlocking joints fixed with screws to metal straps. Plaster on metal lathing provides a good surface ready for decoration. The thick sprayed coatings give an irregular surface that is relatively easy to damage. While less expensive than the hollow protections, they are a wet site process and will need to be hidden as they are not suitable for furnishings and so are more suited to beams than columns. The thin coatings, however, are themselves a paint finish and can therefore be used for either exposed sections. As indicated above these coatings can be sprayed onto the sections before delivery to site, although care must be taken not to damage the surface and the need to touch up the paint should be allowed for in the production programme.

It is only when the frame is fully protected that a true comparison between construction cost and time can be made between structural steel and the in situ or precast reinforced concrete frame. Steel is often assumed to be quicker when observing the erection of the beams and columns but slabs and fire-proofing and even stairs may still have to be completed. These operations will determine when following operations can commence, normally associated with activities concerning the facade and services. For many buildings, when these factors are taken into account, the time and cost may not be significantly different. However, some designs will favour one of the frame options. Column layouts, spans and floor plate shape will all affect the economics directly. Decisions on frame material will include factors other than direct time and cost. Market factors and risk will have to be taken into account along with integration into the overall building design and the proposed technical solutions for other elements of the construction.

Structure choice and overall design

When considering a potential structural solution for a building, it is important to see it as part of the overall building design. In addition to integrating the structure into the overall technical strategies and with the technical solutions for other elements of the construction, the development of the frame solution will have to be in harmony with the overall design of the building. At the most practical level the positions of columns will have to be compatible with room layouts. These will inevitably condition the structural grid. This grid is normally based on the column (or wall) centres which are normally repetitive to gain the best economy for the frame.

Some spaces in buildings can be considered less likely to change over the life of the building. Areas unlikely to change include circulation cores for staircases and lifts plus any shafts for the vertical distribution of services. Even toilets and other wet areas are perhaps less likely to change. The permanence of other spaces will depend on use and the requirements of the client's brief, which will also determine the acceptability of column centres and even wind-stability elements for the frame.

The elevation of the building will interact with the structural layout. The regularity of the structural grid will have to be accommodated in any scale or rhythm associated with the elevational treatment in windows, change of materials or articulation of the face of building. The elevational treatment will affect the provision of panels for wind stability, which is more difficult to achieve if openings are present in the external wall.

As soon as some overall design concept is emerging it is possible to start to identify the sort of structural grid that will facilitate the design. This will in turn point to potential solutions for the frame.

This early identification and analysis of possible structural solutions needs a sound overall understanding of the performance and production of the technologies involved with some ability to establish approximate sizes to judge whether the final detailed design can achieve the design concept. In this respect the selection of the structure is no different from the choice of any other part of the building.

Integration with other building elements

The relationship of the frame with the external enclosing elements, particularly walls, of the building is predominantly one of detailing the physical connection. Here load transfer is associated not only with the support of the wall, often with eccentric edge loadings, but also with the transfer of wind loads. There is also a need to allow for inherent deviations, with potential subsequent relative movements and production-induced deviations associated with the achievable tolerances, in the production and erection of the frame. Other considerations in the detailing of the connection are the environmental concerns such as thermal cold bridging, which may prove difficult to achieve at the detailing stage.

The structural slabs can provide many of the environmental requirements for floors, such as soundproofing and fire resistance for compartments, with very little additional detailing. The floor can also provide the thermal mass for night purging in buildings based on passive environmental thermal design. Other internal enclosing elements such as partitions will impose loads that may influence the frame choice and detailing based on load distribution patterns.

The distribution systems for the services often compete for space in the floor zones, particularly with beams, as do the finishes, both floor above and ceiling below. There will most probably be a need to support services components from the frame but these are often relatively light, although some items of services plant can impose substantial loads.

Environmental concerns associated with sustainable development have had a significant influence on strategies to provide comfortable and healthy environments within buildings. Over the recent past providing such an environment has been achieved with building services often requiring a considerable input of energy obtained from the burning of fossil fuels, now accepted as a major contribution to climate change. Passive design strategies for creating these internal environments involve using the fabric of the building, including the structure, combined with building services systems. Perhaps most easily exemplified by the use of thermal mass in the floors, this is discussed more fully in Chapter 15.

Summary

1. Concrete and steel as strong, quality-controllable materials have allowed the development of the structural skeletal frame at a useful scale for commercial buildings.
2. Skeletal frames are made up of columns and beams with slabs and wind-stability members, although in situ reinforced concrete has allowed the development of plate floors where the horizontally spanning member is a slab without the need for beams.
3. There are limits to the economic spans for beams and slabs that give column centres and frame layouts that can be applied usefully to a variety of building forms, including high-rise structures.
4. In addition to the structural requirements of a frame, durability and fire resistance are major performance considerations.
5. Reinforced concrete is a composite, normally with steel bar reinforcement taking the tensile forces and the concrete the compressive forces.
6. The compatibility of the properties of the concrete and the steel allows for economic designs exploiting the full potential of both materials, and these determine the structural design and the production process.

7. The production process involves formwork and falsework as temporary works into which the reinforcement as mats and cages has to be placed prior to the pouring of the wet concrete. The temporary works can be removed as the concrete cures and develops its full strength.
8. A major element of the cost is the temporary works particularly with in situ construction. Simplification of the sectional shape to simplify formwork and falsework will reduce both construction time and overall costs.
9. Steel has significant strength in both compression and tension and can therefore be used as a single material for beams and columns. It is, however, expensive and the shape of the structural section is significant to concentrate the steel where stresses are greatest, giving the characteristic 'I' and 'H' sections.
10. Structural steel sections are hot-rolled in a rolling mill and structural members fabricated in a workshop prior to being delivered to site to be erected and bolted together to form the frame. The major cost is in the steel sections and fabrication.
11. The slabs for structural steel frames are most likely to be reinforced concrete, either in situ composite construction or precast planks.
12. The choice between frame materials will depend on the overall structural layout determined by the building design and on the choices made on the other elements of construction.

26 Roof Structures

This chapter looks at the options for providing a roof structure over large uninterrupted areas. Originally developed for industrial buildings this is now a requirement of many other commercial uses. There is a considerable variety of forms for these structures. The plane frames are the most economic and widely adopted form, with the two-way spanning flat and curved structures being used for special roof structures often being left exposed. In this chapter these forms are first introduced as broad options that are followed by some construction details for the most widely used forms of lattice girders, portal frame and two-way spanning flat space frames.

Introduction

Many commercial activities today require large areas to be covered with little or no internal interruption from columns. The skeletal frame does not fit this brief well. Minimising internal columns requires these structures to have long spans. Solutions are likely to be more readily available if the loads can be limited. Imposed loadings on roofs are relatively light and dead loads can be limited with lightweight claddings.

Historically the first commercial activities requiring this type of structure would have been industrial for manufacture or warehousing. In the early days of the Industrial Revolution, buildings fulfilling this type of brief would have been multi-storey constructed with external loadbearing walls and substantial timber beams, joists and trusses. This form of construction structurally limited the width of the building to around 12 metres. The inside would have been just one large space with lighting being provided from windows on both sides. These long narrow buildings would have been constructed near to running water as a power source. This general form continued to develop with the introduction of steam power and the use of cast iron and later wrought iron for the internal structure. These eventually developed into the early form of skeletal frames at the end of the nineteenth century with the introduction of concrete and steel, although early examples of skeletal frames can be identified with cast-iron columns and riveted wrought-iron beams.

While the multi-storey industrial building continued to be developed through the nineteenth century, the single-storey building covering a large area was seen as providing a better working condition with roof lighting (originally north light and later monitor lighting). From around the middle of the nineteenth century these single-storey buildings would still, like the multi-storey construction, have had external loadbearing walls and internal columns. Trusses, initially in timber and then in steel, would have spans limited to around 15 metres. The internal space would still have had columns but this was not a problem with the manufacturing methods of that time. It was the introduction in the late 1920s of mass production with flow line methods characterised by the car industry, and the assembly of large machines such as aircraft (shipyards had a longer history of long-span roofs), that demanded the clear internal space that is seen as the major demand today.

The ability to respond to this true demand for long-span structures to roof over large uninterrupted areas was made possible by the increasing availability of steel and the experience from bridges. The early American car plants were designed and constructed with lattice girders with spans up to 100 plus metres.

In more recent times manufacturing in the UK has declined and, while warehousing for distribution centres is still required, the demand for long-span structures has moved to retail and leisure. This move from industrial to direct use by the public has led to a new and more diverse aesthetic. This is supported with larger budgets, particularly for finishes, and has often been integrated with the form and expression of the structure.

There is now a great variety of forms of these roof structures based on bending, compression and tension structures in both slab and grid forms. Many have been tried and a few have become common forms. This chapter will introduce the range but only consider in any detail the more common forms.

Variety of structural form

There have been many ways to present the variety of this structural form. In this text the basic approach is given in Table 26.1. The first types are the plane frames, followed by the two-way spanning flat structures then two-way spanning curved (and folded) structures. The main forms of each are indicated, together with the principal materials from which they can be constructed.

Most of these spanning structures will need a column or mast system to support them above the usable space they are covering. Arches, some shells and domes can spring from the ground as their structural action comes from their upward curved form that creates space under the rise of the structure.

The column as a component has been established in the analysis of the skeletal frame. However, the mast as a means of support is particular to long-span structures. Masts rise above the roof line to hold suspension cables to support the spanning structure. They are commonly associated with nets and fabric structure but can be employed to support beams, girders and even grids and space frames externally to limit spans without the need for internal columns. Another form of support that limits spans is the tree or umbrella support system, which can also be used to distribute shear forces where the spanning structures connect to the columns.

These roof structures are normally associated with single-storey building but can be employed as the top storey to provide a roof to a multi-storey structure. Many airport terminal designs provide examples of this approach.

Chapter 11 introduced these long-span structures and suggested that long is defined as having a span of over 35 metres, with medium being 15 to 35 metres and therefore short being up to 15 metres. While there is no universal acceptance of this definition and it does not

Table 26.1 Types of long-span roof structures

Structure types	Main forms	Potential materials
Plane frames	Lattice girders	Steel and timber
	Portal frames	Steel, timber and concrete
	Arched ribs	Steel and timber
Two-way spanning flat structures	Flat grids	Steel and timber
	Space frames	Steel and aluminium
Two-way spanning curved structures	Solid shells	Concrete and timber
	Grid shells	Steel and timber
	Suspended nets	Steel
	Fabric structures	PVC and PTFE

work for all forms, it does give clues on the cost and type of structure and the construction that is possible. This chapter will include roof structures that can be used for medium as well as 35 metres plus, in some cases to 100 metres or more, but these are very expensive. More limited spans, even if this means some internal support, will always be more economical.

In terms of the cost of these structures the plane frames are normally the most economic, particularly in their most straightforward structural form, and are therefore perhaps the most common. For many of these long-span structures the additional cost is justified in the dramatic forms these structures can make and are therefore often designed to be seen and become part of the aesthetic of the building.

Broad options

Plane frames

The plane frame provides a two-dimensional structure that when repeated along the length of the building at regular intervals provides support for the roof and cladding.

Lattice girders are pin-jointed frames similar to trusses but take the form of a beam having some depth along the whole section, as illustrated in Figure 26.1. Whereas trusses have ceiling ties and rafters that define the pitch of the roof, girders have top and bottom chords with internal members, known as bracing. The lattice girder allows much greater spans than the truss where spans would be limited to

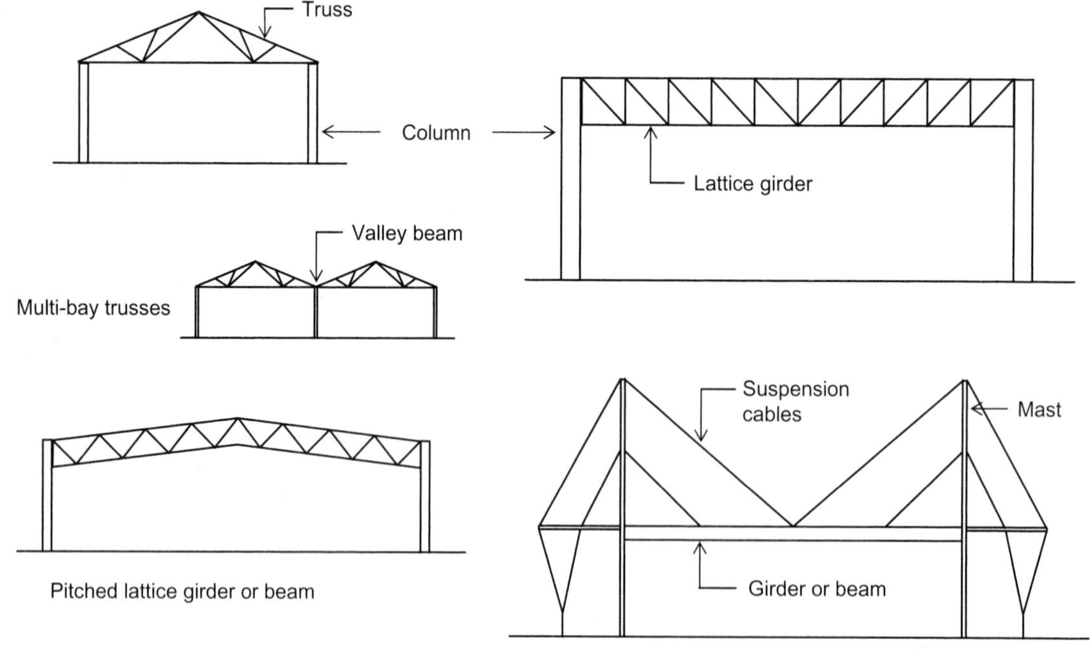

Figure 26.1 Pin-jointed plane frame forms and support.

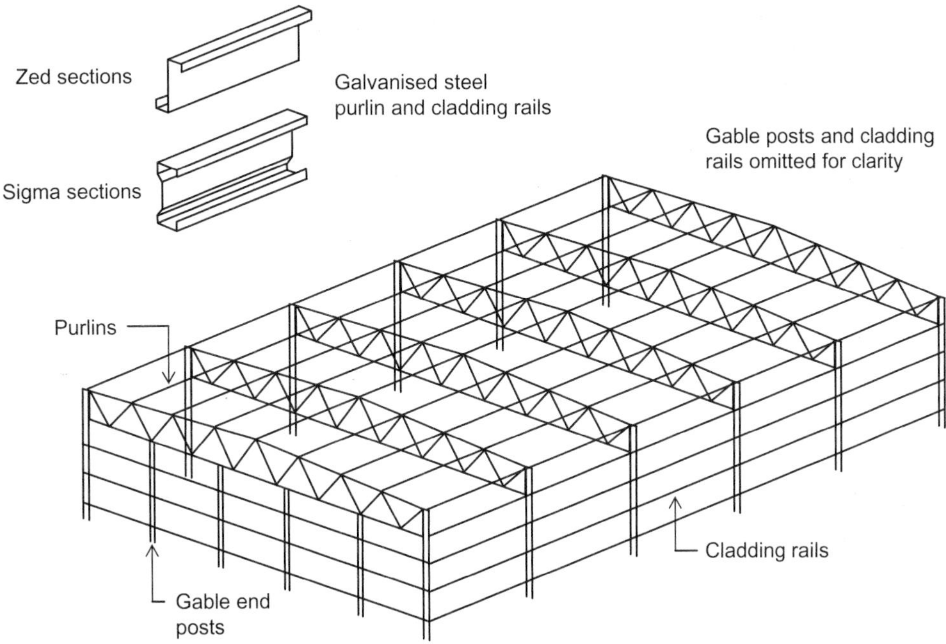

Figure 26.2 Purlins and cladding rails.

around 15 metres. Trusses are therefore not normally classed as long-span structures, although they can produce large areas with multiple spans. The top chord of the lattice girder can have a small pitch to provide drainage for the roof. These spanning forms would normally be supported on columns but can be designed to be suspended from masts.

A characteristic of all these buildings is the use of lightweight claddings and roofing to limit dead loads, although the lower part of the walls is often masonry resting directly on a ground beam. Plane frames will normally be set at between 4- to 8-metre centres and will need additional structural support for this lightweight cladding and roofing. This is normally achieved with purlins for the roof and cladding rails for the walls, although they may be similar structural sections. This arrangement of enclosure support and some common sections is shown in Figure 26.2. As the spans get longer it may be more economical to set the frames at 8 to 12 metres and then use light trusses for the purlins. The external walls then need intermediate columns to take the cladding rails. It is usual practice for all buildings to make the gable end frame with intermediate columns to provide the support for the cladding rails.

These structures are not stable against wind forces and will require bracing. The normal pattern of bracing is shown in Figure 26.3. Bracing is required between the frames to maintain overall stability. This bracing is normally provided at the corners at the gable ends, but in long buildings some internal frames may also need to be braced. To ensure that these braced areas stabilise the whole building it is necessary to stiffen the eaves and gable at the level of the tie or bottom chord with a girder action, to transfer the wind loads to the braced frames. Lattice girders are often deep structures with normal span to depth ratios of around 1:25 and are therefore susceptible to lateral tortional buckling. In most designs there is a reliance on the purlins that carry the roof covering to take axial loads to resist the bending of the top chord. There will then need to be bracing between the

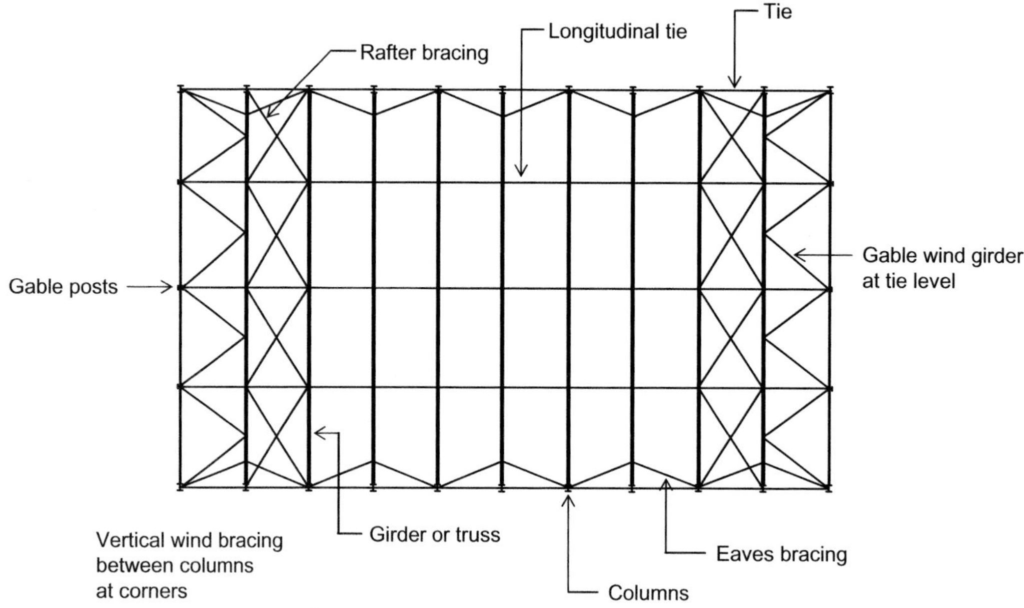

Figure 26.3 Bracing to truss and girder plane frames.

end frames to form a girder action with the rafter or top chords.

There is a wind condition that may cause a failure in these fully stabilised girders. If there is any chance of the wind generating uplift forces that overcome the weight of the structure and the cladding, the girders will experience a reversal of the bending effect. The truss or girders themselves can be constructed with chord and bracing members capable of resisting both compression and tension forces and therefore work under conditions of wind reversal. However, in the reversal of bending the bottom chord is susceptible to lateral tortional buckling but does not receive the support of the purlins and rafter bracing. This lateral instability can be overcome with longitudinal ties at the level of the tie or bottom chord that work in conjunction with the gable end wind girder.

The second type of plane frame is the moment or portal frame. The moment connection between the spanning member (the rafter) and the column reduces the bending forces at the centre of the span. This general form in the three main structural materials is shown in Figure 26.4. The moment connection transfers some of the bending forces into the column. While this knee joint must remain a fully fixed moment connection, the bending forces can be further modified by arranging for pin joints to be provided at the foot of the column (two-pinned frame) or with both the foot of the columns and the centre of the span (crown) to be pinned (three-pinned frame). The introduction of pinned joints will increase the maximum bending forces but in specific circumstances can be used to advantage. The two-pinned portal removes the horizontal thrust from the foot of the frame that has to be resisted by the foundations. This is helpful in some poor ground conditions, although an alternative is to provide a tie between the feet of the frame eliminating the need to make the foot of the frame a pin joint. The release of the crown to form a three-pinned portal makes the structure statically determinant, which has in the past assisted in providing the structural analysis necessary to design a safe structure.

The roof and wall cladding will require purlins and cladding rails for support. There is, however, less need for wind bracing for portal frames, as the connection between column and

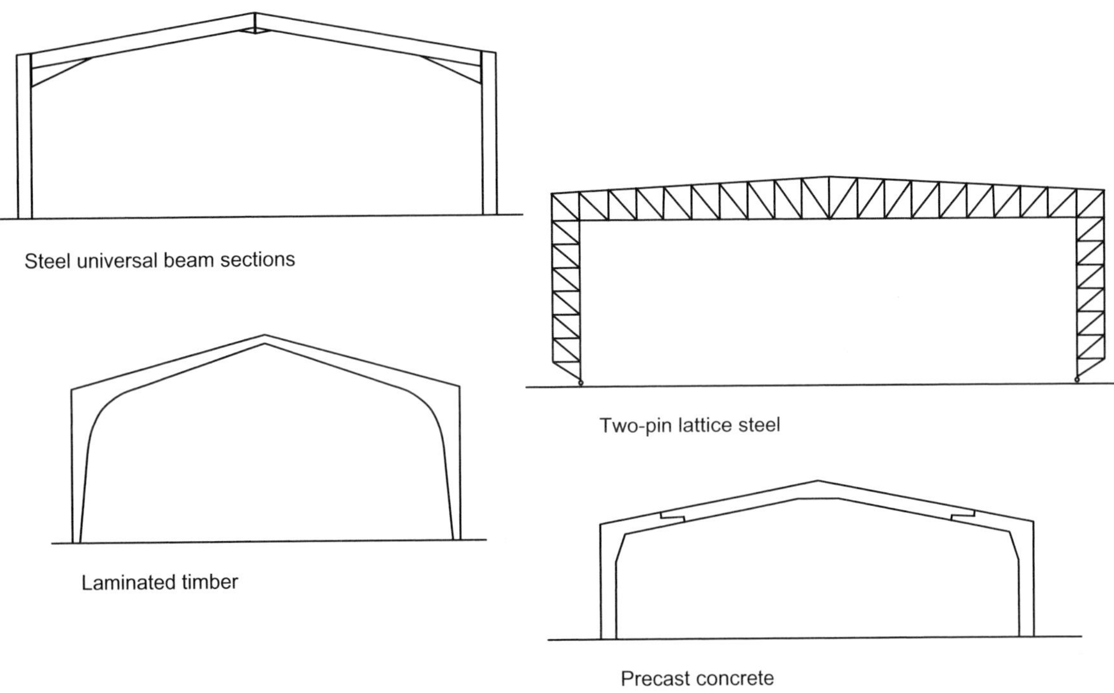

Figure 26.4 Typical portal forms in various materials.

beam is a moment connection that provides wind stability in the line of the frame. They are, however, just as vulnerable between frames, where bay bracing between columns will be required. In the portal frame the rafter may suffer lateral tortional buckling and will rely on the purlins with appropriate gable end bracing to resist this failure. As with lattice trusses if wind reversal is a potential condition then measures must be taken to stabilise the bottom of the rafter.

Another form of plane structural member is the arch rib. Like all arched structures, it is curved to contain the line of force so the member remains in compression from the vertical dead and imposed loads, giving it a potential to achieve great spans or more slender structures. This form does, however, generate a horizontal thrust as well as a vertical load at the support. If the arch springs from a structure above ground, this will have to be buttressed or the arch will have to be tied, normally at the springing points. The arch may spring from the ground and will therefore have no above-ground supporting structure and the horizontal thrust will have to be resolved in the foundations or tied at the base. Like portals, arch ribs can be formed as two-pinned structures with pins at the base or three-pinned with a further pin at the crown.

Two-way spanning flat structures

These flat structures work in bending in much the same way as the lattice girders but, by interconnecting the top and bottom chord systems of structures spanning in both directions, they act as a single two-way spanning structure. In the flat grid the girders remain plane structures intersecting the top and bottom chords with the girder bracing remaining vertical between their respective chord systems, as shown in Figure 26.5. In the space frame the bracing runs between adjacent chord systems, providing a potentially more efficient and stiff structure. In order to keep the bracing at an efficient angle the chord systems have to be relatively close together or the whole frame would get very deep. This also limits the length of the bracing to minimise the

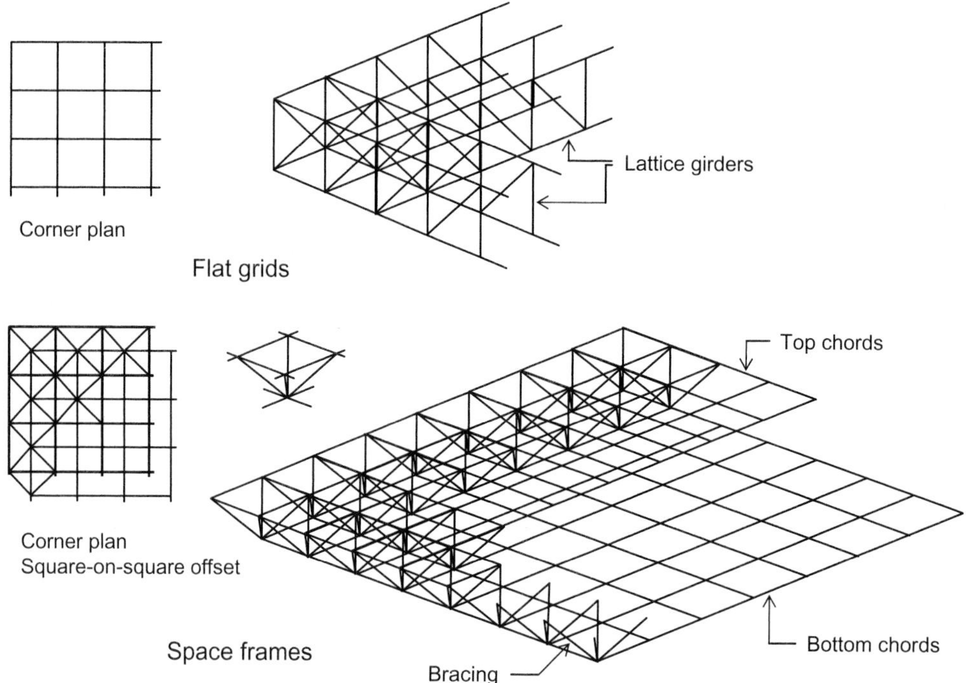

Figure 26.5 Two-way spanning flat structures.

chance of buckling in the slender compression bracing members. In practice this generates chords between around 1.2 and 2 metres apart. To further reduce the length of the bracing and maintain the efficient angle and limit depth of the structure it is also advantageous to offset the top and bottom chord systems. These alternatives generate a number of configurations, but the most common is a pattern known as square-on-square offset and this is the pattern illustrated in Figure 26.5.

The interconnection and geometric form of the space frame provides a stiff structure that is stable against wind forces. However, if supported on columns or masts, the structure is laterally unstable and bay bracing between columns to stabilise the support system will be required. The stiffness gives low deflection and provides effective distribution of point or moving loads. This inherent structural stiffness makes the use of aluminium a possibility. Aluminium as a material for building structures is limited by its strain characteristics. Its high strain rates give large deflections at working stress. However, with the space frames, even aluminium structures can have acceptable deflections and gain the advantage of the lightweight material, significantly reducing the dead loads.

Curved structures

These structures are formed from curved surfaces with geometry that ensures that the forces in the structure are in either compression or tension in at least one direction, normally both. The surfaces are either solid or grids. Like all the tension or compression structures, they induce horizontal forces at the support as well as vertical loading. In structures where complex curves give compressions in one direction and tensions in the other shear forces at the edges have to be resolved in edge detailing.

Shells are rigid solid surfaces in reinforced concrete, or for smaller spans timber can be used. Single curvature shells such as barrel

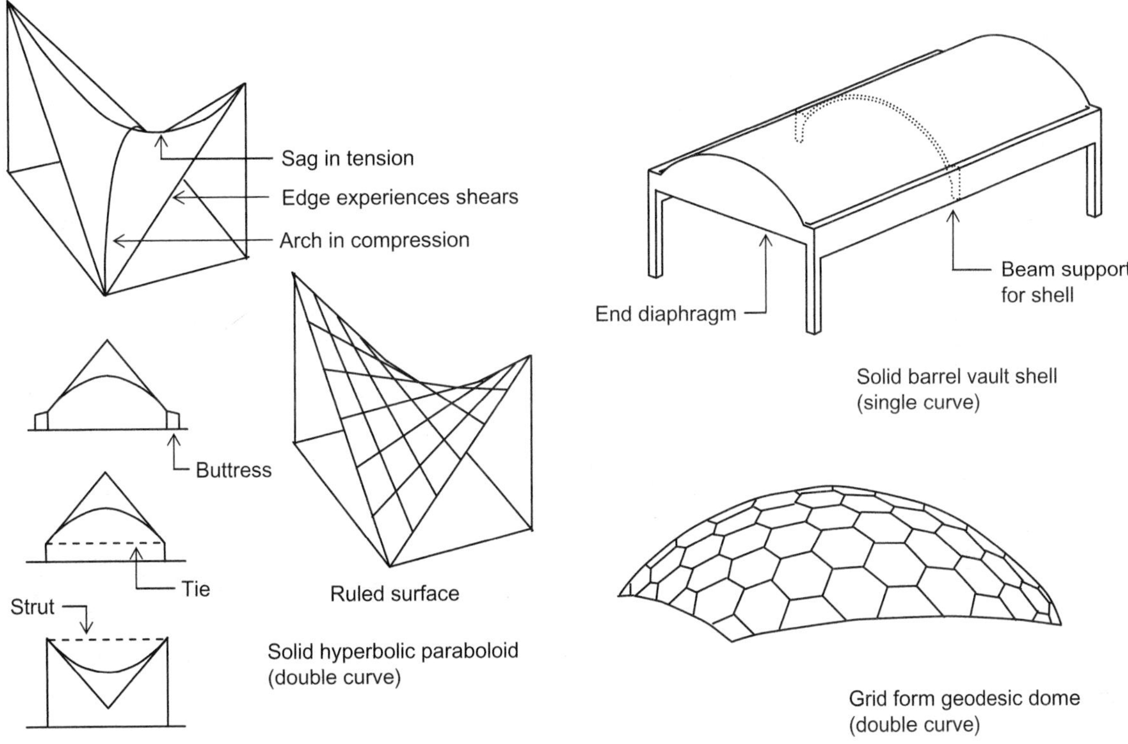

Figure 26.6 Two-way spanning curved structures.

vaults, as illustrated in Figure 26.6, are in compression across a curve that has to be supported by a bending member and this limits the span capability of this form in the bending direction. Double curvature shells will be curved in both directions, forming saddles, hyperbolic paraboloids or domes where cross sections that are sagging will be in tension and those that rise in the middle will be in compression. These forces make the section of these structures very thin except where shears are created at the edges. This is illustrated in the hyperbolic paraboloid shown in Figure 26.6. The horizontal forces from the shell can be resolved in a number of ways. The spreading of the arch can be supported by a buttress but more common would be the tie. An alternative is to support the sagging section with a strut. All these are also illustrated in Figure 26.6.

It may at first appear difficult to form such complex curves, particularly in a material such as concrete. Figure 26.6 indicates that these shapes are made of ruled surfaces, straight lines that can be used to form the temporary works onto which the surface can be formed that produces the hyperbolic shapes.

Shells made of solid surfaces are now not common, but the grid equivalent – where the structure is made from interconnected components carrying the compression and tension forces with a covering then applied either as a support for the waterproof cladding or as a stress skin on the grid – is. These would include steel and aluminium-based geodesic domes and timber grid shell forms. The basic form of the geodesic dome is shown in Figure 26.6.

Shells normally have some compression forces, whereas the tension structures are, as the name implies, entirely in tension. These structures are normally more flexural and less stiff as the absence of compression forces means that buckling is not a failure mode. This flexural nature makes them more difficult to clad and weather unless the structure is the weatherproofing, as with fabric structures. They also require suspension cables for support from either masts or perhaps guy ropes directly to the ground. In steel these would include net structures, of which there are only a few examples but the more widely used forms are the fabric or tented structures. Used as a feature on some buildings they are not normally long spans, although domes can have clear spans that would put them into the long-span category.

Construction details

Lattice girders and beams

Lattice girders and beams are most often made in steel but the section used for the chords and bracing varies. Of the variety of options three are shown in Figure 26.7. For medium spans, lattice beams provide an economic solution. The type illustrated uses cold-rolled sections for the chords and rods for the bracing. These structures are relatively light, which helps reduce dead load and assists handling.

The girders illustrated are made from hot-rolled steel section that allows longer spans to be achieved. Early girders would have used angles and channels connected with bolts to gusset plates but modern girders are more likely to be made from tube or square hollow sections (SHS) or rectangular hollow sections (RHS), as shown in Figure 26.8. With these hollow sections the joints would be welded in the factory and end-plated to provide bolted connections to be made on site. Large girders may have to be delivered to site in sections. Each member at the junction will be end-plated to be bolted on site. Site welding can be used for the connection of these sections, where the appearance of the girder is significant.

Given the susceptibility of girders to lateral tortional buckling, girders can be fabricated

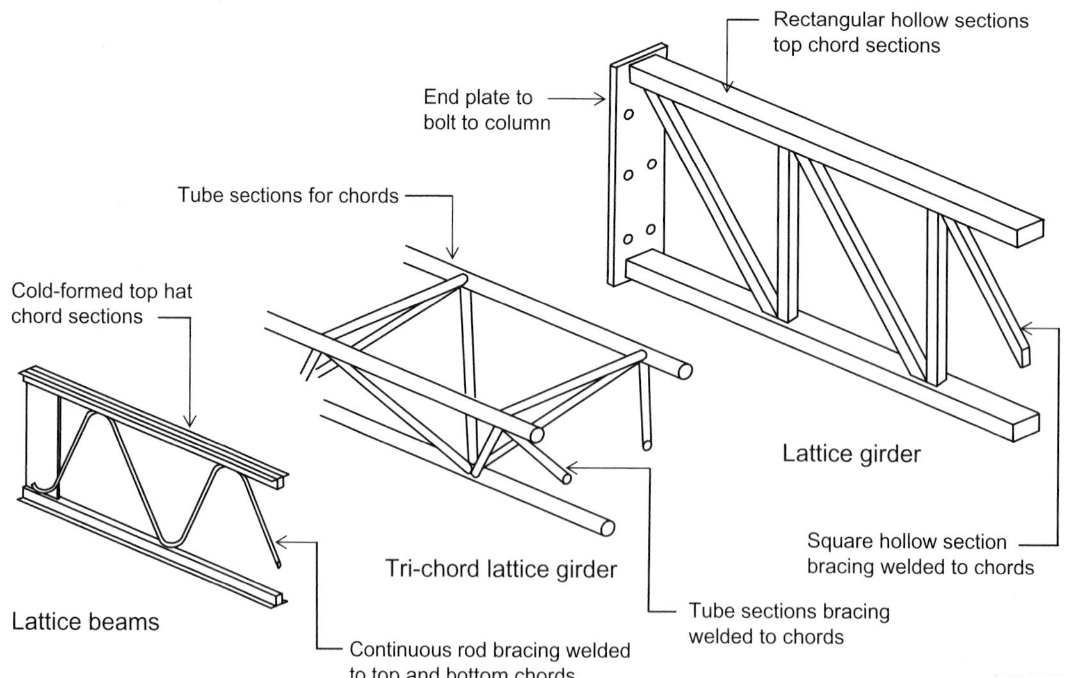

Figure 26.7 Lattice beams and girders.

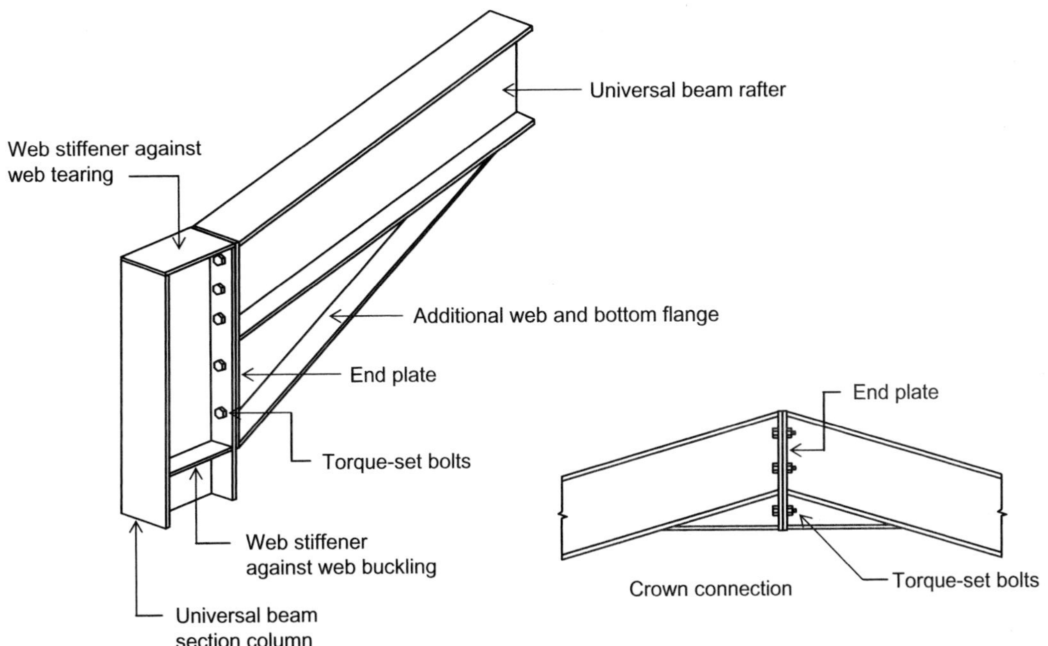

Figure 26.8 Steel portal knee and crown connections.

with twin top chords to give a triangular section to the whole girder, as shown in Figure 26.7. This stabilises the girder and provides closer support for the covering, reducing the span of the purlins, although the connection to the column has to ensure that the girder does not overturn.

The girder can be made of timber where timber sections are bolted together often with shear connectors to increase the resistance of the joint to tearing the timber along the fibres. For longer spans the sections may be laminated, a process where standard cut timber sections are glued together along their length to create larger sections. The process of laminating can produce large sections used as beams and can even be laid up to form curves for portal frames, as indicated in Figure 26.4.

Portal frames

To create the portal action the joint between the column and the roof rafter has to be detailed to be a full moment joint. This can be achieved relatively easily in reinforced concrete, but the sections become large as the spans increase and therefore the applications in reinforced concrete have been limited in the past to small-scale factories and, particularly, agricultural buildings. The most successful material for portals giving a range of uses and spans is steel. Timber that is laminated to make the column and rafter one member also makes an effective structure that can be left exposed (even externally) to become part of the aesthetic of the building.

Steel portals have both the column and the rafter made from universal beam (UB) sections as both are subject to bending. Where the two connect at the knee joint, there is unlikely to be sufficient resistance to bending with a connection made with just the depth of the rafter. A typical knee joint is shown in Figure 26.8. The depth of the connection is increased by end-plating the rafter with plate that extends below the bottom flange of the rolled UB to give a greater area over which to build up the shear resistance at the connection. This also allows an additional web and bottom flange section to be added below the rolled beam to increase the bending capacity as the moments increase at the connection. The moment will cause a rotation in the top of the column that may tear the web at the top or buckle the web at the bottom of the connection. This can be resisted with the web-stiffening plates at the top and bottom (and in some cases middle) of the connection. This connection will depend on the friction between the end-plate on the rafter and top of the column. This requires two unpainted surfaces and bolts that can be set to a torque to ensure the two surfaces are pulled together with a controllable force as part of the erecting procedure. The connections include both welding and bolting representing the factory and site processes in the assembly of these frames (fabrication and erecting). Similar detailing is used at the crown, also shown in Figure 26.8.

The connection at the foot of the portal is very similar to that at the foot of any column (see Chapter 25) but may need more bolts and a thicker base plate if it is to transfer the moment to the foundation. If the foot of the column is to be pinned, the joint will have to be detailed to release the moment.

Ribbed arches

For the spanning member to act as an arch in compression the line of force has to remain within the section. Once the member is in compression, the large spans can be achieved with relatively small sections. Many early cast-iron roofs for major Victorian railway stations used this principle and these can now be reproduced in steel, concrete and timber. The horizontal thrust is often taken on a tie at ceiling level, although these ribs can spring from the ground.

Space frames

As indicated above, the space frame can be generated in many configurations of top and bottom chord arrangements but the most common is the square-on-square offset, as shown in Figure 26.5. Normally in steel, although aluminium systems have also been developed, this simple configuration has been created in a number of different forms of different sectional members and components. Two of these forms are illustrated in Figure 26.9.

The first form is created with a prefabricated unit comprising a square of angles that will form the top chord system welded to bracing rods that meet at a central node ready to receive screw-ended rods to make the bottom chord system. The entire roof can be constructed with these inverted pyramid-shaped components. These units are created in a number of sizes. The 1.2-metre square forms a structure 1.2 metres deep, giving spans of up to around 45 metres if supported on all four sides. Connection to the columns is made to the top chord section via fabricated support brackets.

The second form of this type of structure is based on the node with the means to connect grid members for both chord sets and the bracing. These will be individual members with each end prepared to connect to adjacent nodes. The sections are normally tube or hollow sections and this form can be developed in steel to provide the longest spans for this form of roof but, as with all structures, are more economic over shorter spans. The nodes and the preparation of the ends of the chord and bracing members are a factory process with assembly on site.

With all forms of space frame the assembly has to take into account the fact that the structure is not self-supporting during assembly. It only works as a structure, even to support its own weight, when it is complete. Two approaches can be considered. Sections of the frame can be built on the ground and then lifted into place or a full temporary platform can be erected and the frame built in position, receiving support from the temporary works until the completed sections are connected to the columns. If constructed on the ground, lifting points are likely to be some distance from any position where a single crane could stand. This has led to either using multi-crane lifts or

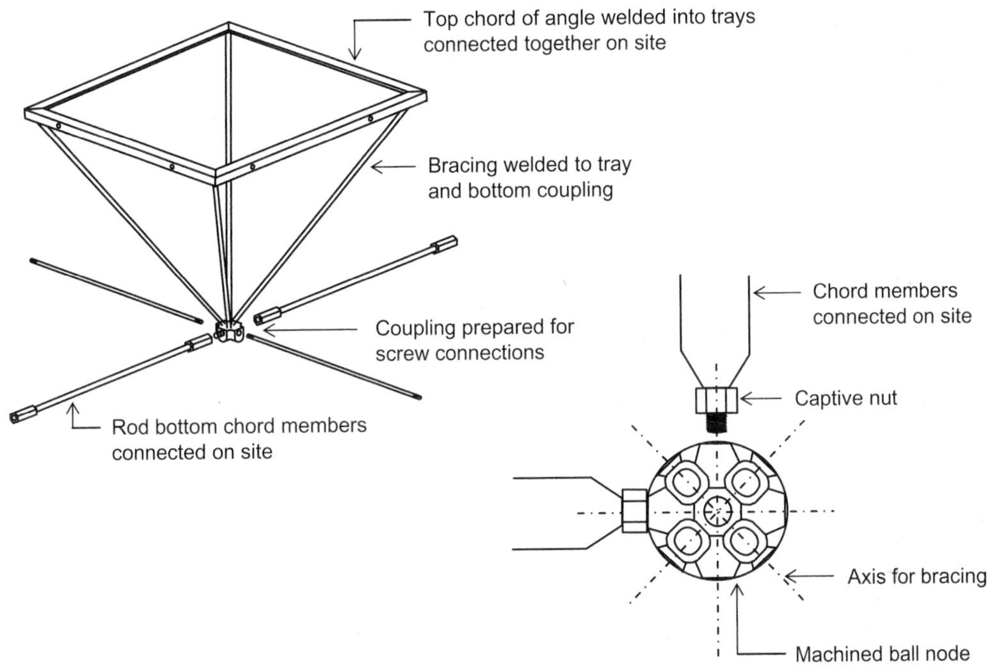

Figure 26.9 Two space frame systems.

jacking the assembled frame from the top of columns, often temporarily, only being released when the permanent connections to the columns (or beams) have been made.

The provisions of a waterproof covering and roof drainage have to be considered for these large areas of flat roof. The frames can be set to an angle but, as these structures are often left exposed, this will leave a sloping underside that may not be acceptable. Another solution is to make the bottom chord system slightly shorter than the top, giving a camber to the roof to create a fall. What this structural form does provide is a good support system for the decking. With top chords even at 2-metre centres, purlins are not required as metal decking can be fixed directly to the top chords and span without further support. Providing a waterproof covering for large areas of flat roof needs to be detailed with care as the direct exposure to the sky creates the potential for large thermal movements. Coverings based on single-layer waterproofing systems with high elasticity values are specified. These are discussed further in Chapter 29.

One other aspect of a large flat roof is the provision of lighting for the interior of the building. The space frame, with its top chord support system, offers good opportunities to provide large areas of glazing anywhere on the roof.

Other structures

As indicated earlier in the chapter there are other structural forms capable of forming a long-span roof. Even within the few forms discussed in this chapter, there are many other variations in configuration and detail. While the most common have been introduced, more research would have to be undertaken to fully explore the options for a specific project involving a roof where the span and the appearance might suggest a more specialised form.

Summary

1. Roof structures to cover large uninterrupted areas are now a common requirement for not only industrial buildings but also retail, warehousing and leisure buildings, as well as part of projects such as airports.
2. Plane frames in the form of lattice girders and portals provide economic structures normally in steel, but laminated timber is also an option. These represent the most economic and therefore the most widely specified form.
3. Two-way spanning structures offer potentially a great variety of forms, both flat and curved, based on both grids and solid structures.
4. More spectacular in appearance, the two-way spanning forms are often used on buildings where they are left exposed to gain the value of the appearance of the structural form.

27 Loadbearing Structural Walls

This chapter considers the potential and the forms of loadbearing walls for commercial buildings. The potential for low-rise, multi-storey buildings as well as applications to large open single-storey buildings to support roof structure will be introduced. Multi-storey options in masonry, concrete and timber walls are considered along with their favoured floor construction to form the whole building. Options to take the loads from single-storey industrial roofs in both masonry and concrete are introduced, where their potential height and length have to be taken into account in the detailing and specification.

Broad options

While most commercial and industrial buildings will have structural solutions based on framed structures, the use of loadbearing structural walls to carry the vertical loads may prove a viable alternative in some circumstances. Multi-storey construction can be successfully achieved with structural walls, as long as the internal spaces (governing the floor spans) are relatively small and repeated over the full height of the building, or at least to a podium level around first-floor level where a framed transfer structure could be envisioned. For wind stability and the prevention of the potential rotational collapse of high buildings the overall width/height ratio of the building needs to be considered to limit slenderness, as well as the need for internal walls to give either a cellular or cross-wall configuration. The cellular layout has combinations of external and internal walls that form boxes on plan, acting as cores throughout the building. Cross walls act as stiffening shear walls, so it is preferable to have cross walls in both directions but floor layouts that show cross walls predominantly in one direction can be stabilised by return ends on the cross walls or with the connection between the wall and the floor, as in the box frame construction described below.

These internal arrangements once established cannot be easily altered during the life of the building and there normally needs to be other performance requirements, such as sound or fire, for the walls that would demand substantial construction even if a frame were specified. Buildings that fit this configuration would include flats, study room accommodation and hotels.

Designs based on the masonry cavity construction with masonry internal walls as introduced in Chapter 19 can achieve buildings in excess of 10 storeys (timber-framed less), but it is also possible to consider walls constructed in reinforced concrete, either in situ or precast for multi-storey construction, and this will be discussed below.

Some low-rise designs may also offer the opportunity to consider a structural wall solution. While the layout constraints are less easily generalised than those for multi-storey, some clues can be suggested to spot the potential for economy in these designs. Designs that are open and light are less likely to be amenable to structural wall solutions. If the walls can be thicker, heavier and require limited openings, typical commercial floor spans can be accommodated. Mixed structures may be considered with some floors and roofs carried on walls and some on frames using columns.

Single-storey long-span roof structures can be supported on walls rather than on columns. Like all loadbearing options, the functions of cladding and structure become combined into the one element and the appearance of the building is likely to be significantly influenced by the choice of this structural form.

Floors and roofs

While there are unlikely to be beams associated with the floor construction for loadbearing walls, the slabs will be similar to those used in framed multi-storey structures. While the joisted floors will probably achieve the structural spans associated with many loadbearing wall projects other than the single domestic dwelling, they are unlikely to provide the sound and fire performance requirements between rooms in multi-storey occupations without significant modification to that of the standard house joisted floor. Almost certainly reinforced concrete slabs would be specified, possibly with precast components. The floor construction is most likely to be chosen to match the wall construction. Block walls may use plank or beam and block floors, while precast walls will almost certainly use precast slab floor units and in situ walls will have in situ floors. These choices match the site handling and operative expertise to give continuity to production processes.

Long-span roof elements can also be similar to those described in the previous chapter. While the moment action of the portal cannot be replicated when structural walls are used, trusses, space frames and solutions such as curved precast shell roof units can be successfully used with wall structures. In these buildings the production match between walls and roof is not as critical as the choice of wall and floor in the multi-storey building. The wall–floor cycle is not repeated with a single-storey building and economy can be best achieved with the appropriate choice of wall and roof structure. However, the junction of wall and roof becomes a significant detail not only to achieve structural performance but also to facilitate the production process.

Multi-storey construction

Multi-storey masonry walls

It is desirable to develop the external cavity wall to exploit its weathering and insulation potential, but the specification and detailing for the structural performance of multi-storey structures will have to be reconsidered as well. This reassessment of strength will have to be extended to the consideration of the loadbearing internal walls. In house construction the strength properties of the brick/block and mortar are unlikely to be the governing factors. Weathering and overall stability and even thermal and sound performance are most likely to be the performance choices that govern the thickness of the wall and the choice of bricks, blocks and mortar. In multi-storey building the masonry will have to be subject to structural design to gain the potential economies associated with efficient wall thickness.

In designed masonry the strength of the brick/block and the mortar combine to give strength to the wall as a whole. These need to be reflected not only in the specification but also in the on-site quality-control procedures for both materials and workmanship. While pre-mixed mortar may reduce the variability of site mixing operations, laying with full beds and curing are crucial to achieving overall wall strengths. Cement mortars will be required but significant proportions of lime can be introduced without significant loss of strength. Lime mixes of around 1:1½:1½ can produce strengths only 95% of a cement-only 1:3 mix. In designed masonry it is possible to achieve up to around 14-storey flats with 300 mm external masonry cavity walls and 175 mm internal block walls.

Stability considerations include the individual wall elements and the overall stability of the building against wind loading and rotational collapse. This stability is achieved by the bonding of the bricks and blocks to form panels and then the bonding of return walls to limit the length of individual panels and to stiffen the whole structure. The connections of the floors can then offer restraint to the walls, influencing the effective height of the panel. As with all masonry construction it is important to ensure that there are no tensile stresses in the structural elements.

The wall panels will be subject to buckling where slenderness determined by the height/thickness ratio will be a major influence on how much load can be taken before buckling. With economic thickness established to ensure strength, the control of slenderness needs to consider the dimensions of the panel, particularly the height. If the floors can provide constraint, the effective height of the wall can be taken to be less than the actual height, and the support provided by return walls will affect the effective length. With buildings for flats or hotel rooms with limited storey heights and frequent internal walls, achieving this stability is not a significant problem but care is needed in the connection of the floor and the wall to provide the lateral restraint. These limits to effective length and height and the lateral restraint of return walls and floors will also stiffen the structure to resist rotational collapse of the whole building. Floors may also have to be across the cavity every second or third storey. While this will spread the load across both skins of the cavity wall, it will create a cold bridge and require detailing if the appearance of continuous brickwork is required across the face of the whole building.

The whole building will also be subject to wind loading. The basic cellular or cross-wall nature of the structure and the enhanced resistance of the floor connections ensure the transfer of the wind loads from the external walls to be resisted by the whole structure. Given a reasonable ratio of width at the base and the height to limit the overall slenderness of the building, the stiffening of the structure will limit deflection. In addition pre-compression of the brickwork by loads from the floors above will limit the risk of tensile forces arising in the masonry from all but the top floors.

Openings will reduce the area of brick- or blockwork taking the vertical loads and will reduce the stiffness of the panels. For these reasons openings may have to be limited both in size and how close together they can be arranged. As openings become taller and closer together the masonry takes on the characteristics of a column rather than a wall panel and, along with the width of the opening, the stiffness of the whole panel is reduced. These are structural design considerations. While localised wall thickenings, called piers, can stiffen walls and masonry can be reinforced or even pre-stressed to overcome these conditions, designs requiring large openings may be better to employ a framed structure.

Multi-storey concrete walls

Concrete wall structures can be achieved in both in situ and precast forms. Although it is possible to construct walls that do not experience tensile forces in plain concrete, nearly all designs will be reinforced.

In situ concrete relies for its economy on the simplicity and repetition of the formwork and falsework, which becomes an option if there are to be dimensionally repetitive spaces vertically throughout the building. In addition cross-wall layouts can provide an opportunity for formwork/falsework systems that do not have to be disassembled between pours and can be lifted to the next level. This requires structures that are open on one side, allowing the striking and lifting of large sections of formwork. Such a structural form that offers all these opportunities for the formwork/falsework is the box frame, which is illustrated in Figure 27.1. The stability of the frame against racking is provided by the connection between the floor and the slab. The illustration shows the cross walls across the building, as this is the normal direction in which walls with high performance specifications for fire and sound are required. However, if the major internal walls are in the length of the building, casting sections that effectively extrude the cross section of the structure before being lifted to the next level can be envisioned.

In box frame construction the wall and floor cast in one operation with a formwork/

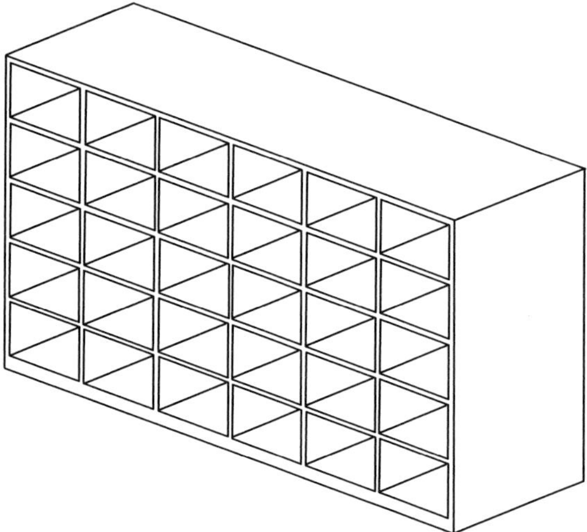

Figure 27.1 In situ box frame cross-wall structure.

falsework unit is called a tunnel form. Kickers are cast on the newly formed slab to receive the formwork that is brought up from below after the wall reinforcement has been fixed. When the slab reinforcement is fixed, the next lift can be poured. Both walls and floors will be around 150 mm thick. Openings can be formed in the cross walls during casting and this leaves the front and back elevations to be treated as any framed building. The end elevations will have relatively few, if any, openings and will need insulating and a facing to provide a fully functioning wall.

Precast panel construction is another option for cellular or cross-wall floor configurations requiring both external and internal structural walls. The use of precast compared to in situ concrete for structural walls will limit the need for falsework and will give potential gains in the quality of finish and the accuracy of construction but increasing transport costs and change in the joint behaviour in performance. Joints between precast panels will have less resistance to racking forces, making the building less able to resist wind forces with the need to provide wind stability, most probably using cores. These joints also make the building more susceptible to disproportional collapse. This is a failure condition where the accidental removal of one member (in this case a wall panel) will cause the building to suffer a collapse disproportionate to the cause. This will require panels to be tied together and back into the floors over and above the needs of the joints to transfer the normal building and wind loads.

Precast structural wall panels are normally matched with precast panel floors often of similar size and weight to give continuity in the production process and the efficient use of common resources. In the second half of the twentieth century this construction was used for flats and took advantage of the ability to provide not only an external finish on the concrete but also a sufficiently fine internal finish to make it ready for decoration. In an era where thermal performance requirements were minimal the concrete offered good fire and sound resistance and could be produced to be weatherproof. The external joints between the panels then had to be water- and wind-proofed to produce a weather-tight wall. This solution is unlikely to be applicable at the beginning of the twenty-first century. Not only have thermal performance requirements increased but also the appearance of concrete panels, however well finished, is unlikely to be a preferred external finish.

If precast panel construction is used, it is likely to provide only the structural element of the external wall. It will still provide good fire and sound resistance but the external appearance, thermal and waterproofing functions will be provided by other elements of the external wall. The internal walls will not require these functions and the quality of finish for internal decoration could still be considered.

For structural needs the walls are likely to be around 125 to 175 mm thick. Openings can be formed in the panels for both windows and doors, but it is difficult to form openings across more than one panel. Joints between panels are normally formed in the thickness of the panels, as shown in Figure 27.2. Edge profiles and pockets on the panels provide for steel loops from adjacent panels to be tied

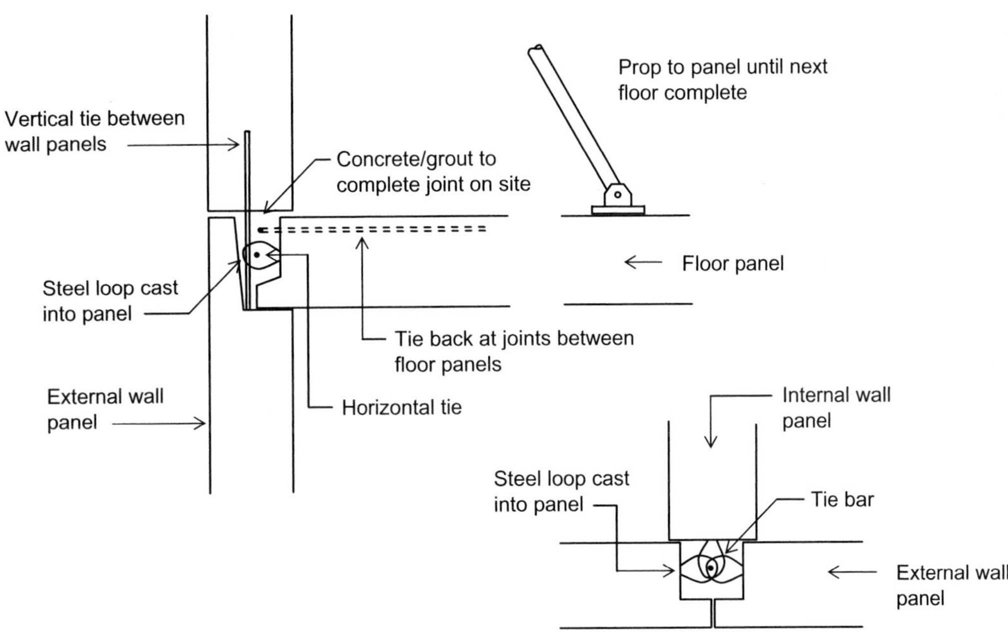

Figure 27.2 Precast concrete panel construction.

together with rods or wires and an in situ filling completing the joint when the panels are in their final position. Wall panels will require propping until the joints are completed for the floor above. Slots have to be formed between the floor panels to receive the reinforcement to provide the horizontal tying which becomes effective when the slot is filled with in situ concrete.

These connections provide for load transfer and resistance to disproportionate collapse. If the walls provide a cellular layout, these connections should give the structure its overall stability; otherwise, in situ concrete or braced structural steel circulation cores could be erected before the precast work begins. However, stacks of precast volumetric units could be specified to provide the stability as the work proceeds. These volumetric units could be circulation cores but might also be conceived as rooms complete with floors. The most likely rooms are the bathrooms that will be located one above the other for servicing. The volumetric solution also offers the opportunity to prefabricate parts or all of the room services and finishes, eliminating a concentrated and lengthy process from the site programme.

Multi-storey timber

The concepts of large panel construction can be achieved in timber. Although currently unable to reach the number of storeys possible with masonry or concrete, it can be considered for flats and accommodation blocks with their limited floor spans offering the opportunity to provide the floors in timber panels or cassette. Panels can be based on the framed panel similar to the details given in Chapter 19. Similar details can also be achieved with steel studs rather than timber.

The development of solid structural panels based on composite boards such as SIPs (structural insulated panels) and solid wood construction can provide an alternative to the framed panel. These are shown in Figure 27.3. Both can be made into panels between around 70 and 250 mm thick but sections of around

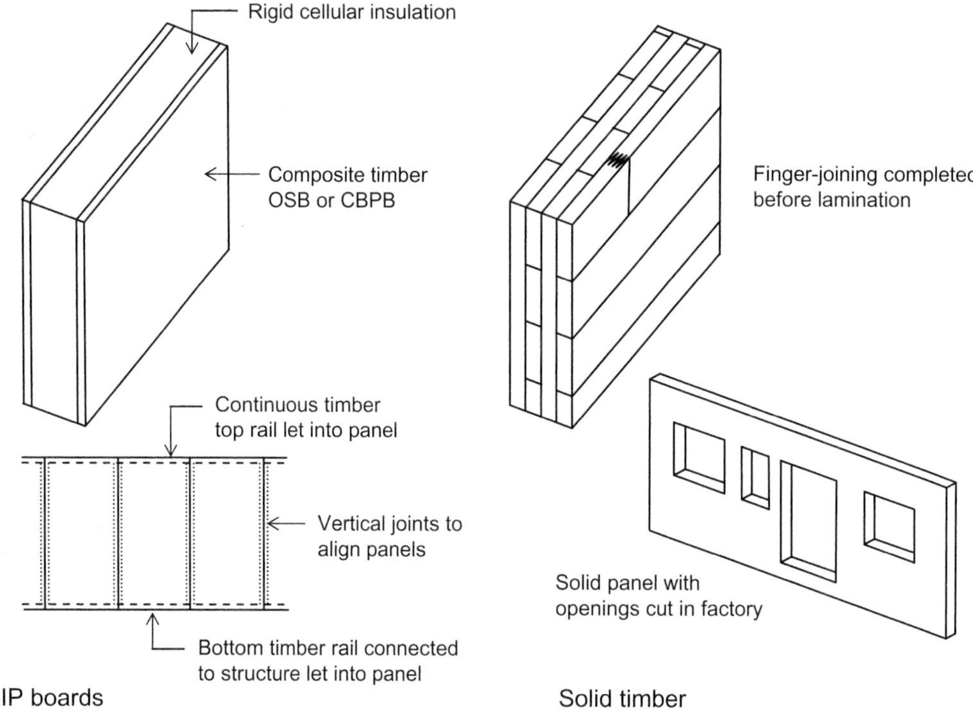

Figure 27.3 SIP and solid timber walls.

100 mm are more common for this form of construction.

SIP panels are two composite timber boards bonded to a core of a rigid cellular insulation. The composite boards are likely to be oriented strand board (OSB) or possibly cement-bonded partial board (CBPB) between 8 and 15 mm thick. The rigid cellular insulation can be self-adhesive polyurethane or polyisocyanurate or boards of polystyrene or mineral fibre that have to be bonded to the facing boards. Panels are normally produced in sheets 1.2 metres wide by either a standard 2.4-metre boards or specifically for storey-height panels. Many of these systems are propriety and the nature of the site joint varies with the manufacture but some location and interlock is required with some preparation of the edges of the panels at the factory. Edge profiles will normally involve some machining out of the insulation to let in timber sections to make the connections required between panels and to other elements of the structure. These panels offer good insulation, even in 100 mm thick panels, but will still require an internal finish and an external cladding with weatherproofing performance. Structurally these panels can take vertical loads for walls both internal and external but cannot take bending. Floors will have to be constructed with joists and boards similar to timber frame detailing. However, the insulating performance of this composite can be used with panels over the rafters at roof level.

Solid wood is produced by laminating using offcuts and other short lengths of timber, finger-jointed and laid up as cross-ply. The laminating process can produce boards up to 4 metres wide and 18 metres long, depending on the capacity of the factory. This material has good bending characteristics so can be lifted flat and used as floor slabs. Openings can be cut out of solid panels, which can then be transported to site. All the panels have square edges providing simple detailing between wall and floor panels, where the joints can be taped for airtightness. Insulation and a cladding to the external face can be provided as on-site operations. This material can be used directly as the internal finish.

The lower weight of the timber panels over the concrete allows more to be transported on each lorry. This has given this type of prefabricated construction the term 'flat-pack', with whole buildings being delivered on a limited number of lorries.

Multi-storey volumetric

Prefabrication can be taken even further with the potential to provide volumetric units. When rooms are repetitive in layout and limited in size (width of around 3 metres) to be transported to the site whole, room units can be stacked on top of each other to create the structural system itself. If the room layouts are sufficiently repetitive, not only on plan but also vertically throughout the building, then whole sections of the building could be considered for construction with volumetric units that also form the structural system. One such combination of room pods is shown in Figure 27.4. Mixed structures where the lower floors may be a framed structure supporting the repetitive rooms on the upper floors may be another combination. This approach has been successfully employed for hotels, student accommodation and flats. This offers the opportunity for high levels of prefabrication and reduced site-erecting time, although project time taking into account design and factory production may not provide an overall time reduction for the client from commissioning to occupation.

The use of concrete for volumetric construction is an option but its weight limits the size of units so other options may be considered. Framed panel walls in timber or steel studs may be considered. It may be that the pods need to be stacked on strengthened corner sections to carry the load. The size and weight of units will determine transport and cranage requirements but the potential rigidity of the pod allows larger internal spaces to be considered, at least for medium-rise buildings. Units can be delivered with open sides that, when connected, provide these larger spaces. This may need the introduction of beams in the structure and temporary bracing for transport. It could be argued that these are no longer loadbearing walls but form a hybrid structure. Levels of services and finishes to be included in the prefabrication may add weight and increase the susceptibility to damage in handling, so care must be taken in devising site sequence and procedures for subsequent activities.

It is not only the level of prefabrication for the internal services, finishes and fittings that needs to be chosen but also the external wall treatment. Units can be produced with an external finish and weathering joints detailed to make the building weather-tight, but it may be preferable to add a facing to the pods to provide an appearance similar to other cladding solutions discussed in Chapter 29, although this will add to site production time.

Single-storey industrial construction

Walls for industrial buildings need to be high and long normally without the lateral support or stiffening effects of internal walls and floors. This makes both element buckling and wind stability major problems for open industrial buildings. Lateral restraint will be provided at the top of the wall by the roof structure, which, when sufficiently braced, will resist the overturning of the wall. The uninterrupted length of walls for these buildings also introduces the need for movement joints for the inherent deviations from both moisture and thermal movements. Options have been developed in both structurally designed masonry and reinforced concrete.

In masonry where tensile forces cannot be resisted there is a need to increase the effective width of the wall to increase both the buckling resistance and to limit the bending forces induced by wind loading on the stabilised wall. Traditionally this has been achieved with piers at regular intervals. This approach can be

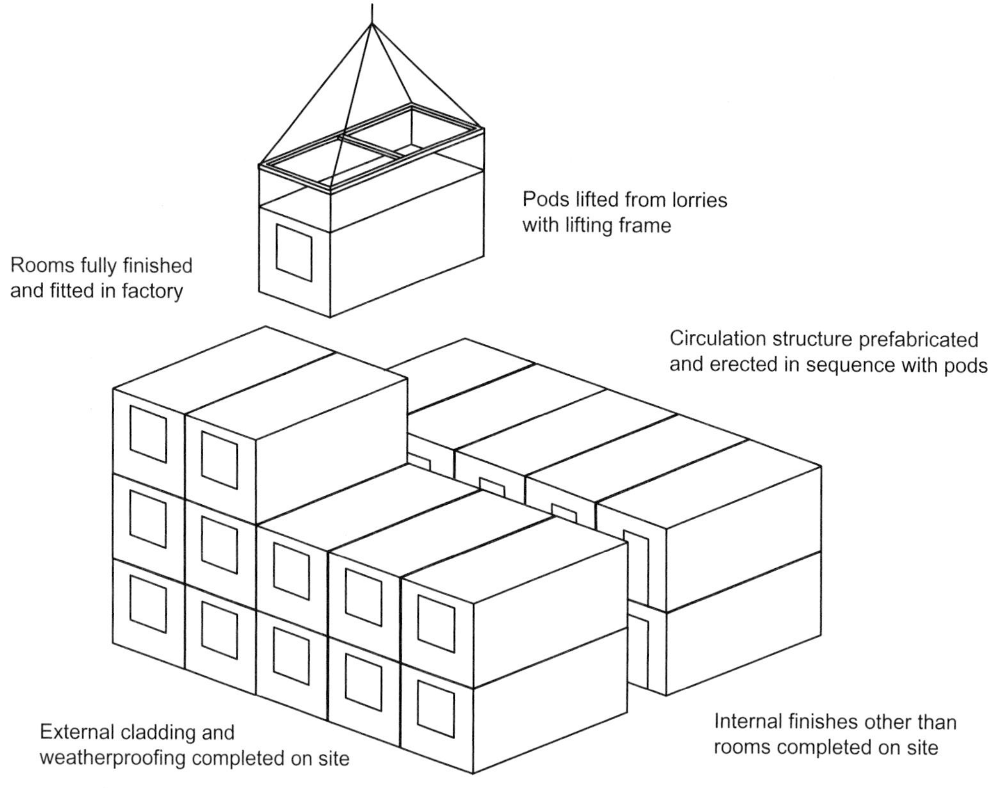

Figure 27.4 Volumetric construction.

adopted for these industrial walls but the piers become more than thickening and need to be much deeper, perhaps between 1 and 2 metres, forming the appearance of fins, and this has led to this approach being known as fin wall construction. This approach is shown in Figure 27.5.

The fins are normally expressed on the outside of the building formed in the outer skin of a cavity wall, which can then adopt standard detailing with ties and insulation. While smaller openings can then be formed in the cavity wall using standard details, the larger openings often required in industrial buildings are normally formed by removing a section of wall between the fins. If the fins form the jamb of the opening, they may have to be thicker to stiffen the edge of the wall at the opening.

The centres of the fins would normally coincide with the spacing of the roof structural elements to provide the bearing (onto a pad stone) and tied connection required to transfer the load and provide the lateral restraint. This detail may have to include a reinforced concrete ring beam to achieve this connection. Movement joints would also have to coincide with the fins, as indicated in Figure 27.5.

An alternative to the fin wall is the diaphragm wall, as shown in Figure 27.6. This construction form increases the width of the wall by increasing the cavity and then ties the walls with cross ribs of brickwork or blockwork to ensure a single structural action against buckling and wind forces. The sizing of the wall is dependent on the modular size of the brick discussed in Chapter 19. The overall thickness of the wall would normally be between 1.5 and 2.5 bricks (approximately 550 mm), although wider walls may be necessary depending mainly on the height of the wall. The centres of the

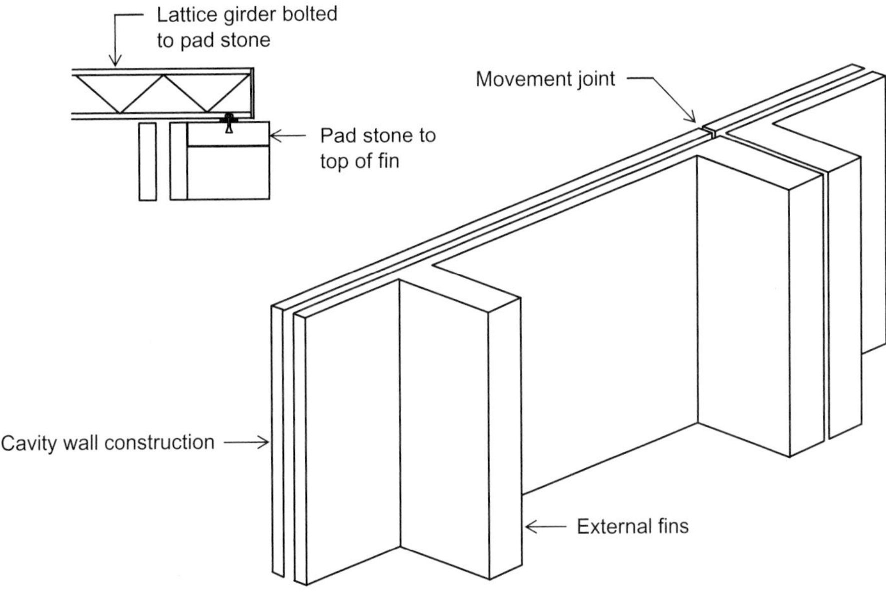

Figure 27.5 Fin wall construction.

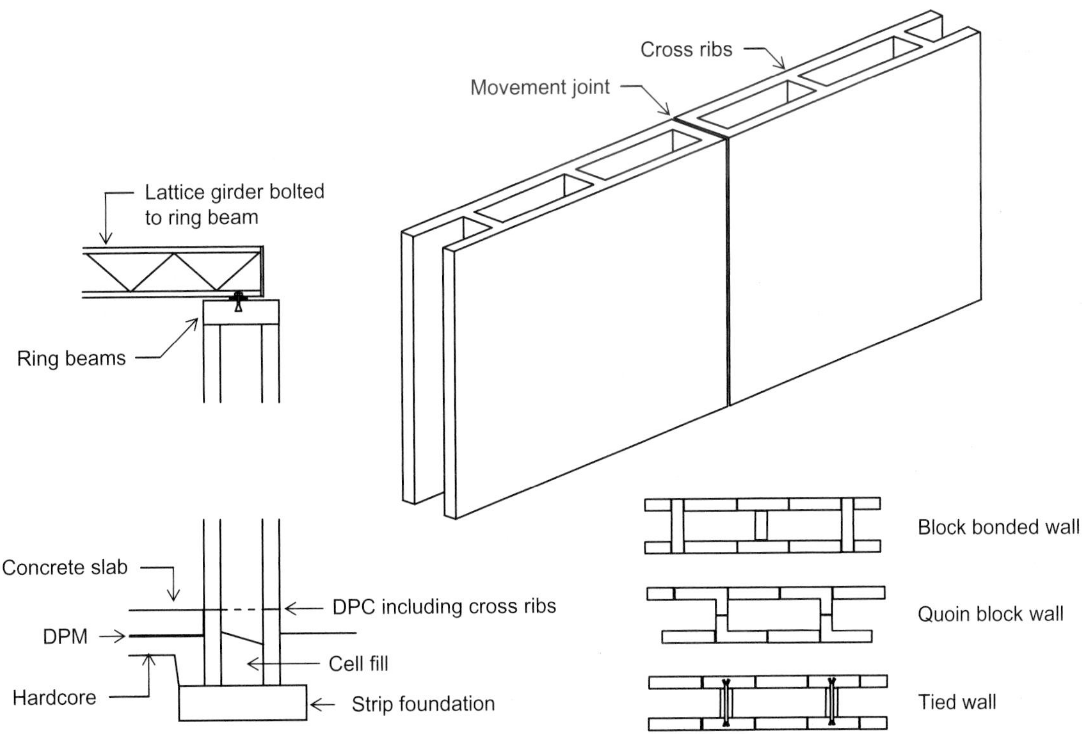

Figure 27.6 Diaphragm wall construction.

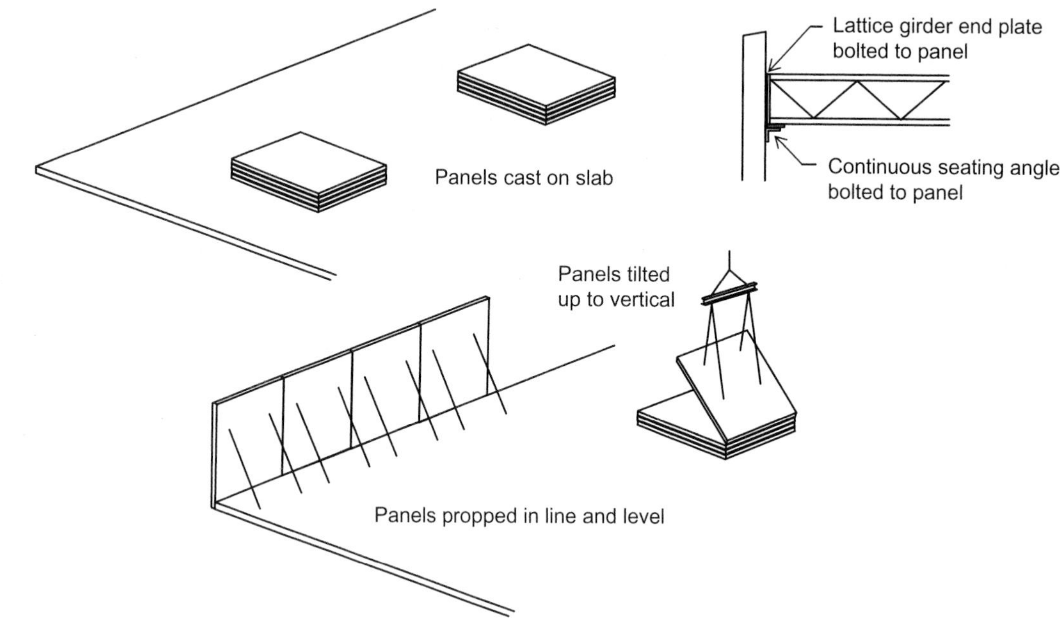

Figure 27.7 Tilt-up construction.

cross ribs also have to be brick modules, normally 4.5 or 5.5 to give whole bricks between the bonding headers that tie the cross ribs to each leaf. Alternative bonding arrangements in blockwork are shown in Figure 27.6. Opening and movements joints now have to be arranged between the cross ribs. This wall will need a reinforced concrete ring beam detailed at the top of the wall to provide the bearing and the lateral restraint connection to the roof structure.

The need for the thickness of the masonry wall is not required by reinforced concrete as this can resist tensile forces. Potentially relatively thin structural walls around 150 to 200 mm can be used as long as the roof and the floor can provide the lateral restraint to overturning at the top and sliding at the bottom of the wall. The structural wall will provide good fire, sound and even some weather protection, but the wall will still need to be detailed to ensure full waterproofing, appearance and thermal performance, introducing other components and materials to form a fully functioning wall.

In situ solutions would need to employ a significant amount of formwork and falsework and, if the full-height wall is to be cast in one pour, care needs to taken in the specification of the concrete for its workability and in the methods used to place and vibrate the concrete in deep narrow shutters. If precast concrete is considered, transport requirements may limit the width of the panels as the wall will need to be cast in single-height units to provide the structural integrity that would be difficult to provide in a horizontal joint between two smaller panels. One system that overcomes these transport difficulties is a process known as tilt-up and is illustrated in Figure 27.7.

Large panels are cast flat on the structural slab adjacent to the position where they are to be erected. The slab becomes the formwork for the face of the panel, with formwork being required to form the edge profile of the panel. They can be cast with insulation and finishes or in stacks of structural wall units, the panel below becoming the formwork for the one above. They can also be cast as continuous walls and then cut into panels at an early age. Lifting points have to be cast in. The precast panels now only have to be moved a few metres after they have been tilted up into the vertical position, where they are lowered onto the levelled foundations. In order to ensure fixity and restraint at the base of the wall there may be an infill strip at the slab edge to tie in the wall, or the panel may be lowered into a rebate in the slab edge or against a prepared metal fixing. Adjustment is necessary to bring the vertical joint between panels into tolerance before final fixing. They then need to be propped in the vertical position until the roof structure, normally including all the roof bracing, is erected to provide the permanent lateral support. To provide this support the roof structure is normally fixed to the face of the panel that extends above the roof line to form a parapet. The ends of trusses may not only be bolted to the panels but also be fixed to a continuous seat angle connecting all the panels together just below the level of the bottom chord of the truss.

The pace of the production process for this system tends to have the panels erected and fixed before all the drying shrinkage has taken place. While the reinforcement within the panel will be designed to resist the tensile forces induced by the shrinkage in the large panels, the joints and the fixings will have to be designed to absorb a proportion of this early-age shrinkage.

Summary

1. Loadbearing wall structures may be considered for many buildings and have been developed for both multi-storey and single-storey industrial buildings.
2. With a limited number of storeys loads can be taken on the walls from floors and roofs even with normal commercial spans, although these cannot then be altered in the life of the building.
3. Multi-storey buildings with internal layouts that provide limited spans and a cellular or cross-wall arrangement of external and internal walls can be considered for loadbearing wall structures.
4. The floor construction has to be matched to the choice of wall to match the production process. While in situ floors would be used with in situ walls, precast concrete floors offer a match with both masonry and precast concrete walls. Timber floors would be used with timber walls.
5. Overall stability can be achieved from the cellular or cross-wall arrangements of the internal walls with appropriate connections between floors and walls, although cores similar to framed buildings can also be considered. Volumetric construction has an inherent overall stability.
6. Disproportionate collapse is a major consideration in the construction of the walls and the design of the connection between wall panels and between panels and the floor.
7. Walls for single-storey industrial buildings are potentially high and long, making buckling and movement a feature of the construction detailing. Increasing the effective width and the provision of movement joints are important in the design.
8. Options are possible in precast concrete panels, but the size of the panel makes transport difficult. This has led to the tilt-up system, where panels are cast on the slab adjacent to the final position of the wall.

28 Structure Below Ground

This chapter identifies three groups of structure as being constructed below ground: foundations, basements and large area industrial floor slabs. For each of these the choice is based on engineering and production processes to promote safety and speed of construction. Both shallow and deep foundations will be introduced, and in particular the procedures that allow even deep foundation to be constructed from ground level. Soil-improvement techniques are also introduced. Basement structure and its waterproofing will be discussed, along with techniques to form the structure before excavation to combine at least some of the temporary and permanent work functions. In large area floor slabs the significance of joints and the way they can be incorporated into the production procedure will be considered.

Introduction

While entitled 'Structure Below Ground', this chapter will cover not only foundations and basements but also ground floors, specifically large area floors for industrial buildings. Ground floors for other commercial buildings will be based on the same detailing as the ground support slab used for domestic construction. Clearly these elements of construction have in common that they are formed in or on the ground and therefore have to interact with the soil in achieving full performance. This in turn means that solutions will be chosen predominantly for engineering and production reasons. Work in the ground is hazardous and can be expensive and therefore solutions are chosen against production processes to suit the site and achieve the engineering objectives. Choosing methods that can be constructed from ground level will have a significant safety and cost advantage, although ensuring quality and performance of unseen construction then has to be considered. Solutions that incorporate temporary works for earthwork support can bring safety advantages and lower the technical and contractual risks involved. Understanding the production and engineering aspects of the work is essential for both designer and constructor. It is not uncommon in complex groundworks to produce production sequence drawings as well as engineering details. The architectural design and environmental performance considerations are limited but where they are relevant in the final choice they will be discussed with the structural issues.

While the key production processes will be explained, there will be little on the engineering aspects. Commonly available options and their relationship to building type and site context will be introduced. This will make some broad initial choices possible but will not allow any depth of analysis. This would only be possible with engineering and production expertise.

Foundation design

The interrelationship between structure, foundations and the soil was introduced in Chapter 20. The analysis of the structure required a clear understanding of not only the distribution, magnitude and direction of the loads but also the stiffness of the construction above ground. For commercial buildings that are predominantly based on frame structures this gives a different set of considerations from house foundations. With the vertical loads in frames carried by columns the loads take on the characteristics of point loadings on the ground. Given the capacity of the frame to create larger building, the loadings are almost certainly likely to be much greater than those from housing. The frame also has different stiffness behaviour to the loadbearing wall.

The stiffness of the frame is mainly dependent on the joints between the structural members. Rigid joints create stiff structures, which, if subjected to differential settlement between columns, redistribute stresses in the frame members. Frames with pin joints subject to differential settlement will accommodate small rotations without changing the stresses in the structure. Wind stability will locally stiffen the structure so some care has to be taken in assuming no additional stresses in these areas. Although unusual, where high differential settlements have to be allowed – in, for example, mining subsidence areas – lightweight structures with sprung bracing may need to be considered to create a fully articulated structure. In these buildings the connections of the enclosure elements may also need to be detailed to accommodate this movement.

Another building configuration that may experience different settlement values is the tall block rising from and forming part of a relatively low-rise development. This is not directly to do with the stiffness of the structures, as both may have similar flexibility characteristics, but the difference in the total magnitude of the loadings and the possibility of different foundations solutions will make matching the predicted settlement difficult. To reduce the risk of movement the junction where these two different scales of structures meet would be designed as a movement joint allowing some relative vertical movement.

The second idea introduced in Chapter 20 was the soil as part of structure. This requires knowledge of the strength and compressibility of the soil, normally expressed as a bearing capacity. In addition it is important to identify any tendency to volume change with change in moisture content or frost action along with any risks for subsidence on the site. It also requires some understanding of the stress distribution in the soil, and this was introduced in Chapter 11. The same information is required for the design of foundations for commercial buildings. However, given the greater variety of loadings and construction forms for framed buildings, the scope of a site investigation may need to be greater but would still follow the basic approach identified in Chapter 20.

The third key idea explored in Chapter 20 was associated with foundation design. It established four conditions that must be satisfied:

1. The load has to be distributed over a sufficient area to limit settlement and avoid collapse.
2. The centre of gravity of the load has to be over the centre of the area of the foundation.
3. The foundation has to be stiff enough to press evenly on the ground over the whole area of the foundation.
4. The foundation has to be founded at a depth below ground to be resting on a sufficiently stable soil mass.

These basic requirements are also applicable to commercial framed buildings and will govern the size and detailing of solutions, as they did for the simple strip foundation.

The final section that may usefully be revisited from Chapter 20 is the scope and process of a site investigation. These are often of more significance on a commercial building, where the scale of building and loadings not only introduce more stress into the ground but production activity may have to be carried out below ground with the need for temporary earthwork support to be designed. This makes the need for information about the site and the soil more critical if risks – technical, financial, and health and safety – are to be reduced.

Foundation types

Shallow foundations

Where soil strengths and loading conditions permit, the simplest foundation for a column is the pad foundation, as shown in Figure 28.1. Normally square, if constructed sufficiently thick it can be made of plain concrete similar to the simple strip footing where the 45° shear determines the depth of pad. It is, however, not unusual to specify a mat of reinforcement in the bottom to give some bending resistance, as this will reduce the potential thickness of the pad as the plan size of the pad increases. The use of reinforcement also makes rectangular pads possible if this better suits the foundation layout.

As the required area under the foundation increases it may be better to take one line of columns and make a continuous column foundation, known as a combined column base, as shown in Figure 28.1. While this looks like a strip foundation, the point loading means that the foundation bends along the strip as well across the foundation. This beam action requires reinforcement to resist the bending action with main steel along as well as across the foundation. This reinforcement requirement is shown in Figure 28.1.

Where perimeter columns are required adjacent to restrictions, it may not be possible to place the column at the centre of the foundation without the foundation encroaching on adjacent construction or beyond a boundary. Under these circumstances it may be possible to combine the loads from other columns. Three variations – the combined base, the cantilever foundation and the balance base – are illustrated in Figure 28.2. While the combined foundation is the simplest arrangement, it still requires some projection of the base beyond the column towards the restriction. The need to

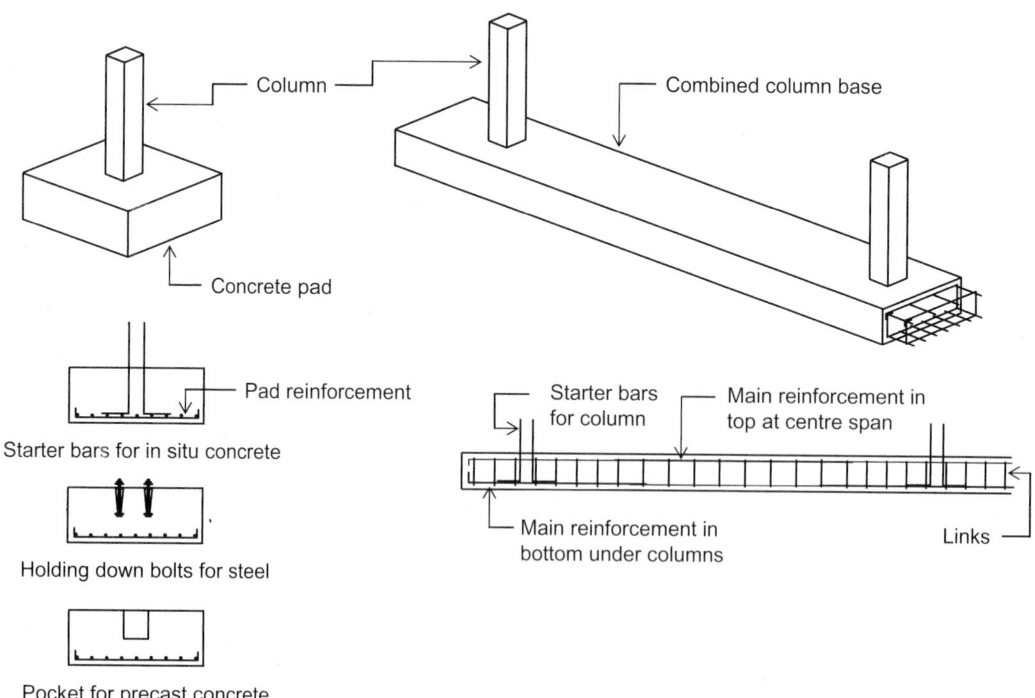

Figure 28.1 Pad and combined column shallow foundation.

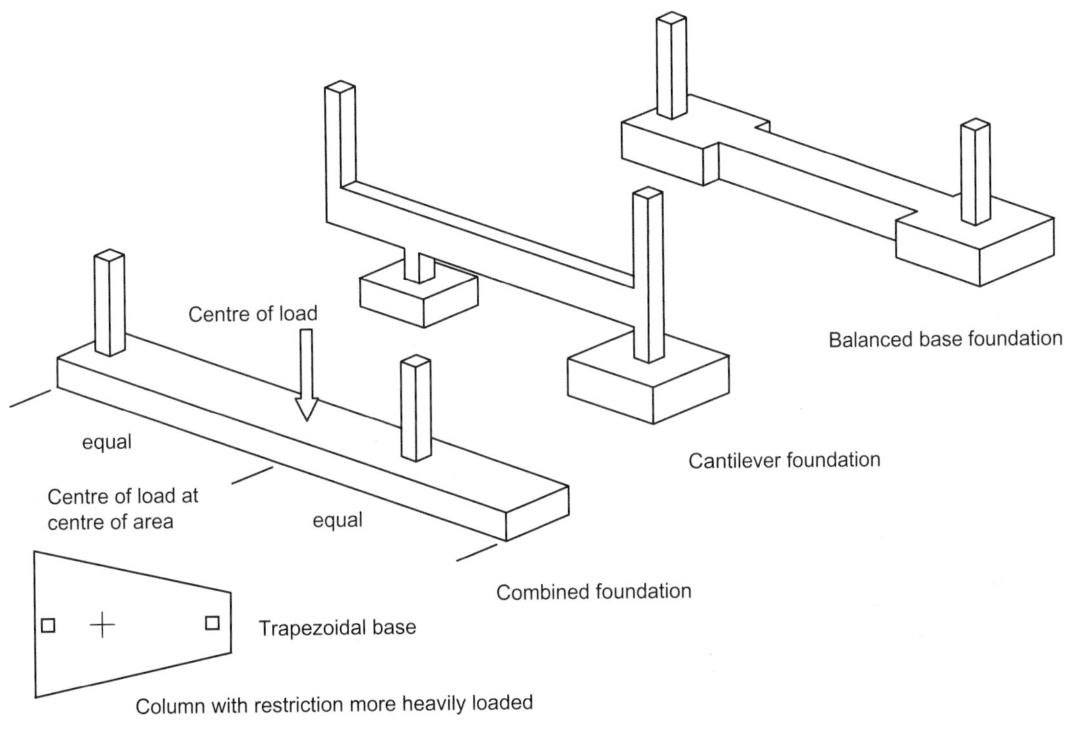

Figure 28.2 Shallow foundation for column loads close to restrictions.

keep the centre of load at the centre of the area of the foundation requires an extension beyond the more heavily loaded column. This is illustrated in Figure 28.2, which also shows an alternative, to design a trapezoidal shape to keep the base within the restrictions. The cantilever foundation allows the column to be built right up to the adjacent building or boundary with the pads well within the footprint of building. However, the combined depth of the cantilever beam and the bases can make the founding level relatively deep. The balanced base may achieve a column sufficiently close to the restriction without the additional foundation depth but will not gain the full advantage of the cantilever foundation.

While it may be possible to construct shallow foundations without earthwork support in most soils, it is unusual to construct the foundations without the use of formwork. The exception may be a simple mass concrete foundation where the soil can be cut to relatively close tolerances (to minimise excessive use of concrete) and will stand for the few hours between excavation, verifying the founding conditions and pouring the concrete. However, even mass concrete foundations may need the starter bars or bolts for column connections to be cast in and these will have to be held in place. This is often best achieved from the formwork. As long as foundations remain relatively shallow the health and safety risks from soil collapse remain low, although the risks of falls and collision with excavating plant remain.

Temporary formwork will be similar to that introduced in Chapter 25, although support falsework may well be improvised depending on the actual site conditions. Permanent formwork may be considered using either blockwork or proprietary sheets of plastic incorporating mesh reinforcement. These plastic sheets distort under the pressure of the wet concrete but the distortion is of little consequence when buried in the ground.

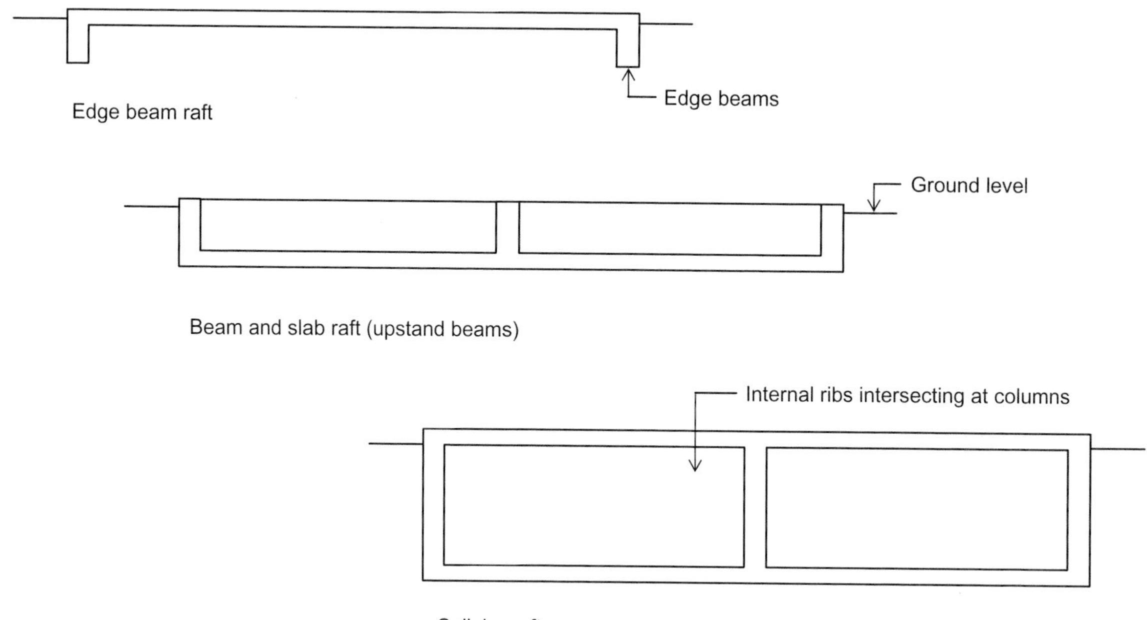

Figure 28.3 Raft foundations for framed buildings.

Raft foundation

As discussed in Chapter 20 an option for a foundation formed close to the surface that can be used on soils of limited bearing capacity is the raft. By effectively designing the whole ground floor to be the foundation the loads are spread over the whole footprint of the building, limiting the pressure on the soil immediately below the raft, although it will stress the soil to greater depths (albeit that the stresses will be relatively low). Designing rafts for framed building with their high-point loadings will mean that for all but the lightest of frames the raft will require a construction form that incorporates a natural stiffness greater than the plane or edge beam rafts detailed in Chapter 20 to gain the even load transfer assumed in a raft design.

This stiffness is normally achieved with a beam and slab construction that is designed with beams across the building as well as around the edge forming a grid with the columns connected at the intersection of the beams. As the soil provides continuous support and the columns create point loading the bending in this slab is the reverse of a suspended floor slab. Deflections in the beams between columns are upwards. For structural efficiency, to use the concrete slab as the compression zone upstand beams can be used. This will, however, require the construction of a floor supported on the beams. This is the arrangement shown in Figure 28.3. Reinforcement will now be very similar to suspended slabs, although the formwork and falsework arrangements will be much simpler.

With increasing need for stiffness the raft can be developed into a cellular form (also shown in Figure 28.3). This will be lead to greater depths, which will need to be founded at lower levels, increasing the need for working below ground. This construction will now need production arrangements more akin to basements than shallow foundation. Indeed, it may be economic to make the raft full basement depth and use the internal space created as floor space in the building. The construction of basements will be discussed later in this chapter.

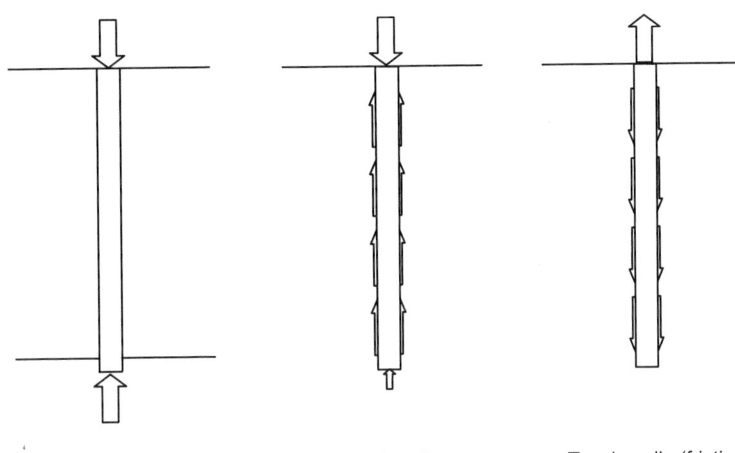

Figure 28.4 Load transfer from piles.

If a raft (or a basement) were founded on soil that was some depth from the surface, the weight of soil removed would compensate for some of the weight of the building, which would mean that the additional pressure on the soil would be less than if the building were founded at ground level. For some lighter buildings with basements this may mean that the soil experiences no additional stress at all, eliminating settlements. Some clay soils, however, heave when the stress of the soil is removed, and this will nullify some of the pre-consolidation effect.

Deep foundation

If sufficient bearing capacity for an economic shallow foundation cannot be achieved near the surface, it may become necessary to consider deep foundations. These will almost certainly take the form of piles, most probably reinforced concrete, but systems employing steel piles filled with concrete have been developed. Load transfer from piles can be end-bearing or use the friction between the pile and the soil. This is illustrated in Figure 28.4. Friction piles can be used to transmit tension forces by resisting being pulled out as well as compressive forces by resisting being forced further into the ground. It is also possible to form some piles at an angle to resist turning moments or resultant forces from complex structures. These are known as raking piles.

The process of forming the pile will be carried out from ground level, eliminating the health and safety risks due to working below ground. This means that the process must ensure the integrity of the unseen construction and there must be some way of judging when an appropriate bearing depth has been reached. On some sites the soil conditions may be sufficiently well established from the soil survey and have been shown to be consistent across the site to provide a design for the length of the pile. There may be a need for bearing capacity to be verified by some production indicators such as set, which is discussed below. It may be necessary to perform a loading test on a trial pile (or group of piles), on the site, to verify design calculations. It may also be arranged for continuity tests to be carried out on each pile after it has been installed. This will not confirm the bearing capacity but check the integrity of the pile over its full length, which, if not achieved, could reduce the potential bearing capacity.

A number of production processes have been developed to form piles, but they fall broadly into two categories:

1. Displacement piles where the driving process compresses the soil to the side of the pile as it enters the ground so no soil is removed

2. Replacement piles where soil is excavated to create a hole into which the concrete is poured

Displacement piles are normally preformed and driven into the ground, although they may be hollow and filled with concrete to complete the working pile. Although these piles can be knocked off line by obstructions in the ground, they do provide a full cross section and can be driven to a set (number of blows to drive a predetermined distance) as a measure of bearing capacity. Fully pre-formed piles ordered to a predetermined length can be used on sites where the bearing depth is well established. If the bearing depth is uncertain or variable, sectionalised piles may be used where sections can be added, as the pile is driven until a set is reached.

Replacement piles have to be excavated to the required depth. Some piles may need lining as the excavation proceeds, to reduce the risk of collapse. The lining is then removed as the concrete is poured. It may only be the top few metres of soil that are unstable, and in these circumstances the lining need only be inserted to retain this unstable layer. For many soils full lining will be necessary to ensure a full cross section is created throughout the length of the pile.

Continuous flight auger (CFA) piles (a replacement system) overcome the need for linings as they retain the soil with the full length of the auger. The installation of this system is shown in Figure 28.5. The cutting flight of the auger is arranged around a hollow stem. When the auger has been screwed into the ground to the appropriate depth, concrete is introduced via the hollow stem as the auger is removed, bringing the excavated soil to the surface. This eliminates the need for the lining and creates the pile in one operation. The excavated soil now needs to be taken off site. The range of CFA piles is between 300 mm and 1200 mm in diameter and can be created up to 30 metres long. For larger diameter piles up to 3000 mm rotary flight auger systems can be used, but these are not continuous flight and the soil has to be brought to the surface in sections. This leaves the bore hole susceptible to collapse and these piles may have to be lined. The pile is then concreted in one operation as the lining is removed. These rotary bored piles start at 600 mm diameter and can be drilled to depths of 70 metres (note this system is not illustrated).

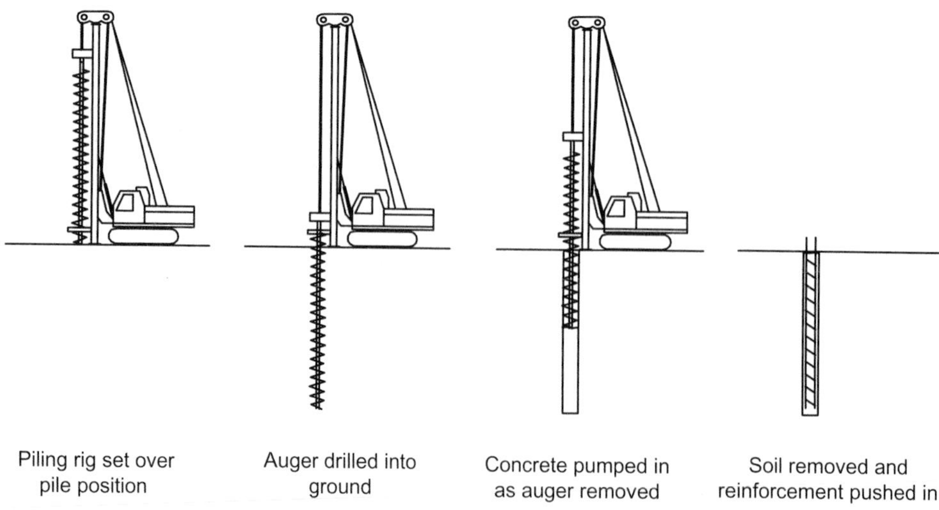

Figure 28.5 Continuous flight auger piling.

Both displacement and replacement piles require piling rigs, but the mechanism for displacement is normally driving with hammers, creating noise and vibration, while for replacement piles a rotary action screws the auger into the ground. Displacement systems installed with rotary rigs to overcome the noise and vibrations are available, including systems that introduce the concrete through a hollow stem as the tool is removed in much the same way as the CFA piles.

Another displacement pile that is not hammered into the ground but screwed in is the steel screw pile, where a hollow steel pile with an auger flight at the end is screwed into the ground and then filled with concrete. The bearing end is now the diameter of the flight and it is possible to remove the pile at a later date by reversing the screw action.

Figure 28.6 shows typical piling plant and site operations associated with pre-formed displacement piles. Precast piles are produced in standard lengths up to around 18 metres, although longer piles can be obtained and sectional systems can create longer piles. Once driven the tops of the precast piles are trimmed to expose the reinforcement. Steel 'H' section bearing piles are also driven with similar plant.

While large diameter rotary bored piles of 1200 mm diameter or greater can be used as a single pile, most piles will be between 400 and 800 mm diameter and will be arranged in groups with a cap connecting them at the top to which the column will be connected.

These pile groups with the cap provide some stability to the top of the piles that, together with a system of ground beams connecting the pile caps, will hold the tops in position. Pile groups also provide some margin of safety if one pile has a lower than anticipated loadbearing capacity. Some typical pile group arrangements are shown in Figure 28.7. When arranging groups, the distance between piles is important if each is to achieve its full loadbearing capacity. For end-bearing piles the piles should be at least twice the diameter of the pile centre to centre; for friction piles this spacing needs to be at least the perimeter of the pile. The connection to the pile cap will be made by taking the reinforcement from the pile into the cap. The top metre of the pile will be removed to expose the reinforcement and ensure the quality of the concrete

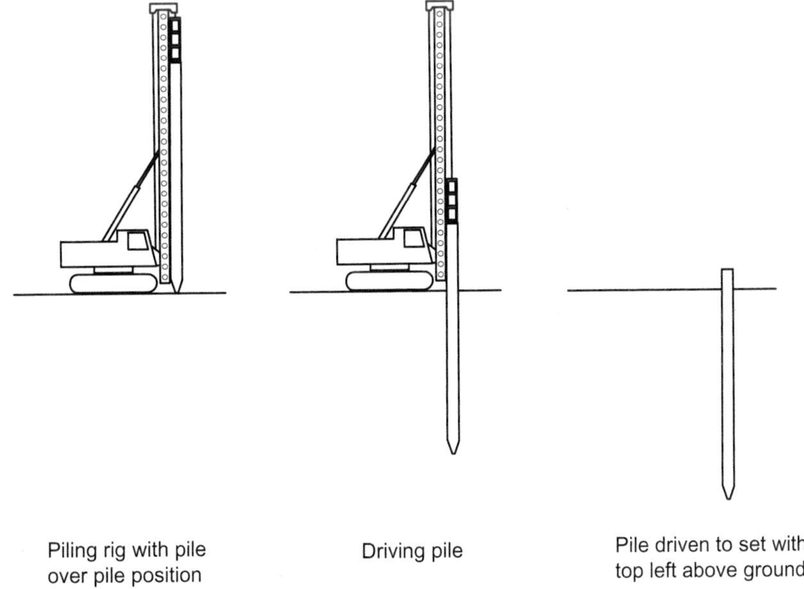

Figure 28.6 Precast driven piling.

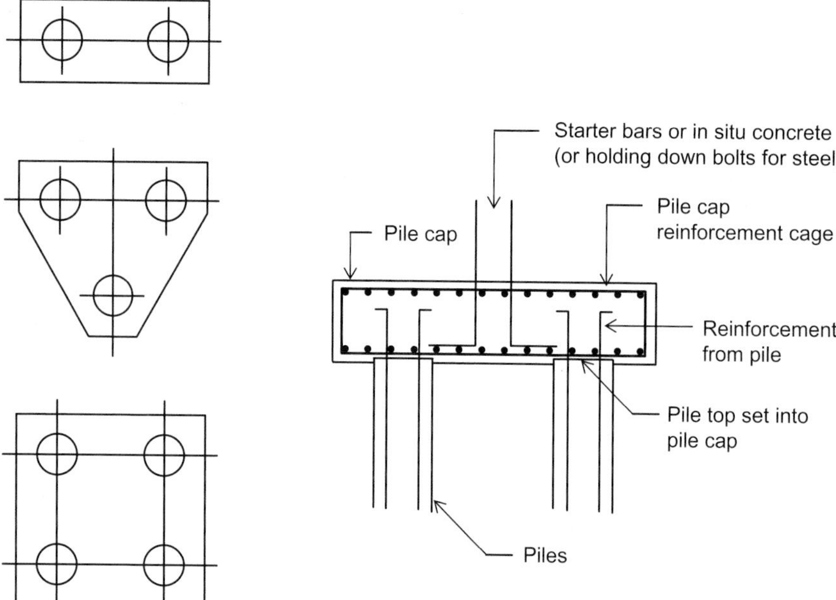

Figure 28.7 Pile cap arrangement.

at the top of the pile. In addition the cap will be cast with the top of the pile itself around 50 mm into the cap to give a mechanical restraint to the top of the pile. This is shown in Figure 28.7. It is then common to connect the pile caps together with ground beams that also stabilise the tops of the piles.

Soil improvement

Soil improvement aims to increase the bearing capacity of the soil. This will be of value where the improvement may allow a less expensive foundation solution. It is also used on recently filled sites where high compressibility will require deep foundations for the column loads, but there is still value in improving the bearing capacity for the ground-floor slab. The role of soil improvement in foundations may be to allow a shallow rather than deep solution or a simpler raft construction. It is not without cost and will take time, so its use needs to be set off against simpler or less expensive foundation and/or slab solutions.

The techniques used fall into three broad categories:

1. Compaction
2. Installation of stone columns
3. Introduction of cementatious materials

Many soils can be improved by densification by removing the air (compaction) or in some cases the water (consolidation) from the voids, bringing the soil particles into closer proximity to improve the bearing capacity. While this can be achieved by loading the area of soil with a depth of overburden, this takes time for the densification to take place, and so the more immediate dynamic consolidation process is more likely to be considered. Dynamic consolidation is undertaken by dropping heavy weights onto the surface of the ground. The resultant indentation is a measure of densification where the mass (typically 7 to 11 tonnes) and shape of the weight (normally cone-shaped) will determine the effective depth of consolidation, depending on the needs of the subsequent foundation work. The densification is achieved with two or three passes with the weight dropped in an overlapping pattern thus reducing the surface level across the whole site.

The introduction of stone columns requires the forming of holes by displacing the soil to compact the soil adjacent to the hole. This is achieved with a vibrating poker that is driven into the ground. Stone (often recycled concrete) is then poured into the hole and the poker reintroduced to drive the stone into the surrounding ground. The hole is then refilled with stone to reinstate the continuity of the improved soil. The technique is known as vibro-displacement and compaction and is used to create a grid of stone columns across the whole site.

The introduction of cementatious materials can be achieved in one of two ways, depending on the depth of improvement required. For surface improvement the soil can be removed, mixed with cement and water to the optimum moisture content and then compacted back in place. For deeper applications grouting techniques can be employed. This requires the injection of cement and water (a grout) to fill pours' fissures and voids that are limiting the bearing capacity of the soil.

Each of these techniques has to be carefully chosen for soil and site conditions, bearing in mind the subsequent engineering works for which the improvement is intended. Such a discussion is beyond the scope of this text, which is intended only to introduce the broad engineering and production options associated with work below ground.

Basements

Unlike for foundations, the choice of basement construction has to consider environmental control (mainly waterproofing) as well as loadings and will have to be constructed by working below ground level. The necessity to work below ground increases the health and safety risks. It also introduces the need for temporary works for access and earthwork support. These two factors have led to the development of techniques that include part of the permanent basement wall that can be completed from ground level that will also provide the structural support for the soil during the construction process. This is an example of how the engineering requirements and the demands of the production process can be integrated at the design stage selecting technical solutions to provide more economical and/or safer production environments.

Basement structures are potentially subject to loads not only from the soil but also from groundwater. This is illustrated in Figure 28.8. On many sites the groundwater will be below the level of basement slab so the walls have only soil pressure to resist. The soil will still be wet with its potential to create dampness in the basement, but there will be no hydrostatic pressure on the structure. On sites where the groundwater table is higher than the basement level both walls and slabs will be subject to hydrostatic pressures.

The high groundwater level not only puts bending forces in the walls and slabs but also induces buoyancy in the basement. While this buoyancy may be overcome by the full weight of the finished building, a partially constructed building may not be of sufficient weight, so the basement may float upwards. Traditionally on these sites temporary groundwater pumping would be used to lower groundwater during construction. This would not only allow work to be carried out in a dry excavation but also, if the pumping were continued until sufficient building was completed, overcome buoyancy problems. An alternative would be to flood the basement to an equivalent level to the groundwater when pumping were stopped, filling the basement as the natural groundwater level returned.

Approaches to waterproofing

Waterproofing basements falls into one of three broad categories:

1. Tanked protection
2. Structurally integral protection
3. Drained protection

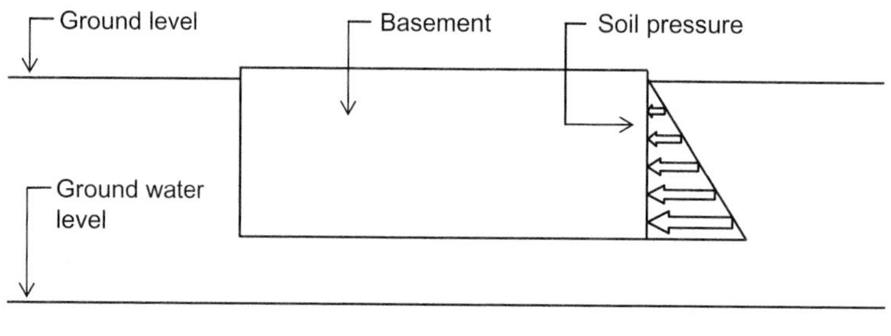

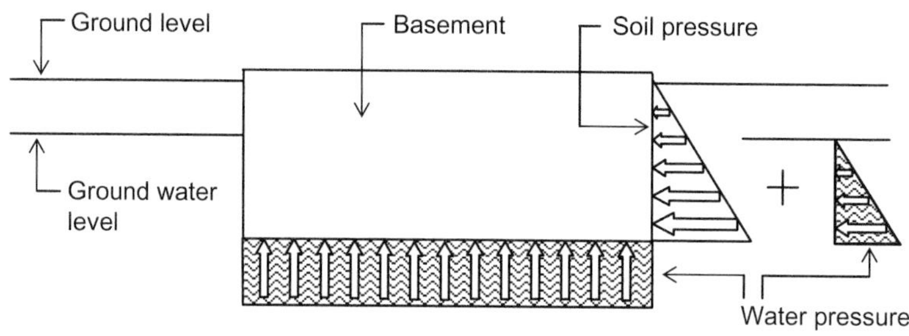

Figure 28.8 Ground forces acting on basements.

Choice will depend on the risk to the basement usage. While a fully dry basement can be achieved by all these methods, the greatest risk of limited performance is associated with the structurally integral method. As the performance level determined by the use increases it may be worth considering either tanked or drained protection methods, even though these methods are more expensive.

Structurally integral protection is based on the naturally low permeability of good-quality structural concrete. However, the waterproofing of the basement will be compromised if the structure develops cracks and the day joints will provide weakness in the waterproofing of the wall if not detailed and constructed with some care. Construction is shown in Figure 28.9. Walls will need to be at least 250 mm thick and be reinforced to limit cracking. Defects in the concrete such as grout loss from formwork causing honeycombing must be avoided. Day joints will need to be cleaned and prepared for the next pour and will need to incorporate a waterstop, such as the one shown in Figure 28.9. Such a concrete structure can be made waterproof but it will not be vapour-proof, and so dampness can still occur within the basement. An external or internal vapour barrier will need to be applied.

Tanked protection will involve the introduction of a continuous waterproof layer to both the slab and walls of the basement. It must be protected from mechanical damage and may have to be supported to stop hydrostatic pressure breaking the bond between the tanking material and the structure. This risk only exists if there is any chance that the basement will become below the water table. Typical construction for both internal and external tanking is shown in Figure 28.10. External tanking requires only limited protection, as all the ground forces hold the tanking layer against the structure, but creates additional production concerns. The basement slab has to be constructed on the tanking material, which will need to be protected by a screed, and

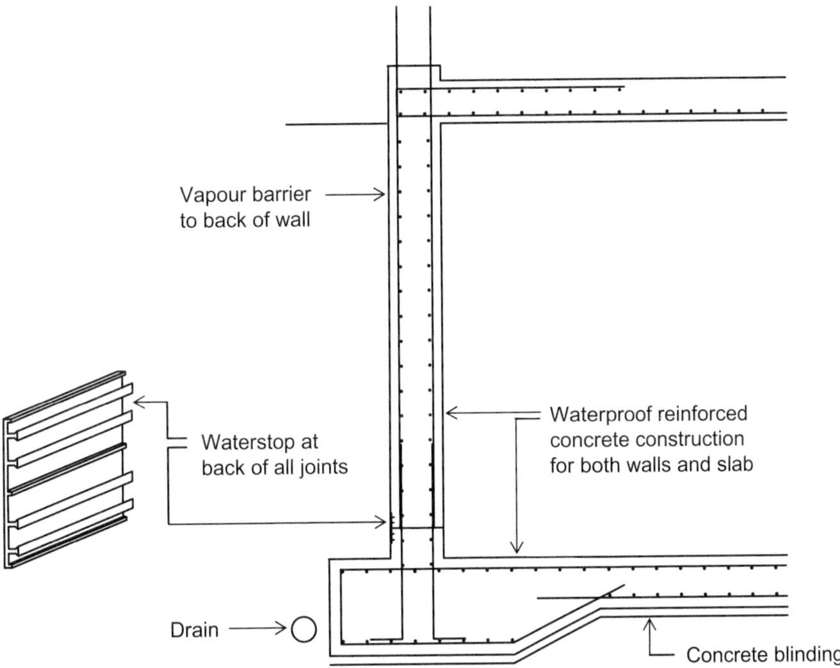

Figure 28.9 In situ reinforced concrete basement.

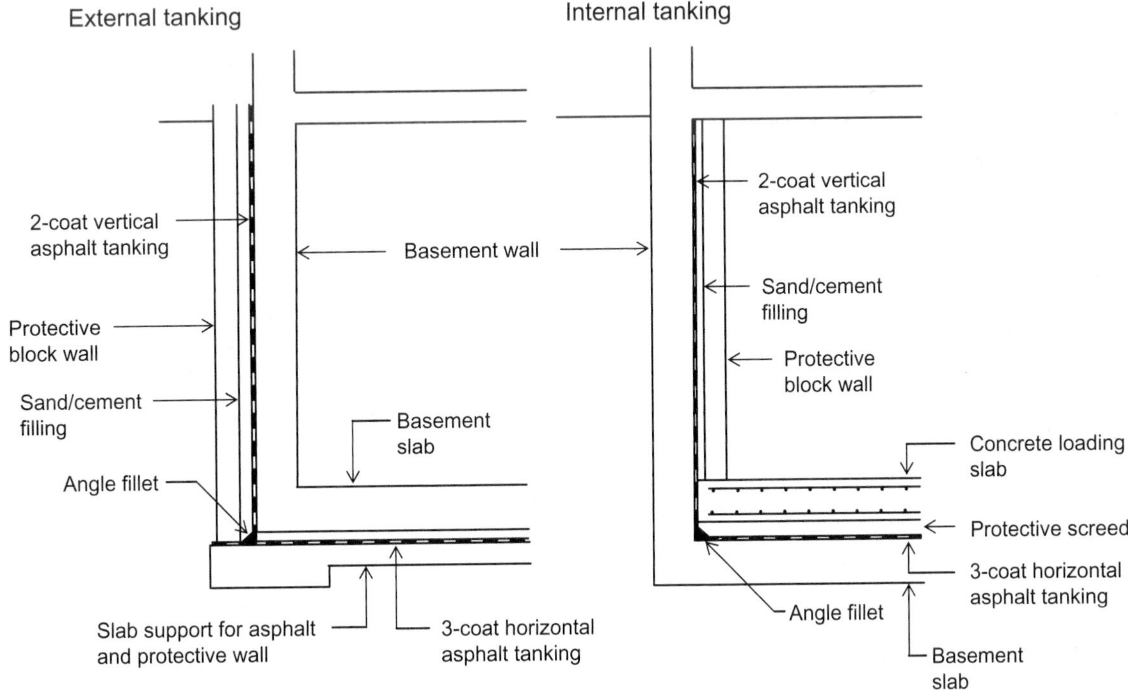

Figure 28.10 Traditional asphalt tanked basement.

working space is essential for the application of the tanking material and the building of a protective skin wall prior to backfilling. Internal tanking is applied to the completed basement within the structure where space and a safe environment can be created more easily. Protective walls and screeds are then required to provide a surface for the finish to the wall. With internal tanking the risk of detachment with water pressure is greater and therefore precautions such as loading slabs may have to be specified. Internal tanking also has to be applied to internal columns, which can introduce a damp path as the tanking cannot be applied under the column at slab level.

Traditionally the waterproof membrane would be asphalt. This is a hot-applied material in three layers, building up to 30 mm thick horizontally and 20 mm thick vertically. Each layer is laid in bays with the joints between bays staggered between layers. Where horizontal layers meet vertical layers, a 50 mm angle fillet has to be applied to ensure a seal between them. Where services penetrate the wall, an asphalt sleeve has to be applied to the pipe and then an additional asphalt pad applied to the wall around the pad and an angle fillet used to seal the joint between the pad and the sleeve around the pipe.

An alternative to asphalt is the cold application of bituminous membranes. These membranes are supplied as sheets rolled with an adhesive surface protected by a release paper. Rolls are around 1 metre wide and up to 18 metres long. Joints have to be lapped and where horizontal and vertical surfaces meet additional reinforcing strips have to be applied before the sheets are laid, being turned to seal the junction. The surfaces of the basement structure have to be clean, dry and free from any small defects that could puncture the membrane. Vertical surfaces then need to be primed to ensure adhesion, which is only achieved by pressing the adhesive side of the membrane to the structure's surface. The bituminous-based materials are more flexible than asphalt and can accommodate small movements.

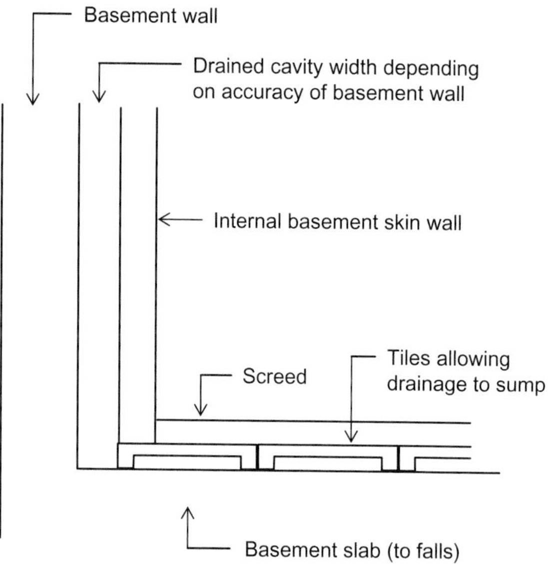

Figure 28.11 Drained waterproof protection for basement.

The third broad option for waterproofing basements is the provision of a drained cavity between the basement walls and slab and between the floor and walls that provide the internal surfaces. This approach is illustrated in Figure 28.11. This solution may be used either when the use of the basement demands high levels of performance to ensure damp-free environments or when the chosen basement wall construction has to be assumed to leak in a small number of areas and therefore damp conditions must be taken to be inevitable. This is the case with the combined temporary/permanent wall solutions introduced below.

While the cavity approach to waterproofing is very reliable as a passive system, as has been proved over the years, with the external cavity wall the draining mechanism often relies on a pump to remove the accumulating water from a sump formed in the structural basement slab. Gravity drainage may be possible but levels often preclude this.

Temporary/permanent wall solutions

The integration of temporary and permanent works for basement walling illustrates the

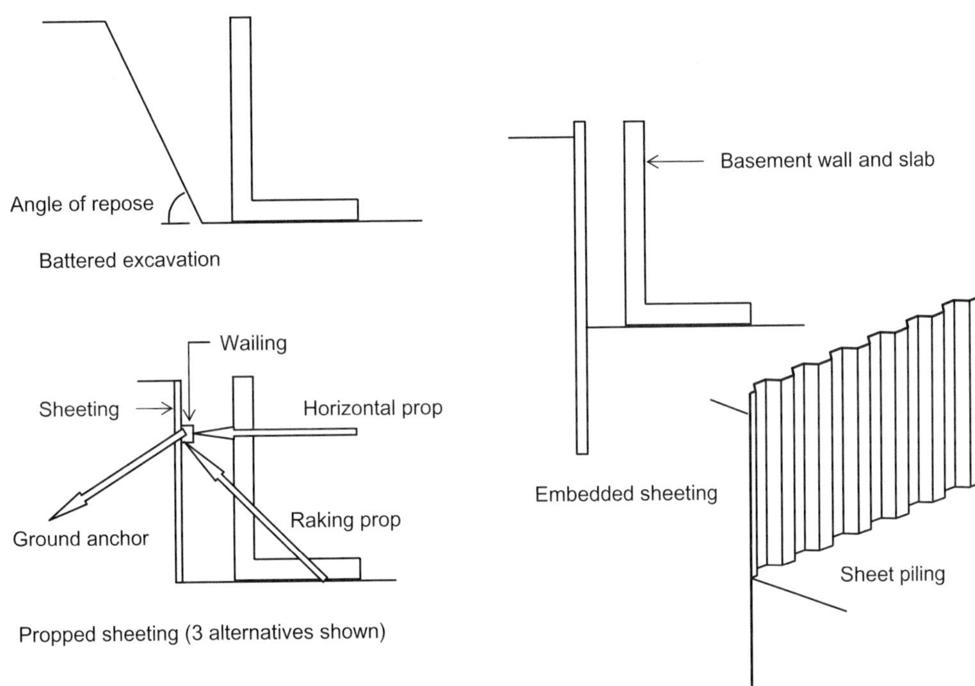

Figure 28.12 Broad approaches to earthwork support.

extent to which engineering and production issues come together in work below ground. The need for a safe working area below ground leads to extensive temporary works. Although beyond the scope of this text to look at this in detail, the broad approaches with some ideas of solutions are given in Figure 28.12. All have safety, cost and time implications, depending on the ground conditions and the proximity of existing or concurrent development. For the methods shown in Figure 28.12 the temporary works, as their name suggests, contribute nothing to the performance of the final wall and will be removed when the permanent wall is capable of resisting the earth and any hydrostatic pressures. Some part of the temporary support walling may not be removed but will be 'lost' as construction proceeds and will not contribute to the permanent wall's performance. The exception to this may be the sheet piling, which can be used as the permanent formwork for a concrete basement wall and its strength can be taken into account in the design of the permanent wall.

Temporary works do not need the accuracy of line and plumb of permanent works or the high requirements of the levels of waterproofing and durability. Obtaining higher performance in these systems would be problematic. However, the idea that temporary works could be incorporated into permanent works provides opportunities to reduce health and safety risks and perhaps reduce overall contract time. The installation and removal of temporary works can be time-consuming and restrict the speed with which the permanent works can progress. Like all temporary works, earthwork support gains economies because components can be reused a number of times during the construction. Schemes that can be phased to reuse temporary works components will reduce costs. However, the ability to open up the entire excavation at one time will considerably improve the opportunities to progress the works, reducing overall time for the contract.

In addition to sheet piling being incorporated into the permanent walls, a number of systems

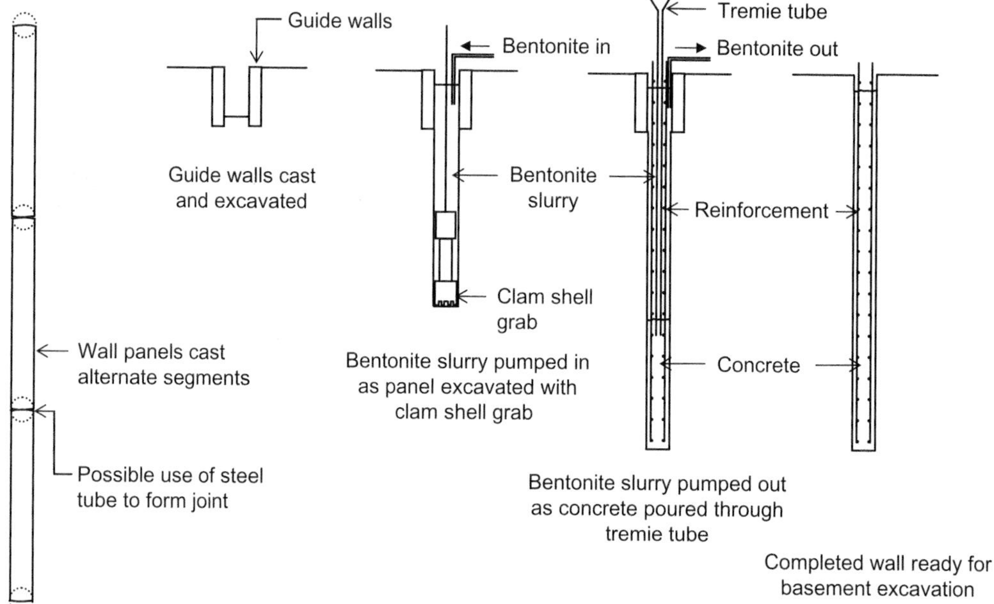

Figure 28.13 Diaphragm walling.

have been developed to provide the actual concrete wall itself from above ground prior to excavation to provide both temporary and an element of the permanent works. They do not achieve the accuracy in line and plumb required for the permanent works, nor can they ensure waterproofing performance, but they will satisfy the durability expectations. These solutions are often combined with cavity tanking to achieve both tolerances and damp-proofing performance for the internal spaces in the basement.

Two of the main options for installing walls from above ground level are diaphragm walling and contiguous and secant piling. In diaphragm walling the wall is constructed in panels between 600 mm to 1500 mm thick and 5 metres long excavated with a specially designed clamshell grab to achieve the deep excavation depths required. The sequence of operations is shown in Figure 28.13. The work commences with the construction of guide walls. The sides of the excavation are supported by bentonite slurry, which is introduced into the trench as the excavation proceeds and then removed and returned for reuse as the concrete is placed. In its most straightforward form, constructing alternate panels gives a straight joint between panels, and these are almost certain to allow water to seep between the panels. The introduction of steel tubes at the ends of the panels that are removed and grouted when the panels are complete helps to seal the joint against seepage. It is also possible to produce more shaped joints to provide interlocks to improve the structural integrity of the wall and its susceptibility to seepage.

In contiguous and secant piling options, the excavation and the concreting for the wall is achieved with a standard piling auger (see Figure 28.5). Jointing varies between contiguous and secant systems, as shown in Figure 28.14. In contiguous systems the piles do not interlock; indeed, there may be a small gap between each pile. In secant walls piles are installed in an alternating pattern where the second set of piles is cut into the first to provide an interlock. This not only improves the struc-

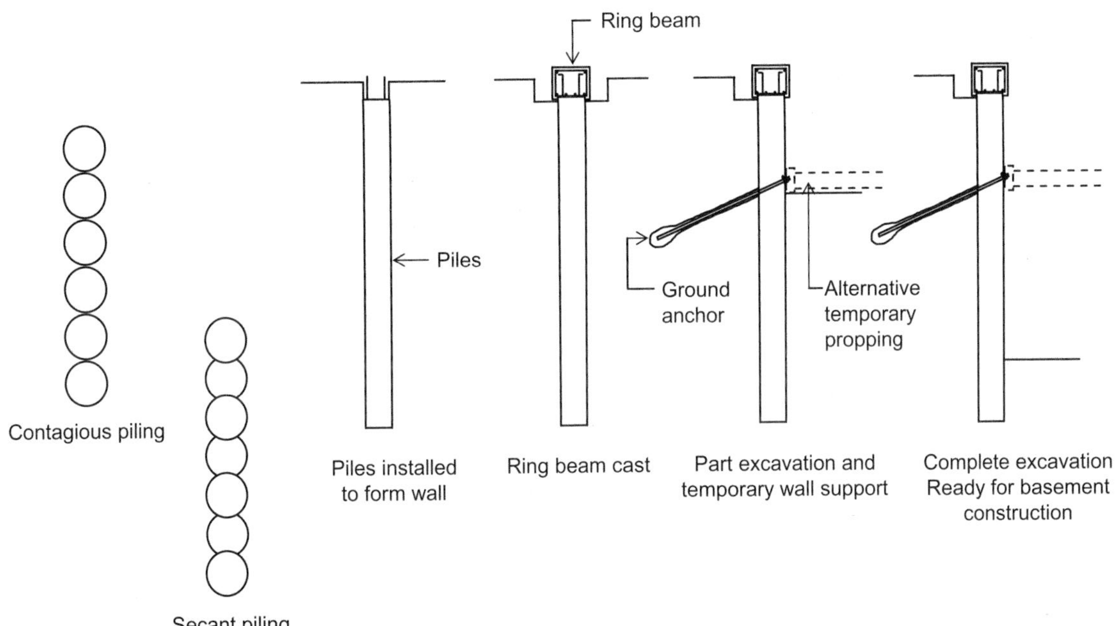

Figure 28.14 Contiguous and secant piled walling.

tural integrity of the wall but also limits the potential for seepage. Both of these are important in some soil conditions for both the temporary works phase of the life of the wall and when it is acting as part of the permanent basement wall. The drilling of the alternate piles requires cutting into the first set of piles. This can be achieved with what are known as soft, firm or hard primary piles. In soft and firm primary piles a retarder is introduced into the mix so that the concrete has limited strength when the secondary set of piles is drilled. Hard primary piles are allowed to develop full strength and then a high-torque cutting attachment has to be used to excavate the second set of piles. These piles will then require a ring beam to be cast across the top of the piles to tie them together to act as a wall.

Both the diaphragm wall and the pile walls are normally designed to be propped up by the internal floors. If the basement is excavated before these floors are constructed, the walls will need support during construction. A typical sequence of excavation is shown for the pile walls in Figure 28.14. If the walls are not embedded sufficiently to act as cantilevers, ground anchors or internal temporary propping will be required to support the walls part way through the excavation.

Another alternative is to cast the floors before excavation to eliminate the need for this temporary support. A hole has to be left in the floor to allow the excavation to be carried out below the floors. This is known as top-down construction. This is only possible if the foundations and internal support for the floor (walls and columns) can also be provided before excavation. This has the potential benefit of allowing the superstructure to commence as the basement excavation proceeds, a process known as concurrent construction, where sub- and superstructure are constructed at the same time. In practice, organising the excavation work so it has little interaction with the construction of the superstructure may be problematic on all but the larger sites.

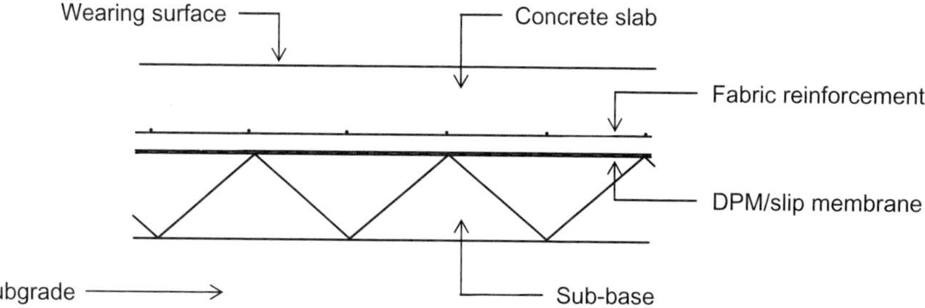

Figure 28.15 Typical industrial concrete ground floor section.

Large area industrial concrete ground floors

The basic floor construction for these large area floors shown in Figure 28.15 appears very similar to that of the ground-supported slab floor for the house outlined in Chapter 17. However, the size of these floors means that the engineering design and the production process are very different. While not strictly work below ground, these floors are introduced in this chapter because, like foundations and basements, the detailing and specification is dominated by engineering and production issues. They also have performance requirements over and above structural strength and durability. These floors have surface performance requirements for appearance and, often more significantly, in their wearing characteristics and flatness tolerances. These performance requirements come from the operational demands of the activity housed in the building. Operational demands often require high loads not only from static equipments but also from wheeled traffic. The development of high-bay warehousing racking systems in particular has made new demands on floor flatness. In addition, the automation of the wheeled traffic, such as wire-guided systems where the wires are buried in the floor, has limited the position of reinforcement so as not to interfere with the signal operating the equipment.

The shrinkage of the concrete in the slab becomes highly significant to the design specification, the detailing and production decisions as the uninterrupted area of floor increases. The shrinkage will induce cracking that, if not controlled and contained to specific designed joint positions, will lead to the slab failure of surface characteristics and structural strength. The method of construction, particularly the specification and spacing of joints, becomes the focus of both engineering design and production methods.

These two aspects of design and production have to be considered together. Over the last half of the twentieth century approaches to both design and production changed dramatically, evolving together to provide higher performance floors in much reduced construction time. This has not so much affected the basic cross section of the floor, as shown in Figure 28.15, so much as had a significant effect on pour sizes, surface finishing and joint design.

When designing joints, a distinction is made between free and restrained movement in order to determine the size of the crack that may form. Both have to restrain vertical movement once the crack has formed. By restraining the movement the crack width is limited and the vertical restraint can be achieved with the slab reinforcement and aggregate interlock. If larger movements are required, the joint must be free to move and so dowel bars have to be introduced (and de-bonded on one side) to resist the potential relative vertical movement that would create a step at the joint.

The idea of joints can be a little misleading in that components to induce cracking are cast into the slab as the pour proceeds. These only become joints when the slab shrinks and cracks. This controls the cracking to the line of well-

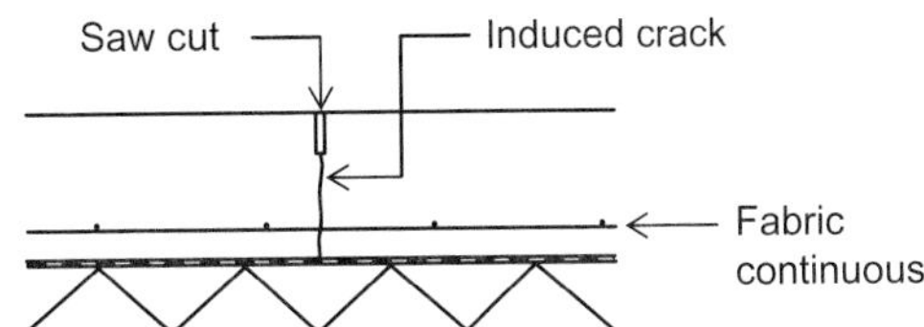

Sawn restrained movement joint

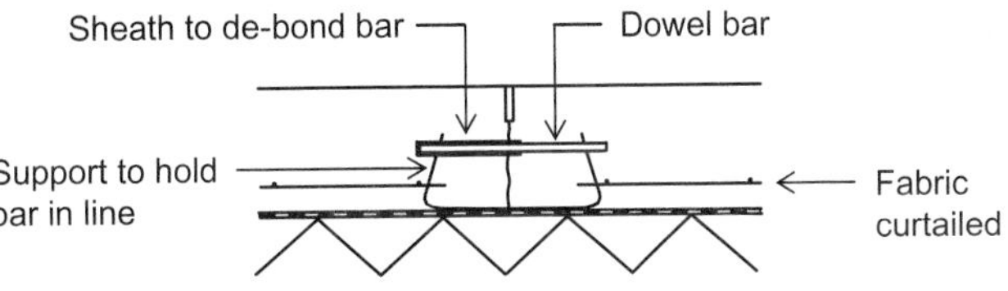

Sawn free movement joint

Figure 28.16 Sawn joints.

defined bays. These are now often known as sawn joints and are illustrated in Figure 28.16. The inducement to crack at the joint is a saw cut around 3 or 4 mm wide and to around a quarter of the slab's depth. The saw cut has to be filled with a material that will accommodate the movement yet give support to the concrete arris formed on either side of the saw cut. This is often a sealant carefully chosen for its movement accommodation factor (MAF) and its hardness to support the arris. This is a compromise as, generally, higher MAF values are achieved at the expense of hardness values. Other filler systems such as highly compressed foam strips and extruded plastic strips have also been developed.

Concrete pours will require breaks with edge formwork, and these joints are known as formed joints. The production process has limits as to the areas that can be levelled and finished and in the volume of concrete that can be laid in one day. Formed joints in the slab can also be designed to be free movement or restrained movement, examples of which are shown in Figure 28.17. Two other joint types are the tied joint, which looks like a formed restrained movement joint, and the isolation joint. Isolation joints are required where the slab meets other elements of the construction such as walls and columns. In these joints restraining vertical movement is to be avoided and therefore isolation details are required. These isolation details are shown in Figure 28.18.

All slabs will require joints (despite one construction method introduced below as 'jointless'), with the choice and distribution depending on the basic construction method chosen. These construction methods are known by the area of slab completed in each pour. Long strip was developed from highway construction and many of the specifications are still based on highway design. In long strip each strip will be between 4 and 6 metres wide. Laid in alternate strips to allow some of the shrinkage across the strip to take place before the infill strip is cast, the edges of the first strips will be formed either with temporary formwork (often 'road forms') or with permanent, precast screed rails, which can be accurately levelled to assist in achieving high flatness specifications. The width of the strip is limited by production equipment. The method of compacting and levelling the concrete is from the side forms. This is, therefore, a major consideration with the wider strips requiring beam screed to limit deflection and mechanically assisted vibration to ensure

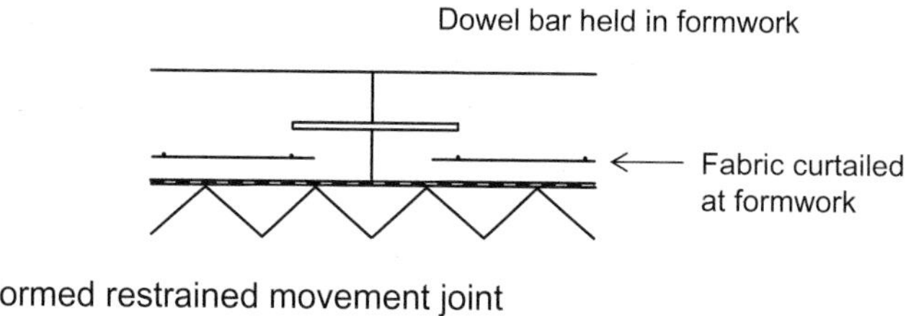

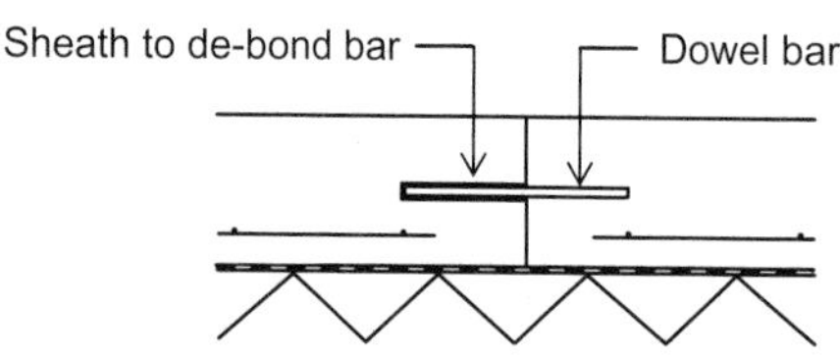

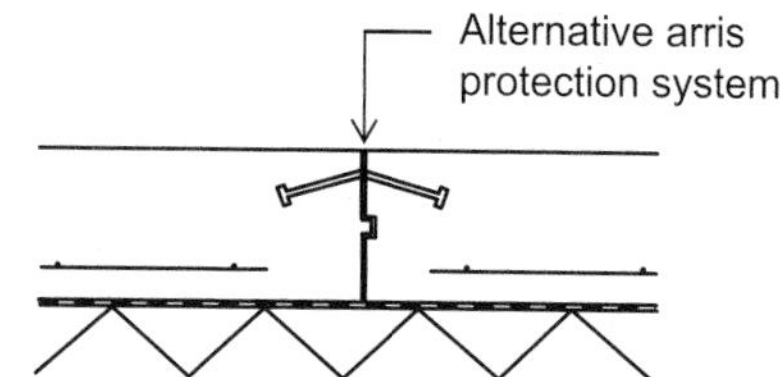

Formed free movement joint

Figure 28.17 Formed joints.

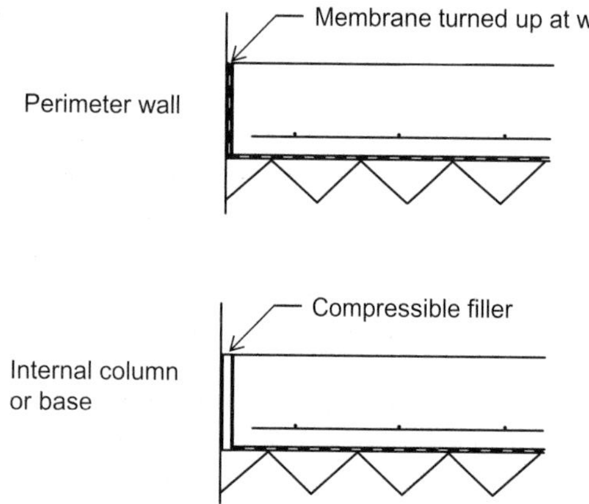

Figure 28.18 Isolation joints.

compaction. Finishing is normally power-floating achieved from the side of the strip limiting the width to reach the middle. Curing membranes to ensure good surface integrity and concrete quality also have to be applied.

With the long strip method the width dictates the number of pours to complete the whole floor and hence the overall construction time. Large area construction, both jointed and jointless, and wide bay construction employ production equipment to achieve laying, compacting, levelling and finishing without the limitation of the width of a strip. This allows much larger areas to be completed in a day and hence reduces the total time to complete the floor. A full introduction to this equipment is beyond the scope of this text but it is highly mechanised and involves high capital investment, limiting the number of companies that can offer the service.

Large area construction allows several thousand square metres to be laid in a single operation in a day. Fixed forms will be used for the edges of the pours, typically around 50 metres apart. These edges will need to be formed joints. In jointed construction sawn joints will be specified within the pour. These will be absent in jointless pours. The ability to eliminate sawn joints is facilitated by the use of steel fibres incorporated in the mix as reinforcement replacing the need for mesh reinforcement. While this will give a continuous wearing surface eliminating the potential for arris deterioration at the sawn joint, it does increase the risk of fine visible cracks appearing on the floor and the formed edge joints have to accommodate all the shrinkage opening up by as much as 20 mm.

It is more difficult to gain the highest standards of flatness required across the whole of a large area pour. One way to overcome this is to adopt wide bay construction with bays up to 15 metres wide constructed with large pour equipment techniques. It is then checked for flatness on the hardening concrete and a high spot removed before fully hardening. The limit in width allows access for the equipment to carry out this so-called bump cutting operation.

Flatness, or more specifically surface regularity, requires specification and measuring techniques designed specifically for these large area floors. These define a number of parameters that need to be fully specified for the surface and to be checked to ensure the operational performance of these floors. For this surface regularity specification two distinct operational conditions are identified. Free and defined movement refer to the pattern of movement of the wheeled equipment envisioned in the operational use of the building.

The discussion so far has covered the slab and its surface characteristics. Figure 28.15 identified other layers in the overall floor construction. The membrane is typically 1200-gauge polythene sheeting lapped 300 mm at the edges. This is a similar specification to the house floor damp-proof membrane (DPM), but it needs to be appreciated that it has another function in the successful performance of these floors. It is identified as a slip membrane. Its purpose is to allow the shrinking concrete to 'slip' on the sub-base, helping to relieve the tensile forces and focusing the cracking at the joints as the movement takes place. The membrane of this specification does, of course, also fulfil the vapour and damp-control requirements. If in addition gas control is required, the specification of the membrane may have to change but the dual function with allowing slip can still be achieved.

This slip function puts an additional requirement on the specification of the sub-base. Normally a fill material, specified as road sub-base fill type 1 or type 2, has to be smooth enough to allow slip. Top surface tolerances on levels should not allow any part higher than specified as this will reduce the thickness of the slab.

The sub-grade or soil on which the slab is to be constructed has a major influence on the required thickness of the slab. On very poor soils it may be economic to carry out soil improvement (discussed above) before completing the slab. On very good soils it will even be possible to specify the slab without the need for a sub-base layer, but this is unusual.

The cross-section in Figure 28.15 shows a nominally reinforced slab representing a mesh reinforcement placed towards the bottom of the slab. In slabs with sawn joints the crack-resisting function of the steel mesh is reduced but still offers continuity across the joints. It is therefore considered that the position of the mesh (traditionally at the top, where cracking is least acceptable) is less important. The replacement of this mesh with steel fibres has been discussed above but two other reinforcement conditions may be required. It is possible to design slabs that are not reinforced, but these are not common in the UK. For slabs with local high loadings or supported on piles, fully reinforced (or even pre-stressed) slabs will have to be adopted.

One last point to make about the production requirements of these floors is that the equipment and techniques that have been developed are based on the assumption that the floor will be laid after the roof and walls have been completed. The floor will be laid inside, protected from the elements. If this is not the case, other methods may have to be chosen. Even hot and drying conditions are not conducive to producing good concrete in large exposed areas without considerable protection, which would not be required for concrete laid inside.

Summary

1. Foundations, basements and large area industrial floors are all dependent on an understanding of the behaviour of the soil and the implications for engineering and production that are often dominant in choosing the methods to be adopted.
2. Shallow foundations will normally be in reinforced concrete and may need earthwork support and almost certainly require formwork.
3. Shallow foundations for framed buildings in good bearing soils can be formed on pads with options to combine the column loads, particularly where loads come to the ground near to restrictions.
4. On poor soils rafts can be considered, stiffened by beams that intersect at the column positions or developed into cellular structures that, if deep enough, can become used as basements.
5. Deep foundations are normally based on piles installed from ground level either by driving a pre-formed pile or tool that displaces the soil or by drilling a hole into the soil and replacing it with concrete.
6. Piles carry loads either by end-bearing or friction and are normally designed in groups with a cap connecting the tops of the piles and ground beams connecting the caps. Large diameter piles (1200 mm plus) may be one pile for each column connected by ground beams.
7. Soil improvement by compaction, installation of stone columns or the introduction of cementatious materials may be worth considering if it allows the use of shallow foundations or perhaps improves the support for the ground-floor slab.
8. Basement walls are subject to soil pressure and, if below groundwater level, hydrostatic and buoyancy forces.
9. Waterproofing of basements can either be designed integral with the structural wall or tanked (external or internal) or with drained protection.
10. Diaphragm walling and contiguous and secant piling can be used to install the walls from the ground level to provide a combination of temporary and permanent wall solutions to create the required safe working environment and potentially reduce construction time.
11. Large area industrial concrete ground floors will be subject to tensile forces as the concrete cures and shrinks, causing cracking. Joints must be arranged to restrict the cracking to specific bays where the crack can be controlled and the joint detailed for the movement.
12. If large pours are to be undertaken, many of these joints will have to be formed with components cast into the concrete to induce the crack and control the movement. These are normally known as sawn joints.
13. Flatness is now a significant aspect of the performance of the floor and this has also influenced the production process and the development of specialist equipment to achieve large area pours.

29 External Enclosure to Structural Frames

The major part of this chapter considers a range of options for the construction of external walls to framed buildings. It first introduces some general considerations that will influence the choice of materials and components and their specification and detailing. Aspects of appearance, performance and construction including establishing structural integrity are discussed as well as issues around site assembly operations. The considerations that influence the choice of roof construction are considered, although only green roofs are considered in detail.

Introduction

To undertake a full analysis of the building enclosure to ensure a choice of construction that will perform and can be built is one of the major challenges in designing and constructing a building. The external enclosure normally comprises two elements: walls and roofs. While these have many environmental performance requirements in common, their influence on appearance and the actual form of construction can be very different. This difference is, however, likely to be less for the lightweight cladding of the industrial buildings that are discussed towards the end of this chapter.

External walls

There are a number of words in common usage for the wall that further indicate the wide range of analysis required in the choice of enclosure elements.

- 'Façade' refers to the appearance of the building, the face that can be seen. It is determined by the context and the design vision for the building.
- 'Skin' refers to the behaviour of the whole wall in modifying and moderating the environment. It is determined by the flows and transfers being mediated that define the functions of the wall and the internal and external conditions that determine the performance.
- 'Wall' refers to the physical construction. It is determined by the requirements of the wall as facade and skin but is also influenced by its need for structural integrity and the production resource required for its economic, timely and safe construction.

It will be necessary to consider the enclosure elements as each of these to ensure the full performance of the final construction.

External walls (and roofs) have to operate in varied and sometimes harsh physical environments being exposed to the external climate. These differences between internal and external conditions will create flows across the wall. The need to satisfy the requirements for low energy and low environmental impact solutions will make demands not only in its operation as skin but also in the choice of materials in the specification of the wall. Part of the exposure condition is to the forces from wind. This wind loading acts on the external wall and then has to be transferred to the structure.

Facade, material and time

It is most likely that the starting point for the choice of facade will involve the identification of a material(s) to be used externally. Each material tends to have a natural component form for the scale and performance of the external wall: concrete in panels, timber in boards, steel as sheets, brickwork as continuous face. This link of material to component form is not, however, fixed. For example, materials such as brick and stone-faced concrete panels can be designed to look similar to laid brick or stonework. The choice of material and its component form for the facade will significantly affect the detailing of the wall as some will need full or intermediate support to ensure structural integrity, while others can be designed to be fixed directly to the building frame. Those that do not provide the integrity of wall will need a backing structure or sub-framing fixed to the building frame to support the finish. This will be discussed more fully below in the section on wall types.

The wall as facade will change with time. The facade will weather. Exposure conditions, both natural and man-made, will potentially change both the appearance and the performance of the materials forming the face of the wall. The wall will experience both soiling and destructive forces from its exposure to the elements. This requires care in the choice of both the materials and the detailing of the wall. For some materials these change can be seen as beneficial with buildings softening and mellowing with time. For others potential staining and discoloration along with deterioration processes can spoil the appearance, detracting from the quality and design intention for the building as a whole. Understanding the potential for these changes becomes a major area of analysis for a proposed solution during the process of developing specifications and details.

These weathering actions will determine the need for maintenance and repair. Discoloration and staining may need cleaning, while deterioration may lead to the need for redecoration or even repair during the life of the building. These should be anticipated and evaluated in the overall adoption of a solution to meet the client's brief and the quality of the design being pursued to deliver that brief in specific context for that building.

At the end of the building's life the assumption is often for demolition, where aspects of recycling of material need to be considered. However, it may be better to allow for dismantling of the facade to allow not only the potential for re-cladding but also the ability to reuse some components. Such decisions do involve assumptions about future demands for materials and building forms, but if these objectives can be achieved with care in detailing, to allow sequence and fixings to accommodate removal and reuse, the opportunity should be considered.

Skin, layers and environmental performance

The external wall has many environmental functions to fulfil in its role as skin. The wall will mediate all the potential flows of energy as heat, light and sound and the physical transfer of moisture and air to maintain the internal conditions that keep us dry, warm and clean and provide levels of light and sound that determine our physical well-being. In addition the wall will provide many aspects of privacy and security that determine our social and psychological well-being.

In fulfilling these functions the wall will provide waterproofing, ventilation and lighting as well as thermal and acoustic properties, all of which directly influence the internal conditions. The sense of connection to the outside world and levels of privacy are provided with views from both inside and outside by areas of glazing. The construction of the wall contributes to security from threats from both society and fire.

The choice of facade material and its component form is unlikely to fulfil all these environmental performance requirements. The performance demands on the wall increase as passive designs are adopted for the generation and control of internal conditions. The engaging of the active fabric to control conditions makes more demands on the position and size of openings probably as windows, and this will affect the appearance of the facade and be influenced by the choice of facade material.

The need for other layers in the wall other than the facade becomes inevitable: insulation layers, fire-proof layers, internal finishes behind the facade and shading and screens in front affecting the appearance of the building. These need to be fixed and jointed to the element in the wall, providing the basic structural integrity that transfers the loads to the building frame.

The analysis of a proposed solution then needs to be made against required performance levels. These will be set by the required conditions in the building and the external climatic and social conditions specific to the context of the building being considered. The ideas associated with both setting conditions and the approaches to achieving environmental performance and the analysis that has to be undertaken have been explored in Part 1 of this book.

Wall, structure, openings and assembly

The choice to adopt a frame structure for a building, as opposed to loadbearing walls, will release the external walls from the need to carry dead and imposed loads, although they will still be subject to wind loads, which will have to be taken into account. This release from a major structural role makes possible a much greater range of construction forms for the walls themselves. This is clear from the observation of any modern cityscape, but closer analysis will show the emergence of a few common forms. Identifying the characteristics of these common forms at the concept design stage suggests a variety of ways of thinking about the options.

Thinking of the wall as facade has identified the face material and the component form that best expresses the vision of the building and its context. When viewing the wall as skin, the sub-elements of glazing and the layers within the wall should be considered in terms of their size and position in order that they achieve requisite levels of performance. But at this stage these are just the parts of the wall required for performance. They need to be arranged as an assembly. This raises the question of structural integrity: which is the element that provides the basic structure and, in turn, to which the other parts can be fixed? Which element allows the transfer of loads (including wind) to the building frame? If none of the elements suggested by the performance as facade or skin can act in this structural capacity, a separate structural sub-system will be required.

This analysis of structural integrity will also include aspects of achieving structural connections to the building frame. The connections will influence the ability of the wall to accommodate movement and achieve fit in the assembly process. There will need to be consideration of both inherent and induced deviations.

The need for structural integrity

Thinking of the wall as an assembly will require the construction to be visualised as a set of components of different materials and different sizes connected together to achieve the requirements of facade and skin. The spatial arrangement of the components and the detailing of the joints and fixings have to be related to the structural support. This idea of the wall being based on components of various materials and sizes but acting as a whole leads to a need to identify an element of the wall that achieves structural integrity.

The self-weight and the wind loading on the external wall have to be transferred to the structure at each floor level and to the columns. This provides a grid of possible support, the dimensions of which are dependent on the form of the structural frame. For buildings where skeletal frames are chosen, floor-to-floor heights are likely to be around 3.5 metres and column centres between 4.5 and 9 metres. If the building is framed with a long-span roof structure, the height of the eaves can be between 4 and 14 metres and the column centres between 6 and 12 metres. Another distinction between skeletal and long-span roof structure is the need to limit dead loads in the enclosure of the long-span roof building, particularly the roof covering. This leads to the use of lightweight systems being used for at least part of the wall, particularly where buildings have high eaves.

One of the elements of the wall has to have the structural integrity to carry all the loads back to the building frame. This integrity function may be achieved by the facade cladding itself or may be one of the other layers providing the skin functions or it may have to be an element with the sole function of providing the structural integrity.

It is possible, therefore, to think about walling solutions as gaining this integrity in one of three ways:

- From the facade components themselves that can span between floor edges and/or columns
- From one of the other skin performance layers that can be used to support the facade components
- From a purpose-designed structural sub-system for support

If panels of reinforced concrete are chosen, there is the possibility to design them to span across the available structural support with no need for structural sub-systems. If clay bricks are chosen, they can be laid on site to form wall panels also capable of spanning between the structural support provided by the frame beams and columns. These two examples bring out another major aspect that influences the choice of the construction, the manufacture and assembly process. Concrete panels are likely to be envisioned as prefabricated off site while clay bricks as an on-site traditional trade process. These are the most likely production options but not the only ones. It is possible to cast the concrete in situ and to pre-form panels of clay bricks, where effects such as stack bonding can be achieved. These production processes have a considerable influence on the connection to the frame.

Another cladding system that includes the components to span between floors is curtain walling. In this system vertical mullions are connected to the building frame to provide support for glazing or panels of a variety of materials which mimic the edge profile of glazing units.

The three examples above can gain their support and restraint from the main building frame members. However, many facade materials and their component forms will not span across the building frame directly and will need to be fixed to sub-systems that can achieve the structural integrity and transfer the loads to the frame.

Many facade materials are best fixed to a full backing panel, as either a solid (masonry) or a framed (timber or steel) system, as the cladding components have little or no spanning capacity of their own. These component forms would include renders, flat sheets, tiles and boarding. The choice between solid and framed backing panels is likely to be influenced by other criteria identified in the analysis of the wall as skin, such as thermal or sound performance, and by issues such as sustainability, production and cost.

Some facade materials, formed into economic and/or design determined component size or shapes, have some spanning capacity but are not sufficient to make use of the structural frame directly. While these could be fixed to full backing walls, components such as sheets or insulated sandwich panels can be fixed to a structural sub-framing system, such as cladding rails. Profiled metal sheets or sandwich panels offer a lightweight solution often chosen for the type of building that may use a long-span roof frame. These sheets need regular support at around 2-metre centres. Lightweight panels in metal or glass reinforced cements (GRC) or plastics (GRP), while being capable of being made into panels that can span directly to the building frame, may be designed in smaller panels to a specific size and pattern for the building. Here cladding rails will be at a specific height, most probably based on the windowsill and head heights. These structural sub-framing systems are normally designed as part of the building frame, not the wall, and will most likely be in steel. Perhaps the most easily identified need for a sub-framing system is for structural glazing where the glass is the only skin layer and does not have the strength to span between frame members and so generates the need for an exposed sub-framing system.

All these systems will be discussed in more detail under the sections below that deal with the detailing and specification of each of these broad options.

The need for lighting and ventilation

In most external walls of a building, as part of the skin functions, there will be need for light and ventilation, normally identified by windows. For many cladding options this will be resolved by providing holes in the wall into which a frame can be fixed with glass in either fixed or opening lights. While this is a good general description of many of the solutions, it is not always accurate. In curtain walling, for instance, the windows are just one of the panel options to fix between the structural mullions. It is also possible to make the window frame of a size that fits inside the structural frame so that it the window becomes a cladding panel. This can be seen in shop fronts and entrance lobbies. It is then possible to envisage a system where these storey-height frames are used to clad a building with panels and windows being fixed in the frames in much the same way as panels and frames are fixed into curtain walling.

Structural glass as an approach eliminates not only the solid wall but also the frame to the glass to give uninterrupted areas of glass often across more than one storey of the building. This has considerable skin environmental design implications, of which a limited ability to open any of the glass as with a conventional window is one. This provides a challenge to the provision of structural integrity. The sheets of unframed glass are unable to span the distances provided by the structural frame, indicating that a structural sub-system is required as identified in the previous section.

Wall manufacture and assembly

The manufacture and assembly of the components of the external wall pose particular challenges to the construction of the building. The wall itself is a major assembly being made of many components all of which have to fit together, although they are often made of many different materials and in different places. The assembly process is carried out at the edge of the building and often at height and so has particular health and safety concerns. The location of the wall at the edge of the building also makes particular demands on handling what can be large and sometimes heavy components.

The choice of components and levels of prefabrication together with the detailing, particularly at any connection to a frame, can significantly affect the need for temporary works and introduce potential health and safety risks. This in turn affects the cost and time elements, and the quality, of the construction and increases the technical risk of premature failure.

Movement and fit will need to be incorporated into the joints and fixings details. The potential variety of materials not only within the wall but also between wall and building frame, given the variety of loading and environmental conditions, increases the risk of differential movement. The analysis of the magnitude of these movements (inherent deviations) becomes part of the design of the detailing of jointing and connections.

It is also in the jointing and connections that manufacturing and assembly tolerances (induced deviations) will have to be accommodated. The provision for adjustment to line, level and plumb to ensure each component is in position within its allotted tolerance is resolved in the joints and fixings. Again, this analysis has to be part of the choice of detailing of the assembly.

In considering production this detailing also determines sequence and the need for temporary works, particularly for working platforms to give the assembly teams safe access and operational conditions. It may become part of the analysis to consider the reverse process, dismantling, in the detailing to increase the potential for reuse and recycling.

Wall options

The major factors influencing the choice of wall have now been introduced as:

- Function as a facade
- Function as skin
- Identification of elements providing structural integrity
- Provision of lighting and ventilation
- Production and assembly issues

There are many common forms of external wall (with their fixings to different frame types) that can potentially resolve the issues listed above. But these options result in numerous details being used on different building at any one time. To try to cover all the available options in a textbook such as this is inappropriate as it could never be comprehensive and would date very quickly. There are, however, a number of broad options that have emerged which can be discussed in terms of their general form and typical detailing. These would include:

- Cavity brickwork
- Cladding panels
- Curtain walling
- Facing materials (including rain-screening)
- Lightweight cladding
- Structural glazing

Cavity brickwork

The form of wall and detailing introduced for loadbearing domestic walls has been successfully transferred to the cladding of framed structures. When cavity brickwork is used to clad multi-storey frames, the loads are transferred to the frame at each floor (sometimes every other floor) and not taken to its own foundations. It gains its stability from the frame and not the return walls of the loadbearing structure.

The choice of brick defines the component form of the facade and the scale and range of colour and bonding patterns that can be achieved, with the most likely material being clay bricks, although it is possible to imagine other combinations of blocks and other materials that can form components of this type. Weatherproofing is ensured by the cavity, and the approach to damp-proof detailing around openings can be applied to the connections to the frame. Detailing and materials choice for insulation and soundproofing are again the same as for domestic construction.

The major difference between cavity brickwork in housing and its use as an enclosure to a frame building is the connection to the structure. Figure 29.1 shows a detail that provides support for the external skin on an angle fixed to the edge beam or edge of the slab. The inner skin sits on the edge of the slab, gaining full support directly from the structure. The cavity is shown passing in front of the beam or slab edge to give continuity to the insulation to eliminate the potential cold bridge through the structure.

Figure 29.1 shows the angle bracket fixed to the slab edge with a cast-in channel. This is a fixing method used for many of the external wall options often in conjunction with angles and brackets. The principle of this fixing is shown in Figure 29.2. The slot formed by the channel allows for adjustment in one direction. Brackets with slotted holes are an effective way to offer adjustment in the other two directions, allowing the fixed component to be adjusted to line, level and plumb. Channels and tee-bolts are available in a variety of materials and duty grades.

Providing support for the brickwork on an angle means that any movement in the frame will move the wall panel resting on the angle, normally at one-storey height. Any movement in the wall will be restrained by the support of the panel above. The potential for movement in frame and wall needs to be assessed. The columns will shorten due to loading, but if the majority of loading is applied before the cladding is built most of this initial movement will have already occurred. If reinforced concrete columns are used, there will also be a drying shrinkage, again tending to shorten the column, but again most of this might have occurred if there had been a sufficient time delay between

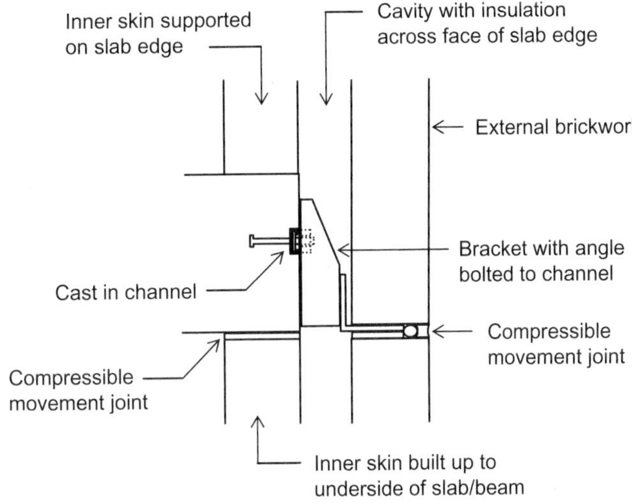

Figure 29.1 Cavity wall support connection to frame.

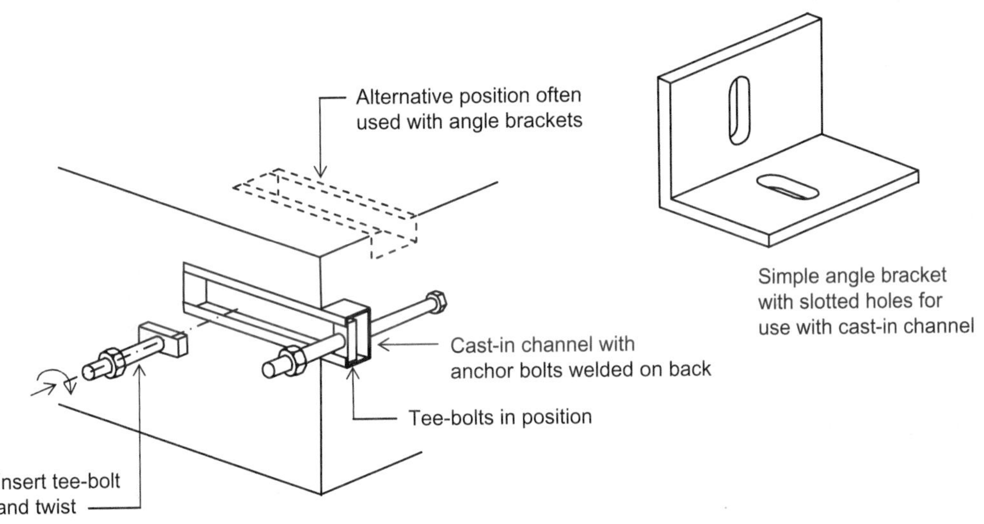

Figure 29.2 Cast-in channel fixings.

casting the column and constructing the brick panel. In reinforced concrete, however, a further phenomenon, structural creep, occurs that continues to shorten the column over time, even though the loading remains constant. All the inherent deviations lead to decreases in the height of the column, but the inherent deviation in clay bricks leads to some increasing in dimensions. While thermal and moisture changes in the climate will directly influence the external skin of the brickwork, both expanding and reducing dimensions, the initial equalisation process, unlike for concrete (which shrinks), is for the bricks to expand.

When the magnitude of these changes in dimensions of the frame and brickwork are identified, they show a clear need for a movement joint at the top of the panel below the support angle for the panel above. It may need to be greater than 10 mm and therefore larger than the mortar joints and will need to be waterproofed with a mastic, compressible pointing, which may have a different colour from the mortar.

In addition to these horizontal movement joints long walls in brickwork may need vertical movement joints to accommodate the expansion and environmental movements within the brickwork itself.

The angle support takes the load from the wall itself to the structure, but this will not transfer the wind loads to the frame. The wall will also need to be restrained from moving away from the structure as it is balanced on such a narrow support, particularly with a movement joint at the top of the panel, even though the wall will have wall ties connecting it to the inner skin. The wall can gain restraint from the beam above and from the columns (not shown) with ties or cramps built into the mortar joints. If the columns are spaced too far apart, wind posts may have to be introduced between columns to support the brickwork panel.

Openings in the brick panels can be formed in the same way as for domestic walls, as can the provision of thermal insulation (as detailed in Chapter 19).

Cavity brickwork can be described as a trade process. The panel of brickwork will be laid on site between each supporting angle (the inner skin between each floor inside the building frame). It will require a traditional scaffold and the materials distribution needs to service a normal 'bumping out' supply of bricks, mortar, ties and insulation to the bricklayer. This materials distribution is often on a greater scale, particularly in vertical height, than the domestic process, with hoists and possibly cranes being employed for this purpose. Access for personnel may also have to be provided by hoist as buildings increase in number of storeys. Given safe access, the health and safety provision for the working area will be provided by the scaffolding and site PPE (personal protective equipment).

In addition to the normal bricklaying process there is the requirement to fix the support angle. Normally detailed to be bolted to anchors cast into the concrete slab or to plates or cleats fixed to the building frame, this attachment process can be carried out from the scaffold if appropriate lifts are arranged.

Cladding panels

Brickwork panels, as introduced in the previous section, can be characterised as a trade process with the wall being built on site. Cladding panels are pre-formed, normally off site, and then have to be lifted and fixed onto the frame. Both forms, however, have their own internal structural integrity and therefore need only support and restraint from the beam and columns of the frame.

Cladding panels are most often constructed in reinforced concrete using steel rod or mesh reinforcement with concrete mixes designed for strength, durability and workability characteristics appropriate to relatively thin section casting, typically around 75 mm webs, and casting for fine detail and weathering qualities. Panels can either hang from or sit on the frame. This choice is often based on the height of the panel. If the panel is full-storey height spanning from beam to beam, it will be more stable if hung from above with a bottom fixing to align and restrain the panel. If the panel is designed to cover the beam edge, perhaps from the window head below to the windowsill above (sometimes known as a spandrel panel), it will sit on the beam. The width of the panels is normally a sub-module of the column centres. They will be restrained at the columns with the panels that do not line up with a column being restrained by the adjacent panels. Typical panel profiles and support and restraint arrangements are shown in Figure 29.3. The support and restraint have not only to transfer the loads but also to allow adjustment to maintain position, line, level and plumb of each panel with its neighbour and across the whole facade. This has to be allowed for in the angle restraint detail shown in Figure 29.3.

These choices for support and restraint will be determined by the size, shape and repetitive

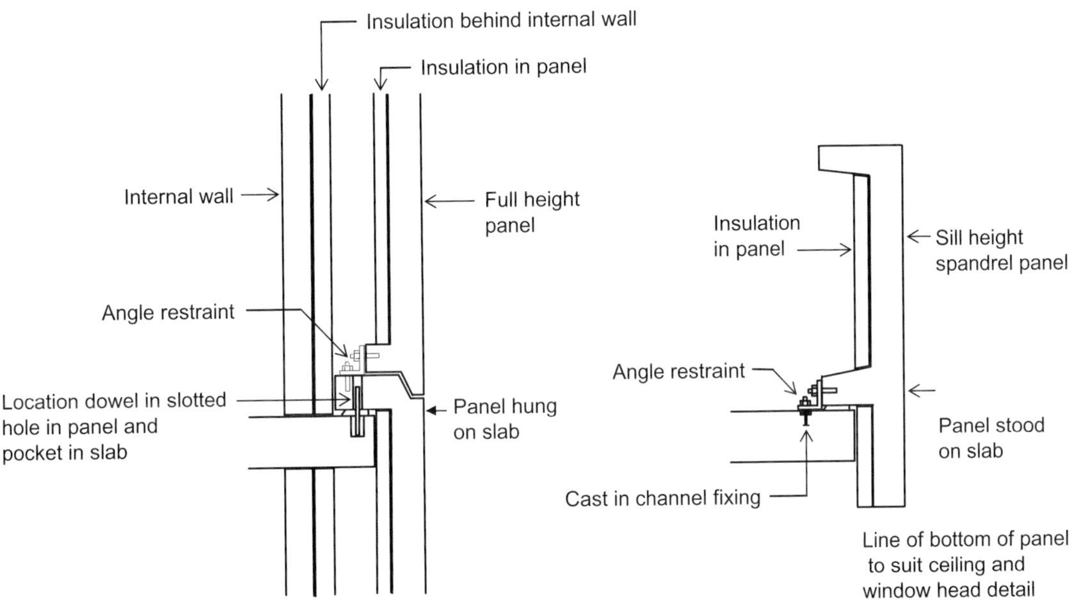

Figure 29.3 Concrete cladding panels connection to frame.

pattern of panels on the facade. This will give a rhythm to the facade determined by the design. Colour and texture can be achieved with shape, pattern and finish on the panel. In this respect there is great variety. Exposed aggregate and relief finishes can be created on the surface of the concrete itself. Other materials can be cast onto the face, such as brick to mimic the brickwork of the cavity wall cladding or stone to provide an alternative construction method to traditional stone-faced buildings.

The joints between panels have to be designed to allow for production deviations (induced) to ensure fit and panel and frame movements (inherent deviations) during the life of the building while remaining weatherproof. These joints can be either closed-sealed or open-drained, as illustrated in Figure 29.4. All these factors influence the width of the joint.

The closed or sealed joint is filled with either a mastic or a gasket that adheres to or forms a seal to the side of the panel and remains flexible through the life of the panel to form a waterproof joint. In the open-drained joints baffles direct any water that does enter the joint down and back out to the surface. These baffles do not stop wind and therefore a fixed wind baffle has to be incorporated into the joint design. Both these joints have required widths of 15 to 25 mm.

The support, appearance and jointing arrangements determine the size and profile of the panel. The thickness of the panels will vary but will be around 75 mm with ribs and thickenings to ensure rigidity and provide fixings and resist shear at the support. This will in turn determine the weight of the panels, which can be considerable and may influence the way they can be picked up and moved from casting to final position on the building. Lifting points for handling will have to be incorporated into the design of the panel and its travelling arrangement by lorry and by crane to ensure a safe approach to the building edge. The detailing of the fixing will determine where the cladding fixers have to stand to guide in the panel and make the fixing and restraints so the panel can be released from the crane. This detail significantly affects the safety risks and the need for temporary working platforms and safety provision. It is possible to detail this fixing to be carried out without scaffold giving a clear approach for the panel to the building edge, but this will require

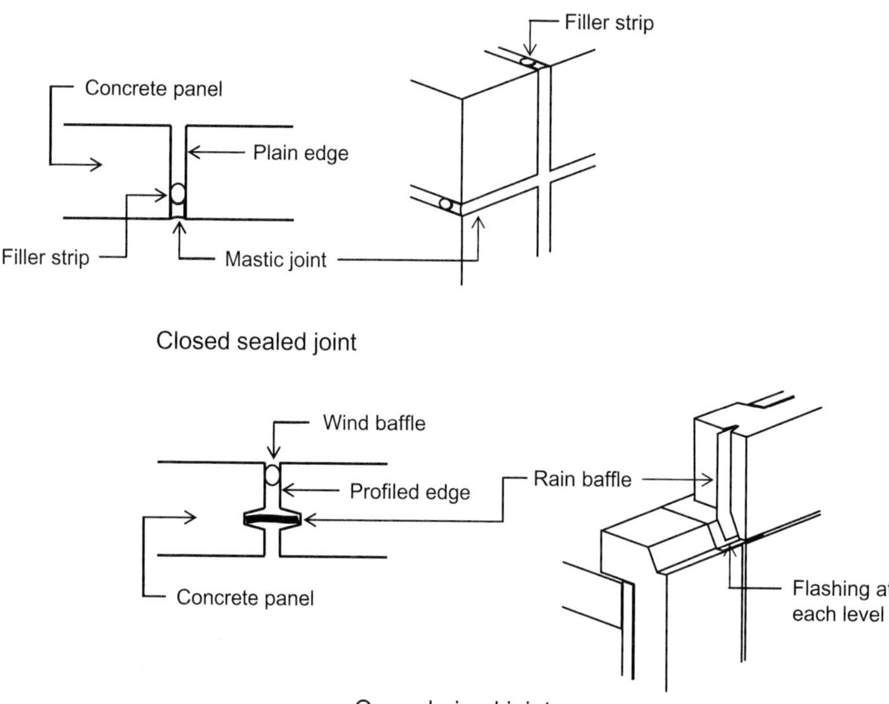

Figure 29.4 Joints between concrete cladding panels.

other edge protection or personal harnesses for the fixing operations.

Openings in the wall can be provided either by incorporation into the full-height panel or by the space between spandrel panels, possibly with full-height panels to cover the columns. The edge profile can be cast to receive window or door frames directly with detailing to incorporate weatherproofing and fixings to the concrete of the panel.

The panel itself will provide a measure of sound and fire resistance, but the overall performance of the wall will depend on the jointing process, which could introduce a weakness and point of failure. The panel also has some thermal mass to attenuate solar gain but has little insulation value. The panels are highly profiled, which gives opportunities for insulation, but again the ribs and edges to form the joints and support details will cause weaknesses in insulation as potential cold bridges. There will need to be an internal skin to provide a basis for the internal finish, and this will need to overcome any weakness in the performance of the panels themselves. The design of the internal skin will have to consider the fire, sound and thermal performance of the whole wall, particularly at the joints and support points. This internal skin could be either block or a framed panel similar to the internal skin of the cavity brick wall discussed in the above section.

Two other materials used to make cladding panels are glass-reinforced cement (GRC) and glass-reinforced plastic (GRP). Both of these have the advantage of being formed in much thinner sections than traditionally reinforced concrete and therefore offer considerable weight reductions for panels of the same size. They also offer much finer surface finish and edge details as well as more versatile profile shapes as the manufacturing process is normally spraying and laying-up respectively in the moulds rather than pouring and vibration in reinforced concrete. They do, however, have greater inherent

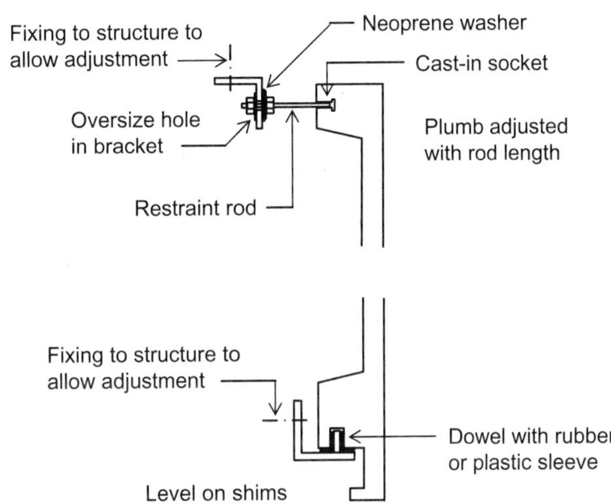

Figure 29.5 Fixings for GRC panels.

movements and can suffer surface crazing and colour variations in manufacturing and variable weathering of the surface if the finish and colour are not carefully selected.

GRC uses alkali-resistant chopped glass fibre reinforcement in a cement-rich mix. Without the need for cover to the glass fibre reinforcement, panel sections need only be around 10 to 15 mm thick. Panels of this thickness are not very stiff and will need ribs around the edge and around any openings and even across the middle of larger panels. These thickenings to the edge will also house the restraint fixings. Continuous support is normally provided at the bottom with restraint at the top. These fixings must allow for accurate positioning as the fine finish of the panel will need to be matched in the precision of the positioning of one panel to the next. Typical connections are shown in Figure 29.5. Another requirement of the connection to the structure is to allow for movement of the panel exposed to environmental changes, as this is greater than in a reinforced concrete panel. This is not only in the oversize or slotted holes; the joint should be provided with neoprene washers in addition to the metal washers to ensure that movements do not stress the corner to the panel, as it is likely to crack. Large panels may not have sufficient stiffness for these edge fixings alone and so may need a steel stud system to be attached to the back of the panel as it is moulded to stiffen the panel and provide fixings back to the structure.

The potential for movement will also affect the jointing options. Sealants or gaskets can be used for closed joints, and open joints can be detailed if there is sufficient depth in the edge rib to provide the profile for the baffle or a rainscreen design could be considered.

GRP panels have the potential to be even lighter and with more pronounced and shaped profiles. The glass reinforcement is laid up in mats into the resin, normally polyester, giving very thin sections of 3 to 6 mm. This makes the panels potentially very flexible and they will need stiffening with edges and ribs that will be moulded into the resin. The edge profile for the jointing and the fixings at each corner will also have to be moulded into the panel in the laying-up process with the resin. Again, the inherent deviation can be significant and this must be allowed for in the joints and fixing connections.

Neither GRC nor GRP materials would score highly on environmental grounds, with the rich-cement mix for GRC and the resin base for GRP, even though they are in relatively thin sections. There are also questions on the disposal of these materials at the end of the cladding's life.

Both GRC and GRP panel systems provide little in the way of environmental control, so an inner wall is required to provide the major skin function. The panels can have limited thermal insulation thickness by filling the back of the panel between the ribs, but the ribs will act as cold bridges. Panels have been designed as sandwich constructions, with insulation glued between two panels. This provides the stiffness required and can give increased insulation thickness without cold bridging. However, these panels may bow as the environmental conditions experienced by the inner and outer faces of the panel will be sufficiently different for differential movement between them to cause distortion in the overall panel.

Curtain walling

Curtain walling is a system based on a structural framework, normally vertical mullions, connected to the building structure, spanning a storey height connected to the edge beam or the edge of the slab. The mullions will normally be around 1800 mm centres, although wider centres can be achieved in some systems. Transom sections, based on the same profile as the mullion, but normally not so deep, are fixed to the mullions to form a series of glazable openings and stiffen the mullions against distortion under wind loading. This arrangement is shown in Figure 29.6. An alternative is to make the transoms the main members providing horizontal support at the slab edge and sill height (not illustrated). The transoms at the sill height can be fixed to the columns but will require intermediate support from stub columns or wind posts.

While it is possible to envision curtain walling being designed in many materials, the most common is aluminium. The structural sections can be extruded as complex profiles, normally based on the box section, and include shapes to accommodate both fixing and weatherproofing details. The section is normally produced in two

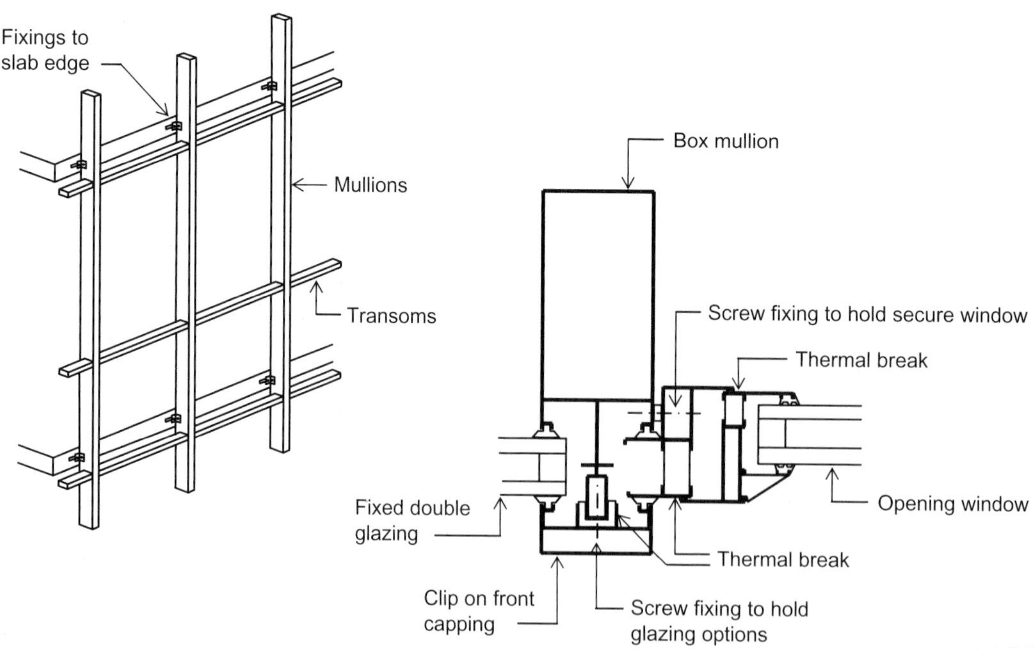

Figure 29.6 Curtain walling.

halves, one structural and one to allow glazing often with clipped capping to finish the outside and hide the fixings. An example is shown in Figure 29.6. Structural sections in aluminium are normally part of commercially available systems that determine the centres of mullions. This is in contrast to the brick and concrete panel options discussed in the sections above, where the panel size and shape can be determined by the design. In curtain walling the dimensional discipline is more rigid. The mullions are normally all set apart to the same centres across the whole facade, or section of facade, clad in this way.

As can be seen from Figure 29.6 the transoms and mullions are designed to receive glazing directly. This does not, however, have to be transparent glass but could be any panel that mimics the edge of a glass glazing unit. Figure 29.6 also shows a window made from aluminium sections specifically designed to work with the glazing profile in the mullions and transoms.

If solid panels are specified, it is possible to achieve some of the skin functions such as insulation and fireproofing. However, this is likely to be limited and greater performance can be achieved if a wall is built behind these panels forming conventional sill and head details inside the building.

The mullion and transom section can significantly affect the appearance of the facade. Most box sections are around 50 mm wide on face and need to be between 100 to 200 mm deep, depending on the span and wind-loading conditions. Normally at least some of the section will stand proud of the glazing and capping arrangement and the finish on the aluminium will affect the appearance of the wall. Finishes such as anodising can provide colour, and so this system in not limited to the natural appearance of the metal. Systems have been devised where none of the box section shows in front of the glazing. The edge of the glazing units are created as frameless and connected to the structural mullion and transom sections using technology from structural glazing.

Connection to the building frame and allowance for movement and adjustment to achieve the required accuracy in alignment has to be considered. Connections are relatively straightforward with plates and brackets secured to the frame to receive the mullion sections at each storey height. This can be achieved with the use of cast-in anchors and brackets, as shown in Figure 29.2.

Facing materials

The three options introduced above provided very different facade treatments but all have in common the structural elements of the wall being provided by the facade components themselves. For many facade treatments the materials to be used and/or the component forms required will not have the structural capacity to be connected directly back to the building frame. These options are often known as facings as the structural integrity of the wall has to be provided by another element of the wall.

This structural integrity can often be provided by a backing wall that will need to be specified to achieve the environmental performance of the wall such as the thermal, acoustic and fire requirements. These facade options will be referred to in this section as facings.

The backing wall can be solid, normally block or framed panels in either timber or steel studding. These options have been discussed in Chapter 19 in walls for housing. In this case, however, the walls will be infill to the building frame. The infill wall will only be required to take the weight of the facing materials and the wind loading back to the building frame. The infill wall will, therefore, require support and restraint detailing similar to the internal skin of the cavity wall option discussed above.

The demand for high levels of insulation in a wall has led to a preferred solution where insulation boards or slabs are required to cover the structure and infill support wall (which itself may be insulated) to eliminate cold bridges. This complete covering of insulation separates the facing from the infill structure that is to provide the structural integrity to the facing components. This limits the choice of insulation material and component form. This is most likely to be boards that will keep their shape, but even these may crush under the direct bearing of the fixing. The fixing of the insulation has to be achieved with large plate or continuous, wide-strip washers behind the mechanical fixings so the force of the tightening of the fixing is spread over a sufficient area of the face of the insulation. The insulation is now stable and the large washer or strip can now be used to fix any battens or support rails required for the facing. If renders in a polymer-based material are being used, they can be applied direct to the insulation if care is taken to match the specifications.

This facing approach can provide support to a very wide range of materials, including:

- Natural and reconstructed stone
- Timber and timber-based boards
- Tiles
- Renders
- Metal

Whichever facing material is chosen, the size, shape and regularity of the component form and type and frequency of fixings will have to be established. These variations in the component form and frequency of fixings will influence the choice of backing wall.

This choice of material and component form will also influence how the wall will achieve the waterproofing function. For the options that are slabs, panels or boards there is the question of the joints between the facing components and how they perform in weatherproofing. Some joints may be designed to provide surface exclusion so all the water runs off the surface of the facing while others may allow water into the joint with a controlled drainage path back to the face of the wall further down the wall. This drainage path can be within the joint or the joint may be completely open allowing a proportion of the water into the space behind the cladding. Both these approaches are generally known as rain-screening.

Rain-screening

As indicated above, rain-screening is not specific to any facade material but is defined by the approach to weatherproofing. It can be used with a number of facade materials, normally in panel or board form, but all will need attention paid to the detailing of the joints between panels.

The system relies on a cavity but, unlike a masonry cavity, the outer facade components have open joints to screen the rain but allow the entry of air. The backing wall provides the air barrier. With pre-formed panels or boards the cavity needs to be around 25 mm. The protection offered by the rain-screen allows the insulation to be fixed on the outside of the backing wall and therefore across the structure, eliminating any cold bridging. The insulation should be vapour-permeable so any vapour is removed from the ventilated cavity and any condensation that does form on the cold external face of the insulation will be evaporated and removed. This allows the whole building to breathe.

Two approaches to rain-screening can be identified:

- Drained and back-ventilated
- Pressure-equalised

Both have open joints around 10 to 20 mm wide between the panels and a cavity. In drained and back-ventilated only simple edge profiles and baffles are used on the panels or boards, so some moisture will enter the cavity. It is now necessary for the material forming the other face of the cavity to have moisture-resistant properties and for the support to the panels to remain durable in wet conditions. However, these wet conditions can be limited by ensuring a clear drainage path out of the cavity at the bottom edge (including openings) and air gaps at the top to ensure a good ventilation rate to remove the moisture as it evaporates, keeping the cavity dry.

Pressure-equalised systems aim to keep the cavity dry by denying entry to the rain. While still open, the joint edge profile is more complex to shelter the open joint and protect it from entry by wind-driven rain. These edge profiles have to be more complex or deeper if the pressure in the cavity is likely to become lower than outside. Care has to be taken with cavity closures to form zones in the cavity that will not be influenced by variations in wind pressures around the building, lowering the pressure in the cavity on the positive pressure side of the building where the rain is being driven onto the cladding face.

If materials are chosen with care, the back-drained and ventilated rain-screen is the simpler system to detail and is applicable to many facing options. This is the choice that will be discussed where appropriate in the sections on materials options that follow below.

Natural and reconstructed stone

While it is possible to build a traditional stone wall inside the frame in much the same way as the cavity brickwork or to face reinforced concrete panels with stone for a similar effect, stone can be fixed in thin slabs as a facing fixed to a backing wall.

For natural stone the size and particularly the thickness of the slabs will vary with the stone type. The size of slabs will depend on the quarry of origin, the cutting techniques and the weight of the panel for handling as well as the type of stone. Broadly stones are either igneous, such as granite; metamorphic, such as marble and slate; or sedimentary, such as limestone and sandstone. These broad, essentially geological, categories are useful as they indicate the stone's strength, density and ability to be polished and its surface weathering characteristics. These in turn affect the thickness that the stone has to be to hold together as a slab and the ability of the edge to be prepared for fixings without the potential for the edge failing locally at the fixings. The thickness will be determined by a recommendation from the quarry from which the stone has been cut. But as a guide, granites are likely to be between 25 and 38 mm, slate and marbles between 15 and 36 mm, limestone between 25 and 75 mm and sandstone between 50 and 75 mm, although in the sedimentary rocks some may have to be as thick as 100 mm.

Improvements in sawing techniques have allowed much thinner slabs to be cut, but these would not take edge fixings or be strong enough to survive on the face of the building. Stone is, however, expensive and the savings in using these thin sections has led to section composites with 5 mm of stone adhered to an aluminium honeycomb making a panel with an overall thickness or around 25 mm. These are very light and can even be used as panels in curtain walling.

The other method of providing a more economic solution is to use reconstructed stone. This is a concrete product using crushed stone and white cement. With a probable thickness of around 40 to 60 mm depending on slab size they are usually lightly reinforced and may have fixings cast in. However, they can be fixed in the same way as natural stone.

Fixings, like most other panel or slab solutions, have to provide support and restraint. Support is only required at each floor level and above each opening so long as each stone is bedded on the stone below and then each stone is restrained at top and bottom (sometimes at the top and bottom of the sides) to the backing wall. Stones are rarely more than 900 mm deep, so there may be four to six stones between supports. The support is likely to be an angle but it may only be short lengths of corbel picking up the corners of each stone. The restraint fixings will also be metal but closer to the surface and long-term small movements may well crack the bedding and side joints, leading to a high risk of damp conditions for the fixings. For this reason the fixings would be stainless steel or a non-ferrous metal such as phosphor bronze. Some possible metal support and restraint fixings showing the stone edge preparation and the securing of the fixing back to the wall are shown in Figure 29.7. These fixings have to allow for tolerances on the backing wall and in the stones themselves with adjustment to bring stones in line and plumb. As with the cavity brickwork solution the movement in the frame needs to be taken into account with a movement

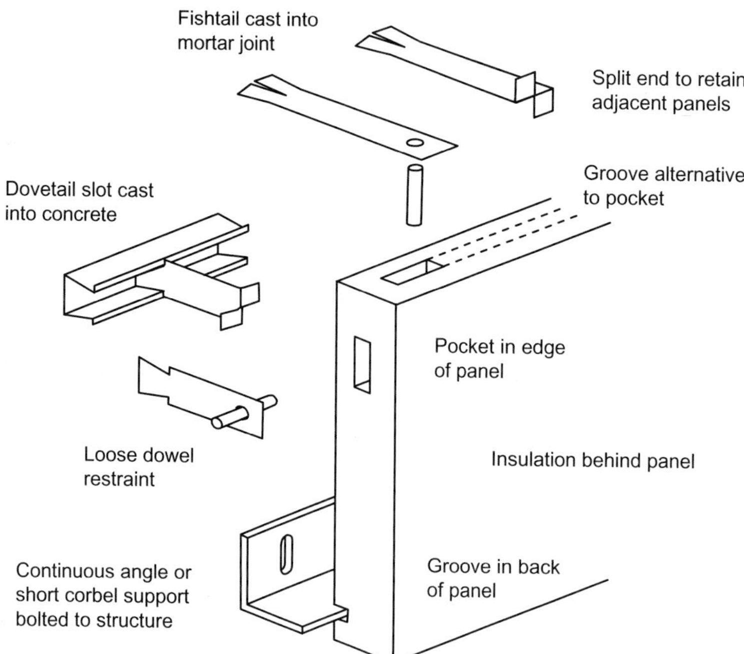

Figure 29.7 Fixings for stone facing.

joint formed under the stones supported at each floor level.

Jointing between stones is normally a mortar made with stone dust, lime and cement, although sand may be specified for sandstones and granites. Joints are likely to be thinner than brickwork at around 5 mm to accommodate the fixings. Jointing can be specified with sealants throughout but flexible mastic will be required at the movement joint that will have to be 10 to 15 mm to accommodate the movement in the sealant without failure.

Timber and timber-based boards

Timber as cladding has long been used as a board material with a variety of profiles and jointing details. Fixings have traditionally been nailing to timber battens that can be attached to a variety of backgrounds, which determines the type of fixing and the need for packing for plumbing dependent on the dimensional accuracy of the backing wall. The weatherproofing has normally been by overlapping or interlocking the boards where allowance for swelling and shrinkage of timber has to be taken into account. Detailing with flashings and cover strips maintain weatherproofing at corners and openings.

Questions of longevity, protective finishes and maintenance with timber products have limited the choice of this material particularly, for larger and taller buildings. However, the introduction of timber-based fibre boards making use of recycled or waste timber in a variety of binding agents, which can be cut into panels as well as boards, is now a viable choice for a greater range of scale of buildings. These board materials have potentially better weathering qualities with less maintenance and offer a greater variety of appearance. The basic construction of the facing component with battens and backing wall is still the approach to this type of cladding, although the joint detailing can be arranged to act as a rain-screen as discussed above.

The performance as skin and the connections to the frame will be determined by the choice of backing wall, which will determine the type of fixings that will be used to attach the battens and may influence the detailing to accommodate movements and the need for detailing to take into account variations in production deviations.

The issues around protective finishes and maintenance and the possible need to replace cladding in the life of the building, affecting lifecycle costs, need to be considered in the original specification, choice of fixings and detailing. Again, there is variety in the basic longevity of the material and the life of the protective coatings, both of which will have a cost implication for the initial and running costs of the building. The combination of material and protective finish will influence the weathering regarding both change of colour and the potential for staining. Timber performs well in any environmental analysis so long as it comes from a managed source, preferably local to the site. It can score well on embodied CO_2 and disposal strategies, which can even include burning for energy production at the end of its life.

While the basic detailing of this facing material is fairly straightforward, the lifecycle needs careful attention if the cladding is to weather well and fulfil its full-life potential.

Tile forms

In the past clay and ceramic tiles have been used as facing materials. On a domestic scale tile hanging using plain roof tiles supported by nailing to battens has been common in much the same way as vertical slate has been used in areas where slate roofing is traditional. When larger civic, more urban developments were undertaken by the Victorians, a facing material known as faience was used. Faience work is tiles of fired clay where individual tiles are modelled to fit together, fixed to a backing wall, to produce highly decorative facades in terracotta.

While, as with timber, there may be a return to these more traditional materials, a return to either of these options for anything other than local use is unlikely given the scale of development and the demands of modern manufacturing processes.

Two forms that have been used for commercial buildings are external wall tiling, used in much the same way as wall tiling in bathrooms, and the development of carrier systems, which support profiled tiles providing a facing often as a rain-screen.

For external wall tiling there is a need for continuous support for bedding the tiles. The choice of tile to remain durable in external conditions and provide the weathering surface can be met with quality ceramic tiles, but the specification of the bedding material would need careful consideration. If large areas of tiling are required, movement may be a concern, as these could lead to stress in the backing materials and loss of adhesion. As a minimum the allowance for movement between the frame and the backing wall will have to be reflected in the tile finish but movement in long runs of tiles horizontally would also need breaking down into smaller panels.

For the profiled tile systems a support framework akin to battening will need to be fixed to the backing wall. A clipping system engages a profiled edge to the tile that is attached to the support rails to provide a dry facing system, where even the joints are open and do not need a mortar jointing material. With these open joints movement will be allowed for in the clips and joints between tiles.

Renders

Render is another traditional facing material that is being used on commercial-scale buildings. Being a wet-applied material, it requires full backing support and relies on adhesion or a mechanical key to remain stable and crack-free. Each render coat will have a required thickness with the first coat matched to the backing support, known as the substrate, to ensure a stable finish. The substrate may be solid or laths. Solid substrates can take a wide range of forms from concrete and brickwork to board materials, including insulation. Laths are now normally metal sheets, cut and expanded to form an open surface. These sheets are then fixed to battens to provide a void behind. The render when applied squeezes through the open surface and, when set, forms a mechanical key with the lathing.

Finish coats need to be matched to the exposure. Traditional finishes have been sand and cement renders, such as roughcast and pebbledash, and these have weathered well in the UK, but the smooth sand and cement renders, often coloured with paints, can discolour and become dirty, particularly if lighter shades are specified. This has led to the development of polymer-based renders often with coloured aggregates and through-colour mixes. The use of polymers such as silicone and acrylics, often as polymer-modified cement mixes gives increased adhesion and flexibility that overcome the major defects in rendering of hollowness and cracking. These renders may also be thinner than the cement renders.

The need for higher levels of insulation with the desire to cover the structure and infill support wall to eliminate cold bridges has led to the development of systems of insulation and renders. These polymer-based renders are, however, more expensive but have superior weathering and movement characteristics and so maintain their appearance over time.

Metal facings

Steel, aluminium, copper, zinc and even gold can be considered facade materials. All these metals are relatively expensive and to cover the large areas of a facade they will be formed from sheets where there is a need to produce details to limit thickness, as this governs the amount of metal used and therefore costs.

Steel and aluminium can be bent and formed into relatively stiff components from thin sections to produce profiled sheets or flat-faced (smooth or low relief pressed patterns) trays. The inherent stiffness of the profile folding means that they only need intermittent support. These sheets or trays can have edge profiles that seal for weatherproofing but can be effectively used with rain-screen designs. Rather than flat sheet or open panels, insulation can be introduced between two metal sheets that are sealed at the edges to form a joint profile to fix the sheets together and form a waterproof joint.

Copper and zinc are softer metals and do not have the inherent stiffness to be bent or formed into stiff panels and therefore have to either cover timber-based panels or be fixed to a continuous background much the same as roofing details. Gold, the most expensive, is applied as leaf, microns thick, to a background, where it forms more of a finish than a facing in its own right.

Given the variety of metals and their support arrangements, a wide range of details have emerged, many of them proprietary systems. It is, therefore, not intended to give details in this text.

As a facing their weathering characteristics are important and in this respect the choice of metal is important. Considerations of weathering involve an analysis of potential colour change and staining as well as deterioration mechanisms. Metals vary greatly in these properties. Gold does not deteriorate or change colour with time. It is part of its value that it exhibits these characteristics. In contrast steel corrodes easily and this changes appearance and performance dramatically. Non-ferrous metals also corrode, and this changes their appearance, but corrosion products protect, considerably slowing down loss of section and therefore performance. Copper is well known for its green patina of copper oxide and is often chosen specifically for this weathered appearance. Understanding these changes and hence the longevity of the metal is a key part of any analysis if metal is to be used on the facade.

Most non-ferrous metals are normally used without additional surface coating to provide a natural colour and weathering pattern. This is not normally the case with steel and aluminium. While the corrosion product of aluminium is stable, it is a dull grey. Steel will need some form of protection unless stainless steel or one of the corrosion-resistant steels, such as corten steel, is specified. Steel-corrosion protection starts with the specification of a zinc coating by the hot-dip galvanising process but a colour coating may still be required, which can assist in corrosion protection. The colour coatings for steel are normally plastic coating known as organic coatings. The most widely used is polyvinyl chloride (uPVC) but this can suffer colour change if deeper colours are used. Coatings such as acrylic polymethylmethacrylate (PMMA) or polyvinyl fluoride (PVF) are more expensive but are tougher and more colour-fade resistant. Aluminium, while more expensive than steel, is naturally more durable and can be painted with acrylic coatings or anodised in a range of coloured finishes.

This section on facings has referred to metals applied to backing walls. These can include profiled sheeting, but these are much more widely used as what is often known as lightweight cladding, normally used with frames for long-span structures where it will be fixed to a structural sub-system normally designed as part of the structure. This walling option is discussed in the next section.

Lightweight cladding

While the deadweight of the enclosure wall is always a concern, the nature of the skeletal frame does not often make this a priority above appearance and environmental performance. Further the location, use and client expectations of a building where the form and layout lead naturally to a skeletal frame will demand certain facade treatments and are likely to include a significant role for glazing. In most buildings employing a skeletal frame storey heights are limited, providing the opportunity for connection back to the frame at intervals of around 3.5 metres. This has led to the variety of solutions discussed above.

For frames creating long-span roofs these demands are usually reversed. There is need to limit the dead loading, particularly the roofing elements of the enclosure. The demands on appearance and requirements for glazing, particularly in the walls, can be limited. Support at floor edges is unlikely to be available as these are single-storey buildings, and eaves heights may reach 14 metres. The most common lightweight systems are based on profiled sheet metal connected to a sub-framing system, normally designed as part of the structure. These sub-framing members are known as cladding rails for the walls. It is not unusual for the wall cladding to be similar to the roof system, where the sub-framing members are known as purlins.

The metal sheets will be either steel or aluminium with coatings and finishes as explained in the section above. They will be cold-rolled to produce a trapezoidal or ribbed profile, giving strength to the section and allowing them to span between cladding rails. The cladding rails (and purlins) will normally be cold-folded steel zed sections spanning between columns (or rafters). The sheeting is fixed directly to the cladding rails with fixings at centres determined by wind forces, which will put negative pressures on the cladding making this the most likely failure mode.

Figure 29.8 shows details of lightweight cladding with insulation behind an inner lining board. For the walls the lining board is fixed to the inside of cladding rails, but for the ceiling the board can be supported on tee-bars similar to the suspended ceiling details shown in the next chapter. Sandwich, or laminated, construction with insulation between two metal sheets with an edge profile to form a waterproof joint can provide insulation in the cladding sheeting itself.

While windows can be formed in this wall, roof lights are more common to provide overall light levels in these single-storey deep buildings, although large doors may be required. The activities around and inside these buildings often means that the lightweight cladding may be specified only above about 6 metres from the ground with a brick or block cavity wall forming the lower part of the enclosing wall. This masonry wall will be built from its own foundation, possibly a ground beam, for support gaining restraint from the columns. If columns

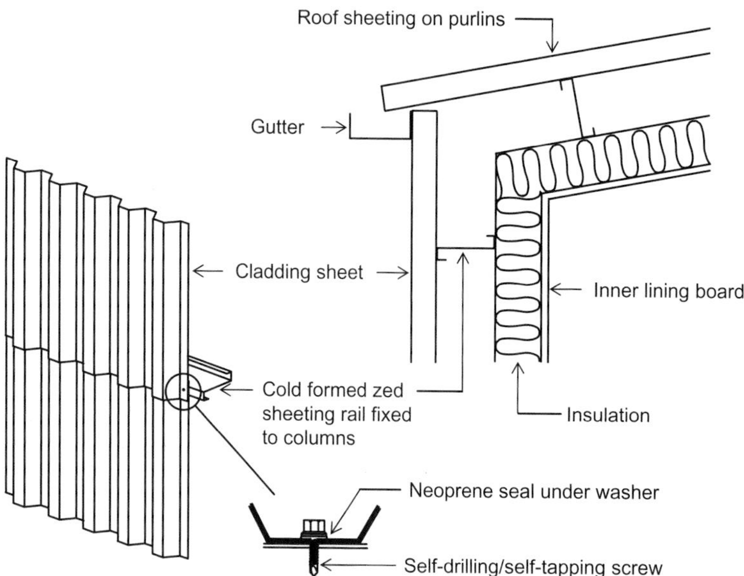

Figure 29.8 Lightweight cladding.

are too widely spaced, wind posts may be required.

Structural glazing

While it is possible to construct a completely glazed wall using storey-height frames, structural glazing provides a frameless wall of glass 20 metres plus high and of unlimited length. For walls of greater than one-storey height the glass needs a sub-framing for support that will be visible and is therefore normally engineered with high-quality structural components. The glass wall will need to transfer the wind loads to this sub-framing and accommodate both deflections and the environmental movements.

In the glass-making process the sheets are allowed to cool slowly (annealing) as this allows the glass to be cut and drilled. It leaves the glass with limited strength and when it does shatter it breaks into the characteristic large dagger shapes that can cause serious, life-threatening injuries. Both its limited strength and its safety risk analysis would eliminate this material as suitable for this application. Such construction is only possible using toughened glass that increases its strength, its security and safety in that when it shatters it forms tiny small squares that minimise injury. For additional security and safety, laminated glass that holds together when shattered, so even further limiting the risk of injury, can be specified. Toughened (also known as tempered) sheets cannot be cut or drilled after toughening but have sufficient strength to provide most of these structural glazed walls in 12 mm, and possibly 15 mm, glass.

The other significant material development that allows this form of construction is the jointing material. Silicone elastomeric sealants are highly extensible and adhere to both organic and inorganic materials and retain these properties for decades, even in external exposure conditions. These clear silicone rubber sealants form joints between the panes at around 10 to 12 mm wide and at the edges, where the sealants can be used directly into rebates of grooves to the surrounding construction. Given the limits on the size of glass sheets (normally the size of the toughening oven), the regular joints provide the flexibility and movements required, although care still has to be taken in fixing the glass to the sub-framing.

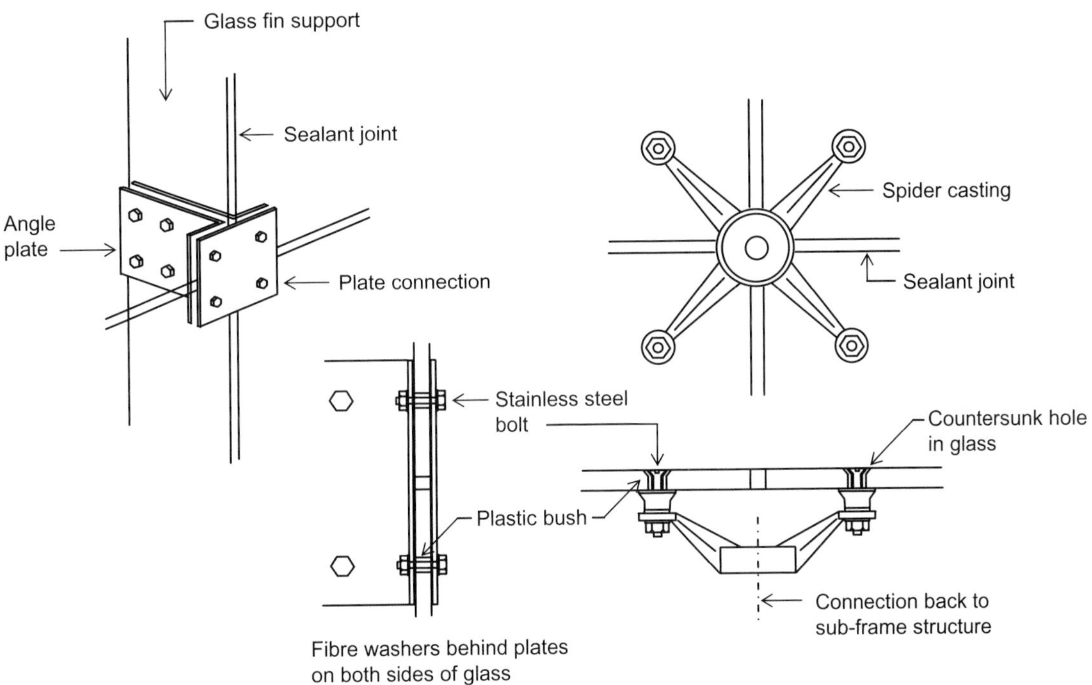

Figure 29.9 Structural glazing.

Before discussing the sub-framing options, it should be pointed out that for single-storey walls it is possible to use the structural silicone to glaze directly into the surrounding structure if movement and fit issues can be overcome, but this is not normally what is envisioned by structural glazing. There are two sub-framing options for structural glazing normally known as:

- Plate connection
- Planar glazing

The plate solution with a glass fin support system is shown in Figure 29.9. The connections are metal plates that sandwich the glass using stainless-steel bolts through oversize holes in the glass with fibre gaskets and plastic bushes to avoid glass-to-metal contact to allow for thermal movement. These plates can be modified to create hinges for doors and even suspension details, where glass panels can be hung from the structure to distribute the weight of the wall.

Planar glazing reduces the visual impact of the connection on the glazing. The connection is based on a spider or star fitting that holds the corner (and edge on larger panes) of the glass away from the supporting sub-framing. An example of a spider bracket is given in Figure 29.9. The form of the sub-framing is now highly variable. It can be based on the glass fin or on a steel structure that can vary from relatively standard tube and box structure to lightweight vertical lattice frames. The use of spider brackets creates the need to accommodate movement at the point of fixing to the glass. Holes are countersunk and a shaped nylon bush prevents metal-to-glass contact and allows some rotation in the fixing as the wall flexes under wind conditions. This connection also has to provide the adjustment to bring the glass sheets into alignment and create the even thickness joint to receive the site-applied silicone sealant. Developments in structural glazing have included the introduction of curved sheets, double-glazed units, opening sections, angled surfaces and even horizontal glazing to form roof areas.

Roof construction

Chapter 18 introduces the functions of roofs and options for waterproof coverings. While it concentrates on pitched roofs for domestic construction, it also provides the introduction for roof options to commercial buildings. Chapter 18 then provides details for the concrete-tiled roof commonly used on housing, but it is not the intention of this chapter to provide similar details of possible roof covering options for commercial buildings. This section will focus on an approach to identifying the broad options for the roof as an enclosure element.

Pitched and flat options

Commercial buildings create the need for larger areas of roofs than houses need. These can be created with pitched roofs, although the context and visual impact of the design may require a flat profile to the roof. Domestic architecture in the UK is dominated by roofs with relatively high pitches, say 20° to 40°, where tiles and slates can be used for the waterproof covering and gutters at the eaves can achieve water collection and disposal. It is also possible to use other waterproof finishes such as sheets or felts and even glazing options using patent glazing to these high-pitched roofs to achieve alternative visual impact and lighting options. However, these relatively high pitches can dominate larger roofs, either with greater rise to the ridge or with multiple spans with internal gutters collecting water, but can be successfully constructed if that is what the design requires.

It is possible to create low-pitched roofs, but the waterproof finishes will need sealing at joints to become a continuous waterproof covering. This is not possible with slates and tiles, but corrugated profiled sheets spanning between purlins similar to the lightweight cladding discussed above can be used, as shown in Figure 29.8. This solution is often used in conjunction with long-span roof structures, where low-profile roofs and low dead loads are required for both wall and roof construction.

For a flat roof the waterproof covering has to be continuous, employing single-ply polymers, metals or asphalt, all of which require full support either on decking (timber or metal) or slab construction. They must be capable of being detailed and worked into the gutters and rainwater pipe outlets for water collection, which now relies on falls that have to be created in the roof fabric. These falls are normally around 1 in 40, taking the water either directly to outlets to rainwater pipes or to gutters formed in the depth of the roof that then take the water to the rainwater pipe outlets. These falls and any gutters have to be formed in the depth of the roof construction.

Forming the falls on a flat roof can be achieved in one of three ways:

- In the structure
- In the screed
- In the insulation

Where the roof structure is created from grid structures such as girders or space frames for long-span roofs, the top chord of the structure can be set to create the fall directly in the decking that will support the insulation and covering. If the roof structure is a slab similar to those designed for the floors, particularly where more complex fall patterns are required, forming the fall in the structure is a less attractive option.

Where concrete slabs provide the support, the finish is not likely to be smooth enough to apply the covering, nor will it be accurate enough to avoid the puddling of water on the roof. Normal construction tolerances for flatness will not allow roofs to drain effectively. To provide both a fixing surface and a fall within tolerance screeds can be applied to create the fall patterns. While these screeds can be lightweight and provide insulation, additional insulation may be specified. The specification of the insulation has led to the third alternative for providing the fall and that is in the insulation boards themselves. If the tolerances on the decking or slab are sufficient, insulation boards can be delivered to site tapered to the fall patterns to be laid directly on the decking or slab construction. They must then be fixed and provide a surface to which to attach the covering so it remains stable under wind and thermal conditions.

The roof will be subject to perhaps the most difficult climatic condition of any part of the external enclosure. Wind loading will create significant uplift forces (these are discussed below). As roof coverings directly face the open sky they will experience radiant exchange, creating surface temperatures significantly different from the air temperature. These are most acute on clear winter nights when the temperature of dark surfaces can drop to –25°C and on clear summer days when direct solar gain can create temperatures of up to 65°C and above (and up to 80°C for some glazing and lightweight coverings over insulation), even in the UK. Light surfaces and protective layers can be considered to limit the thermal extremes but the materials and fixings will still have to accommodate the dimensional changes induced by these temperatures.

Use and access

While in the past pitched roofs might have been used for roof lighting, they were not seen as usable space. Solar collectors and photovoltaic arrays may take advantage of the pitch roof but even these require access to the roof for maintenance. This access is only by authorised personnel who can be trained and equipped to move about the pitch and the roof covering, limiting the risk of both harm to the individual and damage to the covering.

Another consequence of using the roof for lighting and of mounting energy-gathering systems on the roof is orientation. For natural lighting north-facing glazing will give the best even quality of light. For energy-gathering from the sun with solar collectors or PV arrays a south-facing aspect is required, although PV will produce some energy, even from an overcast sky. Quality lighting and energy collection are best mounted on pitched roofs. On flat roofs any roof lights will have to be built up above the level of the roof to weather the opening and give sufficient depth of the construction to shade from direct sunlight which may cause discomfort. Energy-gathering systems are best mounted at an angle even on a flat roof, and therefore some support framework will have to be specified. This support will also penetrate the waterproof covering and details to seal the supports will have to be provided.

Flat roofs offer greater opportunities for usable space. The main distinction remains as to whether access will be for maintenance personnel only or if wider access to the users of the building will be required. Even if access is for maintenance only, often considerably more use is made of flat roofs for services plant and even window-cleaning support provision, and this more frequent access increases the risk of damage. In these circumstances designated walkways may be needed. In some buildings the roof can become another usable space with roof gardens and facilities for other outside activities where the safety of individuals such as edge protection and the wearing qualities of the finishes have to be taken into account.

Loading

The identification of use and access will have an influence on the loading, particularly if the use involves additional dead-loading, such as services equipment or the creation of roof gardens. However, all roofs will be subject to snow and wind loading. Snow may drift and create high loadings behind parapets, but the major characteristic of wind loadings on all roofs is uplift. This loading will lift the covering if its fixing does not offer sufficient restraint. If the covering is fixed sufficiently well, the uplift will be transferred to the roof structure. If the covering does not have sufficient dead weight, it will need to be fixed down to the building structure in the detailing of the fixings.

There is a great range of roof structure from the long-span roof, where dead-loads are kept low, to roof gardens, when loading may even be higher than the floors within the building. Lightweight roofs can suffer complete wind reversal, where the entire dead-load of roof structure is overcome by the uplift forces of the wind. This can occur on simple timber-joisted flat roofs at the domestic scale to complete long-span roof structures on large industrial complexes. It is, however, unlikely on commercial roofs, where concrete slabs form the support structure.

Insulation and condensation

As part of the full consideration of the roof's thermal performance there is one set of circumstances that will significantly influence the detailing and specification of the roof. These circumstances arise during the winter heating period and concern the high potential for condensation to form within the construction below the roof covering. This is particularly acute on roofs where the covering is a continuous membrane because, as well as being waterproof, the layer is also vapour-proof. As the roof construction below the waterproof layer gets colder the chance of condensation becomes greater.

This has led to three identifiable forms of construction, as shown in Figure 29.10:

- Cold roof
- Warm roof
- Inverted warm roof

These options are identified by the position of the insulation relative to the support provided for waterproof covering. Sometimes referred to as a cold deck roof, the cold roof has the insulation below the decking (or slab), leaving the underside of the decking cold and therefore vulnerable to condensation. To control the vapour coming from the room requires either a vapour-control layer or ventilation to remove the vapour from below the deck. Given the difficulties of providing an effective vapour-control layer, ventilation has to be provided above the insulation but below the deck to ensure that the dew point is reduced to that below the temperature of the underside of the deck.

In contrast to the cold roof the warm (deck) roof has the insulation above the deck or slab. This ensures that the deck remains warm and free from condensation but still leaves the waterproof covering cold and condensation will now form on the underside of the covering. In this solution it is possible to use the deck to support an effective vapour-control layer, leaving the cold side of the insulation free from

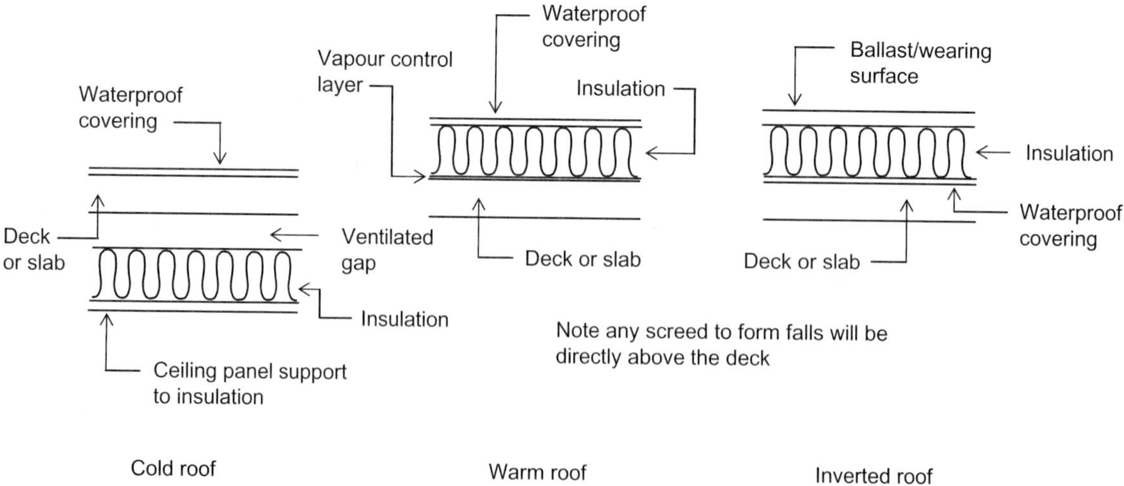

Figure 29.10 Roof insulation options.

vapour transfer, thus eliminating the chance of condensation.

In the inverted roof the waterproof covering acts as the vapour-control layer as the insulation is placed above the covering, ensuring it remains on the warm side of the insulation. This roof construction provides the lowest probability of condensation but clearly has implications for the nature of the insulation material. Most obvious is that it is now external, exposed to the elements and drainage has to be arranged below the insulation. The insulation must not absorb water and must not deteriorate under wet conditions. In addition it is subject directly to wind uplift and, as a lightweight material, will require fixing or ballast to hold it down. If the roof is to be used for access, the insulation will be vulnerable to trafficking and damage from items dropped on the roof. These conditions normally lead to either defined walkways around the roof or the whole roof being paved to allow safe access. The paving will provide the ballast to hold the insulation in place. Insulation and surface paving systems are available, making the inverted roof a viable option even for roofs with full access. It is interesting to note that while inverted roofs need careful consideration of the insulation material, the insulation will protect the covering and will extend the life of the roof as a reliable waterproof element.

Fire safety

The concern with roof construction is the threat from fire from outside the building. Such a threat could jeopardise the safety of the building should the fire penetrate to roof construction and so the performance will have to take into account the whole construction, covering, insulation and deck. There is an additional concern for the surface spread of flame on the exposed face of the roof to limit the chance of fire spread. For most roofs this is the roof covering, but for inverted roofs this would be the paving or ballast.

Roofs will be given a double-letter rating, both A to D. The first letter indicates the penetration classification and the second the spread of flame. As the threat is from an external fire source the allowable designations and the limits to the area of these designations is dependent on the distance to the notional boundary determined by the proximity of adjacent buildings.

Lifecycle and weathering

As with all external enclosure elements appearance will change with time. This was discussed at the beginning of this chapter for external walls in the section 'Facade, material and time'. The same agents of soiling and decay will act on the roofing materials as on the wall. However, it is recognised that the roof exposure is likely to be generally higher than the wall and therefore the destructive agents more aggressive.

Given the wide range of covering materials, it is perhaps not surprising that their life expectancies are very variable. They range from 15 to 20 years for three-layer built-up felt roofs to 100 plus years for metal coverings. They also vary considerably in the capital or initial cost and these are, very roughly, in proportion to their longevity. Initial and lifetime costs are a significant consideration in the choice of roof construction. Generally, failure of roof weathering is disruptive to the operation of the building and puts the fabric of the building at risk from accelerated deterioration. As the roof covering gets closer to failure and its reliability reduces, there will be potential costs of repair. For many roof coverings there is little expectation that they will last as long as other parts of the building fabric, and replacement will have to be undertaken to continue the effective life of the building.

Green roofs

While it is not the intention of this text to give details for decking/slab and waterproof covering combinations, green roofs merit particular mention as they offer some performance characteristics that other options cannot provide.

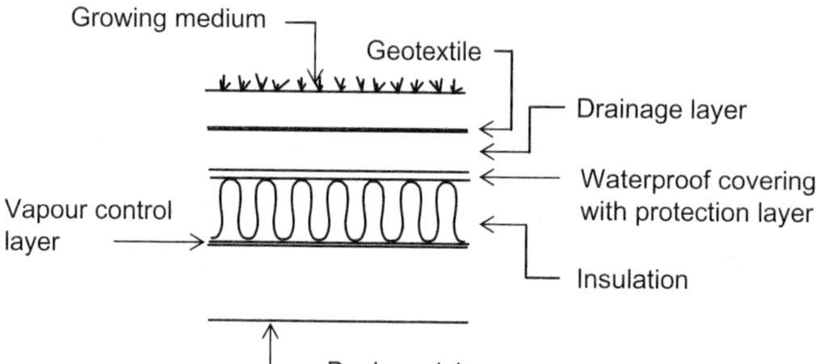

Figure 29.11 Typical green roof section.

Figure 29.11 shows a typical cross section through a green roof.

Up to the waterproof covering the construction is based on normal roof construction, including the provision of a fall. The section shows a warm roof with insulation above the deck, but cold roofs can successfully be used. The waterproofing has to be a high-quality robust layer normally either a rubberised asphalt or single-ply layer thermoplastic sheet membrane. On top of the waterproofing is laid a drainage layer normally of aggregate to take the water to drainage outlets, above which will be an inspection chamber to allow the outlet to be cleaned and cleared. Above the drainage layer is a geotextile filter that will allow water to drain but retain the soil of the growing medium and the roots of the vegetation so the drainage layer will remain free flowing. The growing medium will be dependent on the type of vegetation. The roof-top environment is a difficult growing environment and plants have to be chosen with care. Three levels of vegetation can be identified:

- Extensive green roofs
- Simple intensive green roofs
- Intensive green roofs

The extensive roof is covered with low-growing, low-maintenance plants. They are normally designed for access only. Often known as sedum roofs, the vegetation may also be mosses, succulents or herbs and grasses. The growing medium is from 25 to 125 mm thick. This level of vegetation can be used on pitched roofs, although the specification has to be chosen with care for a pitch above around 1 in 3 to reduce the risk of soil slippage and erosion.

The intensive roofs provide a more landscaped environment. The simple intensive includes lawns and low-growing plants requiring a growing medium of up to 250 mm thick. These plants need maintenance. There may be arrangements for the roof to be more widely accessible but it is more likely to be designed to be overlooked. The intensive green roof is designed to have a high amenity benefit and is usually accessible to be used as a garden. With a growing medium up to 450 mm thick these roofs impose much higher loadings and may need irrigation with the type of plants used on them.

All these options provide benefits to both the thermal response of the building and in the reduced drainage run-off and delayed discharge for all but the most extreme storm events. The fire rating of the roof depends on the thickness of the growing medium and the provision of gravel or slab fire breaks around the perimeter and at around 40-metre intervals on large roof areas. Overall, however, the roof adds to the greening of the urban environment and will help to maintain biodiversity in any location or setting.

Summary

1. The choice and detailing of enclosure elements is one of the major challenges in the design and construction of a building.
2. External walls need to be considered as facade with respect to appearance, skin in terms of environmental performance and wall as the physical construction of an assembly of components.
3. The choice of the facade materials and components needs to be considered not only for their initial appearance but also for changes over time from soiling and deterioration.
4. As skin the environmental performance has to deal with many energy flows, such as heat, light and sound, as well as transfers of moisture and air, to maintain the internal conditions that provide for the well-being of the occupants.
5. There is a need to identify the part of the wall that will provide the structural integrity of the wall that will transfer load to the structure and provide the support and fixings for all the other components of the wall.
6. The assembly of the wall takes place at the edge of the building and at a height requiring temporary works to provide access and a safe working environment.
7. Tolerance and adjustment to allow assembly to line, level and plumb are important particularly in the connection to the structure.
8. Cavity brickwork, cladding panels and curtain walling are systems that, while they have very different appearance, provide the structural integrity to span between the frame structural members in the main facade components.
9. Facings are a range of materials and component forms that all require to be fixed to a backing structure (normally an infill wall) that is supported and restrained by the frame.
10. Lightweight cladding, mainly used on single-storey industrial buildings, is fixed to rails provided as part of the structure.
11. Structural glazing provides a glass wall uninterrupted by framing that will require a sub-structure to support the glass enclosure.
12. Roof structures can be pitched or flat. Flat roofs have to be provided with falls for drainage that can be achieved in the structure, the screed or the insulation. Drainage has to be provided for all roofs.
13. The roof can have a range of needs for access, from the maintenance of the covering by trained personnel to full access for the public as with, for example, a roof garden. The roof as usable space for services equipment or user activities will affect the loadings and the need for walking surfaces and edge protection.
14. Insulation position in the roof can affect the potential for condensation. Three conditions can be identified as cold, warm or inverted roof, each with their own detailing and materials specification.
15. There are options in the choice of roof coverings, which vary widely in their life expectancy and cost.

30 Internal Enclosure

This chapter considers the internal enclosure elements of walls and floors and the finishes they support. The range of functions and the levels of performance that can be expected in commercial buildings from these enclosure elements are identified. There is a brief introduction to applied finishes with more consideration given to the component systems of suspended ceilings, raised floors and partitions that may be chosen for commercial buildings.

Performance

Internal enclosure elements, like external enclosure elements, have different environmental conditions on either side. Their function is then to maintain the required conditions on the occupied side of the division. In internal enclosure both sides are occupied and therefore conditions on both sides are to be controlled and maintained as part of the building design. However, both sets of conditions are normally less aggressive than the external climate, although some industrial processes may need special consideration.

Functions may be fewer and performance less demanding but the three key ideas that were used for external enclosure are still useful. These are facade for appearance; skin as dynamic response to load and environmental conditions; and wall or floor as the construction specification and detailing.

The face of the wall, floor and ceiling is normally known as the finish and is, like any facade, mainly concerned with the choice of material and its performance over time. The material will give colour, texture and quality to the finish plus a wearing surface but will be subject to change with time. Redecoration or even refurbishment may be needed in the future. The wall or floor as skin concerns not just the surface but also the whole element contributing to a range of functions to provide structural integrity and environmental modification. The wall or floor as construction involves the choice of details and specification of the components and materials based on a few broad options.

When considering the performance of the walls and floors, there is initially a need for structural integrity. While walls are not likely to be loadbearing, floors will almost certainly be part of the structural system. The structural floors can, therefore, support the finishes, but non-loadbearing walls need requirements of structural integrity to be considered in their construction. As with external walls, non-loadbearing means not taking dead and imposed load from other structural elements, but this does not mean that they will not receive their own direct loading. Some walls may form the wind stability element of the structure. In addition to direct suspension loadings from fittings attached to the partition, sideways accidental loading from equipment and soft body impacts have to be considered. Stability then includes failure in deflection as well as potential localised damage or collapse. The integrity of the wall is likely to be achieved in much the same way as it is for domestic construction, either as solid blockwork or as framed panels. The panels can be in timber or light steel studs, as discussed in Chapter 19, but solid panels or glazed systems are also common in commercial building and are discussed in more detail below.

Along with structural integrity, the functions of privacy and security have to be considered. Privacy has visual and aural implications that influence glazing and soundproofing performance requirements. The mechanisms involved in sound transmission were introduced in Chapter 10. The mass of the construction together with its rigidity and edged conditions plus any internal cavity with sound-absorbing quilt will determine the transmission loss performance. Many partitions that are site-assembled systems will find achieving these properties difficult. This is not only in the construction of the wall itself but in the jointing and fixings between panels and around the edges. This is in addition to noise transmission paths over and under the partition through the voids in suspended ceilings and raised floors discussed later. Actual sound transmission loss figures achieved on site may well be lower than those achieved under test conditions. Noise transmission through floors normally made of concrete are unlikely to be a problem from airborne sound, but impact sound may be a concern.

Security involves the threat from fire and intruders. Aspects of fire design are covered in Chapter 10, where linings (the finishes) and the fire resistance offered by the element itself refer specifically to internal walls and floors. Security, again discussed in Chapter 10, depends on the nature of the threat. For the threat of theft, security is normally focused on doors and locks, but if additional force is expected the strength of the wall may have to be considered as a loading issue of structural integrity.

Applied finishes

For the internal enclosure elements the choice of finish is normally a major consideration to provide appearance and a wearing surface. Finishes have to be applied to a background or substrate that will provide the structural integrity but will often not have either the tolerances (normally flatness) or, in the case of joists or studs, the continuity to directly apply the final decorative, wearing finish. It is necessary to provide beds (floors) and backings (walls and ceilings). Originally these would have been wet processes with screeds (floors) and renders and plaster (walls and ceilings) being specified. These wet processes are still used where modern materials give quicker drying times but dry processes have also been developed. Dry processes are normally based on boards with chipboard or oriented strand board (OSB) (floors) and plasterboard (walls and ceilings) being the most widely used.

For these applied finishes the building-up of the surface for decoration from the structural substrate needs care in the specification. The number of layers, their thickness and the materials of which they are made will depend on a number of factors.

Each material will have a minimum thickness to maintain its own integrity, depending on the support it gets from the substrate. For floors, sand and cement screed will be around 50 mm, although this can be thinner if the screed can be bonded to the concrete. Flooring grade chipboard and OSB supported at 600 mm centres will need to be 22 mm, or supported at 450 mm centres will need to be 18 mm thick if spanning in one direction with continuous support in the other. The jointing of the edges of the sheets and the centre of the fixings is also important in this specification. For walls, two-coat plaster and plasterboard will be around 12 mm. Plaster can be applied to either masonry or metal lathing, and plasterboard can be screwed to studs or applied to dabs of adhesive mix onto a masonry background. Plasterboard can be specified with a feathered edge for jointing as dry walling or with a square edge for the application of a skim of plaster.

For most construction these thickness and fixing specifications are sufficient to provide a stable surface within tolerance for the application of the decorative and wearing surface. It is also necessary to check that the specification will accommodate movement and that there will be no adverse chemical reaction with either the substrate or the decorative finish that would affect the long-term performance.

There are a wide range of substrates and perhaps an even wider range of surface finishes and each combination will need any bed or backing construction to be chosen for the circumstances, taking into account the total surface area and the wear and tear expected at the finish itself. Further consideration is also required if services distribution and systems involving heating (such as underfloor heating) are to be specified. It is beyond the scope of this text to discuss all of these combinations and even the range of finishes themselves. Both technical and manufacturers' information in this respect is widely available.

Component systems

Component systems have been developed for walls, floors and ceiling elements of internal enclosures. Floor and ceiling systems will use the structural slab for fixings and support but normally provide a void that is often used for services distribution, where accessibility has to be considered. These systems are known as suspended ceilings and raised floors and can be used to generate the floor-to-ceiling height appropriate to the internal space. Wall component systems, normally known as partitions, include their own elements of structural integrity but are not usually capable of being loadbearing.

One aspect that most component systems have in common is modularity or a repeating coordinating size, which was introduced in Chapter 4. The components will be produced in a limited range of dimensions but based on the repeating size that, with common jointing details, allows notional modular space to be enclosed. The basic modular dimension is 300 mm. The modular size of ceiling tiles, for example, being 600 × 600 or 600 by 1200 mm, as are lighting units allowing the coordination of both finishes and services into any suspended ceiling system. The working size of the tiles is less than the modular size to allow for the jointing and tolerances.

Partition systems based on panels will have widths of standard panels at 1200 mm maximum for handling but may have 600 or even 300 mm unit widths to make up lengths with limited make-up pieces. Partitions also need some coordination in the heights of panels. This is not only for overall floor-to-ceiling height but also for doors and solid and glazed panels that make up that overall height. While overall heights normally conform to the module, the heights of part panels may be to the sub-module of 100 mm.

Suspended ceilings

Suspended ceilings may be chosen to reduce room height and provide a decorative surface but are often used to create a void for services, including the support system for lighting fittings and air vents. This use of the void for services demands provision for access. Many systems allow access at any point in the ceiling by raising individual tiles, but using this access often damages the tiles over time, which can spoil the appearance. Some balance needs to be struck between the assumption of full and open access and some planned locations. While access for commissioning may be acceptable when damaged tiles can be replaced, it may be better to limit the need for access for emergency and repair later in the life of the building. If the upgrading of, for instance, wired communication services is anticipated, designated access may also be advisable.

Suspended ceilings are also often used to improve room acoustics with the selection of sound-absorbent tiles. However, they normally offer little extra in the way of reduction in sound transmission and can create the conditions for noise transmission over the top of partitions if these are only taken up to the underside of the suspended ceiling.

Fire design is another area of concern in the choice and detailing of suspended ceilings. Most systems cannot be assumed to add to the fire resistance of the floor elements. If this were required, a joint-less system based on plaster or plasterboard would have to be specified, making access to services very difficult and effectively limiting the use of the void for services distribution. If the ceiling construction has no effective fire resistance, the void becomes a major path for the transmission of smoke, heat and volatile gases. If the void extends over partitions (it cannot extend over a compartment wall), the void may have to be divided by fire barriers. Another issue with fire and suspended ceilings is the surface spread of flame. This may also contribute to volatile gases and dripping hot material that may assist the spread of the fire and limit escape time. Associated with this is the integrity of the system fixings to avoid premature collapse of the system in a fire.

With the increasing use of passive design, where the thermal mass of the structural slab is being used to control thermal conditions, suspended ceilings cannot be used. Methods of providing a finished slab and accommodating services and support for lighting without suspended ceilings have to be found. This also reduces the ability to provide acoustic control as the slab soffit is hard and reflects sound, although this does give a surface that naturally limits the spread of fire.

The most common form of suspended ceiling is based on the frame and panel and is illustrated in Figure 30.1. A lightweight metal framework is fixed to the underside of the slab and, in the simplest form, the tiles, often fibreboard and lightweight, are laid into the framework. This allows access by just lifting the tiles but, as has been indicated, these are fragile and easy to damage. If pressure differences are possible as in lift and entrance lobbies, the tiles may need to be clipped in. In the system illustrated the metal of the framing can be seen. If this is not desirable, concealed fixings can be arranged with grooved tiles and zed section runners or tongue-and-grooved tiles. This makes access more difficult. An alternative that gives a concealed fixing is to use coated metal trays often perforated with a fibre based-bat insert that clips to special spring steel runners.

It is possible to eliminate the runners by forming the metal tray as a strip where the

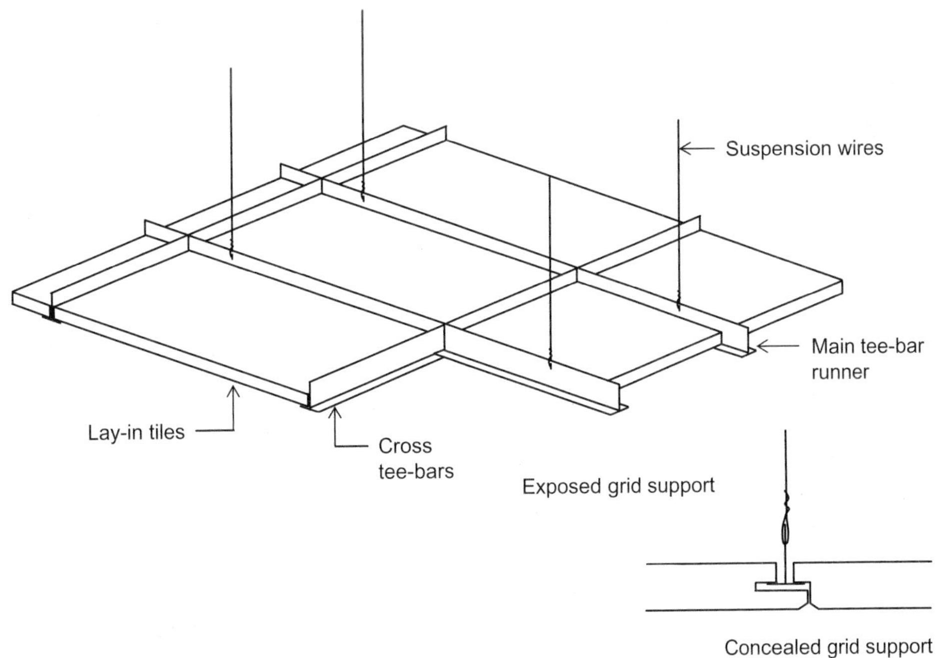

Figure 30.1 Suspended ceilings.

increased bending resistance of the shaped metal tray can span between the bearers. This clearly gives a very different appearance to the ceiling and is often known as a linear strip system after its overall appearance.

One other form of suspended ceiling is the open grid ceiling where the 'tiles' are deep open-box sections that provide no visual barrier if viewed directly from below.

Raised floors

The increasing need for wired services to be distributed widely across the floors, particularly for computer stations in open offices where changes in layout are be anticipated, have led to the development of raised floors. The possibility of a raised floor can lead to decisions to house other services in the floor rather than in the ceiling, particularly if the exposed soffit is required for passive thermal design. Ducting and pipework can be located in a raised floor if sufficiently deep. Indeed, the introduction of heating and cooling air at floor level in displacement systems (see Chapter 31) has some advantages in creating thermal comfort conditions. If airtight, the whole void can be used as a plenum for return air. With the increasing speeds of wireless systems for data transfer some operations may now need less wiring. Less expectation of upgrading these systems during the life of the building reduces the need for either suspended ceilings or raised floors.

Like suspended ceilings, the floors gain support from the structural slab and provide a void to accommodate services and therefore need access. Unlike the ceiling, the floor has a significant imposed load requirement, which makes structural demands on the construction. Floor systems vary depending on the designed imposed loading and this will affect their cost. Cost also varies with the amount of accessibility and the depth of the void beneath the raised floor. There needs to be a careful analysis of the requirements and benefits of a raised floor, indeed, of the need for a raised floor at all.

If only wiring is required to work stations in large open areas, it may be possible to provide ducting in the screed. The grid of ducts could be as close as 1500 mm centres with junction and pulling boxes at the intersections. While this is not strictly a raised floor, it may provide the functions of a raised floor at a more economical capital cost.

If raised floors are required, the choice is between a shallow void, normally less than 100 mm, or a deep floor, which can be up to 300 mm deep. The analysis of the need for access is based on the expected incidence of change in servicing, normally new or modifying wiring and/or of emergency repair. In both cases there will be disruption to the operations in the building and this itself has a cost to the user of the building. If access can be planned and limited, cost can be reduced.

While the choice of floor systems is mainly based on the services to be housed and the need for access to these services, there are other performance considerations to be made. Again, as for suspended ceilings, there are both acoustic and fire implications. Walking on a floor over a void creates the potential for the generation and transmission of noise within the room. This will require a certain level of rigidity and weight with adequate fixings to keep noise to an acceptable level. It may need sound absorption in the void and sound control under partitions between sensitive areas. However, if the floor construction is isolated on resilient pads, the transmission of noise through the slab to the space below can be reduced. For fire resistance the materials of the floor, known as platform floors in the Regulations, may need to be made of incombustible materials. Like suspended ceilings, raised floors do not add to the fire resistance of the element but the void can provide a route for smoke, heat and volatile gases and barriers may be necessary to limit the open areas that may exist in the floor void.

There are implications for the production process in the choice of floors and the services to be installed. Deviations on the flatness of the structural slab have to be matched to the system chosen. Deep floors can tolerate greater deviations as they have adjustment in the supports, but some shallow systems can only follow the

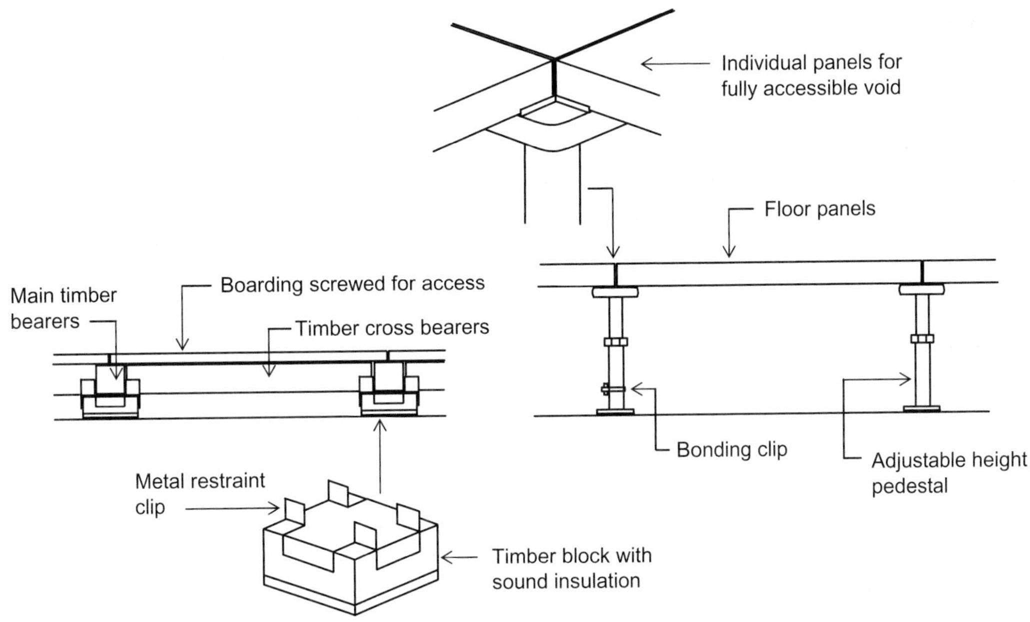

Figure 30.2 Raised floors.

profile of the surface of the slab and so may require packing under the supports. The sequence of the work and the ability of trades to work concurrently will be influenced by the choice of systems, particularly when considered with the work to be completed at ceiling level where temporary access has to be provided from the floor.

Shallow systems are available based on metal decking and timber-battening supports with both partial and full access options. These are essentially designed for wire distribution systems and will include electrical power and data socket-outlet boxes to be cut into the floorboarding prepared for carpet or other specified floor finishes. The metal decking provides support for 22 mm flooring grade boards cut and fitted to provide the access required. The timber systems provide a floating floor with battens at 600 centres to receive 22 mm floorboarding for lighter loadings where fire resistance is not required. The typical details of a timber-batten raised floor are shown in Figure 30.2.

Deep-floor systems to house a range of services are made of pedestal supports and floor panels, which are also shown in Figure 30.2. These are modular systems with the panels normally 600 × 600 mm that provide full access by lifting any tray. Electrical and communication outlet boxes or ventilation grills are cut into the panels. The panels can be specified for light, medium and heavy loading where the composite boards in a metal framing can be up to 40 mm thick. For deeper floors stringers forming a continuous support grid for the trays may be necessary. Stringers will also be necessary if an airtight plenum is to be provided by the void below the floor. The metal components will need bonding to the electrical earth system to protect both occupants and maintenance workers should the system come into contact with any live wiring.

Partitions

Partition systems are more varied than either ceilings or floors as they cover a much wider range of uses and hence performance conditions. Partitions can provide full wall functions between internal rooms with different uses or be partial height divisions within rooms such as toilet partitions. They can be movable in that they can be open and closed, folding along a defined line to create either one or multiple rooms or be transferable to different locations for temporary division. They can vary in appearance, including the presence or absence of glazing. They range from simple block or stud wall constructions with applied finishing to prefabricated composite board pre-finished systems or purpose-built timber-joinery glazed panels.

Partitions are sometimes identified as demountable or relocatable. In many commercial buildings it becomes necessary over time to change partition layouts to accommodate new business needs and operational requirements and efficiencies. These changes are to an extent unpredictable at the time of specification, being dependent on changes in the use of the building in response to future circumstances and even changes of occupation. The degree of demountability required will depend on the frequency of change and the tolerable disruption to the occupier at the time of change. Partition systems with features to accommodate demountability will allow partition layouts to be reconfigured with disassembly being achieved without damage to the partition components and re-erection being achieved with limited dust and noise and therefore interruption to the users' operations. These features are often achieved at the expense of appearance and levels of performance, most often sound transmission reduction. Although it will be possible to reuse the partitions in a simple demount and re-erect procedure, often services and finishes need to be moved, repaired or replaced, making the change more disruptive than just moving partitions.

As periods of change become less predictable and/or less frequent and if these periods of refitting can be achieved by giving contractors complete access to the areas involved, the need for demountable features becomes less. It may be better to provide less expensive systems that are effectively demolished at a time of change and new partitions erected. This will be a dusty and possibly noisy process causing disruption to operational activities. This is a significant decision for many clients when commissioning a building or considering a refit of an existing structure. It is also prudent to consider levels of recycling of the materials specified if the reuse of a whole component system achieved with demountable systems is not possible.

The above discussion on demountability has raised the issue of services integration into partition systems. This is another significant feature of a partitioning system. Services may need to be contained within the partition or be fixed and supported on the face. The major services within the partition may be the wiring distribution to switches and power and communication sockets on the face. In highly serviced areas dado height trunking may be employed, leaving no wires or outlets to be incorporated into the partitions themselves. If only light wiring is required, floor channel ducts and posts and/or door frames can by designated for wiring with cover strips for access to change or modify the service provision. Partition systems vary in their ability to take fixings and bear weight on those fixings. If this function is limited, strengthening bars may have to be incorporated at predetermined locations, but this limits the versatility and reuse of the partition components.

It has already been identified that partitions have very different appearance characteristics. Many are made of panels that are expressed in the appearance of the wall. Here the connecting components or cover-strips have to be evaluated as part of the overall visual experience and the finish provided. Varying levels of the prefabrication of partition systems means that factory-applied pre-finish can be considered, as these components are often one of the last to be installed in relatively clean and protected conditions.

Partition systems can also vary in their ability to provide performance levels for privacy and security functions. Privacy can be both visual and aural, leading to specifications with differing glazing configurations and sound transmission loss values. As indicated above sound transmission loss is as dependent on the detailing above and below the partition and any joints between panels as well as on the properties of the partition itself. Security functions include both fire and intruders, where again the detailing above and below the partition in the voids created by suspended ceilings and raised floors have a significant impact on performance.

Considerations of privacy and security have highlighted the issue of the height of the partition. It is possible to think about partitions being erected from the structural slab to structural slab or alternatively from floor finish to ceiling finish. This choice is particularly significant if suspended ceilings and raised floors are to be specified. This has implications not only for the performance detailing but also for the ability to provide fixings for stability and for the erection sequence with services and other finishes as well as for demountability in future changes in layout. In many buildings the performance requirements of partitions vary from one location to another, and therefore different solutions may be appropriate to suit each set of circumstances. Some internal division may have to be full height, as for compartment walls for fire design or where sound transmission loss has a high performance specification. Some may be less permanent or have lower performance requirements, and it may be better to erect these from the floor up to the underside of the ceiling finish. This will affect the choice of system.

As has been indicated there are a number of construction options available for internal enclosure walls, but they are based on a few broad options:

- Masonry
- Stud
- Frame/sheet
- Frame/panel
- Panel/panel
- Movable panel

These options vary in the way they achieve their structural integrity and the way they are secured to the floor and ceiling. All systems have a height limit depending on their thickness and head and sole (floor) fixings that must be taken into account in the specification.

Masonry and stud construction were introduced in Chapter 19, where solutions for house construction are discussed. Commercial building may have greater performance demands, particularly for fire, and may have greater floor-to-ceiling heights. Both will influence the final specification and detailing but the basic construction will be the same. These construction options would not normally be seen as systems, as they involve traditional construction practices and require applied finishes. They do, however, provide a good solution where high performance is required and limited demountability can be accepted. Many internal enclosure walls are likely to be permanent. Walls around staircases and highly serviced areas such as plant rooms and toilets are examples. These will have higher performance requirements, for example for surface spread of flame or the wearing qualities of the finish. Walls around core activities such as auditoria or production facilities are unlikely to be changed without anticipating major interruption to operational activities and may well have high performance requirements where block or even concrete walls may be an appropriate solution.

Partitions based on frames and panels are likely to be component-based systems, factory-produced and modularised with common jointing and fixing details and the potential for self-finished components. The frame systems introduce a vertical post, normally at 1200 mm centres, to provide, with the head and floor channel, a frame into which factory-produced units can be fitted. The basic format of these framed systems is shown in Figure 30.3. In the frame/sheet systems the units are based on studs for structural integrity with face sheets producing a hollow unit, much the same as a stud partition. Similar to stud partitions, services can be accommodated in the hollow units themselves, as can soundproofing quilts to one side of the cavity. The type, thickness and/or number of sheets can affect the sound and fire performance. In frame/panel systems the panels have their own structural integrity,

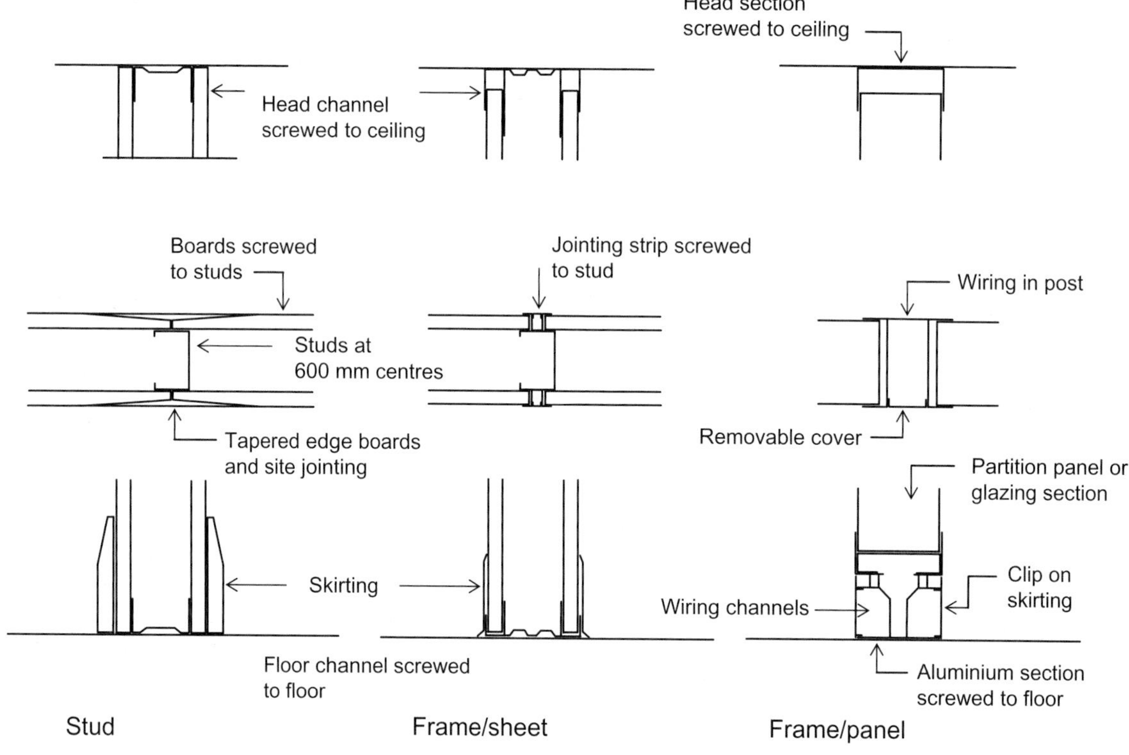

Figure 30.3 Partitions.

based on a solid or honeycombed core, and fit into the frame. Services now normally run in the posts or floor channel duct. In these systems panels will be interchangeable with either half- or full-height glazing units normally based on aluminium glazing sections.

Panel-only systems use the natural structural integrity of the panel to eliminate the need for a post to make up the framing element of the system. The edge of the panel may have to be profiled, and/or jointing clips and cover strips will be required to make the connection between panels. Vertical services may be more difficult to accommodate but often demountability is enhanced with fixings based on floor-jacking facilities. These accommodate deviation and provide a pressure fixing that will allow ease of disassembly with limited numbers of permanent fixings to be made good when dismantled. However, they do need a stable floor and ceiling structure to achieve the pressure fixing. The fire resistance of all these systems is now dependent on the jointing arrangements as well as on the resistance of the panel itself.

Once a panel has its own structural integrity, it can be used free-standing for screening or connected with hinges and fixed top and bottom in flush floor and ceiling tracks that facilitate folding to provide movable panel partitions. These can be arranged to be moved manually; larger, heavier units that can give higher performance ratings can be motorised.

Each system has to be evaluated for the circumstances against the need for flexibility, servicing and appearance as well as its performance for privacy, security and structural integrity functions. The production issues that need to be considered include tolerance and sequencing, which need to take into account the probability of damage if self-finished units are to be used and the possible need for protection in the final stages of construction. All these systems will vary in the way doors and glazed areas can be incorporated and detailed and this may be another factor to take into account in the final choice of a system.

Summary

1. Internal enclosure elements will have different conditions on either side and support a finish so the concepts of facade, skin and physical construction are applicable to these internal elements.
2. Consideration of privacy and security is normally the main concern of these enclosure elements, with visual contact and sound being key elements of privacy, while, for security, an awareness of the risks of fire and intruders is essential.
3. There is a need for structural integrity, and this will normally be provided by the structure for the floor finishes but will have to be incorporated into partition walls.
4. Applied finishes may involve beds and backings to build up from the structural substrate or background to a surface for the decorative finish.
5. Suspended ceilings will be hung from the structural slab to provide a framework of support for tiles, light fittings and possibly other services fittings such as air vents. The void created is normally used for services distribution and will therefore require access.
6. Fire and sound performance will have to be considered in the void, particularly if partitions are fixed to the underside of the ceiling.
7. Raised floors can provide service distribution voids above the structural slab. These raised floors then have to be designed for imposed loadings and have some level of access identified. Sound and fire through the void also have to be considered.
8. The depth of the void is significant in the construction method, giving rise to the distinction between shallow and deep floors.
9. Partitions cover a range of forms of construction, from blockwalls to temporary movable screens.
10. Services and glazing may have to be integrated into the partitions.
11. Demountability or relocatability are concepts in the commercial internal division of space and refer to the ability of the partition to the taken down with the potential for its components to be used in another location. This relocation may also involve services and will leave any fixings to be made good.
12. Masonry and stud partitions provide good performance but are disruptive to remove and will have limited scope for reuse of the materials.
13. Framed and panel systems have jointing and fixing details that facilitate changes in services and the reuse of materials and components as well as prefabrication and quick installation procedures. The frame and panel nature of the system will be evident in the appearance.

31 Services – Scope and Space

This chapter introduces the services element of commercial buildings. It does not cover the engineering or any detailed description of the systems that would feature in the detailed design but the issues that need to be considered in the early stages of design. Initially this involves the identification of the range of services that would be required on the project. It is then necessary to be clear about the space these will need for the plant and equipment involved and the distribution systems to integrate this with the structure, enclosure and usable space.

Introduction

It is not the intention of this text to give any engineering design or much information about systems and equipment for services for commercial building. It is not the intention of this text to describe the services for commercial building in the detail it did for housing in Chapter 21. Performance requirements for services in houses are sufficiently similar for the common solutions described in Chapter 21 to be adopted for most housing developments. With the possible exception of the sizing of the boiler and radiators for a heating system, and some new low-energy systems where common solutions have not yet emerged, engineering calculations are unlikely to be employed for any of the house services.

Many of the basic ideas introduced in Chapter 21 will be the same for commercial buildings but what changes is the scope of services to be incorporated and the way they affect the allocation of space and the layout of the rooms in a commercial building. These are the aspects of services that will be explored in this chapter.

Scope

The scope of building services in commercial buildings can be wider than in housing. Passive design is reducing the need for the more traditional heating and ventilation services, while communications and security are increasing. This is marked by a reduction in ducted distribution and an increase in wiring. Developments in wireless systems in communications and connections between switching and equipments may in the future reverse the trend to wired systems.

There are many ways to think about types of services, often in categories that are useful at different stages when thinking about design. One way, mentioned in the last paragraph, is by distribution systems of ducts, pipes and wires with possible further distinctions between pipes for supply and pipes for drainage that need to be laid to falls or wires carrying power and those carrying data signals where power poses a particular set of dangers to life and a fire hazard in buildings. In Chapter 15, thinking about sustainability, energy consuming services were identified as a category worth considering.

The categories to be used in this chapter are shown in Table 31.1. The first are the environmental services – which include heating and ventilation, with its development into air conditioning – and lighting systems, as these contribute directly to environmental conditions. These are concerned with comfort and contribute to creating a healthy environment. More

Table 31.1 Range of services for commercial buildings.

Category of service	Service
Environmental	Heating
	Ventilation
	Air-conditioning
	Lighting
Public health	Water supply
	Drainage
Power supply	Electrical
	Gas
Security	Fire detection and alarm
	Fire-fighting
	Intruder detection and alarm
	Access control
Operational	Communications
	Movement of people
	Specialist (e.g. medical gases)

closely associated with healthy environmental conditions are what have been called the public health services of water supply and drainage.

Environmental services and much of the operational equipment used in commercial buildings require power, usually electricity being widely distributed in the building, but gas may also be required locally for heating or cooking. These systems also include environmental generation using solar and wind power and provision for heat pumping and biofuels.

Security systems are often a feature of commercial buildings to help create a safe environment, be they fire detection and alarm as required by law or access control and intruder alarms considered necessary by the building operator. Security systems would also include sprinklers and fire-fighting, provision such as hose reels and risers

The final category includes the operational services required to support the activities being carried out in the building. These are the communication services, now dominated by the use of computers and telecommunication, and the services associated with assisting the movement of people. These include lifts and escalators but also the equipment for automatic doors and the systems associated with disabled access.

In some buildings, such as hospitals, the specialist nature of certain activities requires specialist services, such as the distribution of medical gasses and specialist waste systems.

At an early stage in the design it is necessary to identify which of these services will be required and the spatial extent of each. Some estimates of type and size of equipment and the associated distribution systems have to be made to ensure adequate space is provided in the right places. It is a characteristic of services that they have moving parts and can have dangers associated with their failure or misuse. While some aspects such as the outlets and manual control must be accessible to the user, much of the equipment and distribution systems need to be hidden and have restricted access to qualified personnel only. In addition some plant generates noise, and this will influence its location related to other activities in the building. The spatial arrangement of the components of the services systems needs to be integrated into the layout of the building at the same time as the organisation of rooms and circulation spaces to best suit the activities of the users.

Another decision to be made early in the design process is the level of any building management systems (BMS) that are to be provided. Computer-based management systems can fulfil a number of functions. They can take information from sensors and activate the services as part of the control process. This is often associated with the energy-consuming services of heating, lighting and ventilation or air conditioning, where they are sometimes called building energy management systems (BEMS). They can allow for central or even remote intervention via the Internet to switch services on or off or to activate equipment. They can provide information drawn from metering devices either to inform immediate decisions on interventions or to inform future use. They can also monitor user behaviour to provide feedback to make the overall use of the building by the occupants more efficient. These systems may be termed building management information systems (BMIS). Whatever the level of management required, the sensors, metering and means of flow and collection of information have to be incorporated into the building and are becoming services in their own right. It normally requires wiring, although wireless systems can be envisioned.

Space

Each of the services identified may have a variety of options but each will have certain characteristics that determine the needs for their layout and space requirements in the building. This follows the decisions that have to be taken about the services systems that were identified in Chapter 10. Spatial decisions will be based on the requirements for:

- User outlets
- Distribution network
- Items of plant
- Building entry (or exit)
- Control
- Access for maintenance

The location for each of these parts of the services system will need to be chosen and the amount of space required for their installation, maintenance and even replacement identified. This requires not only an understanding of how the user and operator of the building interact with the service but also knowledge of the service itself, how it works and the physical size and characteristics of the parts.

For small passively designed buildings the space required for the services may be as low as 4% of the gross floor area, while for air-conditioned offices this may be as high as 9%. For highly serviced buildings such as sports centres with swimming pools this may rise to between 20% and 30%.

Environmental services

Heating

Space heating requirements provided by the active services in commercial buildings have many options. Centralised systems may be similar to those described for housing in Chapter 21, with boilers, pipes and radiators. It may be part of a full air conditioning system with summer cooling and humidity control as well as winter-heating capabilities, where the air may be conditioned in large central air handling plant to be distributed and returned in ducts through vents in each room. Passive design is making large air conditioning systems unnecessary, but some winter heating is still normally required in commercial buildings where low-carbon systems may be considered to complement to summer passive cooling.

Each heating system makes demands on the space requirements and location of the plant and equipment. The air-handling plant of air conditioning systems is normally on the roof to collect and exhaust air at the top of the building, where there is a good supply of fresh air, whereas if ground source heat pumps are used, plant rooms will need to be on the ground floor. Even the more conventional heating plant using carbon-based fuels may be best on the ground floor. Originally this would have been because of the weight of the boiler plant and fire risk, but it was also convenient to make the fuel connections and allow the future maintenance and replacement of plant easier. This is still true for biofuel boilers that require fuel delivery and storage. Indeed, in the air conditioning option with heating coils in the air-handling plant on the roof the boiler may still be housed on the ground floor providing the heat (probably via water in pipes) to the roof, although low water capacity boilers have considerably reduced the weight of boiler plant, making roof-based boilers possible. Conventional boilers have been getting smaller, thus reducing the size of plant rooms, but some of the new low-carbon heat sources require larger areas to house the heat exchangers.

The above paragraph introduced only a couple of centralised heating options. These were discussed not because of their popularity or efficiency but to show how different systems have different location and spatial requirements. They illustrated differences in the outlets for the user (radiators or air vents), the type of distribution network (ducts or pipes), the size and position of plant (roof or ground floor) and the position of entry to the building (roof and/or ground floor). Other heating systems will have different combinations of location and size of plant and distribution networks. These have to be visualised and understood at an early stage in the design of a building if the detailed design of the services and their operational characteristics are to be achieved in an economic and efficient manner.

Heating systems vary in the way the heat is to be introduced into the room and the mechanism that is expected to distribute it as evenly as possible plus any provision for return air, particularly if it is intended to recover heat from the air before being released to the outside. While some heating systems have a radiant component that heats the occupants (and other objects that then become convectors), most systems employed in commercial buildings rely on room air movements to distribute the heat. This makes heat entry at floor, or at least low, level preferable. This is achieved with underfloor heating, linear skirting heaters or radiators (which actually have a majority convection effect) around the perimeter of the room. Where heating is based on heated air, entry will often be at high level, relying on the entry velocity to encourage air distribution in the room. For displacement systems where there is a low-entry air velocity the air has to be introduced at floor level. These alternative locations significantly influence the design and location of the distribution network.

Whether distribution of the heat is through water in pipes or air in ducts, the distribution network for these centralised systems is going to be widespread throughout the building and will need to be routed to each room, requiring both vertical and horizontal distribution. While the need for widespread routeing is the same for ducts and pipes, the space they require is different because ducts are significantly larger than pipes. Not only is the heat that can be carried in the same volume of water greater than in air but the system calls for the temperature of water to be higher than the air, making the pipes considerably smaller than ducts carrying an equivalent amount of heat.

Vertical service shafts will be required in multi-storey buildings to house the pipes or ducts to each floor. Their size will depend on the type of distribution system and the floor area to be served by each shaft. These shafts will need access to be provided for maintenance at each floor and will constitute a risk of fire spread, and so will need to be detailed for fireproofing, particularly at openings and where services transfer to horizontal distribution at each floor level. This will influence the specification of access door and fire-stopping around the services.

Horizontal distribution options can be seen as different patterns. Perimeter systems, as the name implies, make circuits that serve the sides of rooms, while branched patterns allow distribution to the middle of rooms. These patterns will be influenced by zoning control where circuits and the position of the vertical shafts will influence the actual distribution and size of the networks required.

The other two questions about horizontal distribution concern whether it should be above or below the structural slab and whether the services need to be hidden or exposed. Hidden services below the slab normally run suspended from the slab, behind suspended ceilings. Access is normally achieved by lifting ceiling tiles. Hidden services above the slab need to be housed in the floor screed or in the space provided by raised floors. Details of both suspended ceilings and raised floors are given in Chapter 30. Exposed services can run at skirting level or be suspended from the structural slab. Options are now being influenced by passive design decisions where exposed slab soffits for thermal storage and night-purging may become part of the summer temperature control strategy. In such systems suspended ceilings are not available to hide the services, and services suspended from the slab will influence the flow of air across the soffit and the efficiency of the night-purging phase.

At the other end of the distribution system from the room outlets there will be a boiler or heat exchanger, depending on whether a fuel or heat is brought into the building from outside. This will be a significant piece of plant that will need to be located in a plant room, which should be secure and only accessible to maintenance staff. The plant room will also contain the pumps to circulate the heat, normally in pipes, either directly to radiators or to their equivalent in each room or directly to an air-handling plant if the heat is to be distributed in the air through ducts. The plant room may also be used for plant for other services such as a colorifier for the hot-water system.

Associated with the heating plant will be the entry of the fuel or the heat and, if combustion takes place in a boiler in the building, there will be a need for a flue. Entry of fuel or heat will most probably be at ground level, and on commercial buildings the flue will need to be taken to a high level, probably above the roof. Entry points are best directly into the plant room area, but the flue will need vertical space to be taken to the roof. With efficient boilers flue gases are not excessively hot and so will produce a condensate plume in the cold air at the flue terminal and increases the risk of condensation in the flue; therefore, provision to remove condensate must be provided in the plant room.

Where heat is brought into the building, the plant room will have to house heat exchangers. Heat may be brought into the building in water, perhaps from a central boiler plant or from a combined heat and power (CHP) unit. These systems provide sufficient heat to achieve a flow temperature in the distribution medium, normally water, to run conventional radiators. With the introduction of heat-pumping from low-grade heat sources such as the ground, the potential flow temperatures are low and therefore the use of radiators would not be efficient as they would need to be larger. Low-flow temperatures are better suited to under-floor heating, where the large area of rising heat provides a comfortable environment that, in conjunction with good insulation below any ground floors and the high thermal mass of the floor, can give good internal temperature control even with large changes in external temperatures. Heat can also be obtained from the air-source heat-pumping with room-based fan convector units heating the air directly. These are discussed in the section on ventilation and air conditioning below as they can be reversible to provide summer cooling.

Along with an understanding of the parts of the heating system itself, it is necessary to understand how the system is to be controlled. This will depend not only on the type of heating system but also on the strategy to be employed for the building. It is usual to zone commercial buildings for heating with sensors and valves to control sections of the building. Areas designed for different activities will require different conditions, and these will need separate control. If large areas of the building are for similar activities, as in offices, then control may be by floor and/or by orientation of north- and south-facing facades.

There will need to be connections between sensors and switching/diverting components. These are electrical connections requiring wired services. These may also be connected to a central computer system to manage the whole building and provide information of performance for energy use. These wired systems can become a significant part of the heating system.

Systems will require commissioning. This process takes place at the beginning of the life of the system when it is fully running with all controls and management systems in place. This is not only to test that the parts are working and responding to the sensors but also to balance the flows in the system to ensure each heat emitter gets its share of the distribution medium, be that water or air. This is initially completed with the building empty to correct gross discrepancies but has to be repeated when the building is in use over the whole heating cycle. Access is required for checking flows and setting valves and dampers at commissioning, albeit only for the early few months of the building's life.

The systems discussed above have centralised heating connected to heat emitters via distribution networks. Local or direct heating systems can be considered. In these systems the power or fuel is distributed to the emitters. In domestic systems this includes electric convectors or fan heaters for instant heat or thermal storage to take advantage of off-peak tariff electricity for slow release during the daytime heating period. It also includes gas and solid fuel stoves that need to be connected to flues. In commercial buildings this may prove a better solution for large industrial buildings rather than office type development, unless localised demand in otherwise passively designed systems is identified. For industrial buildings large, floor-standing independent warm-air heaters can heat large areas but are noisy as they rely on high-capacity burners and fans to deliver the air into the space at relatively high velocities. Alternatively radiant heaters with high surface temperatures can be mounted at a high level, relying on individuals being in direct line with the heaters. These are relatively quiet if burning gas and silent if powered by electricity. These systems are often specified where intermittent use requires limited or even no warm-up periods.

Ventilation and air conditioning

Both of these have been introduced in the discussion on heating and much of the plant and the distribution network issues around air-handling plant and ducting have been identified. In buildings incorporating the passive design principles identified in Chapter 15 a few areas, such as kitchens and toilets, may need ventilating and only areas with high unavoidable heat gains may need air conditioning or cooling in the UK. For these buildings there is no requirement for widely distributed ducting systems. Either dedicated systems to areas such as kitchens or vertical collection systems from toilets repeating on each floor (see the section on public health services below) may be required. The exhaust discharge points from these ventilating systems are normally at a high level as it is assumed that the exhaust air may be unpleasant. Fans are also needed to extract the air from the room. These can be noisy and set up vibrations and so may need to be located away from sensitive areas.

In buildings where passive design principles may be difficult to apply mechanical ventilation systems, possibly with some comfort cooling or air conditioning, may have to be considered to serve some or all of the general areas. It should be emphasised that this should be identified at the early stages of the design so that the configuration of the whole building can be revisited to see whether heat gains can be reduced and/or natural ventilation incorporated before mechanical systems are adopted.

Full air conditioning provides ventilation with the introduction of fresh air, filtering of the mixed fresh and return air, humidity control as well as cooling and heating. Centralised systems condition the air in large air-handling plant normally on the roof (as described under the section on heating above). Heated and chilled water are provided to heating and cooling coils in the air-handling plant in the main air flow along with any filtering and humidity control. This air is then ducted to each room and delivered via diffusers often in the ceiling but can be wall-mounted or even at floor level, although air speeds may have to be limited if people would normally come close to inlet diffusers. Return air is then collected via diffusers to ducting that takes the air back to the main plant, where some return air is mixed with fresh and passed into the air-handling plant. Heat recovery may be incorporated into the exhaust air ducting. The whole cycle is powered by fans incorporated into the air-handling system.

The design of the room inlet diffuser can offer localised control. Plain diffusers are used if large area single zone control is all that is necessary. Controls on the central fan and cooling/heater coils plus dampers in the ducting system that are balanced at commissioning deliver the same air to all diffusers. Variable air volume (VAV) diffusers have dampers linked to a room thermostat which closes and opens the damper. This will cause a change of air pressure in the duct and this has to be monitored by a sensor that changes the speed of the main fan to maintain flow to other rooms still on demand. Dual duct systems have diffusers that are mixing boxes supplied with separate cooled and heated air that is mixed, depending on the demand identified by a room thermostat. This clearly has implication not only for the size and number of ducts but also for the management and control provision for the system.

In these centralised systems moving conditioned air from a central plant requires large ducts that have a significant demand on space and compete with the structure and finishes horizontally and with usable space vertically. Partial centralised air/water systems overcome this by providing air from the central handling plant and chilled and heated water from the chiller and boiler plants to the inlet diffuser units in the rooms. The distribution systems now involve smaller ducts but require pipes for the chilled and heated water. These inlet arrangements can be through either induction units, where the supply air draws in some room air and the mixed air passes over the cooling and heating coils, or displacement systems. In displacement systems cooled air is introduced at floor level, normally through raised floors, at low velocity and relatively high temperatures for chilled air so as to avoid local discomfort at the inlet grills. Air movement is generated by the warm air rising from the heat sources in the room (people and equipment) with extraction at high level. For high cooling loads this may have to be used in conjunction with chilled ceilings or chilled beams, where cool surfaces give a radiant cooling effect and cooled air will fall from the chilled surface. This requires some humidity control to avoid condensation.

Cooling systems for localised high heat gain or poorly ventilated areas can be provided by local units. These systems are served by room-based units where only an electrical power supply and a condensate drain are required to run a small refrigeration unit. Completely self-contained units can be fitted into the external wall, but these are used more for retro-fitting than a conscious initial design choice. Two-part, split systems with a room convector unit and an external heat exchange unit can serve more than one room. These are the two halves of a refrigeration system with the evaporator coil cooling the air in the room and the condenser coil giving off heat to the outside. The two units now have to be connected with pipes carrying the refrigerant. Both units will have fans to move the air over the coils, and the outside unit will need to power supply for the compressor and will therefore produce noise. The position of these internal and external units is now more flexible. Internal units can be mounted on either internal or external walls or can be in the ceilings and integrated into finishes.

Multi-split and variable refrigerant flow systems allow multiple room units to be run from one external unit where the condensing coil may have to have water cooling if the cooling load becomes too great for air cooling. Maintenance of these water cooling units is essential as they can become the source of legionnaires' disease that can, if their location is not carefully considered, become a risk to the public.

The basis of these cooling units, the refrigeration cycle, is also known as a heat pump. It can be designed to work in reverse and becomes the basis of winter heating known as an air-source heat pump. The heat pump is also the basis of ground-source heat pumping introduced above in the section on heating.

Controls and the use of a building management system (BMS) are similar to heating systems depending on zoning and the room control involved. There is also a similar need for access for maintenance, replacement and commissioning.

Lighting systems

Lighting systems do not have as many alternatives as heating and air conditioning systems but they are high-energy consumers and are widely distributed throughout the building.

Room lighting is most likely to be provided at ceiling level, be powered by electricity and therefore have a cabled distribution system. Unless wireless systems are to be used, cables will also be required between manual switching and any automatic sensors used to control the lights. The cables will require both vertical and horizontal distribution, which can be either above or below the structural slab. Hiding and access issues are the same, as discussed in heating above. Cables are normally either laid in or clipped to trays suspended from the ceiling, or fixed to the walls of vertical shafts or in ducts housed in floor screeds or attached to exposed walls. The final distribution to fittings may be in conduits that may be buried in walls and/or floors or may be surface-fixed. Trays, ducts and conduits provide support and dictate the routeing of the cables but must also provide the mechanical protection to the cables, where damage can occur to the electrical insulation and cause a safety hazard to either occupants or trained maintenance staff.

There are options in the lighting source and fittings. These will be selected for their light quality and for their ability to be integrated into the finishes, such as suspended ceilings. They will also be selected for their energy efficiency. Energy use will also be significantly influenced by the control mechanisms provided. Manual switching allows individuals to switch lights on when they are required. However, automatic systems to switch lights off can provide energy advantages. Movement and sound sensors to switch off lights in empty rooms or light sensors, to switch off light as natural lighting provides sufficient illumination, can be specified. These may involve additional wiring. Switching is unlikely to be for individual lights, and the arrangements of banks of lights will depend on the use of the room. However, lighting in banks parallel to the windows will allow lights to switch on or off from the windows as natural lighting levels change.

The major equipment associated with lighting (and power, discussed later) is distribution boards. These provide not only circuit isolation but also the safety features of fusing for each circuit. There are normally many circuits for any building, and while there may be only one distribution board for small buildings boards are often provided for each floor or even different zones on each floor in larger developments. All these boards require access by qualified staff but pose a significant safety hazard to the occupants. This will influence the position of these boards and, while the boards provided at each floor are not large (they are more like cupboards), the need for locking them away does call on some reasonable space to be allocated for this equipment.

The incoming electrical supply to a building will normally be underground and must be brought into a secure room, where fusing and isolation from the external supply network will be required. This room can also house the major distribution circuits for both power and lighting and perhaps metering, although this may be part of the BMS provision. While on small buildings this may be combined with the plant room, separate rooms may be considered desirable on larger developments.

Public health services

Water supply and drainage

It is sensible to consider these together as they are normally part of the same system. Clean water is supplied and used and then needs to be taken away from the building without causing dampness or contamination; both of these are public health concerns. Water is required for a limited number of activities and it is usual to try to concentrate these areas together and repeat them on all floors to form a vertical zone that requires the services of water and drainage. This also proves useful in concentrating the ventilation requirements identified in the section above. In this way horizontal distribution is limited to within the rooms themselves. This is particularly helpful for the drainage, where vertical soil vent pipes are relatively large (100 mm) and horizontal pipes have to be laid to falls, making routeing more problematic over long horizontal distances.

There is one other part of the drainage service inside the building that is not associated with disposal of used (so-called foul) water and that is the collection of rain- or surface water from roofs. This is achieved with vertical pipes from the roof to the underground drainage system. In commercial buildings these vertical pipes may be located on the outside of the building, as in housing, but are often sited inside the building, where they need vertical ducts if they are to be hidden but access for maintenance must also be incorporated. Horizontal rainwater pipes can be provided above ground, perhaps in basements, to limit underground drainage under the building with the associated increased risk of damage and potential for blockage. Systems for rainwater harvesting will involve large underground tanks and requires provision for overflow plus pipework back into the building, where it becomes part of the water supply system. Rainwater will not be considered further in this chapter.

There are a number of separate systems involved in water supply. The most obvious is the need for hot as well as cold water to many points of use, such as sinks and showers. Chapter 21 introduced the idea of the need for different quality levels for the cold-water supply and the high level of regulation on the supply and services systems to maintain the appropriate quality right up to the point of use. In most buildings the highest quality is for drinking and food preparation. In domestic installations the drinking water is taken directly from the incoming main, while water for cleaning may come from a pipework system from a storage tank. In addition there may now be a need for a pipework system for harvested rainwater or semi-treated water from a grey water system for flushing toilets. For each of the outlets there is a need to identify the type of supply along with the points of origin for that supply. This will then lead to decisions on the routeing of the pipe and the allocation of space to the distribution network.

The water and drainage systems meet at the user outlet, such as sinks and toilets. These contain the water while it is being used and are normally known collectively as sanitary ware. In wet areas where spillage or cleaning regimes mean that water will not be contained within the sanitary ware floor drainage will be required. Sanitary ware will have provision for connection to both water and drainage services and will keep the water contained for its journey through the building.

In some systems this journey starts with the entry of the mains water that is taken directly in the pipework to the sanitary ware, and the drainage system takes the water directly to an exit point to the underground drainage system. This would be true of domestic cold supply to kitchen food preparation sinks. There is no plant or other large equipment involved. Control is via taps and plugs in the sinks to draw off water and contain the water until it is ready for disposal. Traps below the sink at the beginning of the drainage system control the escape of gas from the drains into the room. Isolation valves on the water supply system and maintenance access have to be provided for emergencies and repair. Other special fittings such as pressure-reducing or water-miser features may be specified.

Plant and equipment associated with water services is, however, necessary for most supply conditions. The main items associated with water supply are storage cisterns (normally called tanks) to ensure the continuity of supply in case of mains failure. These have to be above the outlets they supply as it is the head of water from the cistern that provides the pressure to deliver the water to the taps or toilet cisterns. These are normally on the roof, where both space and the additional loading have to be taken into account. In tall buildings (above around 10 storeys) the pressure created from a tank in the roof at the ground floor would be high. While pressure-reducing valves could be used, intermediate storage cisterns are often used that have to be housed on intermediate floors. These reduce the quantity of water that needs to be stored at any one point, thus distributing the overall structural loading. Water may also need to be pumped up, possibly from large cisterns on the ground floor (loading direct to ground) with smaller cisterns at the higher levels. In these circumstances it is often necessary to have storage cisterns for drinking water sealed against insects and contamination and limited in size to limit the time the water is left standing in the tank before it is drunk. A separate supply network would then be required for the drinking/food preparation water.

Hot water requires more plant and equipment. Instant water heaters that only need a connection from the cold water and a power supply can provide hot water to remote sanitary appliances, but this is not an efficient way to provide hot water to a whole building. These instant water heaters are known as direct systems in so far as the heater directly heats the water as it is demanded in the sink. Again, direct systems are not a sensible way to heat the water for a whole building, as scale deposits will build up on the heater unit. It is better for the water to be heated indirectly, normally from water that returns to the heater unit to be reheated round a closed primary circuit similar to a radiator for heating the room. The indirect heater coil is contained in the storage vessel that is large enough to ensure a reliable supply across the period of occupation with an estimated level of simultaneous demand and estimated reheat times.

These storage vessels take the form of hot-water tanks (normally called cylinders or, for larger installations, colorifiers) that are fed with cold water and use the pressure in the cold-water system to deliver the heated water to the taps. Hot-water systems can create long pipework systems, sometimes known as dead legs. During periods of low demand, the hot water cools in the pipe. This not only reduces the level of service as users have to wait some time for hot water but is a waste of water and energy as users will run off the cold water until the hot water reaches the tap. Insulation (called lagging) can be specified and even electric heating tape can be used to keep water warm, but secondary return circuits may be required to maintain a reasonable supply of hot water, but the pipework will then lose heat continuously.

The cylinders need to be associated with the heat source and this, along with the length of the distribution pipework discussed above, becomes a consideration in the location of the cylinders. One option is to use space in the plant room. This is often close to the boilers that are used to provide the heat source. If the rooms requiring supply are some distance away or there are a number of zones for these highly serviced areas, local hot-water cylinders may be preferable, albeit that they then need a longer primary circuit connected to each cylinder.

Alternative heat sources may change this spatial distribution. An example of this is the use of solar panels on a south-facing roof. The cylinder now needs to be connected to two heat sources: the solar panels and a source that can provide heating in periods when the solar panels are ineffective. This would normally be the boiler or heat exchanger providing the space heating, although independent sources such as electric immersion heaters can be used. Cylinders need to be relatively close to the solar panels to increase the efficiency of the solar system.

Before considering the drainage system, the issue of bonding the electrical system to the copper (electrically conductive) pipework has to be considered. The plumbing system can become live if contact is made with the electrical systems. This is clearly serious, particularly in the wet conditions associated with sanitary appliances. To avoid any danger from this the plumbing has to be connected to the earth system. This was discussed in Chapter 21. This introduces the need to associate a wiring system with the plumbing pipework.

For all hot and cold pipework systems isolation for emergencies and repair needs to be available and accessible to qualified staff for maintenance purposes. This may influence both the number of isolation points and their location. The aim should be to quickly close down limited sections to effect the repair without disrupting the service to other parts of the system.

The above-ground drainage will follow the same principles of gravity flow in either horizontal pipes laid to small gradients or vertical stacks to maintain the self-cleansing velocities and streamline-flow explained in Chapter 21. A major difference in larger installations required for commercial premises is the approach to induced- and self-siphonage in the traps below the sanitary appliances. In the domestic one-pipe system the pressure differences are controlled by specifying the depth of the seal in the trap and then limiting the horizontal runs to vertical stacks and limiting the proximity of two adjacent connections to the stack. This may not be possible in larger installations and the pressures may have to be equalised with a separate venting pipework system connected to the main drainage pipes just behind each trap. This venting pipework runs in parallel (but opposite in gradient) to the drainage pipes connected to the vent stack. They can therefore be accommodated in the space allocated to the drainage pipes.

Nearly all drainage systems can be designed to work on gravity and therefore no items of plant are required. However, where outlets are required for sanitary appliances or floor drainage that is below the point of entry to the underground drainage system some means to lift the sewage has to be provided. This normally takes the form of a holding tank to which the sewage can flow by gravity with a float switch that pumps the tank out when it is full. Access to this tank and provision for dealing with the risk of breakdown will influence the size and position of the tank.

Drainage systems, other than the example of the float switch above, are generally self-regulating and do not need any control. The correct choice of pipe size, gradient, junctions, traps and venting maintains streamline flow and equalises pressures in the system to keep the system working without intervention.

Like all systems, however, there is still a risk of failure. In drainage these are blockage and leaking. Consideration has to be given to access to clear blockages and to carry out repairs. Drainage is often associated with highly serviced areas. There needs to be space around the pipework to enable clearing and repair activities. This often leads to walk-in ducts behind extensive installation often with access from circulation space and not the room containing the sanitary appliances, particularly if cross-contamination is a concern.

Grey and black water systems that collect drainage water provide some processing and then, in the case of grey water, return the water to the building for some supply conditions, such as toilet flushing, and external use (watering and vehicle washing) will require equipment and pipework systems. These are likely to be external to the building and underground. This will influence landscaping not only directly in the siting of the equipment but also in the provision of water features that can become part of the waste-processing system.

Power supply

Many items of equipment used in the operation of a building need power. While in some industrial settings these may include systems such as compressed air, the two widely used power sources are electricity and gas. What all power supplies have in common is potential dangers that can lead to serious injury and death. Much of the design of the equipment is governed by the hazards it poses, but some of the safety and protection requirements are associated with the design of the building itself. An example of this has already been discussed above with the allocation of lockable rooms for major equipment with access for authorised personnel only.

Gas is not normally widely distributed throughout the building. It may be confined to areas such as boiler rooms and kitchens. If entry of the main with its associated metering and isolation arrangements are close to these areas, the distribution pipe runs can be kept to a minimum. If mains gas is not available and bottles or tanks are to be used, the location and protection of these areas, normally outside the building, will need special consideration. Regulations on gas installations are high, and this may affect the routeing of some pipework. Control and fail-safe conditions are normally built into the gas appliances but isolation valves throughout the pipework will also be required with appropriate access requirements, depending on whether the need is for isolation or emergency shutdown, which may even be external to the building.

In nearly all buildings the widely distributed power supply is electricity. One of the major consumers of electrical power is the lighting, and this was discussed above along with distribution and plant requirements for electrical circuits. The major operational power system is the 230 volt supply to socket outlets or spurs to specific pieces of equipment. While these power circuits are the same voltage as the lighting, there is likely to be a greater overall power demand. This will influence cable and fuse sizing, but the notions of circuits, distribution boards and the need to support and protect cables in trays and conduits are the same as for lighting. Sockets and equipment are likely to be at, or more likely below, dado level with switching incorporated with the outlet. This will require cabling in the walls and even the floor. Space for these cables is often provided by floor and partition systems, which are discussed in Chapter 30.

Some operational equipment will need the higher voltage of the three-phase 415 volt supply that is available from the electricity company's supply system. In the industrial context individual pieces of equipment may have the high-power requirements but in many buildings some services systems equipment may also need the three-phase supply. Domestic supply takes only one phase from the supply company's three-phase cable in the road, and this gives the 230 volt service for the building. In commercial building the incoming main will normally be a three-phase supply and the circuits in the building will be split to the single-phase supply to the lighting and socket outlets. This splitting into a single-phase circuit is normally done at the point of entry, although in large buildings and/or a building complex the three-phase supply may be distributed across the site and then later split into single-phase supplies to reduce losses. This has implications for the size and security arrangement for the plant rooms. As well as three-phase distributions for later splitting to single-phase circuits some larger individual pieces of equipment may need a three-phase supply. These cables are larger and heavier than the single-phase cables and need a larger radius when changing line or direction.

Power generation on site will also make demands on space and may be limited by the space available. On large developments in suitable locations mast wind turbines may be considered or combined heat and power (CHP) units chosen. Smaller-scale generation from photovoltaic arrays may be considered. These are often mounted on the roof but integrated systems, where the photovoltaic cells become the facade or roof covering material, can be used where appearance allows. These are expensive cladding units, although there is a return on that investment. This power can be fed into the normal distribution wiring in the building and may even be sold back to the power companies via net metering arrangements.

Security

As indicated in Chapter 10 two of the major threats to a safe environment are from fire and threats from within society. From these threats have arisen two major services provisions, each of which has a detection and an intervention aspect. The detection provision will have sensors and alarm systems, while the intervention provision will try to reduce the potential harm to the occupants in the case of the threat becoming a reality. The two systems are concerned with the threat from fire and intruders. Other threats including natural events or terrorist attack are normally considered in defensive strategies that are built into the fabric of the building, although some aspects of terrorist attack may be considered in the specification of the intruder systems.

Detection systems for both fire and intruders have many characteristics in common, although the actual equipment may be very different. While intruder sensors may be based on movement, fire detection may be based on ionised particles found in smoke, visibility or even heat. The type of detection chosen determines the position of the detector and the number required to give coverage to ensure early warning. While some units, such as domestic smoke detectors, are self-contained with a power source (battery) and alarm sounder, most systems in commercial and industrial building have a central power source and a separate network of alarm sounders. This requires wiring to a control panel to activate the sounders and hold information on the detector that was activated to identify the location of the fire or the intruder. Alarm activation can also be manual with 'break glass' units for fire and panic buttons for intruder alert.

The detection and alarm systems are devised to give early warning to occupants (and potentially scare off intruders) to promote action, normally evacuation. However, both may also activate the calling of the public services of fire and police. These services then need information to deal with the hazard and possibly to then investigate the incident. This information can be provided at the control panel. In fire detection systems this is known as the fire panel and needs to be near an entrance that will be accessible to the fire service to give them information when attending to fight the fire and assist in any subsequent investigation.

The wiring systems are vulnerable during an incident. Fire may damage the wires and intruders may cut wires, rendering the alarm systems inoperable. Wiring to fire systems will be in fire-resistant mineral-insulated cables, and intruder systems may need to be mechanically protected against malicious damage. Emergency lighting and illuminated signs will also need to remain active throughout the potential escape period.

These detection and alarm systems, unlike most other services, may never be activated but must respond reliably when they are. There is a need to test and maintain these systems, and access to carry out these management functions has to be provided.

While the detection systems have many characteristics in common, the intervention services for fire and intruders are very different. It should be noted that many of the intervention measures for both fire and intruders are dealt with in the passive fabric in providing resistance that is specified in combination with the active services. Only the services systems are being discussed in this chapter. It is also interesting to note that the intervention systems for fire and access may be in conflict with each other and with provision for the disabled, particularly in respect to access control. Access for the disabled to upper floors may be by lift, but these should not be used in the event of a fire. Doors may be needed for emergency exits but need to remain locked against intrusion. This requires the use of emergency push bars on the inside of these doors.

Intruder systems normally involve access control associated with the locks to doors. These are normally localised with card swipe, keys or biometric pads outside, and exit switches inside, the room. Both will be adjacent to the door. The system needs only a power supply and possibly some access to data to verify the user.

Fire intervention systems are much more varied being associated with smoke and fire suppression at a number of stages on the fire curve (see Figure 10.6). The equipment for fire suppression starts with the provision of hand-held fire extinguishers that may be distributed around the building with types selected depending on the type of fire risk in each area. This may be supplemented with fire hose reels that will need to be connected to the cold-water supply system. Sprinklers also activate in the early stages of the fire but are automatic and leave the occupants to escape, continuing to control the fire growth during the evacuation process.

Sprinklers are a water distribution system at ceiling level to feed a grid of heads that are opened individually in response to rising temperature. In the early stages only a few heads may open, so the demand for water may be limited. However, as the fire progresses and large areas of heads are activated, the demand for water may be great. To supply sufficient water the minimum size of the pipe to each head needs to be 25 mm with increasing pipe sizes as more heads are serviced. Installation of around 150 heads will require a main of 150 mm to ensure a supply of water to the whole system. Control gear is required at the start of the system. It will also need testing because, like alarms, although the sprinklers may never be used they must be reliable when activated. For very large open areas such as warehouses, where large numbers of heads may be expected to open at once in a short space of time, the public supply may not be large enough. In these cases large storage tanks will be required on site with pumps to provide the water under sufficient pressure through pipes of large enough diameter to maintain a sufficient water supply to all the sprinkler heads. The special risks associated with basements may also require equipment to ensure that the supply to the sprinkler heads remains effective for a sufficient time for evacuation.

Water is not suitable for all fires, such as electrical or oil-based fires, although oil fires can be cooled by a fine mist of water. Another extinguishing medium may be required but these are often hazardous to life as they exclude oxygen and can only be activated when evacuation is complete. Gas and foam systems come into this category. It may be necessary to provide storage and pipework for these systems, although in some cases the extinguishing medium is carried by the fire service and only pipework and an external coupling point for the fire service will be required in the building's design. Foam inlets to high fire risk service areas such as boiler rooms are an example of this type of provision.

Other intervention services provided for the fire service include wet or dry risers in tall buildings so the fire service does not have to deploy hoses through the building. Also for tall building over around 24 metres the provision of a fire lift may be required that often opens to the outside for the use of fire-fighters in the event of a fire. These are both vertical services and need to be housed in ducts or shafts that take space and need careful location on the floor plan.

The interventions systems discussed above are concerned with escape and fire-fighting. One installation concerned with fire prevention is the use of lightning conductors on tall (particularly pointed) buildings. Lightning seeks a route to earth and tall building attracts the strike. In an unprotected building the route taken through the building to earth is unpredictable and may cause direct death or fire. Lightning conductors provide a low-resistance path that will take the discharge safely to earth. The system is a passive system in that it has no moving parts or equipment. A spike at the highest point on the building is connected to a flat copper bar that is normally attached to the outside of the building and is then connected to a copper earth rod driven into the ground. A number of these sets may be required around the building.

Operational services

This final group of services have aspects in common with many of the services already discussed, but each has characteristics worthy of individual consideration. Communications including telephone and data links are wired services widely distributed through most office and other commercial developments. Sockets are required at or below dado level and therefore, like power, require routeing in partitions and floors. They are, however, low voltage and therefore do not have the safety hazards associated with electrical distribution. The wires carry signals that can also be transmitted wirelessly. While speeds of transfer via wireless systems may be different from wired systems, they are increasingly sufficient and/or more convenient to the operational needs of the organisation. This in turn changes the service from a wire network to the housing of wireless hubs and routers that need to be accommodated in the design.

Movement of people includes lifts, escalators and moving pavements. Unlike most other services, these are themselves major pieces of equipment. Their location is dominated by the layout and operation of the building, where their interaction with people in a safe and efficient way is the main criterion in the design of the equipment. They not only need space allocation but also interact with the building structure in the need for openings and shafts and are often associated with high loadings. They often require parts of the structure to be constructed to closer tolerances than are required for other parts that connect to the structure.

This equipment for movement needs power that is normally provided by electric motors and these, along with winding and control gear, will need a motor or machine room with access for authorised personnel only. For lifts these machine rooms can be at the top, above the lift shaft or at ground-floor level adjacent to the shaft. Hydraulic power is an alternative for goods lifts, where speed is not required and heavy loads have to be lifted over limited heights up to around 20 metres. The hydraulic ram is housed in a bore hole below the lift pit and the system then needs a hydraulic pump in a small machine room, normally at ground level adjacent to the shaft. This pump needs a power supply but not the power demands of the motors required for the winding gear for a cable-operated car.

The last category of operation services covers specialist requirements. The details for these have to be determined for any particular building. In the example of medical gases the user outlets will be at bed stations and in operating theatres. The distribution networks will be pipework, often of a smaller diameter than used for water. There are probably no items of plant, and the supply will be from pressurised bottles, which will therefore need to be delivered and stored safely. Control will be at each user outlet, but isolation and shutdown valves will have to be arranged for zones (perhaps each ward) and at the storage location. Access is required regularly for bottle delivery and changing by authorised personnel. The risk of failure is low for most of these gases, although the loss of supply may be significant to the operation of the hospital, so good access for repair and maintenance will be required.

The need for coordination

While it is extremely unlikely that all the services outlined above will be in any one building, most commercial buildings will require some provision for each of the categories. Some of the services may be able to share shafts and ceiling and floor voids but they will all need their own space in the building. In addition to the competition between individual services there will be competition for space with the structure and the fabric of building in floor zones (the zones between ceiling and floor level) and there will be competition for operational space on plan.

Planning locations and checking for, and avoiding clashes with, major plant and primary distribution routes are crucial at the design stage. It is common practice with the more traditional production arrangement for the final routeing of pipes and wires and even the final position of outlets to be left to site personnel. Indeed many services drawings, particularly electrical services, show final connections as representational rather than literal routeing. With increasing prefabrication the planning of services routeing and the position of final outlets has to be carried out in greater detail at the design stage to ensure components provide the space for services. In high levels of prefabrication the services may be incorporated in components in the factory. Now the coordination of services systems, location and space has to be planned to a high degree to ensure that connections can be made between components on site so the services perform efficiently as a system when the building is assembled.

Not only do clashes have to be avoided but also services pass through both enclosure and structural elements of a building. The need for holes and openings can become an issue for the choice of these other elements. Depending on the size of the hole or opening, this may involve drilling a hole in completed construction or forming the opening as the work is completed. When any opening is formed, the question of maintaining the performance of the element in which the hole is formed has to be considered. For example, in structural elements structural integrity has to be maintained; in external enclosure weatherproofing has to be continued; and in internal enclosure elements fire-proofing may have to be ensured. These are only the most obvious examples. When openings are formed, it is essential to check that all performance is maintained, and the detailing and specification of the fabric has to be considered. In locating and routeing services the ease with which openings can be formed and detailed to maintain performance should be considered. The cost of the opening plus the additional risk of failure should be part of the decision-making concerning the services provision.

Services not only pass through the fabric of the building but also are often contained within it. This interaction of the services with other elements of construction involves the routeing of services through floors and walls, where detailing and provision for installation along with access for modification repair or renewal has to be allowed for. In traditional construction this may involve so-called builder's work in casting in conduits or cutting chases in completed construction for the services, but with the use of internal systems for partitions, floors and ceilings provision for services may become part of the systems design. This is discussed in more detail in Chapter 30.

Another interaction of the services with other elements is for fixings and support. Where most distribution networks impose little additional loading, they still need to be fixed. It has to be clear what they can fix to, the type of fixing that will be appropriate and be coordinated to ensure that other services are not assuming that they can fix to the same place. In highly serviced areas this may lead to specific fixing channels or other systems available to all services.

Plant and equipment for services not only need fixing but also may place significant loads on other elements requiring strengthening if not of the whole element at least of the fixing point. This will need the position of the equipment to be established at an early stage to ensure the strengthened section of the fabric is in the right place, as it may have to be installed many weeks before the equipment is even delivered to the site. The passage about lifts and escalators (in the section on operational services above) raised the issue of tolerance and deviations, and this should be queried for parts of other services when looking at the positioning and fixings provision for plant and equipment.

Production sequence and contract time

The section above introduced the idea that services choice has production implications. The choice of services, their location and routeing will have consequences for the production sequence and hence the overall contract time. On some projects time may be a major determinant of design decisions if early hand-over and occupation dates are required. In either case an appreciation of the implications for production time and sequence of the choice of services and their location is important.

Construction operations are often considered in stages, normally known as carcass, first, second and even third fix. Generally this represents a sequence of operations: installing support, basic components and then finishes, respectively. For wired services this would be conduits and tray followed by wiring and distribution boards followed by fittings sockets and switches. Not only is this a basic sequence but it implies that other operations, by other trades, have to take place between these stages in the installation. For services there is also the need to coordinate all builders' work, as discussed above. This is particularly true in highly serviced areas such as toilets. Choice of location and routeing, along with the fundamental decision on production options from traditional to prefabrication, affects this sequence and hence potentially the overall contract time. It also influences the need for temporary works, health and safety risks and the potential for damage and need for protection in the production process.

The combinations of fabric, structure and services along with the choices for layout and subsequent need for routeing services networks in commercial building are almost endless. Each building project will need an analysis of the production sequence and methods, as outlined in Chapter 13, at an early stage in the design. This may lead to changes early enough in the design process to ensure that tolerances (quality), costs and time can be achieved in the final detailing and specification.

Cost of services systems

While this is strictly not an issue of scope and space, it will be a major consideration in the choice of systems. Perhaps more than for any other part of the building the balance between capital and running costs needs to be considered, including the aspects of maintenance and future modification. This should not only be in terms of money but perhaps also in terms of environmental costs, where additional capital costs may not only reduce running costs but also will be seen as good value for money in terms of the social and cultural role of buildings and their standing as investments in a community's future.

Summary

1. A number of building services will have to be considered for a commercial building, although the nature and extent will vary with each project and this needs to be established early in the design process.
2. The range of services falls into five categories: environmental, public health, power, security and operational.
3. The space required for each will vary depending on how widely distributed the user outlets are together with the location of plant, which may be highly centralised. This will generate distribution networks that need routeing with decisions on concealment and access for maintenance and repair.
4. Control is a key aspect of all building services to both turn on and off and regulate the output of working systems. This may be manual or automatic where building management systems (BMS) may be involved.
5. Environmental services include heating and cooling, ventilation and lighting. As potential energy users their required performance has become linked to passive design.
6. There are a large number of alternatives for heating and ventilation encompassing cooling and full air conditioning with different demands on space and position of user outlets, plant and distribution systems.
7. Lighting will be required for night-time occupation and possibly for tasks or even for general lighting if natural lighting cannot be achieved. As electrical circuits they will require distribution boards and switching arrangements to be established.

8. Control of lighting to switch off lights when sensors show a lack of movement or noise and/or sufficient light levels can make a major contribution to energy saving.
9. The water supply and drainage cycle as public health services can be limited within the building by concentrating the wet areas, particularly to vertical stacks.
10. Water usage is an environmental concern and systems for its reduction and reuse within the building before discharge to the drainage system should be considered.
11. Heating for hot water will need to be identified; this is often associated with the heating plant but other heat sources can be considered.
12. Electricity is the most widely distributed power, with gas a more localised use, but both are potentially dangerous and safety is a major aspect of the design, including the location of plant and equipment.
13. Security systems for both fire and intruders will require elements of detection and alarm. Fire will then include systems for containment and fire-fighting to allow escape and facilitate the intervention of the emergency services.
14. Operational services systems are varied from wired and wireless communication systems to mechanical systems for moving people, such as lifts. Some building will need specialist services such as a hospital's requirements for medical gases.
15. When the range and extent of the services systems for any one building have been identified, they will need to be coordinated, as many will need similar distribution patterns requiring some space to be identified for services generally and some specifically for one system.
16. The coordination of space can affect the production sequence and hence the contract time for both site work and any prefabrication being considered more generally for the project.
17. Services systems often have significant running and maintenance as well as capital costs, both financial and environmental.

32 Guide to Further Reading

The previous chapters described a framework for the analysis of a proposed construction solution in order to make a final choice. The text has introduced many ideas and provided some information on how we build. To make use of this approach in practice, and to keep up to date, you will need to refer to a wide range of published information. This final chapter gives many such sources and suggests how you might evaluate them to help your overall understanding of construction and make choices for specific projects in the future.

What this book provides

It has been the aim of this book to introduce a comprehensive framework that will allow you to identify the nature and range of knowledge that is necessary to choose construction solutions. The framework suggests that you will require knowledge of what the solution could look like and then the knowledge to undertake an analysis of how it would behave in operation and how it could be constructed.

It is hoped that if you understand what is involved in making a choice you will not only have some idea of what you need to know to make an informed selection but also have a way of integrating new knowledge as you increase your understanding of construction and gain experience.

In exploring the framework and the process of choice many ideas and concepts, as well as some details and specifications, have been introduced. This represents only a small part of the knowledge required to operate in practice. While your understanding will grow as you gain experience, there is still much published material that you will need to make use of.

Sources of information – authority, validity and relevance

You will need to refer to a wide range of published information for a variety of reasons. You will need to develop your understanding of the ideas related to the basis of the analysis in the earlier chapters of this book, and this can be achieved initially with the use of textbooks. The later chapters introduce detailing and specifications for housing and commercial buildings. This technical detail is best explored with current publications from a wide variety of sources that identify current regulations and recommendations.

Perhaps the best way to start to think about the range of information that you will need is to understand who prepared the information. This will tell you the purpose of the publication and the basis of the information. Who prepared the published material allows you to judge their authority in the subject area and therefore the potential validity of the information.

The publication may be prepared not by an individual but by an organisation. It may be governmental or commercial. There are many such bodies and associations, whose aims include collecting, generating and disseminating information for construction. Some, like the Building Research Establishment (BRE), have broad scope, while others, such as the Concrete Society, are limited. Some are funded by independent sources, such as government, whereas others are funded by vested interest, although this makes the information no less valuable. You will need to identify the scope of their activities and make some assessment of their authority and independence.

You will need to match the information you seek to the purpose you have for it. You need to judge its relevance. While the subject matter may seem to be what you are looking for, two other factors need to be taken into account. First, when was it published? In many cases the date of publication is significant, particularly in the use of current regulations and the most up-to-date recommendations. Second, where was it published? Building is a worldwide activity, but not all knowledge transfers between countries, particularly in the case of common practice where climate and the available resource base may not make the solution appropriate for another location.

Textbooks and research papers

Textbooks bring together established knowledge about a whole subject area, for example materials, structural design, building services, economics, and will help you understand more about the theory and practice of some of the ideas introduced in this text, particularly in Part 1. You may also need to refer to other textbooks on construction technology that will offer a greater range of solutions to help you understand the development of solutions covered in Parts 2 and 3. However, the solutions presented in textbooks might have been current when the books were published, but there will have been subsequent changes in legislation and new knowledge and innovations in materials and components. The detailing and specifications suggested in all textbooks, including this book, will need to be confirmed from more recent sources as the time lapsed from the date of publication increases.

Textbooks are normally written by academics and researchers who also write research papers. Research papers are written to contribute to debates, often when knowledge is uncertain and contested. Their purpose is not usually to provide direct advice or recommendations for making selections for construction solutions but mostly to help you understand the nature of the problem and the issues that may be involved. They are often written for the education and research communities, but in practice new or innovative design may require reference to current debate (or past original research findings), which will need interpreting in a practical context.

Advice and recommendations – information services

Research papers should not be confused with publications by research bodies that produce publications with the purpose of advice and recommendations for use in practice after the debates and experiments associated with the research itself have been completed. Some of the knowledge on which these publications are based will come from experience and studies of observations of past building behaviour where the recommendations have been developed from practice. Such research bodies include:

- The Building Research Establishment (BRE)
- The Construction Industry Research and Information Association (CIRIA)

An example of this would be the BRE *Digests*, which provide a concise review of individual subjects, from mortars to building on shrinkable clay. The BRE provides a wide range of information directly related to the choice of construction, details and specifications. CIRIA generally provides more comprehensive reports.

Another major group of publications, designed to provide direct advice and recommendations based on research and practice, is provided by organisations concerned primarily with one material and the components that can be created from such materials. They publish information and may provide an advice service to designers for specific projects. The organisations that provide this information service include:

- The British Cement Association (BCA)
- The Concrete Society
- The Steel Construction Institute
- The Timber Research and Development Association (TRADA)
- The Brick Development Association (BDA)
- The Lead Sheet Association
- The Clay Pipe Development Association

Many of these associations, while not providing information about individual manufacturers' products, do represent the interests of commercial organisations in so far as there is a need to provide good information to promote the use of a range of products if they are to be safely and successfully incorporated into buildings. They bring together technical information and good practice, including design data, specification and detailing.

While these organisations are specific to a particular material and their components, they often bring together practice for a particular application. For example, the major publications for timber frame construction are provided by TRADA, but the BDA offers advice and recommendations for cladding a timber frame house in brickwork.

Some of these publications can be based directly on practice, with reports on real projects to provide precedent and examples of how real designs have responded to the brief and the site, giving details and sizes of components that will help in making initial suggestions for similar projects.

Yet another group of organisations that provide an information service are the professional bodies. While some of their services will be exclusive to members, many publish information including technical advice on subjects of interest to their membership. These institutions include all the major construction industry professional bodies.

An example of this type of publication worthy of special mention is the design guide published by the Chartered Institute of Building Services Engineers (CIBSE), which provides not only an explanation of the basis and research behind design procedures for building services but also the data to use in design calculations.

Advice and recommendations – manufacturers' information

Many manufacturers provide information about their products. This includes information about not only the range of products, their shape and size, but also the properties of the materials these products use and any manufacturing deviations they have. They will also provide fixing and jointing requirements. This should not be confused with their advertising or brochures. These are the technical design manuals giving design data, detailing and even installation and health and safety information.

Many manufacturers include details and suggestions as to how their products can be incorporated into construction to meet performance requirements, particularly where these are set by national regulations. This can be seen in the information provided by both concrete block and insulation manufacturers giving examples of how to meet current thermal and acoustic regulations.

Manufacturers sometimes provide information on good practice guidance taken from many of the other sources outlined above. Their literature can also often be a good source for references to the appropriate standards and regulations discussed in the next section.

Standards and regulations

It is important to distinguish between regulations made under Act of Parliament, which carry legal obligations in their application and enforcement for all prescribed development, and the standards that may be written into building contracts, which are then enforced through that individual contract.

Generally, the regulations made under Act of Parliament are written in performance terms and do not prescribe how the performance will be achieved. However, the major regulations of concern in this text, the Building Regulations, are published with Approved Documents, which include construction detail and specifications as well as design information. These suggestions for the construction will satisfy the regulation, but not all situations are covered. Other solutions can be used, but the onus is then on the designer to show performance compliance. This is enforced through the process of building control. The Approved Documents are therefore a good source of information on how construction may meet these performance standards.

For most well-established aspects of construction there will be a British and/or European Standard. While these are written to provide a standard that will achieve a performance associated with notional current requirements, their main function is to specify a minimum standard to which a material, component or product should conform. They do not guarantee performance in a specific building. That still has to be judged as part of the final choice. However, these standards do offer advice as to what specification is applicable to a range of situations and as such provide valuable information on choice as well as a means of specifying for contractual purposes.

Because the development of a full Standard takes time and often relies on practice and the experience of the committees who draw the standard up, new products may not have a relevant standard to refer to. In this case British Board of Agrément (BBA) certificates may be available to outline the nature and use of a particular product.

One set of standards of particular interest to Part 2 of this book are set out in the *National House-Building Council (NHBC) Handbook*. Devised to set standards necessary to gain the warranty cover offered by the NHBC, they bring together many of the recommendations, regulations and advice associated with house construction.

Many of the services installations are covered by specific regulations, an example of which would be the Institute of Electrical Engineers' (IEE) regulations for electrical installations.

Magazines

It is important that the choice of construction be made in the current design and practice context. This is achieved by constantly refreshing the information and knowledge you have about current practice and the social, economic and commercial environments that influence the activity of building. This is possible by working in practice but can be supported by the regular reading of a weekly or monthly magazine.

Professional bodies normally provide such a service for their members, but there are a number of commercial magazines that provide not only news but also cover articles on new technical developments and innovative designs and solutions often associated with particular building projects.

Finding information

It is always possible to seek the information you need by using libraries or directly approaching the organisation or institution that publishes the information. There are, however, databases that allow you to search for and locate information normally based on a common theme or purpose. Originally based on paper, many are now available electronically via the Internet. The electronic storage and access has allowed what were originally indexes, possibly with abstracts outlining the content, to be extended to systems providing full text copies of the information. It is necessary to know who maintains these databases, how they select and authenticate the entries and then make the information available. Many academic and commercially available databases can now be used with confidence, but it is still necessary for you to ensure the authority and relevance of each piece of information for the purpose for which you need it.

These databases should not be confused with using a search engine (e.g. Googling) for information. Using the Internet to access known databases or even the organisations' own information direct from their websites is one thing, but using an open search facility may identify sources of a wide range of authority and relevance. If searches are used, they can identify some excellent information, but much more care is needed to evaluate the material, its source and intention.

Using information – corroboration, confirmation and risk

It may be possible in many cases to find an authoritative and relevant publication and take the information and use it in formulating the solution and then to carry out the analysis to minimise the risk of failure. It is often sufficient to just check the new information against your own existing understanding to confirm it makes sense and then to use it to firm up the solution to be adopted.

However, where this risk is high, or your existing understanding is weak, it may be necessary to seek information from more than one source. It may be that the new information does not seem to make sense with your current understanding so you feel it is necessary to confirm what you are being told.

Asking others for their opinion, if you feel they have the experience and expertise to give a considered opinion, may satisfy this process of corroboration. However, it may involve seeking other information from published sources.

You will not be able to cross-check the information given from several different secondary sources if they all quote from the same primary source. However, if the publications to which you are referring were from authoritative sources, they would have evaluated the validity of the information before they adopted it in their publication. This should give you some confidence in using the information. It is up to you to confirm that it is relevant to your current project.

Remember that it is finally your understanding and good sense that will lead you to make a sound choice.

Summary

1. This book provides a framework for the analysis and choice of construction. Carrying out the analysis and the process of choice in practice will require reference to a range of current and authoritative publications relevant to the project being considered.
2. While textbooks and research papers can inform the process of analysis, aspects of detailing and specification have to be current. A range of publications offering current advice and recommendation are available. These are published by organisations that have been established to provide information and by the commercial companies who provide services and products to the industry.
3. Standards and regulations provide information related to minimum performance achievements that may be used and quoted if they are relevant and appropriate to a specific building project.
4. Identifying publications is often possible through the use of indexes and libraries, and increasingly through electronic sources. Care should be exercised if information is accessed through searches using the Internet rather than established indexes and databases.
5. It is necessary to make judgements on the quality and relevance of information and it may be necessary to seek corroboration if the risks inherent in the use of the information are high.

Index